Maximum Entropy and Bayesian Methods

Fundamental Theories of Physics

An International Book Series on The Fundamental Theories of Physics:
Their Clarification, Development and Application

Editor: ALWYN VAN DER MERWE
University of Denver, U.S.A.

Editorial Advisory Board:

LAWRENCE P. HORWITZ, *Tel-Aviv University, Israel*
BRIAN D. JOSEPHSON, *University of Cambridge, U.K.*
CLIVE KILMISTER, *University of London, U.K.*
PEKKA J. LAHTI, *University of Turku, Finland*
GÜNTER LUDWIG, *Philipps-Universität, Marburg, Germany*
ASHER PERES, *Israel Institute of Technology, Israel*
NATHAN ROSEN, *Israel Institute of Technology, Israel*
EDUARD PROGOVECKI, *University of Toronto, Canada*
MENDEL SACHS, *State University of New York at Buffalo, U.S.A.*
ABDUS SALAM, *International Centre for Theoretical Physics, Trieste, Italy*
HANS-JÜRGEN TREDER, *Zentralinstitut für Astrophysik der Akademie der*
  Wissenschaften, Germany

Volume 79

Maximum Entropy and Bayesian Methods

Santa Fe, New Mexico, U.S.A., 1995

Proceedings of the Fifteenth International Workshop on Maximum Entropy and Bayesian Methods

edited by

Kenneth M. Hanson

Dynamic Experimentation Division,
Los Alamos National Laboratory,
Los Alamos, New Mexico, U.S.A.

and

Richard N. Silver

Theoretical Division,
Los Alamos National Laboratory,
Los Alamos, New Mexico, U.S.A.

SPRINGER-SCIENCE+BUSINESS MEDIA, B.V.

A C.I.P. Catalogue record for this book is available from the Library of Congress

ISBN 978-0-7923-4311-0   ISBN 978-94-011-5430-7 (eBook)
DOI 10.1007/978-94-011-5430-7

Printed on acid-free paper

CONTENTS

The Workshop coordinator, Barbara Rhodes, in between the cochairs, Ken Hanson (left) and Richard Silver at the Bradbury Science Museum in Los Alamos.

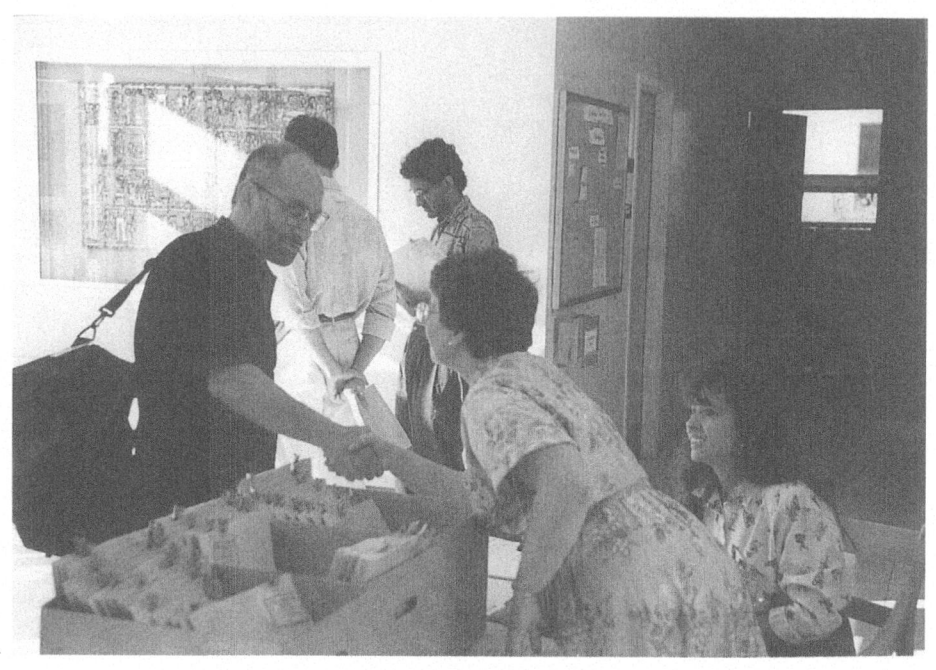

Wlodek Holender gets greeted by Barbara Rhodes and Rose Vigil.

PREFACE

The Fifteenth International Workshop on Maximum Entropy and Bayesian Methods was held July 31-August 4, 1995 in Santa Fe, New Mexico, USA. St. John's College, located in the foothills of the Sangre de Cristo Mountains, provided a congenial setting for the Workshop. The relaxed atmosphere of the College, which was thoroughly enjoyed by all the attendees, stimulated free-flowing and thoughtful discussions. Conversations continued at the social events, which included a reception at the Santa Fe Institute, a New Mexican dinner at Richard Silver's home, and an excursion to Los Alamos that ended with a mixed grill at Fuller Lodge, the main hall of the former Los Alamos Ranch School.

This volume represents the Proceedings of the Workshop. Articles on the traditional theme of the Workshop, application of the maximum-entropy principle and Bayesian methods for statistical inference in diverse areas of scientific research, are contained in these Proceedings. As is tradition, the Workshop opened with a tutorial on Bayesian methods, lucidly presented by Peter Cheeseman and Wray Buntine (NASA AMES, Moffett Field). The lecture notes for their tutorial are available on the World Wide Web at http://bayes.lanl.gov/~maxent/. In addition, several new thrusts for the Workshop are described below.

The Workshop was generously supported by the Center for Nonlinear Studies and the Radiographic Diagnostic Program, both in the Los Alamos National Laboratory, and by the Santa Fe Institute (SFI). This support permitted us to add some new features to the Workshop program. We were able to invite several distinguished speakers to present overviews of some exciting new areas of research: Imry Csiszár (Institute of Mathematical Sciences, Budapest, Hungary) talked about *The maximum-entropy principle and information theory*, Michael Miller (Washington University, St. Louis) discussed *Deformable models*, and Julian Besag (University of Washington, Seattle) presented *The Bayesian inference machine: an introduction to Markov-chain Monte Carlo*. Also, we were able to support the organization of several special sessions, namely *Time Series Analysis*, organized by Mark Berliner (Ohio State University), *Machine Learning*, organized by David Wolpert (Santa Fe Institute and TXN Inc.), *Deformable Models*, organized by Kenneth Hanson (Los Alamos National Laboratory), and *Data Analysis of Physics Simulations*, organized by Richard Silver (Los Alamos National Laboratory).

We wish to express our appreciation of the unflagging efforts of the staff at the Center for Nonlinear Studies, especially those of Barbara Rhodes and Rose Vigil, who coordinated the conference. Their conscientious attention to all details received countless compliments from the Workshop attendees.

We also thank Mary Louise Garcia (Los Alamos National Laboratory) for her expert composition of these Proceedings.

Kenneth M. Hanson and Richard N. Silver
Los Alamos National Laboratory

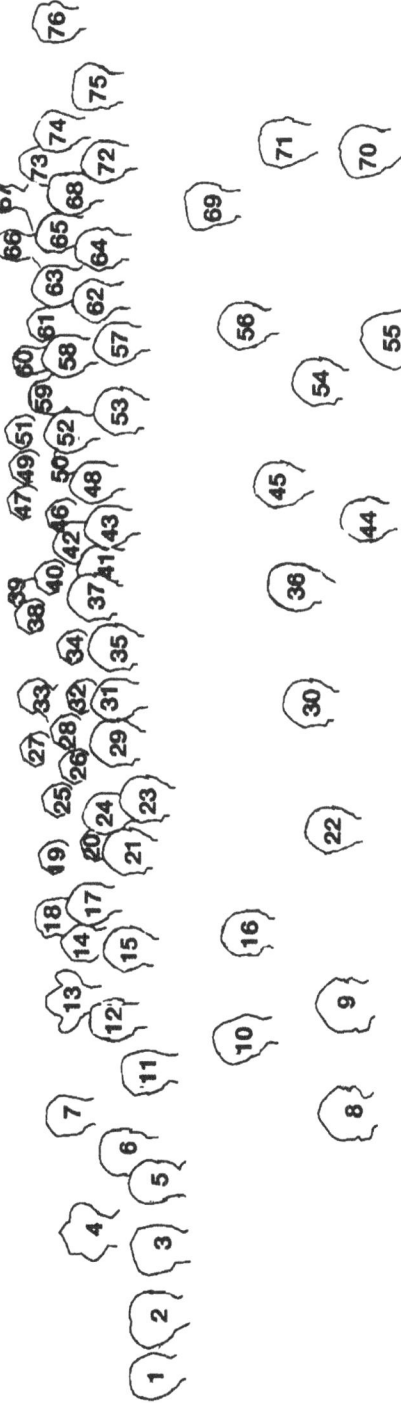

Legend for group photograph 1. Fritz Froehner, 2. Glenn Heidbreder, 3. Rafael Laboissiere, 4. Clint Scovel, 5. Danny Keren, 6. Imry Csiszár, 7. Michael Miller, 8. David Wolpert, 9. Gordon Hogenson, 10. Richard Jordan, 11. Jim Gee, 12. John Lafferty, 13. Tom Mottershead, 14. Anthony Crider, 15. Richard Puetter, 16. Richard Pitre, 17. Tom Metcalf, 18. Tim Gosnell, 19. William Dibble, 20. Chung-Yi Hong, 21. Tony Vignaux, 22. Gustavo Deco, 23. Ray Hawkins, 24. Jonathan Oliver, 25. Vijay Balasubramanian, 26. Roger Bilisoly, 27. John Stutz, 28. Jonathan Shapiro, 29. Ali Mohammad-Djafari, 30. Everett Larson, 31. Anand Ramaswami, 32. Anand Rangarajan, 34. Gene Gindi, 35. Richard Silver, 36. Tad Ulrych, 37. Gary Erickson, 38. Guthrie Miller, 40. Marcel Nijman, 41. Peter Cheeseman, 42. Larry Bretthorst, 43. Sibusiso Sibisi, 44. Vladik Kreinovich, 45. Myron Tribus, 46. Larry Schwalbe, 47. Greg Cunningham, 48. Jim Press, 50. Wlodek Holender, 51. Benoit Dubertret. 52. John Skilling, 53. Volker Dose, 54. Bob Snapp, 55. Rose Vigil, 56. Bob Fry, 58. Miao-Dan Wu, 59. Bill Fitzgerald, 60. Pierre Boulanger, 61. Bob Wagner, 62. Guillaume Evrard, 63. Julian Besag, 65. Saul Youssef, 66. Roland Preuss, 67. Iftah Gideoni, 68. David Montgomery, 69. Rainer Fischer, 70. Barbara Rhodes, 71. Paul Fougere, 72. Vinay Kashyap, 73. Sergei Rebrik, 74. Sri Sastry, 75. Ken Hanson, 76. Ray Smith

PARTICIPANT LIST

Vijay Balasubramanian
Princeton Univ.
Physics Dept., Jaowin Hall
Princeton, NJ 08544 USA

vijayb@phoenix.princeton.edu

Howard N. Barnum
Univ. of New Mexico
Center for Advanced Studies
Albuquerque, NM 87131-1156 USA

hbarnum@tangelo.phys.unm.edu

Mark Berliner
NCAR
PO Box 3000
Boulder, CO 80307-3000 USA

mb@stat.mps.ohio-state.edu

Julian E. Besag
Univ. of Washington
Dept. of Statistics
Box 354322
Seattle, WA 98195-4322 USA

julian@stat.washington.edu

Roger Bilisoly
DX-13, MS-P940
Los Alamos National Lab.
Los Alamos, NM 87545 USA

bilisoly@rio.lanl.gov

Lisa Borland
Physics Dept.
Univ. of California, Berkeley
Berkeley, CA 94720 USA

lisa@gojira.berkeley.edu

Pierre Boulanger
National Research Council of Canada
Inst. for Information Technology
Montreal Rd., Bldg. M-50
Ottawa, Ontario, K1A OR6Canada

boulanger@iit.nrc.ca

Larry Bretthorst
Washington Univ.
Dept. of Chemistry
1 Brookings Dr.
St. Louis, MO 63130 USA

larry@bayes.wustl.edu

Wray L. Buntine
Heuristicrats Research Inc.
1678 Shattuck Ave.
Suite 310
Berkeley, CA 94709-1631 USA

wray@heuristicrat.com

John P. Burg
Entropic Processing Inc.
10011 North Foothill Blvd.
Cupertino, CA 95014 USA

Peter C. Cheeseman
NASA Ames Research Center
MS 269-2
Moffett Field, CA 94035 USA

cheeseman@pluto.arc.nasa.gov

Anthony W. Crider
NIS-2, MS-D436
Los Alamos National Lab.
Los Alamos, NM 87544 USA

acrider@nis.www.lanl.gov

Imre Csiszar
Mathematical Inst.
Hungarian Academy of Sciences
Realtanoda-u 13-15
H-1053 Budapest, Hungary

h1141csi@huella.bitnet

Gregory Cunningham
DX-13, MS-P940
Los Alamos National Lab.
Los Alamos, NM 87545 USA

cunning@lanl.gov

Gustavo Deco
Siemens AG Corporate Research
ZFE T SN 4 Otto-Hahn-Ring 6
81739 Munich, Germany

gustavo.deco@zfe.siemens.de

William E. Dibble
Dept. of Physics and Astronomy
Brigham Young Univ.
Provo, Utah 84602 USA

larson@acoust.byu.edu

Volker Dose
Max-Planck- Inst. fur Plasmaphysik
Postfach 1533
D-85740 Garching, Germany

Benoit L. Dubertret
Laboratoire de Physique Theorique
3, rue de l'Universite
Univ. of Strasbourg
67084, Strasbourg, France

benoit@lpt1.u-strasbg.fr

Frederic Dublanchet
Laboratoire des Signaux et Systemes
Plateau de Moulon
Gif-sur-Yvette Cedex 91192, France

dublanchet@lss.supelec.fr

Phillip R. Dukes
Dept. of Physics & Astronomy
Brigham Young Univ.
Provo, Utah 84602 USA

dukesp@dirac.byu.edu

Gary J. Erickson
Seattle Univ.
Dept. of Electrical Engineering
Seattle Univ.
Seattle, WA 98122-4460 USA

erickson@seattleu.edu

Guillaume Evrard
GRAAL, Univ. Montpellier II
cc 072, Place Eugene Bataillon
Montpellier Cedex 05
F-34095, France
evrard@graal.univ-montp2.fr

Peter S. Faynzilberg
Carnegie Mellon Univ.
Graduate School of Industrial Admin.
Schenley Park
Pittsburgh, PA 15213 USA
petersf@andrew.cmu.edu

Rainer Fischer
Max-Planck Inst. Fur Plasmaphysik
P.O. Box 1533
D-85740 Garching
Germany
rrf@ipp-garching.mpg.de

William J. Fitzgerald
Univ. of Cambridge
Dept. of Engineering
Trumpington Street
Cambridge, CB2-1P2
United Kingdom
wjf@eng.cam.ac.uk

Paul F. Fougere
29 Randolph Rd.
Hanscom AFB,MA 01731-3010 USA
fougere@plh.af.mil

Fritz Froehner
Forschungszentrum Karlsruhe
INR, P.O. Box 3640
D-76021 Karlsruhe
Germany
inr355@dkakfk3.bitnet

Robert L. Fry
Johns Hopkins Univ.
Applied Physics Lab.
Johns Hopkins Rd.
Laurel, MD 20723 USA
robert.fry@jhuapl.edu

James C. Gee
Univ. of Pennsylvania
Grasp Laboratory
Dept of Computer & Info. Sci.
Philadelphia, PA 19104 USA
gee@grip.cis.upenn.edu

Iftah Gideoni
67 Yavne Rd.
Rehovot 76343
Israel
iftah@physics.ubc.ca

Gene Gindi
SUNY Stony Brook
Dept. of Radiology
Stony Brook, NY 11794 USA
gindi@clio.rad.sunysb.edu

Timothy R. Gosnell
MST-10, MS-E543
Los Alamos National Lab.
Los Alamos, NM 87545 USA
gosnell@lanl.gov

James E. Gubernatis
T-11, MS-B262
Los Alamos National Lab.
Los Alamos, NM 87545 USA
jeg@kaiser.lanl.gov

Kenneth M. Hanson
DX-13, MS-P940
Los Alamos National Lab.
Los Alamos, NM 87545 USA
kmh@lanl.gov

Raymond Hawkins
John Stafford Trading
Suite 3904
440 South Lasalle St.
Chicago, IL 60605 USA
cumph@aol.com

David Heckerman
Microsoft
One Microsoft Way, 9S/1
Redmond, WA 98052-6399 USA
heckerman@microsoft.com

Glenn Heidbreder
11509 Hemingway Drive
Reston, VA 22094-1240 USA
heidbreder@aol.com

Leon Heller
P-21, MS-D454
Los Alamos National Lab.
Los Alamos, NM 87545 USA
heller@biophysics.lanl.gov

Gordon Hogenson
T-13, MS-B213
Los Alamos National Lab.
Los Alamos, NM 87545 USA
ghogenso@u.washington.edu

Wlodek Holender
Dept. of Communication Systems
Lund Univ.
P.O. Box 118
S-221 00 Lund, Sweden
wlodek@tts.lth.se

Chung-Yi Hong
Washington Univ.
Chemistry Dept. , Box 1134
One Brookings Dr.
St. Louis, MO 63130-4899 USA
hong@wuchem.wustl.edu

Richard K. Jordan
Center for Nonlinear Analysis
Dept. of Mathematics
Carnegie Mellon Univ.
Pittsburgh, PA 15213-3890 USA
rjordan@andrew.cmu.edu

Vinay L. Kashyap
Univ. of Chicago
AXAF Science Center
5640 S. Ellis AAC
Chicago, IL 60637 USA
kashyap@ockham.uchicago.edu

Daniel Z. Keren
Dept. of Mathematics and
  Computer Science
Univ. of Haifa
Haifa, 31905
Israel
dkeren@mathcs2.haifa.ac.il

Vladik Y. Kreinovich
Univ. of Texas
Dept. of Computer Science
El Paso, TX 79902 USA
vladik@cs.utep.edu

Rafael Laboissiere
Inst. de la Communication Parlée
46 ave. Felix Viallet
F-38031 Grenoble Cedex
France
rafael@icp.grenet.fr

John Lafferty
School of Computer Science
Carneige Mellon Univ.
5000 Forbes Avenue
Pittsburgh, PA 15217 USA
lafferty@cs.cmu.edu

Everett Larson
Dept. of Physics and Astronomy
Brigham Young Univ.
Provo, Utah 84602 USA
larson@acoust.byu.edu

John Lewis
Interval Research Corp.
1801 Page Mill Rd., Building C
Palo Alto, CA 94304 USA
zilla@interval.com

David Madigan
Univ. of Washington
Dept. of Statistics
Box 354322
Seattle, WA 98195 USA
madigan@stat.washington.edu

Robert McKee
DX-13, MS P940
Los Alamos National Lab.
Los Alamos, NM 87545 USA
mckee@lanl.gov

Thomas R. Metcalf
Inst. for Astronomy
2680 Woodlawn Dr.
Univ. of Hawaii
Honolulu, HI 96822 USA
metcalf@akala.ifa.hawaii.edu

Guthrie Miller
ESH-12, MS-K483
Los Alamos National Lab.
Los Alamos, NM 87545 USA
guthrie@lanl.gov

Michael Miller
Dept. of Electrical Engineering
Campus Box 63130
Washington Univ.
1 Brookings Dr.
St. Louis, MO 63130 USA
mim@ee.wustl.edu

Ali Mohammad-Djafari
LSS-CNRS
Lab. des Signaux et Systems
Supelec, Plateau de Moulon
Gif-sur-Yvette Cedex 91192
France
djafari@lss.supelec.fr

David C. Montgomery
Dept . of Physics & Astronomy
Dartmouth College
Hanover, NH 03755-3528 USA
david.c.montgomery@dartmouth.edu

Charles T. Mottershead
AOT-1, MS-H808
Los Alamos National Lab.
Los Alamos, NM 87545 USA
mottershead@lanl.gov

Marcel Jan Nijman
Univ. of Nymegen
Geert Grooteplein 21
CPKI-231
6525 EZ Nymegen
The Netherlands
marcel@mbfys.kun.nl

Jonathan Oliver
Dept. Computer Science
Monash Univ.
Clayton, Vic 3168
Australia
jono@cs.monash.edu.au

Hanbin Pang
Dept. of Physics
Univ. of Cincinnati
Cincinnati, OH 45221 USA
pang@physunc.phy.uc.edu

Richard Pitre
2471 Windbreak Dr.
Alexandria, VA 22306 USA
pitre@n5160d.nrl.navy.mil

James Press
Dept. of Statistics
Univ. of California
Riverside, CA 92521-0138 USA
jpress@ucracl.ucr.edu

Roland T. Preuss
Inst. fur Theoretische Physik
Universitat Wurzburg Am Hubland
97076 Wurzburg
Germany
roland@physik.uni-wuerzburg.de

Richard C. Puetter
Center for Astrophysics & Space
  Sciences
Univ. of California, San Diego
9500 Gilman Drive
La Jolla, CA 92093-0111 USA
rpuetter@ucsd.edu

Anand Ramaswami
Physics Dept., Box 1105
1 Brookings Drive
Washington Univ.
St. Louis, MO 63130 USA
amr@wuchem.wustl.edu

Anand Rangarajan
Dept. of Computer Science
Yale Univ.
51 Prospect St.
New Haven, CT 06520-8285 USA
rangarajan-anand@cs.yale.edu

Sergei P. Rebrik
Univ. of California
Dept. of Physiology
513 Parnassus Ave.
Room S-762
San Francisco, CA 94143-0444 USA
rebrik@phy.ucsf.edu

David B. Rosen
New York Medical College
Dept. of Medicine (Munger Pavillion)
(Grasslands Reservation)
Valhalla, NY 10595 USA
d.rosen@ieee.org

Srikanth Sastry
Physical Sciences Lab.
Div. of Computer Research and Tech.
Building 12A, Room 2007
National Institutes of Health
Bethesda, MD 20892 USA
sastry@helix.nih.gov

Larry Schwalbe
XHM, MS-F663
Los Alamos National Lab.
Los Alamos, NM 87545 USA

schwalbe@lanl.gov

sps2@mrao.cam.ac.uk

Jonathan Shapiro
Santa Fe Institute
1399 Hyde Park Road
Santa Fe, NM 87501 USA

jls@santafe.e

Sibusiso Sibisi
Univ. of Cambridge
Cavendish Lab.
Manngley Rd.
Cambridge C33 DHE
United Kingdom

Richard N. Silver
T-11, MS-B262
Los Alamos National Lab.
Los Alamos, NM 87544 USA

rns@loke.lanl.gov

John Skilling
Univ. of Cambridge
Cavendish Lab.
Madingley Rd.
Cambridge CB3 OHE
United Kingdom

sps2@mrao.cam.ac.uk

Charles Ray Smith
US Army Missile Command
AMSMI-RD-MG-RF
Redstone Arsenal, AL 35898-5253
USA

Robert Snapp
Computer Science & Electrical
  Engineering Dept.
Univ. of Vermont
Burlington, VT 05405 USA

snapp@emba.uvm.edu

David M. Stupin
ESA-MT, MS-C914
Los Alamos National Lab.
Los Alamos, NM 87544 USA

stupin@lanl.gov

John C. Stutz
NASA Ames Research Center
MS 269-2
Moffett Field, CA 94035-1000
USA

stutz@ptolemy.arc.nasa.gov

Irina Tchoumatchenko
Laforia-CNRS
Univ. of Paris VI
4 Place Jussieu
75252 Paris Cedex 05
France

irina@laforia.ibp.fr

Henry Tirri
Dept. of Computer Science
P.O. Box 26
Univ. of Helsinki
Fin-00014
Finland

tirri@c.s.Helsinki.Fl

Myron Tribus
350 Britto Terrace
Fremont, CA 94539 USA

mtribus@netcom.com

Ruey S. Tsay
Univ. of Chicago
Graduate School of Business
1101 E 58th Street
Chicago, IL 60637 USA

rst@gsbrst.uchicago.edu

Tadeusz Jan Ulrych
Dept. of Geophysics & Astronomy
Univ. of British Columbia
Vancouver V6T 1Z5
BC, Canada

ulrych@geop.ubc.ca

George Anthony Vignaux
Inst. of Statistics & Operations
  Research
PO Box 600
Victoria Univ.
Wellington, New Zealand

tony.vignaux@vuw.ac.nz

Robert F. Wagner
CDRH-FDA
12720 Twinbrook Parkway
MS-HFZ-142
Rockville, MD 20857 USA

rfw@fdadr.cdrh.fda.gov

Scott A. Watson
DX-11, MS-P940
Los Alamos National Lab.
Los Alamos, NM 87545 USA

scottw@lanl.gov

Mike West
Duke Univ.
Inst. of Statistics & Decision
  Sciences
Durham, NC 27708-0251 USA

mw@isds.duke.edu

David A. Wilkinson
Chevron Petroleum Tech. Co.
1300 Beach Blvd.
La Habra, CA 90631-6374 USA

vdwlk@chevron.com

David H. Wolpert
Santa Fe Institute
1399 Hyde Park Rd.
Santa Fe, NM 87501 USA

dhw@santafe.edu

Miao-Dan Wu
Dept. of Engineering
Univ. of Cambridge
Trumpington St.
Cambridge CB2 1PZ
United Kingdom

mdw@eng.cam.ac.uk

Saul Youssef
SCRI
Florida State Univ.
Tallahassee, FL 32306-4052 USA

youssef@scri.fsu.edu

Randy J. Zauhar
Tripos Inc.
1699 S. Hanley Rd.
St. Louis, MO 63144 USA

zauhar@tripos.com

WORKSHOP PRESENTATIONS NOT INCLUDED IN THESE PROCEEDINGS

BAYESIAN METHODS AND THE MAXENT PRINCIPLE
P. Cheeseman and W. Buntine

A BAYESIAN APPROACH TO ACOUSTIC-TO-ARTICULATORY
INVERSION IN SPEECH: DETECTING THE POSITION OF THE
VELUM FROM SPEECH SOUNDS
R. Laboissiere

BAYESIAN SPECIFICATION OF MULTIVARIATE TIME SERIES
R. Tsay

BAYESIAN MODEL AVERAGING FOR ACYCLIC DIRECTED GRAPHICAL
MODELS
D. Madigan

ISSUES IN SELECTING EMPIRICAL PERFORMANCE MEASURES FOR
PROBABILISTIC CLASSIFIERS
D. Rosen

DEFORMABLE MODELS
M. Miller

THE BAYESIAN INFERENCE MACHINE: AN INTRODUCTION TO MARKOV-
CHAIN MONTE CARLO
J. Besag

INFERRING NONLINEAR DYNAMICS OF STOCHASTIC SYSTEMS IN
CONTINUOUS AND DISCRETE-TIME
L. Borland

GEODESICS AND OPTIMAL MEASUREMENTS FOR THE INFORMATION
METRIC ON QUANTUM MECHANICAL STATES
H. Barnum

TOWARD THE GOAL OF OBTAINING AN OPTIMAL THEORY OF
LOGICAL INFERENCE FOR APPLICATION TO THE INTERPRETATION
OF PHYSICAL DATA
E. Larson

RECONSTRUCTION OF THE PROBABILITY DENSITY FUNCTION IMPLICIT IN OPTION PRICES FROM INCOMPLETE AND NOISY DATA

R.J. HAWKINS
111 Chestnut Street, #807, San Francisco, CA 94111[§]

M. RUBINSTEIN
Haas School of Business
University of California at Berkeley, Berkeley, CA 94720[¶]

AND

G.J. DANIELL
Department of Physics
The University, Southampton, SO17 1BJ, UK[‖]

Abstract. We present a novel synthesis of MaxEnt and option pricing theory that yields a practical computational method for numerically reconstructing the probability density function implicit in option prices from an incomplete and noisy set of option prices. We illustrate the potential of this approach by calculating the implied probability density function from observed S&P 500 index options.

Key words: options, state-contingent prices, image reconstruction, image recovery, MaxEnt, maximum entropy

1. Introduction

Financial options are contracts that confer the right, but not the obligation, to buy or to sell a prespecified amount of an asset (e.g. a stock, bond, commodity, index, etc.) at a prespecified price at, or during, a prespecified time in the future. For example, a 6-month European[1] "call" option on IBM stock "struck" at $50 would entitle the owner to buy IBM stock in 6-months time at $50. If, at that time, the stock price exceeds $50, say $60, the option owner can buy at $50 and sell at $60: a $10 profit. If, on the other hand, the stock price is less than $50,

---

[§]Email: cumph@aol.com
[¶]Email: rubinste@haas.berkeley.edu
[‖]Email: gjd@phys.soton.ac.uk
[1]The term "European" indicates that the option can only be exercised on the last day of the period specified in the option contract.

1

K. M. Hanson and R. N. Silver (eds.), Maximum Entropy and Bayesian Methods, 1–8.

the option owner is not obligated to buy and can allow the option to "expire". If we denote the strike price of $50 by k, then we can write an expression for the "payoff" of the call option as $\max(x - k,\ 0)$ where x denotes the stock price.[2] The price of the option today can be calculated from the payoff function of the option by (i) multiplying the payoff function by the risk-neutral conditional probability density function (PDF) relating today's stock price to the stock price 6-months hence, (ii) integrating the result over all possible terminal stock prices, and (iii) multiplying the result by a discount factor:

$$c(k,\ x,\ t,\ t',\ r) = e^{-r(t'-t)} \int_{-\infty}^{+\infty} \max\left(x' - k,\ 0\right) p(x,\ t,\ |x',\ t') dx' \qquad (1)$$

where r denotes the continuously-compounded interest rate over the period $t' - t$ and $p(x,\ t\ |x',\ t')$ is the risk-neutral conditional probability density that the stock price will be x' at time t' if the stock price is x at time t.[3]

The traditional approach to the problem of option pricing is to posit that the stock price x evolves in time according to a stochastic differential equation (SDE)

$$dx(t) = \mu[x(t),\ t]dt + \sigma[x(t),\ t]dW(t)\ , \qquad (2)$$

where $\mu[x(t),\ t]$ is a drift coefficient, $\sigma[x(t),\ t]^2$ is a diffusion coefficient, and $W(t)$ is the Weiner process. A Fokker-Plank equation for the risk-neutral conditional probability density is then derived by constructing a risk-free self-financing portfolio consisting of a linear combination of the option, stock, and zero-coupon bonds.[4] Popular - analytically tractable - solutions include (i) geometric Brownian motion $dx = \mu x dt + \sigma x dW$ which yields the well known Black-Scholes [4] equations for option prices, (ii) for arithmetic Brownian motion $dx = \mu x dt + \sigma dW$, and (iii) for the "square-root" process $dx = \mu x dt + \sigma\sqrt{x}dW$. While this approach has worked well in the past we are, with increasing frequency, encountering markets where the expressions for option prices that result from these SDEs are at odds with observed option prices. The most common example of the is seen in the so-called "volatility smile"[5]: if one takes observed option prices differing only in k and inverts the analytic expressions for the option prices to obtain a value for σ one often finds that σ is a function of k when it was originally assumed that σ was independent of k. One remedy for this problem has been to propose functional forms for $\mu[x,\ t]$ and for $\sigma[x,\ t]$ and to fit the resulting expressions to option prices. Alternatively, on can - and we do - view this as problem of inference

---

[2] This has been a very brief introduction to one particular type of option. A complete discussion of options can be found in a number of texts including those by Cox and Rubinstein [1] and Hull [2].

[3] The prefactor $\exp[-r(t' - t)]$ denotes the amount of money you would need to invest today at r to receive a dollar $t' - t$ hence.

[4] The Fokker-Plank equation for the risk-neutral conditional probability density differs from the Fokker-Plank equation that follows from the SDE given above as discussed, for example, in Gardiner [3]. The reader is referred to Hull [2] for a full discussion of the derivation.

[5] This term comes from the picture that sometimes results when one plots the implied volatility as a function of k. A discussion of the volatility smile and of ways of dealing with it can be found in a recent paper by Kuwahara and Marsh [5].

The motivation for this paper is our observation that the integral equation for the price of a European option is formally identical to that of a blurred image. In the language of image reconstruction, observed option prices are a blurred image of the discounted conditional probability density[6]: the discounted conditional probability density is blurred by the option payoff formula[7]. Consequently, it should be possible to obtain the PDF from observed option prices using the MaxEnt approach to numerical image reconstruction.

Broadly speaking the approaches to reconstructing the PDF that have appeared in the literature to date fall into 2 categories[8]. First are those that begin with a particular model of reality (e.g. stochastic process or explicit functional form) and use observed option prices to fix free parameters in the resulting PDF: the well-known implied volatility, the work of MacBeth and Merville [9], Shimko [10], Derman and Kani [11], and Dupire [12] are examples of this approach. Second are those that view the reconstruction of the PDF as a problem of inference: the work of Ross [13], Banz and Miller [14], Breeden and Litzenberger [15], Jarrow and Rudd [16], Longstaff [17], the Lagrange approach of Rubinstein [18], and the present paper fall into this category. While both categories play an important role in the development of a consistent approach to option pricing, we would argue that the inferential approach with its greater flexibility is better suited as the first step in the understanding of financial markets. By providing a complete description of the PDF at a given point in time, an inferential study provides the benchmark against which dynamical models can be tested. Furthermore, in providing this benchmark, it is crucial that the PDF provided by the inferential analysis be as free of method-induced artifact as possible. Thus, MaxEnt is the method of choice.

The use of information theory in general and of Maximum Entropy in particular for the solution of deconvolution problems in economic theory, econometrics, and financial economics has appeared previously and is discussed and summarized in the work of Kapur [19], Zellner [20], Stutzer [21], and Sengupta [22]. Our work builds on this, but differs in that our application of maximum entropy is a novel synthesis of the powerful image reconstruction technology afforded by MaxEnt with the practical need to obtain the PDF for option pricing.

---

[6]Alternatively, we could say that the option prices are a blurred image of the state-price density since the state price density is the discounted conditional probability density.

[7]Indeed, the plethora of formal and practical similarities that we found when comparing the problem of reconstruction in option pricing theory to the problem of reconstruction in astrophysics prompted our "appropriation" of the title of the paper by Gull and Daniell [6] where an early application of MaxEnt in astrophysics appeared. A discussion of the use of MaxEnt in a variety of image reconstruction contexts can be found in the paper by Gull and Daniell [7].

[8]These categories, "ontology" and "epistemology", were first developed and discussed in the context of MaxEnt by Gull [8].

2. Theory

For purposes of illustration, let us consider the expression for the price of a European call option rewritten with a minimum of notation

$$c(k) = e^{-rt} \int_{-\infty}^{+\infty} \max(x - k,\, 0)\, p(x) dx \,, \tag{3}$$

We take $c(k)$, k, r, and t to be known, and ask what PDF is associated with the observed $c(k)$? The observed call value $c(k)$ can also be viewed as the mean value of the function $e^{-rt} \max(x - k,\, 0)$, and it is from this viewpoint that a natural connection to MaxEnt can be easily made: given a set of observed mean values of known functions together with the standard errors of these observed values, we want a PDF that yields the observed values, is positive, and is normalized to unity. As shown by Jaynes [23] in his original work on MaxEnt, the PDF that includes the information provided by the observed mean values but is, otherwise, *assumption-free,* is the one that maximizes the entropy

$$S = - \int_{-\infty}^{+\infty} p(x) \ln p(x) dx \,, \tag{4}$$

subject to the constraints $p(x) \geq 0$, $\int_{-\infty}^{+\infty} p(x) dx = 1$, and that the observed call values be reproduced. If the call values were observed without error then we could apply the original equations of Jaynes [23] as described by Stutzer [21] and Sengupta [22]. Observed option prices are, however, noisy. Consequently we must include a measure of the error in the observation into the reconstruction. The measure of the error in the reproduction of the observations that is most common in the use of MaxEnt is the chi-square error statistic

$$\chi^2 = \sum_{m=0}^{M} \left(\frac{c_{calc}(k_m) - c_{obs}(k_m)}{\sigma(k_m)} \right)^2 \,, \tag{5}$$

where M is the number of observed call prices, $c_{calc}(k_m)$ is the calculated call price at the mth strike price, k_m is the mth strike price, $c_{obs}(k_m)$ is the observed call price at the mth strike price, and $\sigma(k_m)$ is the standard error of $c_{obs}(k_m)$.

   In our discussion so far the imposed constraints have been limited to the observed option prices. In the limit of no option price information this leads to uniform MaxEnt PDF which is generally not consistent with our prior knowledge of the problem. It is possible, and may be desirable, to include more constraints. We can conveniently fold our prior information concerning the PDF into our definition of the entropy as follows:

$$S = - \int_{-\infty}^{+\infty} p(x) \ln \left(\frac{p(x)}{g(x)} \right) dx \,, \tag{6}$$

where $g(x)$ can be the lognormal density or whatever density is most likely in the absence of any option price information. Now, in the absence of any option price information the MaxEnt PDF is $g(x)$ rather than the uniform density.

As mentioned above this paper was motivated by the similarity between the integral equation for a blurred image and the integral expression for the price of a European option $v(k)$:

$$v(k) = \int_{-\infty}^{+\infty} h(x,\ k)\pi(x)dx\ ,\tag{7}$$

where $h(x\ k)$ is the known payoff function at expiration and $\pi(x)$ is the state-price density function. The connection to a blurred image can be seen as follows. If $h(x,\ k)$ is an "elementary claim" in the sense of Breeden and Litzenberger [15] (i.e. a payoff of unity if and only if $x = k$, and of zero otherwise)[9] then the observed option prices would provide a "clear" view of the state prices: $v(k) = \pi(k)$. In general, however, option payoff functions are wider in x than "elementary claims" and the observed option prices are made up of contributions from state-prices over a range of x just as a blurred image is made up of contributions of an image from a range of spatial locations. In terms of state prices, the theory as previously developed follows except that the Lagrange function is formed without the normalization to unity. The MaxEnt state-price density is now obtained by maximizing the Lagrange function

$$L = - \int_{-\infty}^{+\infty} \pi(x) \ln \left(\frac{\pi(x)}{f(x)} \right) dx - \frac{1}{2}\alpha\chi^2\ ,\tag{8}$$

where $f(x)$ is the state-price prior.

We highlight in passing three major features of the MaxEnt PDF that make it particularly useful for option pricing theory. First, we are guaranteed that the PDF will be positive as the entropy is not defined for nonpositive numbers. Second, the MaxEnt PDF is the most uniform density that is consistent with the observed data. While proving this is beyond the scope of our paper[10] it is a key feature of MaxEnt that recommends it over other methods. Thus, any structure that appears in a MaxEnt reconstruction is *required* to reproduce the observed option prices. Finally, from a practical standpoint it is quite easy for prior information about the behavior of options into the reconstruction of the PDF. Indeed, one can incorporate prior knowledge concerning both the option *and* the asset upon which the option is written in a natural and systematic framework.

3. Applications

We have used the iterative MaxEnt algorithm developed by Skilling and Bryan [25] to reconstruct the PDF associated with S&P 500 index options as observed from 1987 to 1992. As a prior we used the lognormal PDF based on the average implied volatility of the two nearest-the-money call options. We used mid-market values as the observed option prices and our proxy for the standard deviation was

---

[9]In our continuous-time formalism the payoff function would be the Dirac delta function $\delta(x)$ and Eq. 7 would become $v(k) = \int_{-\infty}^{+\infty} \delta(x - k)\pi(x)dx = \pi(k)$.

[10]Discussion of this can be found in Buck and Macaulay [24] and references therein.

one-half of the bid-offer spread. The level of the index was included as it is the present value of a payout-protected call struck at zero. The algorithm was stopped when $\chi^2 = M$; (the number of observed prices) as it is the expected value for a χ^2 distribution. While a detailed examination of the temporal evolution of the PDF implied by the nearly 3000 reconstructions that we have carried out is beyond the scope of the present paper and will be discussed in a later communication, we will examine three of these reconstructions shown in Fig. 1.

We begin with the PDF reconstructed from S&P 500 index option prices posted six hours after the opening on January 30, 1987 that is shown in Fig. 1a. In all panels of Fig. 1 the lognormal prior is depicted by the dashed line while the solid line shows the MaxEnt PDF. Option prices during this period were remarkably lognormal and the MaxEnt density is essentially the same as the lognormal prior. Interestingly, this "lognormal" state is also seen after the stock market crash of 1987. This is illustrated in Fig. 1b where we show the lognormal and the MaxEnt PDFs corresponding to the S&P 500 index options prices posted seven hours after the opening on December 17, 1987. The PDF has clearly shifted to lower levels, but remains essentially lognormal. The PDF changes substantially following the events of October 1989. The PDF implied by the S&P 500 index option prices posted three hours after the opening on January 2, 1990 is shown in Fig. 1c and differs greatly from the lognormal prior: indeed, a bimodal shape is seen. This type of PDF is seen frequently, but not exclusively, following the Fall of 1989, and this represents a substantial change in the dynamics of this market. The S&P 500 has changed from a relatively lognormal market to one in which the implied PDF can, and does, make frequent, extended, and large departures from lognormality. Clearly, adequate risk management in this market requires a more nimble approach than that offered by traditional PDF theories.

4. Summary

The reconstruction of a probability density function (PDF) from incomplete and noisy option data is a problem of deconvolution also known as image reconstruction. We have, for the first time, applied a technique of image reconstruction based on Maximum Entropy - MaxEnt - to the reconstruction of the PDF of S&P 500 index options. Like others, we have found that PDFs reconstructed from S&P 500 index option prices posted after the Fall of 1989 can display large departures from lognormality. In contrast to previous methods MaxEnt provides the most uniform PDF consistent with the observed data, and it thus provides a benchmark against which PDF models can be tested. These features, combined with the ease with which prior information can be included in the calculation of the PDF recommend MaxEnt highly for this and other inverse problems in financial economics.

5. Acknowledgments

We thank F.A. Longstaff for useful discussions and W.J. Keirstead for computational assistance. RJH thanks J. S. Kallman for introducing him to the Maximum Entropy Method, D. R. Heatley and S. Kim for insightful comments on an earlier

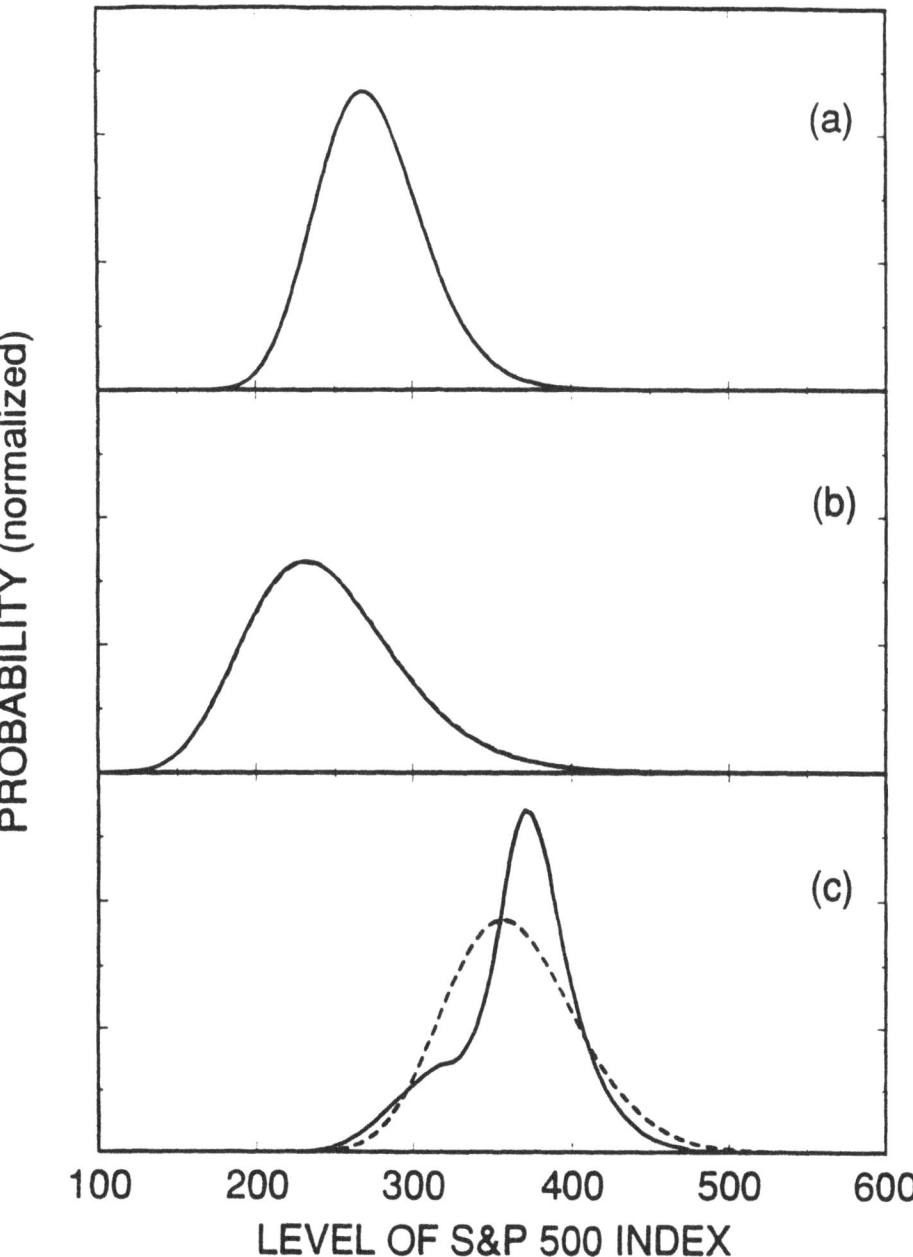

Figure 1. Normalized MaxEnt reconstructions of the probability density function for S&P 500 index options based on observed option prices. The dashed lines are the lognormal prior based on the implied volatility calculation described in the text. The solid lines are the MaxEnt reconstructions. Panel 1a was calculated using option prices from January of 1987. Panel 1b was calculated using option prices from December of 1987. Panel 1c was calculated using option prices from January of 1990. While all panels show both lognormal and MaxEnt results, the difference between them cannot be resolved in panels 1a and 1b.

verson of this paper, J. M. Haile for useful discussions, and W. H. Chute for his
encouragement.

References

1. J. C. Cox and M. Rubinstein, *Options Markets*, Prentice-Hall, New Jersey, 1985.
2. J. C. Hull, *Options, Futures, and Other Derivative Securities*, Prentice-Hall, New Jersey, 2 ed., 1993.
3. C. W. Gardiner, *Handbook of Stochastic Methods for Physics, Chemistry, and the Natural Sciences*, Springer-Verlag, New York, second ed., 1985.
4. F. Black and M. Scholes, "The pricing of options and corporate liabilities," *Journal of Political Economy*, **81**, pp. 637–659, 1973.
5. H. Kuwahara and T. A. Marsh, "Why doesn't the Black-Scholes model fit Japanese warrants and convertible bonds?," *Japanese Journal of Financial Economics*, **1**, pp. 33–65, 1994.
6. S. F. Gull and G. J. Daniell, "Image reconstruction from incomplete and noisy data," *Nature*, **272**, pp. 686–690, 1978.
7. S. F. Gull and G. J. Daniell, "Maximum entropy method in image reconstruction," *IEEE Proceedings Part E*, **131**, pp. 646–659, 1984.
8. S. F. Gull, "Some misconceptions about entropy," in *Maximum Entropy in Action: A Collection of Expository Essays*, B. Buck and V. A. Macaulay, eds., pp. 171–186, Oxford University Press, New York, 1992.
9. J. D. MacBeth and L. J. Merville, "Tests of the Black-Scholes and Cox call option valuation models," *Journal of Finance*, **35**, pp. 285–300, 1980.
10. D. Shimko, "Bounds of probability," *RISK*, **6**, pp. 33–37, 1993.
11. E. Derman and I. Kani, "Riding on the smile," *RISK*, **7**, pp. 32–29, 1994.
12. B. Dupire, "Pricing with a smile," *RISK*, **7**, pp. 18–20, 1994.
13. S. A. Ross, "Options and efficiency," *Quarterly Journal of Economics*, **90**, pp. 75–89, 1976.
14. R. W. Banz and M. Miller, "Prices for state-contingent claims: some estimates and applications," *Journal of Business*, **51**, pp. 653–672, 1978.
15. D. T. Breeden and R. H. Litzenberger, "Prices of state-contingent claims implicit in option prices," *Journal of Business*, **51**, pp. 621–651, 1978.
16. R. Jarrow and A. Rudd, "Approximate option valuation for arbitrary stochastic processes," *Journal of Financial Economics*, **10**, pp. 347–369, 1982.
17. F. Longstaff, "Martingale restriction tests of option pricing models," tech. rep., University of California, Los Angeles, 1990. Working Paper, version 1.
18. M. Rubinstein, "Implied binomial trees," *Journal of Finance*, **49**, pp. 771–818, 1994.
19. J. N. Kapur, *Maximum Entropy Methods in Science and Engineering*, Wiley Eastern Limited, New Delhi, 1989.
20. A. Zellner, "Bayesian methods and entropy in economics and econometrics," in *Maximum Entropy and Bayesian Methods*, J. W.T. Grandy and L. H. Schick, eds., pp. 17–31, Kluwer Academic, Dordrecht, 1991.
21. M. Stutzer, "State prices and Gibbs states," tech. rep., Dept. of Finance, Carlson School of Management, University of Minnesota, 1992. Working paper.
22. J. K. Sengupta, *Econometrics of Information and Efficiency*, Kluwer Academic, Dordrecht, 1993.
23. E. T. Jaynes, "Prior probabilities," *IEEE Transactions on Systems Science and Cybernetics*, **SSC-4**, pp. 227–241, 1968.
24. B. Buck and V. A. Macaulay, eds., *Maximum Entropy in Action: A Collection of Expository Essays*, Kluwer Academic, Dordrecht, 1992.
25. J. Skilling and R. K. Bryan, "Maximum entropy image reconstruction: general algorithm," *Mon. Not. R. Ast. Soc.*, **211**, pp. 111–124, 1984.

MODEL SELECTION AND PARAMETER ESTIMATION FOR EXPONENTIAL SIGNALS

ANAND RAMASWAMI
Department of Physics, Washington University
1 Brookings Drive,
St. Louis, Missouri 63130

AND

G. LARRY BRETTHORST
Department of Chemistry, Washington University
1 Brookings Drive,
St. Louis, Missouri 63130

Abstract. In this paper, probability theory is applied to the problem of estimating the decay rate constants in data that are known to contain sums of exponentials. In many instances the number of exponential components is unknown. One might naively believe that the correct strategy, to estimate the parameters, is to first determine the number of exponentials and then estimate the parameters from a model containing that number of exponentials. However, this is not what the rules of probability theory indicate should be done. Probability theory indicates that the best parameter estimates are obtained from the probability for the parameter of interest independent of the number of exponentials in the data. This probability density function is a weighted average. It is a sum over the probability for the parameter given the number of exponentials weighted by the probability that this number of exponentials is the correct value. When the number of exponentials in the data are well determined, this reduces to the problem of determining the number of exponentials and then estimating the parameters given that number of exponentials. However, when the data fail to strongly support a single value for the number of exponentials the result from probability can differ significantly from this intuitive procedure. In this paper a sketch of these calculations is presented, and numerical examples are used to illustrate the calculations.

Key words: Exponential, Model Selection, Parameter Estimation

K. M. Hanson and R. N. Silver (eds.), Maximum Entropy and Bayesian Methods, 9–14.
© 1996 *Kluwer Academic Publishers.*

1. Introduction

Exponential models are ubiquitous throughout all branches of science and engineering. Thus, there is considerable interest in the measurement and analysis of multiexponential data. In chemistry, the concentrations of reactants often follow exponential kinetic schemes. Radioactive nuclear decay is exponential and in nuclear magnetic resonance (NMR) the rates of polarization and decay of magnetization are also generally exponential in nature. In all of these examples, the values of the amplitudes and decay rate constants contain the information of interest. Because such importance is attached to the amplitudes and decay rate constants, it is desirable to have available mathematical methods with strong rigorous foundations.

In this paper probability theory as generalized logic will be used to solve this problem. The problem of estimating the decay rate constants independent of the number of exponentials will be used to illustrate the techniques needed. Other problems, such as estimating the amplitudes, and noise variance are solved similarly. To estimate the decay rate constants, first, a data set $D \equiv \{d_1, \ldots, d_N\}$ is postulated. This data has been sampled from a time series $y(t)$ at discrete times t_i ($1 \leq i \leq N$); uniform sampling is not assumed. The time series $y(t)$ is assumed to be the sum of two terms, a signal plus noise:

$$d_i = y(t_i) = f(t_i) + e_i \quad (1 \leq i \leq N), \tag{1}$$

where the signal $f(t)$ is taken to be of the form

$$f(t_i) = \sum_{j=1}^{m} B_j \exp\{-\alpha_j t_i\} \tag{2}$$

where e_i represents the value of the noise at time t_i, B_j is the amplitude of jth exponential, α_j is the decay rate constant, and m is the number of exponentials. From Eq. (2), the multiexponential model is invariant under permutations of the labels of the decay rate constants and amplitudes. This invariance manifests itself in the joint posterior probability density for the decay rate constants and can be visualized graphically as multiple peaks of equal probability corresponding to exchange of the decay rate constants. Therefore, a convention must be adopted that distinguishes one component from another. Here the exponential rate constants will be ordered such that $\alpha_1 < \alpha_2 < \cdots < \alpha_m$.

2. Estimating The Decay Rate Constant Independent Of The Number Of Exponentials

In probability theory all of the information relevant to estimating a decay rate, α_j, is summarized in a probability density function. For α_j this is denoted by $P(\alpha_j|DI)$: the probability that decay rate "j" had value α. This quantity is computed from the joint probability for α_j and the number of decay rate constants,

m, by using the sum rule:

$$P(\alpha_j|DI) = \sum_{m=1}^{max} P(\alpha_j m|DI) = \sum_{m=1}^{max} P(m|DI)P(\alpha_j|mDI) \qquad (3)$$

where $P(m|DI)$ is the probability for the number of exponentials given the data and the prior information, and $P(\alpha_j|mDI)$ is the probability for the jth decay rate constant given the number of exponentials, the data, and the prior information. We define max as the upper bound on the number of signal components (which we take as three for practical computational reasons). To obtain the optimal estimate of the decay rate constant requires three steps. First, the probability for the number of exponentials, $P(m|DI)$, must be computed. Second, the probability for the decay rate constant given the number of exponentials, $P(\alpha_j|mDI)$, must be computed. Last, the sum must be computed. From a practical standpoint this last step need only be done if the data do not strongly indicate one particular value of m.

To apply Eq. (3) one must obtain both the probability for the number of exponentials and the probability for the parameter given the number of exponentials. Thus, we have a model selection problem and a parameter estimation problem. Both of these types of problems are well studied in probability theory. For specific examples see [1,2]. For a tutorial on parameter estimation see [3] and for a tutorial on model selection see [4]. Here we use the results described in those papers to illustrate how to estimate parameters independent of the model order.

Before the problem is addressed one might ask: What does it mean to estimate the low decay rate constant, α_1, the medium decay rate constant, α_2, or the high decay rate constant, α_3 given a one exponential model? Here we adopt the convention that if there is a single exponential model the low, medium, and high decay rate constants are all the same. Similarly for a bi-exponential model we defined the medium and high decay rate constants to be the same.

3. Example

The data set used in the example is shown in Fig. 1(a). The probability for the number of exponentials was computed for this data, Fig. 1(b). The probability for the number of exponentials is equally split between one and two exponentials. Note that while this probability distribution indicates there is probably one or two exponentials present, three is completely ruled out. The probability for a third exponential is approximately 10^{-5}.

To estimate the low decay rate constant we must compute the probability density function for the low decay rate constant given the one, two and three exponential models. These probability density functions are shown in Figures 2(a), 2(b), and 2(c) respectively. The probability for the low decay rate independent of the number of exponentials, the weighted average, is shown in Figure 2(d). Similarly, the probability for the middle decay rate constant, Fig 3(d), is computed from the probability for the middle decay rate constant given the number of exponentials is one, two, or three in Figs 3(a), 3(b), and 3(c) respectively. Finally,

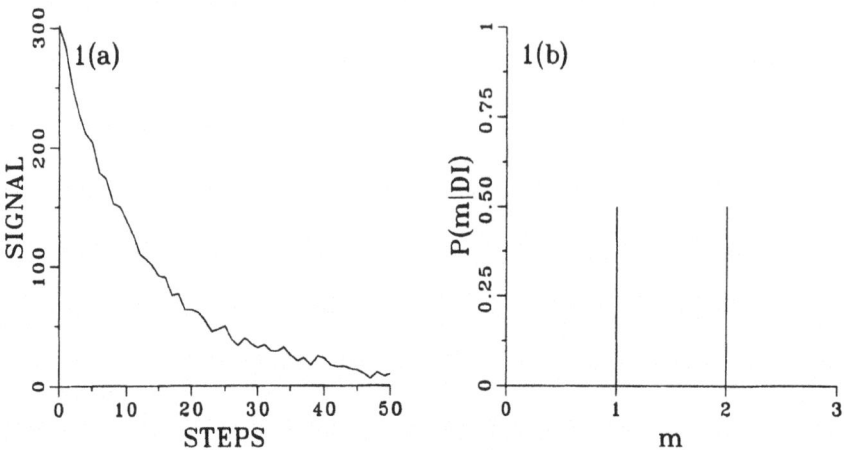

Figure 1. Figure 1(a) is simulated data. It consists of a bi-exponential signal plus Gaussian noise. The exponentials have amplitudes of 100 and 200, and decay rate constants of 0.05 and 0.1 respectively. The noise variance was carefully chosen so that the posterior probability for the number of exponentials given the data, Fig 1(b), would not strongly favor either a one or two exponential model.

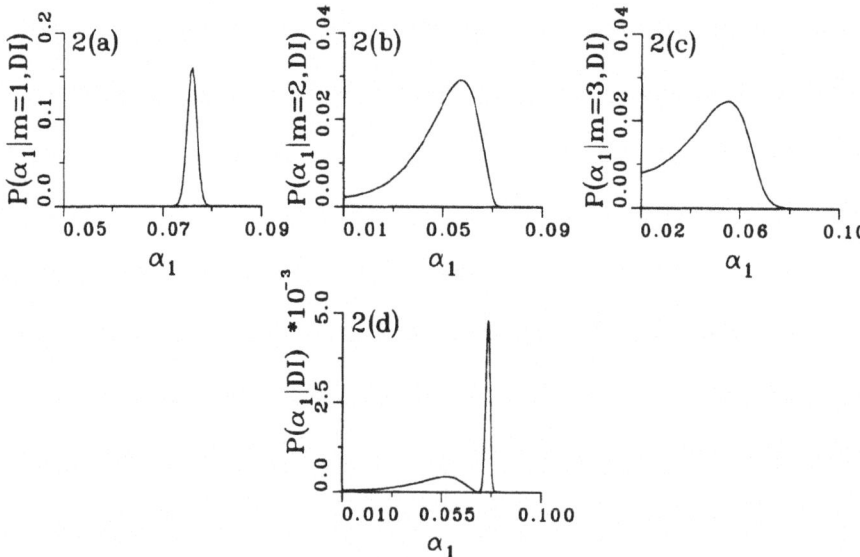

Figure 2. Figure 2(a,b,c) shows the probability distribution for the low decay rate constant given one, two, and three exponential models respectively. Figure 2(d) show the probability distribution for the low decay rate constant independent of the number of exponentials. From Eq. (3) this is a weighted average of Panels 2(a), 2(b), and 2(c); the

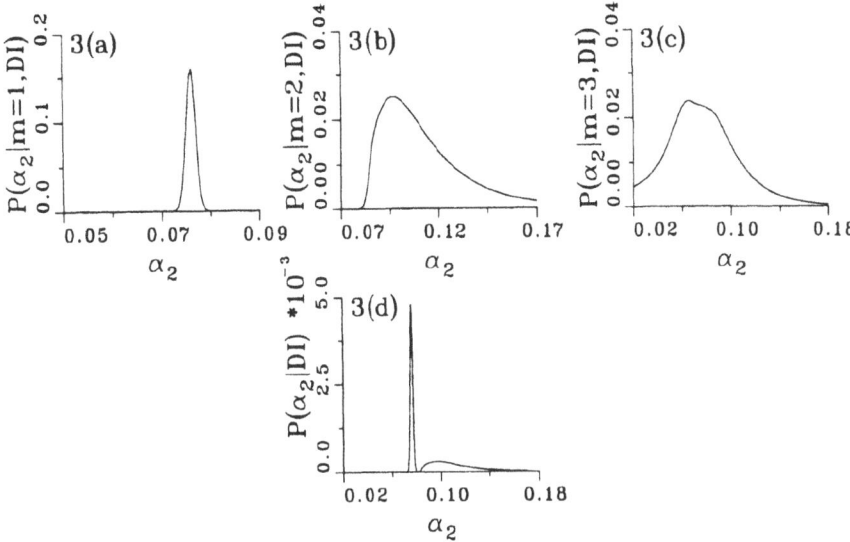

Figure 3. Figure 3(a,b,c) shows the probability distribution for the medium decay rate constant given one, two, and three exponential models respectively. Figure 3(d) show the probability distribution for the medium decay rate constant independent of the number of exponentials. From Eq. (3) this is a weighted average of Panels 2(a), 2(b), and 2(c); the weights are 0.499, 0.501, and 10^{-5}.

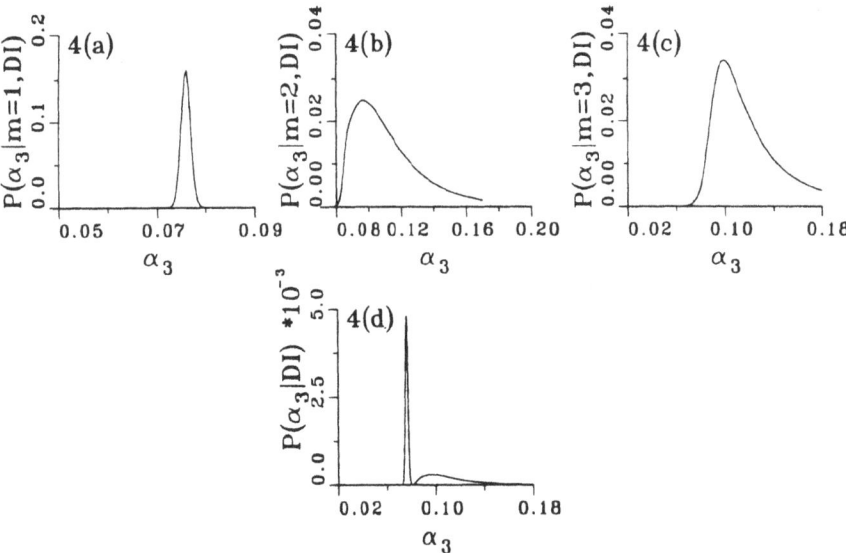

Figure 4. Figure 4(a,b,c) shows the probability distribution for the high decay rate constant given one, two, and three exponential models respectively. Figure 4(d) show the probability distribution for the high decay rate constant independent of the number of exponentials. From Eq. (3) this is a weighted average of Panels 2(a), 2(b), and 2(c); the weights are 0.499, 0.501, and 10^{-5}.

the same procedures may be applied to the problem of estimating the high decay rate constant, Fig 4. Notice that here the probability for the high decay rate constant, Fig 4(d), is the same as the probability for the middle decay rate constant, Fig 3(d). The reason for this is that there is no evidence in the data for more than two exponentials, so the probability for the high decay rate constant is the same as the probability for the middle decay rate constant. This is just what one would expect given the convention we adopted.

4. Summary

Probability theory indicates that all of the information relevant to estimating the decay rate constants independent of the number of exponentials is summarized in a probability density function. This probability density function is a weighted average of the probability for the decay rate constant of interest given that one knows the number of exponentials. The weights are just the probability that the number of exponentials was correct.

References

1. Bretthorst, G. Larry., *Bayesian Analysis. II. Signal Detection and Model Selection*, J. Magn. Reson. **88**, pp. 552-570 1990.
2. Gull Stephen F., *Bayesian Inductive Inference and Maximum Entropy*, G.J. Erickson and C.R.Smith (Eds), in "Maximum Entropy and Bayesian Methods in Science and Engineeringl" Vol 1, pp. 53-74 Kluwer Academic Publishers 1988.
3. Bretthorst, G. Larry., *An Introduction to Parameter Estimation Using Bayesian Probability Theory*, P. Fougere (Ed.), in "Maximum Entropy and Bayesian Methods" p. 53, Kluwer Academic Publishers 1989.
4. Bretthorst, G. Larry., *An Introduction to Model Selection using Probability Theory as Logic*, G. Heidbreder (Ed.), in "Maximum Entropy and Bayesian Methods" Kluwer Academic Publishers, in press.

HIERARCHICAL BAYESIAN TIME SERIES MODELS

L. MARK BERLINER
National Center for Atmospheric Research
& Ohio State University
NCAR, P.O. Box 3000, Boulder, CO 80307-3000, USA. [†]

Abstract. Notions of Bayesian analysis are reviewed, with emphasis on Bayesian modeling and Bayesian calculation. A general hierarchical model for time series analysis is then presented and discussed. Both discrete time and continuous time formulations are discussed. An brief overview of generalizations of the fundamental hierarchical time series model concludes the article.

Key words: Dynamical model, Fokker-Planck equation, Markov process, Prediction, Stochastic differential equation

1. Intorduction

1.1. THE BAYESIAN VIEWPOINT

Much of the Bayesian viewpoint can be argued (as by Jeffreys and Jaynes, for examples) as direct application of the theory of probability. In this article the suggested approach for the construction of Bayesian time series models relies on probability theory to provide decompositions of complex joint probability distributions. Specifically, I refer to the familiar factorization of a joint density into an appropriate product of conditionals.

Let x and y represent two random variables. I will not differentiate between random variables and their realizations. Also, I will use an increasingly popular generic notation for probability densities: $[x]$ represents the density of x, $[x|y]$ is the conditional density of x given y, and $[x, y]$ denotes the joint density of x and y. In this notation we can write "Bayes's Theorem" as

$$[y|x] = [x|y][y]/[x]. \tag{1}$$

Equally important to probability theory and to Bayesian modeling is the relation

$$[x] = \int [x|y][y]dy. \tag{2}$$

[†]Email: berliner@ucar.edu. This work was supported by the NSF under Grant #DMS93-12686.

K. M. Hanson and R. N. Silver (eds.), Maximum Entropy and Bayesian Methods, 15–22.

Of course, relationships such as (1) and (2) hold for conditional densities. For example, if w is a third random variable, then

$$[x|w] = \int [x|y, w][y|w]dy.$$

Furthermore, a variety of nestings, hierarchies, and other relations among various conditional densities are possible. Finally, appropriate factorizations of joint densities are useful. For example,

$$[x, y, w] = [x|y, w][y|w][w] \tag{3}$$

is a familiar result in probability theory.

Equation (3) is particularly relevant to the discussion here. Specifically, it is the basis of *Hierarchical Models*. As modelers, faced with complex structures and a variety of random quantities to be modeled, Bayesians and other "stochastic modelers" break the modeling process of a large collection of variables into the pieces, following (3), and model the required conditional distributions. Hierarchical models have a long history in Bayesian statistics. Some discussion and references may be found in [1] and [2].

Markovian Models form a quintessential example of conditional modeling. In large scale applications of generalizations of (3), various patterns or structures in the conditional distributions are considered. Consider a time series of data, $x_1, \ldots, x_n$. We can write the joint distribution of these values as

$$[x_1, \ldots, x_n] = [x_n|x_{n-1}, \ldots, x_1][x_{n-1}|x_{n-2}, \ldots, x_1] \ldots [x_1]. \tag{4}$$

A common assumption about these conditionals goes something like "the distribution of x_t given 'all' the past x_s's only depends on a restricted subset of the *recent* past." A one-step Markov model is that for each t, $[x_t|\text{the past}] = [x_t|x_{t-1}]$, leading to the joint distribution

$$[x_1, \ldots, x_n] = [x_n|x_{n-1}][x_{n-1}|x_{n-2}] \ldots [x_1].$$

Markovian reasoning has been applied beyond the time series setting. The key observation is that one is free to index a countable collection of random variables in any convenient, meaningful fashion. The modeler may then directly apply (4) and formulate the resulting conditionals. A primary example involves modeling of spatial correlation in spatial statistics and image analysis. See [3] for discussion and references. The Markovian step involves the intuition that the value of a variable at a location, conditional on the values at appropriate collections of other locations, actually depends only on the values in a subset of "nearby" locations. Users of *Markov random field* models use this sort of reasoning, though the usual construction is not a direct use of (4). Hence, some care is taken to insure that the resulting specifications do yield a true joint distribution. *Markov meshes*, special cases of Markov random fields, are constructed by direct use of (4). See [4]. More generally, hierarchical reasoning offers an organized approach to spatio-temporal modeling, but this topic is beyond the scope of this article.

1.2. OUTLINE

Section 2 describes representations of an archetypal hierarchical model for time series. The model is presented in three stages. A casual way to think about these stages is:

Stage 1. [data|process, parameters].
Stage 2. [process|parameters].
Stage 3. [parameters].

In the time series context, the time evolution of the process of interest is primarily modeled in Stage 2. Both discrete and continuous time models are considered. The reader will note relationships between the models described and the so-called Kalman filter formulation; see [5]. In Section 3 a very brief discussion of some natural extensions of the basic models is given.

The purpose of this article is to present notions of modeling strategies useful in time series. For the most part, I will present the models with no concern for computational complexity. The models presented are indeed being developed today. Modern research involving Markov chain Monte Carlo offers a general approach to the approximation of Bayesian results in complex settings. See [6] for some discussion. Also, I will focus on "time domain" modeling; discussion of Bayesian spectral analysis may be found in [7]. Finally, this article is not intended to be a review of Bayesian time series; the reader is referred to [8], [9], and [10] for discussions and further references.

2. Hierarchical Models

2.1. DISCRETE TIME

Assume that a stochastic process, $x_0, x_1, \ldots$, is under study. To allow for measurement error, we allow that the process is observed indirectly as follows:

Stage 1. *Distribution of the Observables.* Assume that a set of n data values, $y_{s_1}, \ldots, y_{s_n}$, are observed. The first stage describes the structure of the conditional distribution of the data, given the underlying x process and any parameters of the model. A common choice is of the form

$$[y_{s_1}, \ldots, y_{s_n} | \{x_t\}_{t \geq 0}, \theta_1] = \prod_{i=1}^{n} [y_{s_i} | x_{s_i}, \theta_1]. \tag{5}$$

The subscripts are intended to allow for a wide variety of sampling procedures. For example, in principle Bayesian analysis is unconcerned with issues of "equally spaced" observations. Next, I have assumed that the modeler wishes to allow for model parameters, represented by θ_1. A common example involves a regression formulation in which

$$y_{s_i} = G(x_{s_i}, \eta) + e_{i+1}, i = 1, \ldots, n, \tag{6}$$

where G is a regression function, η represents unknown regression coefficients, and the errors, e_i are uncorrelated, mean zero, random variables with some fixed density. Let the variance of these variables be denoted by v^2. In this context, we have $\theta_1 = (\eta, v^2)$. Another useful error model is a mixture of two distributions, one of which is comparatively longer tailed than the other, thereby presuming to allow for "outliers." The mixing probability can be incorporated into θ_1.

Stage 2. *Structure of the x Process.* A natural model for the evolution of the x process is a dynamic, typically Markovian, model. For example, a one-step Markov model involves the conditioning formula

$$[x_{t+1}|\{x_o, \ldots, x_t\}, \theta_2] = [x_{t+1}|x_t, \theta_2], \tag{7}$$

where θ_2 is a vector of parameters associated with the Stage 2 model. A common structure for this stage involves an autoregressive model,

$$x_{t+1} = F(x_t, \beta) + z_{t+1}, t \geq 0, \tag{8}$$

where F is a dynamical function, β represents unknown regression coefficients, and the z_t are mean zero, random variables. These variables are often suggested to represent unmodeled environmental effects, "noise," and uncertainty concerning the functional form F. Mixture models for the distribution of the effects may also be appropriate. In general, θ_2 is the vector of parameters composed of β and any parameters in the modeler's specification of the distribution of the z_t's. Finally, a prior for the initial condition is proposed: $[x_o]$. (This distribution may depend on the parameters.) Note that the one-step Markov model is merely an example. Higher order time dependencies can of course be modeled. Also, a high order Markovian model can often be written as a one-step, Markov model via *state space representation.*

Stage 3. *Prior on Parameters.* As a final stage for the model, we construct a distribution for the "parameters" introduced above: $[\theta_1, \theta_2]$.

Note that, the presentation of hierarchical models typically involves Markovian like reasoning, but without explicit reference. For example, the Stage 1 distribution described above is actually

$$[y_{s_1}, \ldots, y_{s_n}|\{x_t\}_{t \geq 0}, \theta_1, \theta_2] = \prod_{i=1}^{n} [y_{s_i}|x_{s_i}, \theta_1],$$

but the model is that, given the x process and θ_1, the distribution of the data does not depend on θ_2. The specification of the components of these three stages yields a bona fide joint distribution for all the quantities modeled.

Direct computation, that is, probability theory, yield conditional distributions of interesting quantities given the observed data. The main object is

$$[\{x_t\}_{t \geq 0}, \theta_1, \theta_2|y_{s_1}, \ldots, y_{s_n}]. \tag{9}$$

Filtering and interpolation (inference for the x process at times corresponding to observation times and between observation times), backcasting or retrospection (inference for the x process at times before the first observation time), and prediction or forecasting (inference for the x process at times after the last observation time), are based on

$$[\{x_t\}_{t\geq0}|y_{s_1},\ldots,y_{s_n}] = \int\int[\{x_t\}_{t\geq0},\theta_1,\theta_2|y_{s_1},\ldots,y_{s_n}]d\theta_1d\theta_2. \qquad (10)$$

(I wrote the above formula as if θ_1 and θ_2 are continuous random variables. The adjustments to the representation in cases involving discrete components are familiar.) I repeatedly used the word "inference" rather than estimation to emphasize the Bayesian view that the conditional distribution of the quantity of interest ought to be the focus. Of course, practical limitations often force summaries of these distributions, though care, including consideration of decision theoretic issues, should be taken.

2.2. CONTINUOUS TIME

A natural starting point for extending the hierarchical model to continuous time is the replacement of (8) with a stochastic differential equation model. (I will not discuss assumptions used to make sense of all the points raised here. See [11].) In particular, consider the model

$$dx = f(x, \beta)dt + \sigma(x, \alpha)dW, \qquad (11)$$

where dW represents white noise and f and σ are suitable functions, so that solutions to the equation make (Itô) sense. Define θ_2 to be the collection β and α.

"Kolmogorov Forward" or "Fokker-Planck" analysis based on (11) can be related to Bayesian calculations. The Fokker-Planck analysis solves the following problem: Assume that the initial value of the process described by (11) is a random variable, x_o, with specified density, $[x_o]$. Find the density, $p(x, t|\theta_2)$, of the x process at time t. The result is that $p(x, t|\theta_2)$ is an appropriate solution to the initial value problem

$$\frac{\partial p}{\partial t} = .5\frac{\partial^2}{\partial x^2}(\sigma^2p) - \frac{\partial}{\partial x}(fp), \qquad (12)$$

subject to the initial data $p(x, 0|\theta_2) = [x_o]$.

Suppose we can solve the Fokker-Planck equation. Assuming data collection as described in (5), the quantity $p(x, s_1|\theta_2)$ may be viewed as the (conditional on θ_2) prior density for $x(s_1)$. Combining the model, $[y_{s_1}|x(s_1), \theta_1]$, and $p(x, s_1|\theta_2)$ via Bayes's Theorem, we can obtain the conditional posterior density

$$[x(s_1)|y_{s_1}, \theta_1, \theta_2] \propto [y_{s_1}|x(s_1), \theta_1]p(x(s_1), s_1|\theta_2). \qquad (13)$$

This object then serves as the initial data for the Fokker-Planck equation for the conditional density of $x(t), t > s_1$. We can proceed sequentially as more data is collected. Let

$$D_i = \{y_{s_k} : 1 \leq k \leq i\}.$$

Then, $[x_{s_i}|D_i, \theta_1, \theta_2]$, serves as the initial data for finding the conditional density of $x(t), t > s_i$.

Example: *The Langevin Equation and the Ornstein-Uhlenbeck Process*. The special case of (11),

$$dx = -\beta x dt + \sigma dW, \tag{14}$$

is easily (Itô) integrated. The parameters, $\beta > 0$ and $\sigma > 0$ form θ_2 in this case. Assume that the prior for the initial condition is a normal (Gaussian) distribution with mean $\mu(0)$ and variance $\tau^2(0)$. Under these assumptions this second stage prior leads to a conditionally Gaussian process.

At the first stage, we assume that the n observations, $(y_{s_1}, \ldots, y_{s_n})$, are conditionally independent, normal random variables with means $x(s_i), i = 1, \ldots, n$, and variances, v_i^2.

Next, the Fokker-Planck equation can be solved (or other methods can be called upon) so that the sequential updating can be implemented. Specifically, consider time s_1. Analysis (see [11], pp. 358, 367-68) yields the prior on $x(s_1)$ is Gaussian, with mean

$$\mu(0) \exp(-\beta s_1) \tag{15}$$

and variance

$$\phi^2 = \tau^2(0) \exp(-2\beta s_1) + (\frac{\sigma^2}{2\beta})(1 - \exp(-2\beta s_1)). \tag{16}$$

Next, we combine this with the data point y_{s_1}, as prescribed in (13). The required calculation is familiar in Bayesian analysis ([1], pp. 129-30). The result is that $[x(s_1)|y_{s_1}, \theta_1, \theta_2]$ is a normal density, with mean

$$\mu(s_1) = \{\phi^2/(v_1^2 + \phi^2)\}y_{s_1} + \{v_1^2/(v_1^2 + \phi^2)\}\mu(0)\exp(-\beta s_1), \tag{17}$$

and variance

$$\tau^2(s_1) = \{v_1^2\phi^2\}/\{v_1^2 + \phi^2\}. \tag{18}$$

We can then continue sequentially as more data is collected by recursing $\mu(\cdot)$ and $\tau^2(\cdot)$, being careful to remember to use lengths of time intervals, $s_{k+1} - s_k$, appropriately, including the definition of the ϕ^2 function at each iterate. □

Calculations for dealing with the parameters in the model are direct, in principle. We would sequentially update the distributions, $[\theta_1, \theta_2|D_i]$ of the model parameters via Bayes's Theorem. Based on these distributions, we can compute the quantity,

$$[x(s_i)|D_i] = \int \int [x(s_i)|D_i, \theta_1, \theta_2][\theta_1, \theta_2|D_i]d\theta_1 d\theta_2.$$

The forward analysis described above yield sequential Bayesian predictive and parameter inference analyses. However, for filtering, interpolation, and backcasting based on the full data, one would need the conditional distributions of the x process at all required time points of interest given the full data set.

Implementation of the above analyses is formidible from a computational view. First, the Fokker-Planck equation is seldom tractable enough to be useful in the sense described above. Second, the application of Bayes's Theorem is also typically numerically intensive. Even in the Langevin/Ornstein-Uhlenbeck prior example, updating with priors on β and σ would be difficult. A variety of approximations are available. If the Fokker-Planck analysis appears to be not useful, an interesting possibility is to use a discrete time approximation to the continuous time model. See [12] for an example.

Another issue involves the assertion that an observation is based precisely on the exact value of the underlying process at a specified instant. In many settings uncertainty in the times of observation arise, see [13]. Second, many data collection techniques involve the observation (with error) of weighted time integrals and transforms of the underlying process. In principle, such data can be modeled by appropriate extensions of (6). Finally, in some circumstances it may be appropriate to consider "analog" or continuously sampled data.

3. Extensions

A variety of extensions, including the use of additional stages in hierarchies, are possible. I only allude to a few of these, primarily in the discrete time formulation, but note that this is an active area of research.

An obvious extension to the models described permits time varying parameters, $\theta_1(t)$ and $\theta_2(t)$. Formally, this is no extension at all, since we could append $\theta_1(t)$ and $\theta_2(t)$ to the definition of the variable x modeled in Stage 2. Formalities aside, it may often be sensible from a modeling viewpoint to separate out parameters from process variables. For example, the parameters might be viewed as "slowly varying," compared to the x process.

A more general direction for modeling time varying structure is to suggest that switching from one model paradigm to another occurs. A natural suggestion is to introduce a process variable, say I, where $I = 1, 2, \ldots,$ or K. Stage 2 is then extended as follows:

$$[x_{t+1}|\{x_o, \ldots, x_t\}, \{I_o, \ldots, I_t\}, \theta_2] = [x_{t+1}|x_t, I_t, \theta_2]. \qquad (19)$$

The variable I then indicates which of K models,

$$x_{t+1} = F_I(x_t, \eta(I)) + z_{t+1}^I, t \geq 0, \qquad (20)$$

and or parameters, $\eta(.)$, is in effect.

The indicator process's evolution itself is then modeled. (The Bayesian use of "mixture models" is of direct use in the sort of models described; see [13].) Two basic approaches are common. First, a (typically, one-step) Markov chain, is used. There is a growing literature on such hierarchical models under the name "Hidden Markov Models." (The "hidden" modifer refers to the fact that the indicators I_t are in fact not typically observed.) A second main class of models involve a variety of "state-dependent" models for the evolution of the indicator process. A key reference in this regard is [14]. In many settings it is natural to believe that the modeling of the dynamics of the x process, say by (8), by defining the autoregressive function relating x_{t+1} to x_t locally depending on the value of x_t. Further, the error process can too be modeled similarly. (H. Tong has been instrumental in making such reasoning, called the "threshold principle," popular in time series analysis; See [15]) Note that we may use the notation of the hidden Markov model, except that the indicator I_t at time t is a function of the state x_t, rather than an automonous process.

The reader may have detected the fact that many of the models described above are not necessarily associated with Bayesian time series. The reader may also have detected that the observation is irrelevant. Good, useful models can be incorporated readily as stages of Bayesian models. Many of us would argue that the hierarchical Bayesian approach therefore subsumes classical modeling in a fashion that both extends the range and enhances the interpretability of time series models.

References

1. J. O. Berger, *Statistical Decision Theory and Bayesian Analysis*, Springer-Verla, New York, 1985.
2. J. M. Bernardo and A. F. M. Smith, *Bayesian Theory*, John Wiley & Sons, Inc., New York, 1994.
3. Noel Cressie, *Statistics for Spatial Data*, John Wiley & Sons, Inc., New York, 1991.
4. K. Abend and T. J. Harley and L. N. Kanal, "Classification of binary random patterns," *IEEE Trans. Inform. Theory*, **IT-11**, pp. 538-544, 1965.
5. R. J. Meinhold and N. Singpurwalla, "Understanding the Kalman filter," *Amer. Statist.*, **37**, pp. 123-127, 1983.
6. J. Besag and P. Green and D. Higdon and K. Mengersen, "Bayesian computation and stochastic systems," *Statist. Sci.*, **10**, pp. 3-66 (with Discussion), 1995.
7. G. L. Bretthorst, *Bayesian Spectrum Analysis and Parameter Estimation*, Springer-Verlag, New York, 1988.
8. M. West and J. Harrison, *Bayesian Forecasting and Dynamic Models*, Springer-Verlag, New York, 1989.
9. A. Pole and M. West and J. Harrison, *Applied Bayesian Forecasting and Time Series Analysis*, Chapman-Hall, New York, 1994.
10. J. C. Spall(Ed.), *Bayesian Analysis of Time Series and Dynamic Models*, Marcel Dekker, New York, 1988.
11. A. Lasota and M. C. Mackey, *Chaos, Fractal, and Noise*, Springer-Verlag, New York, 1994.
12. C. M. Scipione and L. M. Berliner, "Bayesian inference in nonlinear dynamical systems," 1993 Proc. of the Section on Bayesian Statist. Sci., American Statistical Association, Washington, D.C., 1993.
13. M. West, *Bayesian time series: Models and computations for the analysis of time series in the physical sciences* in *This Volume*.
14. M. Priestley, *Non-Linear and Non-Stationary Time Series Analysis*, Academic Press, New York, 1988.
15. H. Tong, *Non-Linear Time Series*, Oxford University Press, New York, 1990.

BAYESIAN TIME SERIES:

Models and Computations for the Analysis of Time Series in the Physical Sciences

MIKE WEST
Institute of Statistics & Decision Sciences
Duke University, Durham, NC 27708-0251[†]

Abstract. This articles discusses developments in Bayesian time series modelling and analysis relevant in studies of time series in the physical and engineering sciences. With illustrations and references, we discuss: Bayesian inference and computation in various state-space models, with examples in analysing quasi-periodic series; isolation and modelling of various components of error in time series; decompositions of time series into significant latent subseries; non-linear time series models based on mixtures of auto-regressions; problems with errors and uncertainties in the timing of observations; and the development of non-linear models based on stochastic deformations of time scales.

Key words: Dynamic linear models, Markov chain Monte Carlo, Mixture models, Non-linear auto-regression, State-space models, Stochastic time deformations, Time series decomposition, Timing errors

1. Introduction

This article catalogues several aspects of time series modelling and analysis of relevance in exploration of univariate time series structure in the physical sciences. Specifically, I briefly review and reference variants of dynamic linear models, or state-space models, that have been in vogue in Bayesian forecasting for years, mentioning practical aspects of handling missing data values, additive measurement and sampling errors, time series outliers, and time series decomposition to investigate latent component sub-series of consequence. Markov chain Monte Carlo (MCMC) methods of posterior simulation in state space models are discussed, and used in an analysis of a deep sea oxygen isotope record related to patterns of variability is historical climate change. This kind of analysis will be of wider interest in connection with exploring (particularly) quasi-periodic structure in time series in physical sciences more generally. Decomposition of the oxygen series into latent components is illustrated in connection with the isolation and investigation of key

[†]email: mw@isds.duke.edu, web: http://www.isds.duke.edu
Invited paper for the *XV*[th] *Workshop on Maximum Entropy and Bayesian Methods*, Santa Fe, New Mexico, July 31-August 4 1995. Research supported in part by NSF under grant DMS-9311071.

K. M. Hanson and R. N. Silver (eds.), Maximum Entropy and Bayesian Methods, 23–34.

sub-cycles driven by Earth-orbital dynamics. Connections are drawn with other component state-space structures. Practical issues of time series analysis subject to uncertainties about the timing of observations, such as arise throughout the physical sciences, are discussed and approaches to incorporating timing errors in time series analysis briefly reviewed. This leads into a wider area of potential for future research in Bayesian time series, in connection with rather general models of stochastic time deformations. Finally, related work in non-linear auto-regression using semi-parametric mixture models from Bayesian density estimation is mentioned. These models are of potential use in identifying systematic non-linearities, such as smooth threshold auto-regressive structure but essentially arbitrary other forms, in observed series.

2. Dynamic Linear Models: The Normal State-Space Framework

DLMs, or linear state-space models, have been central to Bayesian time series analysis and short-term forecasting for almost thirty years ([1]). Applications in socio-economic forecasting and control engineering have exploited modelling flexibly in contexts of time-varying and non-stationary time series structure. More recently, advances in computation have begun to extend the framework to allow coherent inference on variance components, model parameters, non-Gaussian errors models, and other model features, and are opening up new application areas as a result. Some brief review and illustration is given here.

Consider a scalar time series x_t, observed at equally spaced time points $t = 1, 2, \ldots$, assumed to follow the DLM

$$x_t = F'z_t + \nu_t, \quad \text{and} \quad z_t = Gz_{t-1} + \omega_t \qquad (1)$$

with the following components and assumptions: $z_t = (z_{t,1}, \ldots, z_{t,p})'$, the $p \times 1$ state vector at time t; F is a column $p-$vector of known constants or regressors, fixed and constant for all t; ν_t is the observation noise, distributed as $N(\nu_t|0, v_t)$ for known v_t; G is a known $p \times p$ matrix, the state evolution matrix, fixed and constant for all t; ω_t is the state evolution noise, or innovation, at time t, distributed as $N(\omega_t|0, W_t)$ for a known variance matrix W_t; the noise sequences ν_s and ω_t are independent and mutually independent; and, finally, an initialising state vector has a specified normal prior $N(z_0|m_0, C_0)$ based on an initial information set X_0.

Standard normal/linear distribution theory provides well-known linear filtering algorithms for easily computing the derived normal distributions of collections of state vectors and future observations conditional on subsets of the observed series. Key examples include the *on-line* priors $p(z_t|X_{t-1})$ and posteriors $p(z_t|X_t)$, *k-step predictives* $p(x_{t+k}|X_t)$ and *lag-k filtered* posteriors $p(z_{t-k}|X_t)$, where $t = 1, 2, \ldots$, $k = 1, 2, \ldots$, and $X_t = \{X_0, x_1, \ldots, x_t\}$ for all t.

More general dynamic models have time-varying F and G elements, though the case of constant elements is quite far ranging. Until recently, problems of learning elements of v_t, W_t, F and G have been addressed in various ad-hoc ways, though new MCMC methods (see below) provide a major breakthrough.

3. Two Examples DLMs with Auto-Regressive Components

Suppose $x_t = F'z_t + \nu_t$ where

$$
F = \begin{pmatrix} 1 \\ 1 \\ 0 \\ 0 \\ \vdots \\ 0 \end{pmatrix}, \quad
G = \begin{pmatrix} 1 & 0 & 0 & 0 & \cdots & 0 & 0 \\ 0 & \phi_1 & \phi_2 & \phi_3 & \cdots & \phi_{p-1} & \phi_p \\ 0 & 1 & 0 & 0 & \cdots & 0 & 0 \\ 0 & 0 & 1 & 0 & \cdots & 0 & 0 \\ \vdots & \vdots & & & \ddots & \vdots & \vdots \\ 0 & 0 & 0 & \cdots & \cdots & 1 & 0 \end{pmatrix}, \quad
\omega_t = \begin{pmatrix} \eta_t \\ \epsilon_t \\ 0 \\ 0 \\ \vdots \\ 0 \end{pmatrix}, \quad (2)
$$

with $\eta_t \sim N(\cdot|0, u)$ independently of $\epsilon_s \sim N(\cdot|0, w)$, for all t, s. Then $W_t = W$ with entries all zero but for $W_{1,1} = u > 0$ and $W_{2,2} = w > 0$. Denote the first and second elements of z_t by μ_t and y_t, respectively; then μ_t is a simple, first-order polynomial trend (or random walk) process, $\mu_t = \mu_{t-1} + \eta_t$, and y_t is a standard, zero-mean AR(p) process, $y_t = \sum_{j=1}^{p} \phi_j y_{t-j} + \epsilon_t$. Thus x_t is observed as a trend plus auto-regression, subject to measurement and/or sampling errors ν_t. This example is used below; many others appear in [1,2]. Note that μ_t could be replaced by other DLM forms, such as more complex trends and/or regression terms, with direct extension, using the notion of model superposition ([1]).

Another class of auto-regressive component DLMs is developed in [3] and applied in various geological time series analyses in [4]. Such models combine two or more distinct AR(2) components, each exhibiting time-varying periodic structure, but now with persistent, i.e. unit root components. This provides an approach to Bayesian spectral analysis in which several sustained sinusoids of distinct periods are subject to stochastic changes in amplitudes and phased. In the case of, for example, just two such components plus a locally constant trend, such a model has the structure of (2) but now with

$$
F = \begin{pmatrix} 1 \\ 1 \\ 0 \\ 1 \\ 0 \end{pmatrix}, \quad
G = \begin{pmatrix} 1 & 0 & 0 & 0 & 0 \\ 0 & \beta_1 & -1 & 0 & 0 \\ 0 & 1 & 0 & 0 & 0 \\ 0 & 0 & 0 & \beta_2 & -1 \\ 0 & 0 & 0 & 1 & 0 \end{pmatrix} \quad (3)
$$

Models with more than two components are obvious generalisations. As with the defining parameters $\phi = \{\phi_1, \ldots, \phi_p\}$ in the model class (2), the corresponding β_j parameters here appear as primary elements of the state evolution matrix G. One consequence of this is that the structure of Bayesian analyses via simulation methods are simplified in important practical ways, as is illustrated in [3-5].

4. Latent Structure and Time Series Decomposition

A central objective in time series analysis is the decomposition of an observed series into estimated latent components, and the exploration and interpretation of such components. In the DLM context, latent component structure is also central in model building and development ([1], chapters 5 and 6), and the developments in that area lead to simple but useful methods of posterior component analysis, as is now elaborated upon in the context of the example provided by the auto-regressive DLM. This is based on recent developments in [5], and is closely linked to the developments of *similar* and *canonical models* in the DLM framework ([1], chapter 5). In this special model, the eigenvalues of G are the reciprocal roots of the auto-regressive characteristic polynomial. Assume, as is commonly the case, that these roots are distinct and non-zero, occurring as q pairs of complex conjugate pairs and $s = p - 2q$ real and distinct values; write the complex eigenvalues as $r_j \exp(\pm i\omega_j)$ for $j = 1, \ldots, q$, and real values as simply r_j for $j = 2q+1, \ldots, p$. Then the latent AR(p) process has a representation

$$ y_t = \sum_{j=1}^{q} c_{tj} + \sum_{j=2q+1}^{p} c_{tj} \tag{4} $$

where the c_{tj} are latent subseries corresponding to the eigen-structure. For each $j = 1, \ldots, q$, c_{tj} is a quasi-periodic AR(2) process over time t, a modulated sinusoid with stochastically varying amplitude and phase, fixed period $\lambda_j = 2\pi/\omega_j$, and modulus r_j; for each $j = 2q+1, \ldots, p$, c_{tj} is an AR(1) process of modulus r_j. The innovations "driving" these latent component processes are highly dependent.

A major point is that this component representation is computable given a specified ϕ vector determining the ϕ_j elements of G and a specified set of values of the y_t series (see [5,6]). Hence, in any analysis delivering posterior inferences about ϕ and the y_t series, inferences about the latent subseries c_{tj} follow. This is most direct in analyses based on posterior simulation, in which posterior samples of ϕ and the y_t are generated, for then the corresponding posterior samples of the c_{tj} terms may be directly computed, averaged to produce approximate posterior means of the components or summarised in other ways. Exploring the time trajectories of the components can usefully elucidate latent structure in the raw x_t series, as is illustrated below.

Note the connections of this decomposition with the structure of the related models (3). There the equivalent decomposition is explicitly the sum of two (or more) periodic components of time-varying amplitudes and phases but with no damping whatsoever.

5. Measurement Error Models

One important feature of state-space modelling is the explicit recognition of sources of error – measurement errors, sampling errors, outlier and recording errors, and so forth – directly corrupting observations on the underlying physical process; these are the purely *observational error* terms ν_t, explicitly distinct from *process* or *system noise/innovation* terms in ω_t. Biases and distortions of inferences arise in using models that confuse and ignore this distinction. For example, measurement error corrupting a pure AR process leads to shrinkage of estimated coefficients towards zero – posterior distributions favour AR parameter value closer to zero – if such error is not modelled; this is a standard errors-in-variables effect. Error processes dominated by outliers are a naturally a serious practical concern (e.g. [7],[1] chapters 10 and 12).

Traditional extensions of the basic normal error assumption to more realistic forms include outlier accommodating models such as the normal mixture distribution $\nu_t \sim (1 - \pi)N(\nu_t|0, v) + \pi N(\nu_t|0, k^2 v)$. This admits a background level of routine measurement error, with variance v, together with occasional extremes, or outliers, via the inflated variance component with $k > 1$ with some (small) probability π ([8,6,1] and [9] section 16.7). The background variation, measured by v, may be small, even negligible, in some applications ([6]); indeed, a degenerate mixture $\nu_t \sim (1 - \pi)\delta_0(\nu_t) + \pi N(\nu_t|0, q)$ is the default additive outlier model in the "robust" time series fitting functions in S-Plus ([9]). However, the extent to which basic observational errors may impact on inference can only be assessed by modelling and estimating them, and, indeed, routine measurement and sampling errors are often quite substantial and can be of enormous significance in physical sciences ([4]). Hence the non-degenerate version tends to be preferred.

Extension of routine DLM analyses to incorporate these kinds of practically important error models are now feasible using MCMC simulation methods, and these kinds of modelling extensions were important stimuli in the early work on simulation models in state-space approaches ([10,11,6]). Similar mixture models are also of interest in the evolution equation of DLMs in connection with modelling stochastic changes in state vectors that are more abrupt and marked than as expected under the "routine" implied by the basic normal distributions of the ω_t terms ([1] chapters 11 and, particularly, 12). In fact, the early developments of Bayesian forecasting using DLMs were largely predicated on applied problems in commercial forecasting that led to the development of mixture errors models and multi-process DLMs ([12] and [1] chapter 12). Though the concepts and various methods of approximation to resulting analyses have been around for many years, only now, using simulation methods, can we implement "exact" Bayesian analyses under these kinds of realistic error modelling assumptions, and we are at the point where such analyses are set to become more-or-less routine in applied work ([10,11,6]).

6. Posterior Computations for Bayesian Time Series Analysis

Following early work in [13,14,10,11], the development of MCMC simulation methods in state space frameworks is growing rapidly. Note that there are ranges of related developments outside the state-space framework, typified by [15], for example; more references appear in the review of [16]. MCMC methods now permit Monte Carlo inference in the kinds of models exemplified above using variants of basic Gibbs and more general Metropolis-Hastings algorithms to generate approximate posterior samples of collections of state vectors and DLM parameters. Structure of a basic algorithm is briefly illustrated in the auto-regressive component DLM of (2); see [3-5] for further details and other models.

For concreteness, take the model (2). Mixture observational error models are incorporated by assuming the ν_t to be independent normal, $N(\nu_t|0, vh_t)$, with a constant scale factor v and individual variance weights h_t. For instance, the basic contamination model assumes the h_t to be drawn independently from $p(h_t) = (1-\pi)\delta_1(h_t) + \pi\delta_{k^2}(h_t)$; other assumptions are possible ([8,6]). Initial information is summarised through independent priors $p(u)$, $p(v)$ and $p(w)$ for the variance components; independence is not necessary though will often be assumed. For each $n > 0$, define $Z_n = \{z_0, \ldots, z_n\}$, the complete set of state vectors over times $t = 0, \ldots, n$, in a notation consistent with the earlier definition of the observation set $X_n = \{X_0, x_1, \ldots, x_n\}$.

MCMC proceeds by iteratively sampling from collections of conditional posterior distributions for quantities of interest. These are the AR parameters ϕ, the set of state vectors Z_n, the variances v, u, w, and the weights $H_n = \{h_1, \ldots, h_n\}$. For any subset θ of these quantities, write θ^- for all remaining parameters and variables; e.g. $\phi^- = \{Z_n, v, u, w, H_n\}$. Various Gibbs sampling algorithms may be constructed to iteratively sample from conditional distributions $p(\theta|X_n, \theta^-)$ for appropriate choices of θ; one such algorithm, detailed in [5], has the following basic structure.

(a) Sampling $(Z_n|X_n, Z_n^-)$

Fixing model parameters and variance components in Z_n^-, standard normal DLM theory applies and delivers useful normal distributions. In particular, following [10] and [11], most efficient algorithms are based on reverse sampling through the sequence: $p(z_n|X_n, Z_n^-)$ followed by, for each $t = n-1, n-2, \ldots, 1, 0$ in turn, $p(z_t|X_n, Z_n^-, z_{t+1})$ where conditioning values z_{t+1} are, at each stage, the most recent values sampled. Each distribution $p(z_t|X_n, Z_n^-, z_{t+1})$ is computed as follows. Filter forward through time to compute the usual sequence of Kalman filter-based moments of conditional distributions, $(z_t|X_t, Z_n^-) = N(z_t|m_t, C_t)$, for each t. Note that elements $2, \ldots, p-1$ of the state vector z_t are known if z_{t+1} is known; they are simply the elements $3, \ldots, p$, of the latter. Hence, given z_{t+1}, replace entries $2, \ldots, p-1$ of z_t accordingly. Sample the remaining two elements (μ_t, y_{t-p+1}) from the resulting bivariate normal distribution, namely $N(\mu_t, y_{t-p+1}|X_t, z_{t+1}, Z_n^-)$; this may be done efficiently, the precise algorithmic details are laid out in appendix in [5].

This complete process results in a sequence $z_n, z_{n-1}, \ldots, z_0$ that represents a sample from the posterior distribution $p(Z_n|X_n, Z_n^-)$, as required.

(b) Sampling $(\phi|X_n, \phi^-)$

This reduces to a standard problem of computing and sampling the posterior for AR parameters based on observed data $y_1, \ldots, y_n$ plus essential initial values $y_{-(p-1)}, \ldots, y_0$, from the AR process. For example, a standard reference prior produces a normal posterior, though other priors might be used.

(c) Sampling variance components

Fixing Z_n leads to current, fixed values of the errors and innovations ν_t, ϵ_t and η_t for all $t = 1, \ldots, n$. These provide sequences of independent "observations", namely $N(\nu_t|0, vh_t)$, $N(\epsilon_t|0, w)$ and $N(\eta_t|0, u)$, that have known means and inform on the variances v, w and u, respectively. Independent inverse gamma priors for the three variances are conjugate, though sampling the resulting posteriors, $(v|X_n, v^-)$, $(w|X_n, w^-)$, and $(u|X_n, u^-)$, is straightforward with other priors. Priors that are finite range uniforms on the corresponding standard deviations are used in [3-5].

(f) Sampling $(H_n | X_n, H_n^-)$

Given H_n^-, the h_t are conditionally independent and so sampled from individual posteriors proportional to $p(h_t) h_t^{1/2} \exp(-\nu_t^2 / 2vh_t)$ with $\nu_t = x_t - F'z_t$, as above.

Repeatedly iterating through these conditionals leads to a Markov chain sampler on the full space of uncertain state vectors and model parameters whose stationary distribution is the full joint posterior. Posterior samples of the elements of z_t translate directly into posterior samples of the underlying smooth trend μ_t together with samples of the latent AR(p) process y_t, over all t. Parameter samples produce direct approximations to posterior margins. Issues of establishing convergence results for MCMC schemes, especially in connection with mixture error models, are discussed in [10] and [11], for example.

7. An Illustration

An analysis of an oxygen isotope record is briefly summarised as an illustration of state-space analysis with latent auto-regression and a mixture error model, implemented via simulation analysis as just summarised. These kinds of analysis, though by no means yet standard, are accessible computationally and should become routine.

In [4,5] I discuss data and modelling questions arising in studies of the forms of quasi-periodic components of recorded geological time series in connection with investigating patterns of historical climate change. The latter reference discusses one deep-ocean oxygen isotope series used in previous analyses (e.g. [17]). A further such series is explored here; the upper frame in Figure 1 plots the data, which is representative of several oxygen isotope series from cores from various geographical locations, and measures relative abundance of $\delta^{18}O$ timed on an approximately equally spaced 3kyr scale. Further background details appear in [17,5] and references therein. The series stretches back roughly 2.5Myr. Time-varying periodicities are evident; the nature and structure of these periodicities is of some geological interest, especially in connection with the $\sim$ 100kyr "ice-age cycle" ([17]).

Analyses with models based on (2) have been explored, and one such is reported. This uses a locally constant trend plus a latent AR(20) component, and a mixture observational error model with $k = 10$ and $\pi = 0.05$. A traditional uniform reference prior is used for ϕ, and priors for the variances v, u and w are based on finite range uniform priors for the standard deviations. After burning-in the simulation iterations, a posterior sample of 5,000 draws is saved and summarised. Based on the approximate posterior means of ϕ and the elements of Z_n, I evaluate the decomposition (4) for the latent y_t process, representing the decomposition into key quasi-cyclical and other components. Parts of the decomposition appear in Figure 1, displaying the approximate posterior means of acyclic trend, observational errors, the three dominant three quasi-cyclical components (ordered by wavelengths), and, finally, the sum of the remaining components of the latent auto-regression. Posterior samples for the periods λ_j of the quasi-periodic components support values around 110kyr, 42kyr and 23kyrs; more precisely, the approximate posterior quartiles are 105.0-110.0-115.0, 40.8-42.2-43.6 and 22.3-22.9-23.4, respec-

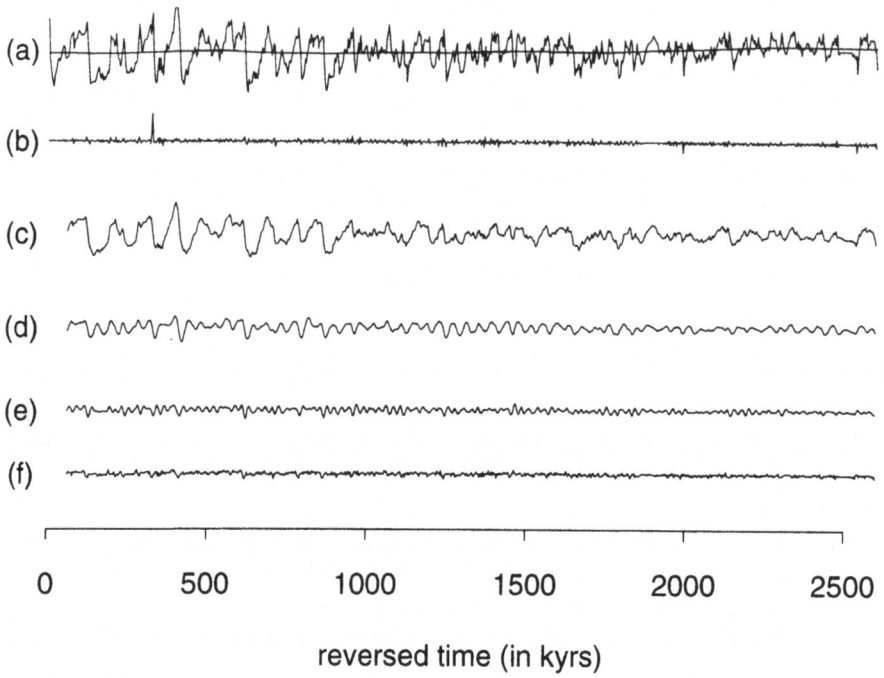

Figure 1. Estimated oxygen data decomposition based on AR(20) state space model analysis. The components displayed are: (a) the series with estimated smooth trend; (b) the estimated observation errors; (c), (d) and (e) the first three quasi-cyclical components, of periods approximately 110kyr, 42kr and 23kyrs respectively; and (f) the sum of the remaining components of the latent auto-regression. The raw series in (a) is the sum of the trend plus all components displayed in (b)-(f).

tively. The latter two are directly interpretable in terms of responses to forcing mechanisms, namely the obliquity of the Earth's orbit (period around 41kyr), and the precession of the Earth about its axis (period around 19-23kyr). The so-called 100kyr "ice-age cycle" is of major current geological interest in connection with debates over its genesis, roughly 1Myrs ago, and, particularly, to questions of whether or not the onset was gradual and inherent or the result of a significant structural climatic change. (e.g. [17] and references therein).

There are several marked observational outliers apparent, though the level of baseline measurement error appears rather small on the scale of the raw data; approximate posterior quartiles for $\sqrt{v}$ are approximately 0.091-0.099-0.0985, those for $\sqrt{u}$ are 0.007-0.009-0.011, and those for $\sqrt{w}$ are 0.187-0.202-0.224.

The changing form, in terms of primarily the amplitude, of the key component of 110kyrs or so in period, is evident, and this seems to be the major factor in determining the change in appearance of the raw series at around 0.8-1.0Myrs ago. The other components have more sustained amplitudes. In addition, the estimated innovations sequence ϵ_t (not displayed) has a consistent form over the full stretch of the the data, with no clear indication of changes in variance. Thus the geological focus is on marked changes in in the nature of the major, longer term oscillations of period between 100-120kyrs. This ties in with the onset of the ice-age cycle of period in this range; from the figure, a rather marked change in the form of this component is suggested. Notice that the model neither accommodates nor predicts real structural change in the series, assuming a fixed auto-regression over the entire time period; nevertheless, the changes in the key quasi-cyclical components are evident from this decomposition and graphical display. Further analyses focussed on this issue, refitting separate models to subsets of the series before and after the apparent change-point, confirm the suspected change-point but do not materially impact on the estimated decomposition as already displayed.

8. Random Observation Times: Timing Errors and Uncertainties

Many application areas suffer from problems of errors and uncertainties in the timing of time series observations. In [4], for example, the x_t series is a geochemical quantity, from deep lake sediment, that is a proxy indicator of local climatic conditions of interest in studying climatic change over time. The data are timed by measures of depth in sediment cores, and then mapped to calendar ages by a complex and uncertain process involving carbon-14 calibration. By detailing and modelling this calibration process, a rather elaborate model for the true but uncertain times of observations is constructed, and this provides a class of prior distributions for the times. Analyses then incorporate the uncertain times along with the model parameters and state variables. The study in [4] describes this, and focuses particularly on how limited the data is as a result of realistic measures of the timing uncertainties.

Simple measurement errors, and truncation or rounding errors in timing, may be subject to assessment using similar approaches. Bayesian inference, implemented via MCMC analysis, as illustrated in this application, is feasible; the conceptual basis is as follows. Generally, denote the observed data series by x_i, for $i = 1, 2, \ldots$, and write t_i for the true, uncertain timing of x_i. If $t_i = i$ we are back to the known, equally spaced case. Given a specific time series model, write Θ for all model parameters and state variables, and T_n for the set of times of the n observations. The model implies a likelihood function $p(X_n|T_n, \Theta)$, and usual analysis proceeds conditional on T_n to induce and explore posteriors $p(\Theta|T_n, X_n)$. Now, with a defined prior $p(T_n|\Theta)$ representing timing errors, uncertainties and other prior information about the timings, we couple conditional analyses (n.b. priors for T_n may depend on the model parameters Θ, as represented in the notation here, though sometimes prior independence with simplify the prior to $p(T_n|\Theta) = p(T_n)$). Assume we can sample Θ vectors from $p(\Theta|T_n, X_n)$ for any specified timings; further assume also that the corresponding conditional posterior $p(T_n|\Theta, X_n) \propto p(T_n|\Theta)p(X_n|\Theta, T_n)$ may be simulated for any specified Θ. Then

MCMC analysis proceeds by iteratively resampling these two conditional posteriors in the usual manner. Some specific models $p(X_n|T_n, \Theta)$ and priors $p(T_n)$ are studied in [4], where various Gibbs and Metropolis-Hastings algorithms for these sampling exercises are developed and applied.

One issue to note is that these problems require time series models for arbitrary spacings between observations. In the DLM framework, this can often be achieved by refining a nominal time scale to a finer, equally spaced scale, as is traditional in dealing with unequally spaced and randomly missing data ([1], chapter 10). An alternative involves embedding the series in an underlying continuous time model ([18]); work is in progress to develop Bayesian analyses in various such contexts.

9. Random Observation Times: Time Deformation Models

The conceptual and technical developments of dealing with uncertain observation times are opening a novel area involving inference on stochastic time deformations. This builds on basic ideas in [18]. which demonstrates that, under certain deterministic deformations, traditional linear models can be mapped into models with characteristics similar to some common non-linear models, such as ARCH models and threshold AR models. The basic approach assumes an underlying, linear time series model for a process $y(s)$, over "operational time" $s > 0$, and that the observed series is a discretely sampled version $x_t = y(s(t))$ where $s = s(t)$ represents the transformation between operational and real-time, or the (inverse of) the time "deformation" function. The connections with modelling uncertainties in timings are evident, at least at a technical level, and the machinery developed there is being extended to provide a a framework for Bayesian inference on time deformation functions. Current studies include *stochastic* time deformations as an approach to modelling *chirp* effects in otherwise harmonic processes, and the modification of traditional Bayesian spectral methods as a result. Various classes of priors, including non-parametric models, are being explored for time deformation functions. Applications include modelling irregular waveform patterns of long series of EEG recordings, in collaboration with Duke psychiatrists, and will be reported elsewhere; some very preliminary discussion appears in [6].

10. Mixture Models and Non-Linear Auto-Regression

A rather different but, in some sense, more traditional approach to non-linear time series modelling is initiated in [19]. The interest here is in identifying and inferring departures from linearity in an auto-regressive context; though not yet developed in a state-space framework, the practical issues of dealing with observational errors will lead to such extensions.

The basic idea is simple. A non-linear AR(p) model for x_t is determined by specifying a class of conditional distributions $p(x_t|x_{t-1}, \ldots, x_{t-p}, \Theta)$, for all t and where Θ denotes model parameters. Building on flexible classes of Dirichlet mixture models developed for semi-parametric Bayesian density estimation, [19] develops models in which these defining conditionals are themselves flexible mixtures;

specifically, mixtures of normal distributions

$$p(x_t|x_{t-1},\ldots,x_{t-p},\Theta) = \sum_{j=1}^{k} w_j N(x_t|b_j, B_j) \tag{5}$$

where, for suitable k,

- the regression functions b_j are each linear in $x_{t-1},\ldots,x_{t-p}$, but they differ across j, and the component variances B_j vary with j but do not depend on conditioning past x_{t-j} values; and
- w_j is a multivariate kernel factor, implying higher conditional weight on mixture components that best "support" the current conditioning values of the "state variables" $x_{t-1},\ldots,x_{t-p}$.

Our approach to inference, described in [19], is largely predictive. Posterior computations deliver approximate posteriors for model parameters (k, Θ) and then the model is interpreted predictively, exploring features of the resulting sets of conditional predictive distributions for future data – just averages of mixtures (5) with respect to the posterior. Technically, posterior computations involve variants of Gibbs and Metropolis-Hastings MCMC algorithms invented for Bayesian density estimation in mixture models ([20]), with extensions to the time series context and also to cover various hyper-parameter problems.

There are obvious connections with "standard" kernel methods – the conditional mean function $E(x_t|x_{t-1},\ldots,x_{t-p},\Theta)$ is a kernel (auto-)regression function – but the Bayesian, model-based foundation provides much more. For example, all aspects of parameter uncertainty are naturally reflected in the analysis, and problems of smoothing parameter estimation are subsumed under posterior inference on corresponding variance parameters and hyper-parameters. In terms of modelling flexibility, the development of a set of full conditional mixture distributions for future x_t as the "state" $x_{t-1},\ldots,x_{t-p}$ varies provides for adaptation to what may be substantial patterns of variation across the state space, as well as in just the conditional mean. Examples in [19] include cases where conditional distributions in some regions of $x_{t-1},\ldots,x_{t-p}$ space are unimodal, but in other regions are clearly bi- or multi-modal. In such cases, conditional predictive means (the "Bayesian kernel regression functions") can poorly summarise location of conditional distributions; they are inherently smooth, continuous functions of the state $x_{t-1},\ldots,x_{t-p}$. By contrast, bifurcating traces of conditional modes are, then, much more informative, capturing the important, distinct components of conditional structure when the mean obscures it by smoothing. In this sense, the models are capable of capturing both abrupt threshold effects (as in TAR models, [21], chapter 3), and smooth transitions (as in STAR models, [21], section 3.3.3), as well as ranges of other non-linearities. Modelling entire distributions also allows for adaptation to changing patterns of variation in spread and other features of conditional predictive distributions as the series evolves. Some example in [19] illustrate these features. Models there include mixtures of quasi-cyclical AR(2) components, designed to allow for and explore rather evident departures from linearity in periodic series; this specific class of state-dependent models naturally allows for variation

in the period/wavelength parameters of quasi-cyclical processes as a function of
the past values of the series, the state variables. Further model developments and
exploration in applications are the subject of current study.

References

1. M. West and P. Harrison, *Bayesian Forecasting and Dynamic Models*, Springer-Verlag, New York, 1989.
2. A. Pole, M. West, and P. Harrison, *Applied Bayesian Forecasting and Time Series Analysis*, Chapman-Hall, New York, 1994.
3. M. West, "Bayesian inference in cyclical component dynamic linear models," *J. Amer. Statist. Assoc.*, **92**, 1995 (to appear).
4. M. West, "Some statistical issues in Palæoclimatology (with discussion)," in *Bayesian Statistics 5*, J. Berger, J. Bernardo, A. Dawid, and A. Smith, eds., Oxford University Press, Oxford, 1995 (in press).
5. M. West, "Time series decomposition and analysis in a study of oxygen isotope records," *ISDS Discussion Paper 95-18, Duke University*, 1995.
6. M. West, "Modelling and robustness issues in Bayesian time series analysis (with discussion)," in *Bayesian Robustness 2*, J. Berger, F. Ruggeri, and L. Wasserman, eds., IMS Monographs, 1995 (to appear).
7. B. Kleiner, R. Martin, and D. Thompson, "Robust estimation of power spectra (with discussion)," *J. Roy. Statist. Soc. (Ser. B)*, **41**, pp. 313–351, 1979.
8. M. West, "Robust sequential approximate Bayesian estimation," *J. Roy. Statist. Soc., (Ser. B)*, **43**, pp. 157–166, 1981.
9. Statistical Sciences, *S-PLUS Guide to Statistical and Mathematical Analysis (Version 3.2)*, StatSci, a division of MathSoft, Inc., Seattle, 1993.
10. S. Frühwirth-Schnatter, "Data augmentation and dynamic linear models," *J. Time Series Analysis*, **15**, pp. 183–102, 1994.
11. C. K. Carter and R. Kohn, "On Gibbs sampling for state space models," *Biometrika*, **81**, pp. 541–553, 1994.
12. P. Harrison and C. Stevens, "Bayesian forecasting (with discussion)," *J. Roy. Statist. Soc., (Ser. B)*, **38**, pp. 205–247, 1976.
13. C. M. Scipione and L. M. Berliner, "Bayesian statistical inference in nonlinear dynamical systems," in *Proceedings of the Bayesian Statistical Science Section*, American Statistical Association, Washington, DC, 1993.
14. B. P. Carlin, N. G. Polson, and D. S. Stoffer, "A Monte Carlo approach to nonnormal and nonlinear state-space modelling," *J. Amer. Statist. Ass.*, **87**, pp. 493–50, 1992.
15. R. McCulloch and R. Tsay, "Bayesian analysis of autoregressive time series via the Gibbs sampler," *J. Time Series Analysis*, **15**, pp. 235–250, 1994.
16. M. West, "Bayesian Forecasting," in *Encyclopedia of Statistical Sciences*, S. Kotz, C. Read, and D. Banks, eds., Wiley, New York, 1995 (in press).
17. J. Park and K. Maasch, "Plio-Pleistocene time evolution of the 100-kyr cycle in marine paleoclimate records," *J. Geophys. Res.*, **98**, pp. 447–461, 1993.
18. J. H. Stock, "Estimating continuous-time processes subject to time deformation: An application to postwar U.S. GNP," *J. Amer. Statist. Ass.*, **83**, pp. 77–85, 1986.
19. P. Müller, M. West, and S. MacEachern, "Bayesian models for non-linear auto-regressions," *ISDS Discussion Paper 94-30, Duke University*, 1995.
20. M. West, P. Müller, and M. Escobar, "Hierarchical priors and mixture models, with application in regression and density estimation," in *Aspects of Uncertainty: A tribute to D. V. Lindley*, A. Smith and P. Freeman, eds., Wiley, New-York, 1994.
21. H. Tong, *Non-Linear Time Series*, Oxford University Press, Oxford, England, 1990.

MAXENT, MATHEMATICS, AND INFORMATION THEORY

I. CSISZÁR
Mathematical Institute of the Hungarian Academy of Sciences
H-1364 Budapest
P.O. Box 127
Hungary[‡][§]

Abstract. This is a mathematically oriented survey about the method of maximum entropy or minimum I-divergence, with a critical treatment of its various justifications and relation to Bayesian statistics. Information theoretic ideas are given substantial attention, including "information geometry". The axiomatic approach is considered as the best justification of maxent, as well as of alternate methods of minimizing some Bregman distance or f-divergence other than I-divergence. The possible interpretation of such alternate methods within the original maxent paradigm is also considered.

Key words: Bregman distance, entropy, f-divergence, information divergence, information geometry, large deviations, maxent axioms, maxent in mean, prior distribution.

1. The scope of MAXENT

Maxent is a method to infer a function $p(x)$ defined on a given set $\mathcal{X}$ when the available information specifies only a feasible set F of such functions. Originally coming from physics (Boltzmann, Gibbs), maxent has been promoted to a general method of inference primarily by Kullback and Jaynes, [1],[2].

Typically but not always, the feasible set is determined by linear constraints on p, i.e.,

$$F = \{p : \int a_i(x)p(x)\lambda(dx) = b_i, \ i = 1, \ldots, k\}, \tag{1}$$

for some given functions $a_i(x)$ and constants $b_i, i = 1, \ldots, k$. Here and in the sequel, λ denotes a given σ-finite measure on (a σ-algebra of subsets of) $\mathcal{X}$. The reader may assume with little loss of generality that $\mathcal{X}$ is finite or countably infinite and

---

[‡]This work was supported by the Hungarian National Foundation for Scientific Research, Grant T016386.

[§]Email: csiszar@math-inst.hu

K. M. Hanson and R. N. Silver (eds.), Maximum Entropy and Bayesian Methods, 35–50.
© 1996 *Kluwer Academic Publishers.*

λ is the counting measure (then integrals become sums) or $\mathcal{X}$ is a subset of a finite dimensional space and λ is the Lebesgue measure.

The function p to be inferred is often a probability density or mass function. Then we will speak of Problem (i) or (ii) according as the only available information is $p \in F$, or also a *default model* $q(x)$ is given, such that $p = q$ would be inferred were $q \in F$. We will speak of Problem (iii) when a non-negative $p(x)$ not necessarily a probability density (such as a power spectrum or an intensity function) has to be inferred, knowing only a feasible set F and a default model $q(x)$.

The maxent solution to Problem (i) is that $p^* \in F$ whose entropy

$$H(p) = -\int p(x) \log p(x) \lambda(dx) \tag{2}$$

is maximum. For Problems (ii) and (iii), the maxent solution is that $p^* \in F$ whose information divergence (I-divergence) from q, i.e.,

$$D(p \parallel q) = \int [p(x) \log \frac{p(x)}{q(x)} - p(x) + q(x)] \lambda(dx) \tag{3}$$

is minimum. If $\lambda(\mathcal{X}) < \infty$, and $p_0(x) = \text{constant} = 1/\lambda(\mathcal{X})$ is the uniform distribution on $\mathcal{X}$, then $H(p_0) = \log \lambda(\mathcal{X})$ and

$$H(p) = H(p_o) - D(p \parallel p_0) \tag{4}$$

for every probability density p. Hence the maxent solution to Problem (i) is the same as that to Problem (ii) with default model $q = p_0$.

I-divergence has a variety of other names (and notations) such as Kullback-Leibler information number, relative entropy, information gain, etc. It is a (non-symmetric) measure of distance of p from q, i.e., $D(p \parallel q) \geq 0$, with equality iff $p = q$ (interpreted as $p(x) = q(x)$ for λ-almost all x). The funtion p^* minimizing $D(p \parallel q)$ subject to $p \in F$ is called the I-projection of p onto F.

It should be emphasized that a maxent solution as defined above does not always exist. On the other hand, if a solution exists it is unique, providing the feasible set F is convex. For more details, cf. Section 5.

2. Information – theoretic heuristic

The amount of information provided by the outcome x of a drawing from a finite or countably infinite $\mathcal{X}$, governed by a probability mass function p, is defined to be $-\log_2 p(x)$ bits. One motivation for this definition is Shannon's

Theorem 2.1: Any uniquely decodable code that assigns a binary codeword of length $l(x)$ to each $x \in \mathcal{X}$, has average length

$$\Sigma p(x) l(x) \geq -\Sigma p(x) \log_2 p(x). \tag{5}$$

The lower bound can be approached closer than 1 bit by a code with $l(x) = \lceil -\log_2 p(x) \rceil$.

Often, $-\log_2 p(x)$ is referred to as the ideal codelength of x for p; the right hand side of (5) is the ideal average codelength. Notice that the average length (under p) of an ideal code for some $q \neq p$, with $l(x) = -\log_2 q(x)$, exceeds the ideal average codelength for p by $\Sigma p(x) \log_2(p(x)/q(x))$.

In information theory, the entropy of a probability mass function p and the I-divergence of p from q are usually defined as $-\Sigma p(x) \log_2 p(x)$ and $\Sigma p(x) \log_2 \frac{p(x)}{q(x)}$. The difference from our definitions (2) and (3), with λ=counting measure, is only a constant factor, i.e., the choice of unit. Thus $H(p)$ measures the average amount of information provided by the outcome of a random drawing governed by p , which may also be interpreted as a measure of uncertainty about that outcome before observing it, or of the amount of randomness represented by p. $D(p \parallel q)$ is an information theoretic distance of p from q. It measures how less informed is about the outcome of a random drawing one who believes this drawing is governed by q than one who knows the true p. Alternatively, it is a measure of information gained when learning that the true distribution is p rather than q.

If $\mathcal{X}$ is a subset of R^k and λ is the Lebesgue measure, the entropy (2) can not be directly interpreted as a measure of average amount of information or degree of randomness; e.g., it may be negative. Still, if the outcome of a drawing governed by a density p is observed with a prescribed (large) precision, say each of its k components is rounded to N fractional binary digits, the probability of any particular observed outcome x will be $p(x)2^{-kN}$ with good approximation, supposing p is smooth. Thus the amount of information provided by this observation will be $-\log_2 p(x) + kN$ bits, and the average amount of information will be $H(p)$ + constant. Thus a density with larger entropy still represents larger randomness. The intuitive interpretation of I-divergence can be similarly extended to the continuous case, indeed, its extension is more straightforward than that of entropy.

For Problem (i), maxent may be regarded as an extension of the *principle of insufficient reason*, i.e., that in case of complete ignorance $p(x)$=constant should be assumed. Of course, this makes sense only when $\lambda(\mathcal{X}) < \infty$, and in that case the uniform distribution on $\mathcal{X}$ represents maximum randomness (formally, (4) shows that the uniform distribution has maximum entropy). If instead of complete ignorance we know that $p \in F$, but nothing else, maxent still suggests to adopt the most random feasible p. The information theoretic meaning of I-divergence provides a reasonable heuristic background of maxent also for Problem (ii). The same, however, can not be said about Problem (iii).

3. Large deviations and maxent

Maxent is closely connected with the subject of large deviations in probability theory, which, in turn, is closely related to information theory. The simple large deviation results stated below are key ingredients of information theory, cf. Csiszár and Körner [3]. However, they have been known in statistical physics much earlier, dating back to Boltzmann. A proof can be given by straightforward calculation using Stirling's formula; an even simpler proof appears in [3], p.30.

Theorem 3.1: Given a finite set $\mathcal{X}$ of size $|\mathcal{X}|$, let $N_n(\hat{p})$ denote the number of n-tuples $(x_1, \ldots, x_n) \in \mathcal{X}^n$ with a given empirical distribution $\hat{p}$, where

$$\hat{p}(x) = \frac{1}{n} \text{ (number of indices i with } x_i = x).\tag{6}$$

Then

$$N_n(\hat{p}) = \exp[nH(\hat{p}) - r_n(\hat{p})],\tag{7}$$

with

$$0 \leq r_n(\hat{p}) \leq |\mathcal{X}| \log n.\tag{8}$$

Corollary: If $x_1, \ldots, x_n$ are drawn independently from $\mathcal{X}$, governed by q, the empirical distribution will be $\hat{p}$ with probability

$$\exp[-nD(\hat{p} \parallel q) - r_n(\hat{p})].\tag{9}$$

Suppose now that $\hat{p}$ is known to belong to a closed, convex feasible set F, and let $p^* = \text{argmax}_{p \in F} H(p)$. Then, providing F contains empirical distributions arbitrarily close to p^* if n is sufficiently large, the Theorem implies that among the n-tuples $(x_1, \ldots, x_n)$ with empirical distribution $\hat{p} \in F$, all but an exponentially small fraction will be in an arbitrarily small neighborhood of p^*, if n is large. Similarly, the Corollary implies that if $x_1, \ldots, x_n$ are drawn independently from $\mathcal{X}$, governed by q, the conditional probability on the condition $\hat{p} \in F$ that $\hat{p}$ will be in an arbitrarily small neighborhood of $p^* = \text{argmin}_{p \in F} D(p \parallel q)$, is exponentially close to 1, if n is large. These facts represent very strong arguments for maxent inference, providing the probability mass function to be inferred is an empirical distribution. This covers many applications of maxent in physics.

Theorem 3.1 and its corollary also imply that if all $(x_1, \ldots, x_n) \in \mathcal{X}^n$ are a priori equally likely, or if $x_1, \ldots, x_n$ are independently drawn governed by q, the conditional distribution of x_1 (or of x_k for any fixed k) on the condition that $\hat{p} \in F$, will be arbitrarily close to the maxent solution p^* as above. More general results of this kind, for general rather than finite $\mathcal{X}$, were proved by Campenhout and Cover [4] and Csiszár [5]. They may be used to argue that maxent is "right" more generally than for inferring an empirical distribution, [6]. It should be added, however, that such arguments do not seem to cover the majority of applications of maxent, in particular, they do not apply to Problem (iii).

4. Maxent and Bayesian statistics

Given a family $\{p_\vartheta(x), \vartheta \in \Theta\}$ of probability densities on $\mathcal{X}$, where $\vartheta = (\vartheta_1, \ldots, \vartheta_k)$ is a parameter vector taking values in a given set $\Theta \subset R^k$, for Bayesian inference about ϑ one needs a prior distribution on Θ. If the family is reparameterized as $\{p_\varphi(x), \varphi \in \Phi\}$, where φ is a one-to-one function of ϑ, one may as well want a prior on Φ. A natural invariance requirement is that the latter be the transform of the prior on Θ under the mapping $\vartheta \to \varphi$.

It has been suggested that the prior should be selected by maxent. In the absence of constraints on feasible priors, this amounts to selecting the uniform distribution on Θ. Unfortunately, in addition to other shortcomings, this method of prior selection does not meet the invariance requirement.

Another method, motivated by information theory, might be to select that prior π under which the average amount of information obtained about ϑ from observing x, known as mutual information, is maximum. By one of several equivalent definitions, this mutual information equals the expectation $E_\pi D(p_\vartheta \parallel \bar{p})$, where $\bar{p} = E_\pi p_\vartheta$ is the marginal density of x. The maximum mutual information is mathematically equivalent to channel capacity in information theory.

The latter method does satisfy invariance but, unfortunately, it often yields a prior concentrated on a finite subset of Θ (always if $\mathcal{X}$ is a finite set). As an improvement, one might take that prior under which a sequence of observations $x_1, \ldots, x_n$, conditionally independent given ϑ, provides maximum information about ϑ in the limit $n \to \infty$. As first established by Bernardo [7] and later rigorously proved by Clarke and Barron [8] (under suitable regularily conditions), this leads to the so-called Jeffreys prior, with density proportional to the square-root of the determinant of the Fisher information matrix $J(\vartheta) = [J_{ij}(\vartheta)]$,

$$J_{ij}(\vartheta) = \int \frac{\partial \log p_\vartheta(x)}{\partial \vartheta_i} \frac{\partial \log p_\vartheta(x)}{\partial \vartheta_j} p_\vartheta(x) \lambda(dx). \tag{10}$$

It may be reassuring to have recovered a prior that had been derived earlier by other considerations, even though this prior is not beyond criticism, cf. Bernardo and Smith [9].

Let us briefly discuss a rather debatable connection of maxent to Bayesian statistics. In many applications, the feasible set is supposed to be defined by linear constraints, cf. (1), but the known values of the constants are subject to errors. In other words, for the unknown p and the known values b_i, the errors

$$e_i = b_i - \int a_i(x) p(x) \lambda(dx) \ , \quad i = 1, \ldots, k \tag{11}$$

are small but not necessarily 0. One way to infer p in this case in the spirit of maxent is to subtract from $H(p)$ or $-D(p \parallel q)$ a regularization term penalizing large values of the errors, and maximize the so obtained functional of p, without constraints. Typically, the regularization term is chosen as $\alpha \Sigma_{i=1}^k e_i^2$, where e_i is defined by (11) and $\alpha > 0$ is a regularization parameter.

This very reasonable technique is sometimes given a Bayesian interpretation which, in the opinion of this author, is questionable even for those who fully accept Bayesian philosophy. Namely, maximizing $H(p) - \alpha \Sigma e_i^2$ or $-D(p \parallel q) - \alpha \Sigma e_i^2$ is interpreted as maximizing

$$\log (\text{prior density of } p) + \log (\text{joint density of } b_i, \ldots, b_k \text{ given } p) \tag{12}$$

which amounts to interpreting the regularized maxent solution p^* as a MAP (maximum a posteriori probability) estimate of p. Supposing that the e_i's are independent Gaussian random errors with common variance σ^2, the regularization term $-\alpha \Sigma e_i^2$ gives the second term in (12) multiplied by $2\alpha\sigma^2$. To make $H(p)$

or $-D(p \parallel q)$ give the first term in (12) multiplied by the same factor, an *entropic prior* proportional to $\exp(\alpha^1 H(p))$ or $\exp(-\alpha^1 D(p \parallel q))$ is assigned to p, where $\alpha^1 = (2\alpha\sigma^2)^{-1}$. An immediate objection is that in the general case what is meant by a prior density of p ? What is the underlying measure in the space of functions on $\mathcal{X}$? If $\mathcal{X}$ is finite and p represents the empirical distribution of a sequence $x_1, \ldots, x_n$, a prior density for p proportional to $\exp(\alpha^1 H(p))$ or $\exp(-\alpha^1 D(p \parallel q))$ is a reasonable approximation, on account of (7), (9), specifically with $\alpha^1 = n$. In this case, the MAP interpretation of the regularized maxent solution p^* is justified, but only when the regularization parameter α specifically equals $(2n\sigma^2)^{-1}$.

5. Mathematical properties. Information geometry

Recall Problems (i), (ii), (iii) stated in Section 1, and their maxent solution defined there. Here we discuss some mathematical results about this solution, including the problem of its existence, when $\mathcal{X}$ and the given measure λ on $\mathcal{X}$ are arbitrary. Suppose first that F is defined by linear constraints, cf.(1). Given the functions $a_i(x)$ in (1) and the default model q, let Θ denote the set of those vectors $\vartheta = (\vartheta_1, \ldots, \vartheta_k)$ for which

$$Z(\vartheta) = \int q(x) \exp \sum_{i=1}^{k} \vartheta_i a_i(x) \lambda(dx) \tag{13}$$

is finite; for Problem (i), $q(x) = 1$ is set. For Problem (i) or (ii), consider the exponential family of densities

$$p_\vartheta(x) = Z(\vartheta)^{-1} q(x) \exp \sum_{i=1}^{k} \vartheta_i a_i(x), \quad \vartheta \in \Theta. \tag{14}$$

For Problem (iii), $p_\vartheta(x)$ is defined similarly but without the factor $Z(\vartheta)^{-1}$.

Theorem 5.1   If $p_\vartheta \in F$ for some $\vartheta \in \Theta$ then this p_ϑ is the maxent solution p^*. If $p_\vartheta \notin F$ for all $\vartheta \in \Theta$ then the maxent solution does not exist, providing

$$\{x : p(x) > 0\} = \{x : q(x) > 0\} \text{ for some } p \in F. \tag{15}$$

Next, for any convex feasible set F write

$$H(F) = \sup_{p \in F} H(p) \tag{16}$$

for Problem (i), and

$$D(F \parallel q) = \inf_{p \in F} D(p \parallel q) \tag{17}$$

for Problems (ii), (iii). The following theorem and corollary comprise results of Csiszár [10], [5] and Topsoe [11]; Problem (iii) was not considered there, but the proofs easily extend to that case.

Theorem 5.2.   For either of Problems (i), (ii), (iii), providing $H(F)$ resp. $D(F \parallel q)$ is finite, there exists a unique function p_0, not necessarily in F, such that for each $p \in F$

$$H(p) \le H(F) - D(p \parallel p_0) \tag{18}$$

respectively

$$D(p \parallel q) \ge D(F \parallel q) + D(p \parallel p_0). \tag{19}$$

If F is of form (1) and condition (15) holds then $p_0 = p_\vartheta$ for some $\vartheta \in \Theta$.

Corollary. (a) A maxent solution p^* exists iff $p_0 \in F$, in which case $p^* = p_0$, thus p^* is unique. (b) For any sequence of functions $p_n \in F$ with $H(p_n) \to H(F)$ resp. $D(p_n \parallel q) \to D(F \parallel q)$, we have $D(p_n \parallel p_0) \to 0$. (c) A sufficient condition for $p_0 \in F$, i.e., for the existence of the maxent solution, is the closedness of F in $L_1(\lambda)$. If F is of form (1) and (15) holds, the weaker condition that Θ is an open subset of R^k is also sufficient.

An example of non-existense of the maxent solution is Problem (i) for the set of densities on the real line

$$F = \{p : \int x p(x) dx = 0, \ \int x^2 p(x) dx = 1, \ \int x^3 p(x) dx = 1\}.$$

The integral (13) with $q(x) = 1$, $a_1(x) = x$, $a_2(x) = x^2$, $a_3(x) = x^3$ can be finite only if $\vartheta_3 = 0$, then the densities (14) (with $q(x) = 1$) are Gaussian. As no Gaussian density belongs to F, the non-existence follows from Theorem 5.1. In this example, the p_0 of Theorem 5.2 will be the Gaussian density with mean 0 and variance 1.

It follows from Theorem 5.1 that if the maxent solution exists for Problem (ii) or (iii) with F defined by linear constraints, then eq. (19) holds with equality. In other words, if the I-projection p^* of q onto F exists then

$$D(p \parallel q) = D(p \parallel p^*) + D(p^* \parallel q) \ \text{ for each } \ p \in F. \tag{20}$$

Actually, this holds even without the hypothesis (15).

The identity (20) is an analogue of the Pythagorean Theorem of Euclidean geometry, I-divergence regarded as an analogue of squared Euclidean distance. Indeed, if p^* were the Euclidean projection of a point q onto a plane F and D would denote squared Euclidean distance, (20) would be just the Pythagorean Theorem for the right-angled triangle pqp^*. Notice that if F is an arbitrary convex set, and the I-projection p^* of q onto F exists, (20) always holds with $\ge$ rather than $=$, cf. (19); this, too, is analogous to a result in Euclidean geometry with D=squared distance.

One simple but important consequence of the Pythagorean identity (20) is that if $F_1 \subset F$, the same $p = p_1^*$ attains both $\min_{p \in F_1} D(p \parallel q)$ and $\min_{p \in F_1} D(p \parallel p^*)$, providing either minimum is attained. Thus I-projection has the

Transitivity property: The I-projection onto $F_1 \subset F$ of the I-projection of q onto F equals the I-projection of q onto F_1, if F is defined by linear constraints.

For arbitrary F_1 and F_2 , both defined by linear constraints, let p_1^* be the I-projection of q onto F_1 and p_2^* the I-projection of p_1^* onto F_2. In general, p_2^* will differ from the I-projection of q onto F_2. However, taking the I-projection p_3^* of p_2^* onto F_1, then the I-projection p_4^* of p_3^* onto F_2, etc., we obtain a sequence $\{p_n^*\}$ that converges to the I-projection of q onto $F = F_1 \cap F_2$, again in analogy with Euclidean geometry. The geometric view turned out very useful in proving the convergence of this and related iterative algorithms, [10], [12], [13] (to be precise, a convergence proof of the I-projection iteration is available for the case when $\mathcal{X}$ is finite).

Simplest to determine I-projection is when the functions $a_1(x), \ldots, a_k(x)$ that define F by (1) are indicator functions of disjoint sets $A_1, \ldots, A_k$ with $\cup A_i = \mathcal{X}$, i.e., the constraints on p are that $\int_{A_i} p(x)\lambda(dx) = b_i, i = 1, \ldots, k$. This is because of the obvious

Scaling property: The I-projection of any q onto F as above is given by $p^*(x) = \lambda_i q(x)$ if $x \in A_i$, where $\lambda_i = b_i / \int_{A_i} q(x)\lambda(dx)$.

If the $a_i(x)$ are indicator functions of non-disjoint sets A_i, such as in the problem of inferring a bivariate probability mass function $p(x, y)$ when only its marginals $p_1(x)$ and $p_2(y)$ and a default model $q(x, y)$ are known, one may apply the above method of iterative I-projections. This amounts to *iterative scaling*, an algorithm that had been used in various fields well before Kullback pointed out that it converged to the maxent solution.

Let us mention one more "geometric" result, not directly related to maxent. It says that channel capacity may be regarded as "information radius", and plays a key role in the theory of universal data compression, [14].

Consider an arbitrary family of densities on $\mathcal{X}$, represented as $\{p_\vartheta(x), \vartheta \in \Theta\}$ as in Section 4, but now Θ may be any set, not necessarily finite dimensional. Let C denote the supremum of the mutual information $E_\pi D(p_\vartheta \parallel \bar{p})$ (where $\bar{p}(x) = E_\pi p_\vartheta(x)$) with respect to the choice of the prior π. Further, for any density q on $\mathcal{X}$, let

$$r(q) = \sup_{\vartheta \in \Theta} D(p_\vartheta \parallel q) \qquad (21)$$

denote the "radius" of the smallest "I-divergence ball with center q" that contains the given family.

Theorem 5.3: C is finite iff $r(q)$ is finite for some q. In that case

$$C = \min_q r(q) \qquad (22)$$

where the minimum is attained for a unique q^*. Moreoover, the maximum mutual information is attained for a prior π iff the corresponding marginal of x equals q^*, i.e., $E_\pi p_\vartheta(x) = q^*(x)$.

6. Other entropy functionals

In the literature, methods similar to maxent but based on other "entropy functionals" have also been suggested. In spectrum analysis, "maximum entropy" usually means maximizing $\int \log p(x) dx$, the Burg entropy of the power spectrum p or, if a non-constant default model q is given, minimizing $\int (\log \frac{q(x)}{p(x)} + \frac{p(x)}{q(x)} - 1) dx$, the Itakura-Saito distance of p from q.

The author is indebted to John Burg [15] for having pointed out that he does not consider $\int \log p(x) dx$ as an entropy functional on its own right. His idea in developing maximum entropy spectral analysis was to maximize the (Shannon) entropy rate of a stochastic process under spectral constraints, and he was using Kolmogorov's formula for the entropy rate of a Gaussian process in terms of its power spectrum.

Without entering philosophy, let us formally define ([16], [17], [18]) the f-entropy of a function $p(x) \geq 0$ as

$$H_f(p) = - \int f(p(x)) \lambda(dx), \qquad (23)$$

where $f(t)$ is any strictly convex differentiable function on the positive reals, and a corresponding measure of distance of p from q as

$$B_f(p,q) = \int [f(p(x)) - f(q(x)) - f'(q(x))(p(x) - q(x))] \lambda(dx). \qquad (24)$$

As strict convexity implies $f(s) > f(t) + f'(t)(s - t)$ for any positive numbers $s \neq t$, $B_f(p, q)$ is a distance in the same sense as I-divergence is, cf. Section 1. It is called Bregman distance, [19].

As examples, take $t \log t$, $- \log t$, or t^2 for $f(t)$. Then $H_f(p)$ will be the Shannon entropy, Burg entropy, or the negative square integral of p. $B_f(p, q)$ will be I-divergence, Itakura-Saito distance, and squared $L_2(\lambda)$-distance, respectively.

Theorem 5.2 remains valid if $f(t) = t \log t$ is replaced by another f, except that the $L_1(\lambda)$-closedness of F will be sufficient for $p_o \in F$, and hence for the existence of $p^* \in F$ attaining $\max_{p \in F} H_f(p)$ or $\min_{p \in F} B_f(p, q)$, only if f has the following property:

$$\inf_{t \geq 1} (f'(Kt) - f'(t)) > 0 \quad \text{for some} \quad K > 1, \qquad (25)$$

[20]. Notice that $f(t) = t \log t$ does have this property but $f(t) = - \log t$ does not. Indeed, the maximum of Burg entropy need not be attained even if $\mathcal{X} = [0, 1]$ and F is defined by linear constraints with continuous functions $a_i(x)$, [18].

The Pythagorean identity (20) with D replaced by B_f, for p^* minimizing $B_f(p, q)$ subject to $p \in F$, where F is of form (1), will always hold if

$$\lim_{t \to 0} f'(t) = -\infty. \qquad (26)$$

Then, of course, the transitivity property stated in Section 5 for I-projections, will also hold for "B_f−projections".

Notice that if f does not satisfy (26) then it can be extended to a (strictly convex, differentiable) function defined on the whole real line. With such as extension, $B_f(p,q)$ can be defined for p and q taking negative values, as well. Now, in (1) it has been tacitly assumed that the feasible functions p are non-negative, but if not needed for $B_f(p,q)$ to be defined, this assumption might be dropped. Then the analogue of (20) will hold unconditionally, but with p^* that may take negative values, as well. This highlights why least squares (minimization of L_2 distance), a preferred method of inference for real-valued functions, is not recommended for inferring functions constrained to be non-negative.

Another family of distances for non-negative functions, with properties somewhat similar to I-divergence, are the f-divergences

$$D_f(p \parallel q) = \int q(x) f(\frac{p(x)}{q(x)}) \lambda(dx), \tag{27}$$

with f as above, and additionally satisfying $f(1) = f'(1) = 0$. Originally, these distances were defined for probability densities, without the additional conditions on f ([21][22]).

As examples, take $t \log t - t + 1$, $-\log t + t + 1$ or $(\sqrt{t} - 1)^2$ for $f(t)$. Then $D_f(p \parallel q)$ will be the I-divergence $D(p \parallel q)$, the reversed I-divergence $D(q \parallel p)$, and the squared Hellinger distance $\int(\sqrt{p} - \sqrt{q})^2 \lambda(dx)$, respectively.

Results similar to those in Theorem 5.2 can be proved also for f-divergences in the role of I-divergence, [20]. We only mention the simple fact that projections defined by D_f share the scaling property of I-projections, stated in Section 5.

7. Axiomatic approach

Whereas maxent is no doubt an attractive method, it is hard to come up with mathematical results that distinguish it unequivocally. For certain applications, mostly for those in statistical physics, the large deviation arguments in Section 3 provide a very strong justification. Apparently, no similarly convincing arguments are available for the majority of applications, in particular for those to Problem (iii).

One difficulty is that the problem of inferring a function from insufficient information, stated in Section 1 in a deliberately vague form, does not easily admit a mathematical formulation, with well defined optimality criteria. Under such circumstances, an axiomatic approach appears most promising: Start from certain properties that a "good" method of inference is supposed to have, and investigate whether the postulated properties uniquely characterize a distinguished method, or else what alternatives come into account. Such an approach was first put forward by Shore and Johnson [23]. The currently available best axiomatic results are those of Csiszár [24]. Below some of these are briefly reviewed, restricted to Problem (iii), for simplicity.

In this Section we assume that $\mathcal{X}$ is a finite set of size $|\mathcal{X}| \geq 3$, and restrict attention to strictly positive functions on $\mathcal{X}$.

By an inference method for Problem (iii), we mean any rule that assigns to every feasible set F defined by linear constraints, and any default model q, an inferred function $p^* \in F$, denoted by $p^*(F, q)$. It should be emphasized that it is *not* postulated that p^* is the minimizer of some measure of distance of p from q. This does follow from the regularity and locality axioms below, but it is to be proved, which represents a major step of the approach.

Regularity axiom: If $F_1 \subset F$ and $p^*(F, q) \in F_1$ then $p^*(F_1, q) = p^*(F, q)$.

Some technical assumptions additionally postulated in [24] as part of this axiom are omitted here.

Locality axiom: If the constraints defining F can be partitioned into two sets, the first resp. second involving functions $a_i(x)$ that vanish for $x \notin \mathcal{X}_1$ resp. for $x \in \mathcal{X}_1$, where $\mathcal{X}_1$ is some subset of $\mathcal{X}$, then the values of $p^*(F, q)$ for $x \in \mathcal{X}_1$ depend only on the first set of constraints and on the values of q for $x \in \mathcal{X}_1$.

The regularity axiom formalizes the intuitive idea that if the inference based on some knowledge happens to be consistent also with some additional knowledge then the new knowledge provides no reason to change that inference. The locality axiom means that if the available knowledge consists of pieces pertaining to disjoint subsets of $\mathcal{X}$ then on each of these subsets, the inference must be based on the knowledge pertaining to that subset.

We recall the transitivity and scaling properties of I-projection, cf. Section 5. These might be used as axioms, requiring them to hold for our $p^* = p^*(F, q)$ in the role of I-projection. Instead of the scaling properly, we will consider a weaker version of it.

Weak scaling axiom: If $F = \{p : p(x_1) + p(x_2) = t\}$ for some x_1, x_2 in $\mathcal{X}$ and $t > 0$, then $p^* = p^*(F, q)$ satisfies $p^*(x_i) = \lambda q(x_i), i = 1, 2, \lambda = t(q(x_1) + q(x_2))^{-1}$.

An even weaker postulate is the

Semisymmetry axiom: For F as above, if $q(x_1) = q(x_2)$ then $p^*(x_1) = p^*(x_2)$.

Theorem 7.1: The regularity, locality, transitivity and weak scaling axioms are satisfied iff $p^*(F, q)$ is the I-projection of q onto F. If weak scaling is replaced by semisymmetry, the possible inference rules will be those where $p^*(F, q)$ minimizes the Bregman distance $B_f(p, q) = \Sigma(f(p(x)) - f(q(x)) - f'(q(x))(p(x) - q(x)))$ for some f satisfying (26). If transitivity is dropped but weak scaling is retained, $p^*(F, q)$ will be the minimizer of the f-divergence $D_f(p \parallel q) = \Sigma q(x) f(\frac{p(x)}{q(x)})$, for some f satisfying (26).

The maxent solution (i.e., I-projection) can be uniquely characterized also by postulating, in addition to regularity and locality, a consistency property for inferring a bivariate function from its marginals, similar to the "system independence" axiom of [23] (this, however, appears less compelling for Problem (iii) than for Problem (ii) treated in [23]).

Another natural postulate might be the *scale-invariance* of inference, i.e., that if $b_1, \ldots, b_k$ in the definition (1) of F and the default model q are multiplied by a constant, the effect on $p^*(F, q)$ be just multiplication by the same constant. This is automatically fulfilled by f-divergence minimization. On the other hand, we have

Theorem 7.2 Minimizing a Bregman distance gives scale invariant inference iff $B_f(p,q)$ corresponds to one of the following functions:

$$f_1(t) = t\log t, \quad f_0(t) = -\log t, \quad f_\alpha(t) = -\frac{1}{\alpha}t^\alpha, \quad 0 < \alpha < 1 \text{ or } \alpha < 0.$$

Among the alternatives to maxent, the most promising might be the minimization of a Bregman distance as in Theorem 7.2. Recall that $f_0(t)$ gives the Itakura-Saito distance; of course, $f_1(t)$ gives I-divergence. Practical success with $f_\alpha(t)$, with a small positive α, has been reported by Jones and Trutzer [17].

The axiomatic approach may help to recognize those situations when maxent should not be used. E.g., the regularity axiom is inappropriate when such a function should be selected from the feasible set F that "best represents" the functions in F. Then a more adequate method than maxent might be to select that $q \in F$ which minimizes the information radius $r(q) = \sup_{p \in F} D(p \parallel q)$, cf. (21) (barycenter method, Perez [25], suitable for F as in (1) only if $\mathcal{X}$ is finite.)

The locality axiom is inappropriate if the inferred function p^* is required to be smooth in some sense. This is more typical in the continuous case, but a discrete version of smoothness may also be a natural requirement. One way to deal with this problem is to modify maxent by a regularization term that penalizes departures from the desired smoothness, similarly to regularization for constraints subject to errors, cf. Section 4. Of course, it may well be that both kinds of regularization are needed.

8. Maximum entropy in the mean (MEM)

Space permits to give but a flavor of MEM, an extension of maxent developed by French researchers, [26], [27], [28], [29]. Though MEM is not restricted to non-negative functions, we keep this restriction here. As a more serious simplification, we restrict attention to functions on a finite set of size m. This preempts significant features of the theory but still affords remarkable conclusions, some apparently new. For convenience, the functions considered for MEM inference will be regarded as m-vectors (with non-negative components), $u = (u_1, \ldots, u_m) \in R_+^m$.

The MEM idea is to visualize the vector to be inferred as the expectation of some probability distribution on R_+^m. Given a reference measure λ on R_+ (typically the Lebesgue measure, or the counting measure on a countable subset of R_+), associate with each $u \in R_+^m$ the set of all probability densities with respect to λ^m whose expectation is equal to u:

$$E(u) = \{p : \int x_i p(x_1, \ldots, x_m)\lambda(dx_1)\ldots\lambda(dx_m) = u_i, \ i = 1, \ldots, m\}. \qquad (28)$$

To infer u knowing only its membership to a feasible set $F \subset R_+^m$, MEM suggests to take the feasible set of densities $\tilde{F} = \cup_{u \in F} E(u)$, determine the maxent density $p^* \in \tilde{F}$, and declare its expectation u^* to be the solution.

One easily sees from Theorem 5.1 that if λ=Lebesgue measure then

$$\max_{p \in E(u)} H(p) = \sum_{i=1}^{m} (\log u_i + 1), \qquad (29)$$

attained for $p(x_1, \ldots, x_m) = \Pi_{i=1}^{m} u_i^{-1} \exp(x_i/u_i)$. The MEM solution u^* is attained by maximizing (29) subject to $u \in F$, which is the same as maximizing Burg entropy.

Suppose next that a default model $v = (v_1, \ldots, v_m)$ with positive components is also given, and let $q(x_1, \ldots, x_m)$ be a corresponding default model for densities, again with respect to λ^m. Then one can proceed as before, taking for p^* the I-projection of q onto $\tilde{F}$. As any density in $E(v)$ might be chosen as q, the MEM idea does not lead to a unique inference method, even when λ is fixed. Rather, any assignment $v \mapsto q_v \in E(v)$ gives rise to a different method. We will consider assignments of product form

$$q_v(x_1, \ldots, x_m) = \prod_{i=1}^{m} q_{v_i}(x_i) \qquad (30)$$

where $\{q_t(x), \; t > 0\}$ is a given family of one-dimensional densities, with expectations

$$\int x q_t(x) \lambda(dx) = t. \qquad (31)$$

For any choice of q, if the I-projection p_u^* of q onto $E(u)$ exists, the MEM solution u^* will be the minimizer of $D(p_u^* \parallel q)$ subject to $u \in F$. For $q = q_v$ as in (30), Theorem 5.1 implies that if

$$p_{\vartheta_1, \ldots, \vartheta_m}(x_1, \ldots, x_m) = \prod_{i=1}^{m} (Z_{v_i}(\vartheta_i))^{-1} q_{v_i}(x_i) \exp(\vartheta_i x_i) \qquad (32)$$

belongs to $E(u)$ then $p_u^* = p_{\vartheta_1, \ldots, \vartheta_m}$. Here

$$Z_t(\vartheta) = \int q_t(x) \exp(\vartheta x) \lambda(dx) \qquad (33)$$

is the moment generating function of the distribution with density $q_t(x)$. The condition $p_{\vartheta_1, \ldots, \vartheta_m} \in E(u)$ is equivalent to

$$\frac{\partial}{\partial \vartheta} \log Z_{v_i}(\vartheta)|_{\vartheta = \vartheta_i} = u_i, \; i = 1, \ldots, m. \qquad (34)$$

Thus, assuming that the equations (34) have solutions $\vartheta_1, \ldots, \vartheta_m$, we obtain that

$$D(p_u^* \parallel q) = \sum_{i=1}^{m} (\vartheta_i u_i - \log Z_{v_i}(\vartheta_i)). \qquad (35)$$

Two types of choice of the family $\{q_t(x), t > 0\}$ deserve special attention.

(i)   Let $q_1(x)$ be given, with expectation 1, and let

$$q_t(x) = (Z_1(\vartheta_t))^{-1} q_1(x) \exp(\vartheta_t x), \quad t > 0 \tag{36}$$

where $Z_1(\vartheta)$ is defined by (33) and ϑ_t is determined to make (31) hold, i.e.,

$$\frac{d}{d\vartheta} \log Z_1(\vartheta)|_{\vartheta = \vartheta_t} = t. \tag{37}$$

It is assumed that (37) has a solution ϑ_t for every $t > 0$, this also ensures that the equations (34) have solutions $\vartheta_1, \ldots, \vartheta_m$. Let

$$f(t) = t\vartheta_t - \log Z_1(\vartheta_t) \tag{38}$$

be the Legendre transform or convex conjugate of the convex function $\log Z_1(\vartheta)$. It is a convex function of $t > 0$ with $f'(t) = \vartheta_t$. A trite calculation shows that in case of (36) the right hand side of (35) equals the Bregman distance $B_f(u, v)$. Thus MEM with the assignment $v \mapsto q_v$ defined by (30), (36) leads to Bregman distance minimization.

(ii)   Given an infinitely divisible distribution on R_+ with expectation 1, let $Z_1(\vartheta)$ be its moment generating function. Then for each $t > 0, (Z_1(\vartheta))^t$ is the moment generating function of a distribution with expection t. Let $\{q_t(x), t > 0\}$ be the family of densities of these distributions (with a suitable choice of λ). Since now $Z_t(\vartheta) = (Z_1(\vartheta))^t$, we obtain that the right hand side of (35) equals the f-divergence of u from v, with f as in (38). Thus MEM with the assignment $v \to q_v$ defined by (30) with the present $\{q_t(x), t > 0\}$ leads to f-divergence minimization.

Examples.   1. Let λ be the counting measure on the natural numbers, and $q_1(x) = (ex!)^{-1}$. Then $Z_1(\vartheta) = \exp(e^\vartheta - 1), \vartheta_t = \log t, f(t) = t \log t - t + 1$. Now the family $\{q_t(x), t > 0\}$ consists of the Poisson distributions in both cases (i) and (ii) above, and MEM leads to regular maxent.

2.   Let λ be the Lebesgue measure on R_+, and $q_1(x) = e^{-x}$. Then $Z_1(\vartheta) = (1 - \vartheta)^{-1}(\vartheta < 1), \vartheta_t = 1 - \frac{1}{t}, f(t) = t - 1 - \log t$. Now the family $\{q_t(x), t > 0\}$ consists of the exponential densities $t^{-1} \exp(x/t)$ in case (i), and of the Γ densities $\gamma_t(x) = x^{t-1} e^{-x}/\Gamma(t)$ in case (ii); thus with these choices in (30), MEM leads to minimizing Itakura-Saito distance resp. reversed I-divergence, cf. Section 6.

3.   Let λ be the Lebesgue measure on R_+ plus a unit mass at 0, and consider a compound Poisson distribution with λ-density $q_1(x)$ given by

$$q_1(0) = e^{-\tau}, \quad q_1(x) = \sum_{k=1}^{\infty} \tau^k e^{-\tau} \gamma_{k/\tau}(x)/k! \quad (x > 0). \tag{39}$$

Here $\gamma_t(x)$ denotes Γ-density as above, and $\tau > 0$ is a parameter. Then $Z_1(\vartheta) = \exp\{\tau[(1-\vartheta)^{1/\tau} - 1]\} (\vartheta < 1), \vartheta_t = 1 - t^{-\tau/1+\tau}, f(t) = t - (1+\tau)t^{1/1+\tau} + \tau$. Thus in case (i), MEM leads to minimizing a Bregman distance as in Theorem 7.2, with $0 < \alpha < 1 (\alpha = (1 + \tau)^{-1})$, and among the f-divergences whose minimization we are lead to in case (ii) is the Hellinger distance $(\tau = 1)$, cf. Section 6.

It is remarkable that via MEM, several alternatives to maxent can be arrived at within the maxent paradigm. Still, the implications of this remain debatable, except for those cases when the MEM idea can be justified by a physical model.

References

1. S. Kullback, *Information Theory and Statistics*, John Wiley and Sons, "New York, 1959.
2. E.T. Jaynes (R.D. Rosenkrantz ed.), *Papers on Probability, Statistics and Statistical Physics, Reidel, Dordrecht, 1983.*
3. I. Csiszár and J. K{orner, *Information Theory: Coding Theorems for Discrete Memoryless Systems*, Academic Press, New York, 1981.
4. J.M. Van Campenhout and T. Cover, "Maximum entropy and conditional probability," *IEEE Trans. Inform. Theory*, **27**, 483-489, 1981.
5. I. Csiszár, "Sanov property, generalized I-projection and a conditional limit theorem," *Ann. Probability*, **12**, 768-793, 1984.
6. I. Csiszár, "An extended maximum entropy principle and a Bayesian justification (with discussion)," *Bayesian Statistics 2*, J.M. Bernardo et al., pp. 83-89, North-Holland, Amsterdam, 1985.
7. J.M. Bernardo, "Reference posterior for Bayesian inference (with discussion)," *J. Roy. Statist. Soc. B*, **41**, 113-147, 1979.
8. B. Clarke and A.R. Barron, "Jeffreys' prior is asymptotically least favorable under entropy risk," *J. Statist. Planning and Inference*, **41**, pp. 37-60, 1994.
9. J.M. Bernardo and A.F.M. Smith, "Bayesian Theory," John Wiley and Sons, New York, 1994.
10. I. Csiszár, "I-divergence geometry of probability distributions and minimization problems," *Ann. Probability*, **3**, pp. 146-158, 1975.
11. F. Topsoe, "Information theoretical optimization techniques", *Kybernetika*, **15**, pp. 7-17, 1979.
12. I. Csiszár and G. Tusnády, "Information geometry and alternating minimization procedures," *Statist. Decisions*, **Suppl. 1**, pp. 205-237, 1984.
13. I. Csiszár, " A geometric interpretation of Darroch and Ratcliff's generalized iterative scaling," *Ann. Statist.*, **17**, pp. 1409-1413, 1989.
14. L.D. Davisson and A. Leon-Garcia, "A source matching approach to finding minimax codes," *IEEE Trans. Inform. Theory*, **26**, pp. 166-174, 1980.
15. J. Burg, "Personal Communication," 1995.
16. C. R. Rao and T. K. Nayak, "Cross entropy, dissimilarity measures, and characterization of quadratic entropy", *IEEE Trans. Inform. Theory*, **31**, pp. 589-593, 1985.
17. L. Jones and V. Trutzer, "Computationally feasible high-resolution minimum-distance procedures which extend the maximum-entropy method," *Inverse Problems*, **5**, pp. 749-766, 1989.
18. J. M. Borwein and A. S. Lewis, "Partially-finite programming in L_1 and the existence of maximum entropy estimates," *SIAM J. Optimization*, **3**, pp. 248-267, 1993.
19. L.M. Bregman, "The relaxation method of finding the common point of convex sets and its application to the solution of problems in convex programming," *USSR Comput. Math. and Math. Phys.*, **7**, pp. 200-217, 1967.
20. I. Csiszár, "Generalized projections for non-negative functions", *Acta Math. Hungar.*, **68**, pp. 161-185, 1995.
21. I. Csiszár, "Eine informationstheoretische Ungleichung und ihre Anwendung auf den Beweis der Ergodizit{at von Markoffschen Ketten," *Publ. Math. Inst. Hungar. Acad. Sci*, **8**, pp. 85-108, 1963.
22. S.M. Ali and S.D. Silvey, "A general class of coefficients of divergence of one distribution from another," *J. Roy. Statist. Soc. Ser. B*, **28**, pp. 131-142, 1966.
23. J. E. Shore and R. W. Johnson, "Axiomatic derivation of the principle of maximum entropy and the principle of minimum cross-entropy," *IEEE Trans. Inform. Theory*, **26**, pp. 26-37, 1980.
24. I. Csiszár, "Why least squares and maximum entropy? An axiomatic approach to inference for linear inverse problems," *Ann. Statist.*, **19**, pp. 2032-2066, 1991.
25. A. Perez, "Barycenter" of a set of probability measures and its application in statistical decision, *Compstat Lectures*, pp. 154-159, Physica, Heidelberg, 1984.
26. J. Navaza, "The use of non-local constraints in maximum-entropy electron density reconstruction," *Acta. Crystallographica*, **A42**, pp. 212-223, 1986.
27. D. Dacunha-Castelle and F. Gamboa, "Maximum d'entropie et problème des moments,"

Ann. Inst. H. Poincarè, **4**, pp. 567-596, 1990.

28. F. Gamboa and G. Gassiat, "Bayesian methods and maximum entropy for ill posed inverse problems," *Ann. Statist.*, Submitted, 1994.

29. J.F. Bercher G. LeBesnerais and G. Demoment, "The Maximum Entropy on the Mean, Method, Noise and Sensitivity", in *Proc. 14th Int. Workshop Maximum Entropy and Bayesian Methods*, S. Sibisi and J. Skilling, Kluwer Academic, 1995.

BAYESIAN ESTIMATION OF THE VON MISES CONCENTRATION PARAMETER

D.L. DOWE, J.J. OLIVER, R.A. BAXTER AND C.S. WALLACE
Department of Computer Science,
Monash University, Clayton, Vic. 3168, Australia[†]

Abstract. The von Mises distribution is a maximum entropy distribution. It corresponds to the distribution of an angle of a compass needle in a uniform magnetic field of direction, μ, with concentration parameter, κ. The concentration parameter, κ, is the ratio of the field strength to the temperature of thermal fluctations.

Previously, we obtained a Bayesian estimator for the von Mises distribution parameters using the information-theoretic Minimum Message Length (MML) principle. Here, we examine a variety of Bayesian estimation techniques by examining the posterior distribution in both polar and Cartesian co-ordinates. We compare the MML estimator with these fellow Bayesian techniques, and a range of Classical estimators. We find that the Bayesian estimators outperform the Classical estimators.

Key words: Bayesian estimation, Minimum Message Length (MML), von Mises distribution, invariance to parameter transformations

1. Introduction

The von Mises distribution, $M_2(\mu, \kappa)$, is a maximum entropy distribution. It corresponds to the distribution of an angle of a compass needle in a uniform magnetic field of direction, μ, with concentration parameter, κ. The concentration parameter, κ, is the ratio of the field strength to the temperature of thermal fluctations. This distribution is a circular analogue of the Gaussian distribution, to which it converges for large κ and small σ.

Circular distributions and the von Mises distribution in particular are of interest in a wide range of fields, such as biology, geography, geology, geophysics, medicine, meteorology and oceanography[3], and protein dihedral angles[10].

We consider a range of estimators for the parameters of the von Mises distribution, both Classical and Bayesian, including a Bayesian estimator [9] obtained using the information-theoretic Minimum Message Length (MML) principle [11],

[†]Email: {dld, jono, rohan, csw}@cs.monash.edu.au

K. M. Hanson and R. N. Silver (eds.), Maximum Entropy and Bayesian Methods, 51–59.
© *1996 Kluwer Academic Publishers.*

and a variety of Bayesian estimators obtained by examining the posterior distribution in both polar and Cartesian co-ordinates. This work raises questions about the effect of parameterisations on Bayesian inference (e.g., [1,4]). We find that the Bayesian estimators outperform the Classical estimators on a series of simulations.

2. The Likelihood Function

The von Mises distribution has density function $f(\theta) = \frac{1}{2\pi I_0(\kappa)} e^{\kappa \cos(\theta - \mu)}$ where θ is in a range of 2π and $I_p(\kappa)$ is the modified Bessel function:

$$I_p(\kappa) = \frac{1}{2\pi} \int_0^{2\pi} \cos(p\theta) e^{\kappa \cos \theta} d\theta \quad \text{and so} \quad I_0(\kappa) = \frac{1}{2\pi} \int_0^{2\pi} e^{\kappa \cos \theta} d\theta \quad (1)$$

For data $D = \{\theta_1, \ldots \theta_N\}$, the likelihood is: $\quad p(D|\kappa, \mu) = \prod_{i=1}^{N} \frac{1}{2\pi I_0(\kappa)} e^{\kappa \cos(\theta_i - \mu)}$

and the negative log-likelihood is: $\quad L = N \log(2\pi I_0(\kappa)) - \kappa \sum_{i=1}^{N} \cos(\theta_i - \mu) \quad (2)$

3. Estimation Methods

We consider and compare several Bayesian and Classical approaches to estimating the concentration parameter κ of the von Mises distribution. We consider the Bayesian MAP estimate (the posterior mode) and the MML estimate. We consider the following Classical estimates: Maximum Likelihood, Marginalised Maximum Likelihood [7], and an estimator proposed by N.I. Fisher [3].

Whereas we were able to obtain an analytical closed form estimator of μ, all of our estimators for κ had to be obtained by numerical methods.

3.1. ESTIMATION OF μ

To find the maximum likelihood estimator for μ, with $x = \sum_{i=1}^{N} \cos(\theta_i)$, $y = \sum_{i=1}^{N} \sin(\theta_i)$, and $R = \sqrt{x^2 + y^2}$ we have that $(\cos \hat{\mu}, \sin \hat{\mu}) = (\frac{x}{R}, \frac{y}{R})$. Because of the uniform nature of our prior on μ (see Section 4.1), the MAP estimates for μ and the MML estimate for μ agree with the maximum likelihood estimator for μ.

4. Bayesian Estimators

We note that the Bayesian estimators require a prior distribution. We consider two prior distributions for κ to see if the Bayesian methods are sensitive to the choice of prior. We also note that the Bayesian MAP estimate requires a choice of parameterisation, an issue we discuss in Section 4.2.

4.1. PRIOR DISTRIBUTIONS

We assume a prior $h_\mu(\mu) = \frac{1}{2\pi}$ on μ independent of κ. We consider two priors [9]

on κ
$$h_2(\kappa) = \frac{2}{\pi(1 + \kappa^2)} \quad \text{and} \quad h_3(\kappa) = \frac{\kappa}{(1 + \kappa^2)^{3/2}}$$

4.2. THE MAP ESTIMATE

It is generally known that the mode of the posterior (and hence the MAP estimate) is not invariant under non-linear parameter transformations [2]. Therefore, we consider these estimates in both polar co-ordinates (κ, μ), and Cartesian co-ordinates $(X, Y) = (\kappa.\cos(\mu), \kappa.\sin(\mu))$, since in some sense, both these representations can be considered "natural". Oliver and Baxter [6] illustrated the manner in which the mode moves given the h_3 prior and the data:

$$D = \{\theta_1, \theta_2, \ldots, \theta_{10}\} = \{279°, 143°, 307°, 153°, 35°, 203°, 325°, 45°, 20°, 74°\}$$

The MAP estimate in polar co-ordinates is $(\kappa = 0.53, \mu = 27.8)$, while the MAP estimate in Cartesian co-ordinates is $(x = 0.22, y = 0.12)$, which is equivalent to $(\kappa = 0.25, \mu = 27.8)$ as shown in Figure 1.

Since we are considering two priors and two parameterisations, we find that there are four MAP estimates for κ. These estimates are the values of κ which maximise the following expressions:

$$\text{MAP}_{h2}^{\kappa,\mu} = \max_\kappa \ \log(h_2(\kappa)) - N\log(2\pi I_0(\kappa)) + \kappa \sum_{i=1}^{N} \cos(\theta_i - \hat{\mu})$$

$$\text{MAP}_{h3}^{\kappa,\mu} = \max_\kappa \ \log(h_3(\kappa)) - N\log(2\pi I_0(\kappa)) + \kappa \sum_{i=1}^{N} \cos(\theta_i - \hat{\mu})$$

$$\text{MAP}_{h2}^{x,y} = \max_\kappa \ \log(\frac{h_2(\kappa)}{\kappa}) - N\log(2\pi I_0(\kappa)) + \kappa \sum_{i=1}^{N} \cos(\theta_i - \hat{\mu})$$

$$\text{MAP}_{h3}^{x,y} = \max_\kappa \ \log(\frac{h_3(\kappa)}{\kappa}) - N\log(2\pi I_0(\kappa)) + \kappa \sum_{i=1}^{N} \cos(\theta_i - \hat{\mu})$$

We note that $\text{MAP}_{h2}^{x,y}$ is not a sensible estimator of κ since $\frac{h_2(\kappa)}{\kappa} = \frac{2}{\kappa\pi(1+\kappa^2)}$ diverges as $\kappa \to 0$. Hence, $\text{MAP}_{h2}^{x,y}$ always gives $\kappa = 0$ independent of the data.

5. The MML Estimate

MML is a Bayesian point estimation method proposed by Wallace et al. [11]. For the MML estimates, we use the prior distributions from Section 4.1. Unlike the MAP estimate, we do not need to consider the parameterisation for the MML estimate, since the MML estimate is invariant under 1-1 differentiable transformations of the parameter space (see [8], [11, p245] or [6, Section 5.4]).

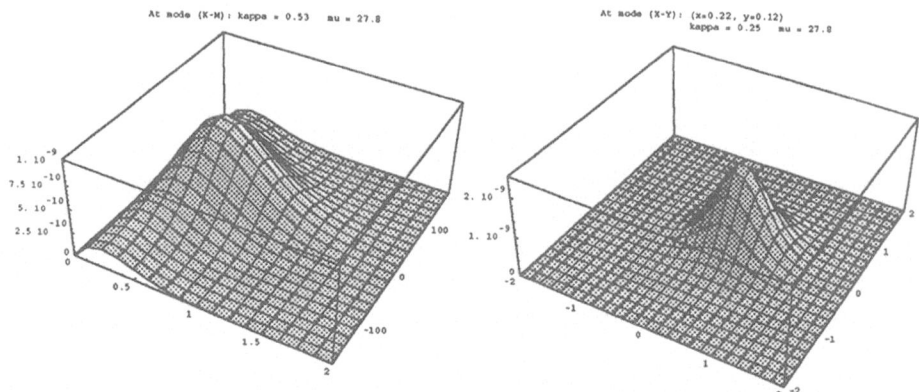

Figure 1.   The Posterior Density in Polar and x-y Co-ordinates

The MML estimate is the value of (κ, μ) which minimises the expression [11]:

$$MessLen(\mu \;\&\; \kappa \;\&\; D) = -log_2 \frac{h(\kappa, \mu) \, p(D|\kappa, \mu)}{\sqrt{det(F(\kappa, \mu))}} + \text{Constants},$$

where we interpret the term $\frac{1}{\sqrt{det(F(\kappa, \mu))}}$ as a volume of uncertainty. Minimising the Message Length is then equivalent to maximising:

$$\text{Exp} = \frac{h(\kappa, \mu) \, p(D|\kappa, \mu)}{\sqrt{det(F(\kappa, \mu))}} \qquad (3)$$

In Appendix A, we derive the Fisher matrix for the von Mises distribution [9]:

$$F(\mu, \kappa) = \begin{bmatrix} E(\frac{\partial^2 L}{\partial \mu^2}) & E(\frac{\partial^2 L}{\partial \mu \partial \kappa}) \\ E(\frac{\partial^2 L}{\partial \kappa \partial \mu}) & E(\frac{\partial^2 L}{\partial \kappa^2}) \end{bmatrix} = \begin{bmatrix} \kappa N A(\kappa) & 0 \\ 0 & N A'(\kappa) \end{bmatrix}$$

where $A(\kappa)$ is defined to be (see Mardia [5]): $A(\kappa) = \dfrac{I_1(\kappa)}{I_0(\kappa)} = \dfrac{I_0'(\kappa)}{I_0(\kappa)}$ $\qquad (4)$

The determinant of the Fisher matrix is: $det(F(\mu, \kappa)) = \kappa N^2 A(\kappa) A'(\kappa)$

and thus the expression to be maximised is: $\text{Exp} = \dfrac{h(\kappa, \mu) \, p(D|\kappa, \mu)}{N \sqrt{\kappa A(\kappa) A'(\kappa)}}$

We maximise this numerically. We note there are potential problems in doing this as $\kappa \to 0$, since $det(F(\mu, \kappa)) \to 0$ as $\kappa \to 0$. This is discussed in Appendix B.

6. Results

We tested the estimation techniques by running the following simulations; we generated N angles from a von Mises distribution with concentration parameter

	$N = 2$				$N = 5$		
	MAE	MSE	MKL		MAE	MSE	MKL
		$\kappa = 0.0$				$\kappa = 0.0$	
MaxLik	243.3	9.0×10^6	241.9	MaxLik	1.0712	2.2207	0.3599
Schou	121.6	2.2×10^6	120.7	Schou	0.5426	1.2513	0.2002
NF	24.43	9.0×10^4	23.86	NF	0.5626	0.7443	0.1509
MML_{h2}	0.2762	0.1038	0.0256	MML_{h2}	0.3421	0.4092	0.0745
$\text{MAP}^{\kappa,\mu}_{h2}$	0.5889	0.486	0.1152	$\text{MAP}^{\kappa,\mu}_{h2}$	0.5947	0.5618	0.1239
MML_{h3}	0.5068	0.3644	0.0873	MML_{h3}	0.5784	0.6569	0.1303
$\text{MAP}^{\kappa,\mu}_{h3}$	1.0622	1.2614	0.2866	$\text{MAP}^{\kappa,\mu}_{h3}$	0.8892	0.9488	0.212
$\text{MAP}^{x,y}_{h3}$	0.3724	0.1827	0.0448	$\text{MAP}^{x,y}_{h3}$	0.4599	0.3078	0.0727
		$\kappa = 0.5$				$\kappa = 0.5$	
MaxLik	758.2	1.4×10^8	675.4	MaxLik	0.8330	2.8412	0.3787
Schou	379.2	3.6×10^7	337.4	Schou	0.7709	1.8765	0.2537
NF	75.83	1.4×10^6	67.16	NF	0.5696	0.7528	0.1674
MML_{h2}	0.2097	0.0700	0.0599	MML_{h2}	0.4247	0.5881	0.1173
$\text{MAP}^{\kappa,\mu}_{h2}$	0.3465	0.1605	0.1269	$\text{MAP}^{\kappa,\mu}_{h2}$	0.3866	0.393	0.1387
MML_{h3}	0.2948	0.1146	0.1049	MML_{h3}	0.4034	0.6369	0.1510
$\text{MAP}^{\kappa,\mu}_{h3}$	0.6046	0.506	0.2713	$\text{MAP}^{\kappa,\mu}_{h3}$	0.4893	0.5215	0.2007
$\text{MAP}^{x,y}_{h3}$	0.1888	0.0547	0.0714	$\text{MAP}^{x,y}_{h3}$	0.2736	0.1416	0.086
		$\kappa = 1.0$				$\kappa = 1.0$	
MaxLik	206.3	4.0×10^6	145.7	MaxLik	1.1863	5.7859	0.4593
Schou	103.3	1.0×10^6	72.55	Schou	1.1153	3.8451	0.3647
NF	20.70	40203	14.25	NF	0.6635	1.2661	0.2047
MML_{h2}	0.6774	0.4846	0.1526	MML_{h2}	0.8235	1.3712	0.2101
$\text{MAP}^{\kappa,\mu}_{h2}$	0.3727	0.2279	0.1615	$\text{MAP}^{\kappa,\mu}_{h2}$	0.5405	0.7788	0.1589
MML_{h3}	0.4125	0.2659	0.1534	MML_{h3}	0.6815	1.2755	0.2065
$\text{MAP}^{\kappa,\mu}_{h3}$	0.3478	0.1627	0.2413	$\text{MAP}^{\kappa,\mu}_{h3}$	0.4445	0.744	0.1772
$\text{MAP}^{x,y}_{h3}$	0.5700	0.3642	0.1464	$\text{MAP}^{x,y}_{h3}$	0.4639	0.2875	0.1149
		$\kappa = 10.0$				$\kappa = 10.0$	
MaxLik	1263	4.5×10^7	96.03	MaxLik	14.79	1126	0.6755
Schou	629.3	1.1×10^7	47.65	Schou	11.28	683.7	0.5045
NF	126.7	4.5×10^5	9.4915	NF	7.3595	1565	0.3161
MML_{h2}	9.5083	90.41	1.1481	MML_{h2}	6.2488	149.7	0.2589
$\text{MAP}^{\kappa,\mu}_{h2}$	8.9035	79.28	0.8111	$\text{MAP}^{\kappa,\mu}_{h2}$	5.522	47.63	0.2204
MML_{h3}	9.0301	81.55	0.8706	MML_{h3}	6.1960	149.2	0.2557
$\text{MAP}^{\kappa,\mu}_{h3}$	8.4152	70.83	0.6289	$\text{MAP}^{\kappa,\mu}_{h3}$	5.975	131.2	0.2108
$\text{MAP}^{x,y}_{h3}$	9.3715	87.83	1.0584	$\text{MAP}^{x,y}_{h3}$	7.604	58.00	0.417

TABLE 1. Results for $N = 2$ and $N = 5$

κ, and with $\mu = 0$. We then applied the estimation methods previously discussed, namely (A) the Maximum Likelihood estimator (MaxLik), (B) the Marginalised Maximum Likelihood estimator (Schou) [7], (C) an estimator proposed by N.I. Fisher (NF) [3], (D) the MML estimator [9] with the h_2 and h_3 priors (MML_{h2}

and MML$_{h3}$), and (E) the three sensible MAP estimators discussed (MAP$_{h2}^{\kappa,\mu}$, MAP$_{h3}^{\kappa,\mu}$ and MAP$_{h3}^{x,y}$).

Tables 1 and 2 give the mean absolute error (MAE), the mean squared error (MSE), and the mean Kullback-Leibler distance (MKL) for each of the above methods averaged over 1000 simulations. We did not present the NF estimator in Table 2, since it is defined to be the Maximum Likelihood estimator for $N > 15$.

	$N = 25$				$N = 100$		
	MAE	MSE	MKL		MAE	MSE	MKL
	$\kappa = 0.0$				$\kappa = 0.0$		
MaxLik	0.3625	0.1705	0.0417	MaxLik	0.1766	0.0390	0.00971
Schou	0.1509	0.0867	0.0197	Schou	0.0732	0.0174	0.00435
MML$_{h2}$	0.1554	0.0503	0.0124	MML$_{h2}$	0.0810	0.0127	0.00318
MAP$_{h2}^{\kappa,\mu}$	0.3266	0.1412	0.0347	MAP$_{h2}^{\kappa,\mu}$	0.1743	0.0381	0.0095
MML$_{h3}$	0.3076	0.1242	0.0305	MML$_{h3}$	0.1685	0.0356	0.00886
MAP$_{h3}^{\kappa,\mu}$	0.4880	0.2571	0.0629	MAP$_{h3}^{\kappa,\mu}$	0.2512	0.0667	0.0166
MAP$_{h3}^{x,y}$	0.3056	0.1246	0.0306	MAP$_{h3}^{x,y}$	0.1664	0.0346	0.0086
	$\kappa = 1.0$				$\kappa = 1.0$		
MaxLik	0.3008	0.1523	0.0447	MaxLik	0.1359	0.0293	0.0101
Schou	0.3116	0.1613	0.0460	Schou	0.1367	0.0295	0.0101
MML$_{h2}$	0.3263	0.1638	0.0457	MML$_{h2}$	0.1402	0.0308	0.0102
MAP$_{h2}^{\kappa,\mu}$	0.2753	0.1316	0.0394	MAP$_{h2}^{\kappa,\mu}$	0.1353	0.0296	0.0100
MML$_{h3}$	0.2841	0.1272	0.0398	MML$_{h3}$	0.1352	0.0287	0.0099
MAP$_{h3}^{\kappa,\mu}$	0.2582	0.1242	0.0387	MAP$_{h3}^{\kappa,\mu}$	0.1312	0.0285	0.0099
MAP$_{h3}^{x,y}$	0.2664	0.1194	0.0372	MAP$_{h3}^{x,y}$	0.1353	0.0295	0.0100
	$\kappa = 10.0$				$\kappa = 10.0$		
MaxLik	2.5336	13.81	0.0479	MaxLik	1.1340	2.1041	0.0103
Schou	2.3778	11.93	0.0448	Schou	1.1179	2.0167	0.0101
MML$_{h2}$	2.2158	9.3569	0.0409	MML$_{h2}$	1.1007	1.8976	0.0098
MAP$_{h2}^{\kappa,\mu}$	2.3792	9.6396	0.0428	MAP$_{h2}^{\kappa,\mu}$	1.1463	1.9549	0.0100
MML$_{h3}$	2.2129	9.3462	0.0409	MML$_{h3}$	1.1004	1.8970	0.0098
MAP$_{h3}^{\kappa,\mu}$	2.3769	9.6302	0.0427	MAP$_{h3}^{\kappa,\mu}$	1.1457	1.9521	0.0100
MAP$_{h3}^{x,y}$	2.4508	9.0782	0.0443	MAP$_{h3}^{x,y}$	1.1404	1.9049	0.0100

TABLE 2.   Results for $N = 25$ and $N = 100$

7. Conclusions and Discussion

From the results given in Section 6, we draw the following conclusions. Firstly, the Bayesian methods for point estimation outperformed the classical point estimators, very convincingly for small N. An extensive comparison of the MML estimator and the Classical estimators was performed in [9].

Secondly, the Bayesian methods were not overly sensitive to the choice of prior. Since $h_3(\kappa) = 0$ at $\kappa = 0$, and $h_2(\kappa) > 0$ for $\kappa = 0$, the Bayesian results using the

h_2 prior were superior to the results using the h_3 prior for small κ, and vice versa for large κ.

Thirdly, the MML estimator was competitive with the Bayesian MAP estimators. For example, when using the h_3 prior for κ, we find that typically the results using MML_{h3} were in between the results of $\text{MAP}_{h3}^{\kappa,\mu}$ and $\text{MAP}_{h3}^{x,y}$; rarely was MML_{h3} the worst of the three, and sometimes it was the best of the three. In addition, the MML scheme avoids the issue of choice of parameterisation. The results using the MAP estimate in Cartesian coordinates were superior to the results using polar coordinates (the obvious parameterisation) for small κ, and vice versa for large κ.

The authors therefore advocate the MML method, but note again that the Bayesian estimators outperformed the Classical methods.

Acknowledgments

This work was supported by Australian Research Council (ARC) Postdoctoral Research Fellowship F39340111 and ARC Grant A49330656.

References

1. J.A. Achcar and A.F.M. Smith. Aspects of reparameterization in approximate Bayesian inference. In *Bayesian and Likelihood Methods in Statistics and Econometrics*, pp. 439–452. Elsevier Science, Amsterdam, 1993.
2. J.O. Berger. *Statistical Decision Theory and Bayesian analysis*. Springer-Verlag, 1993.
3. N.I. Fisher. *Statistical Analysis of Circular Data*. Cambridge University Press, 1993.
4. J.A. Hills and A.F.M. Smith. Parameterization issues in Bayesian inference. In *Bayesian Statistics 4*, pp. 227–246. Oxford University Press, London, 1992.
5. K.V. Mardia. *Statistics of Directional Data*. Academic Press, 1972.
6. J.J. Oliver and R.A. Baxter. MML and Bayesianism: Similarities and differences. Technical report TR 206, Dept. of Computer Science, Monash University, Clayton, Victoria 3168, Australia, 1994. Available on the WWW from http://www.cs.monash.edu.au/~jono.
7. G. Schou. Estimation of the concentration parameter in von Mises-Fisher distributions. *Biometrika*, **65**, pp. 369–377, 1978.
8. C.S. Wallace and D.M. Boulton. An invariant Bayes method for point estimation. *Classification Society Bulletin*, **3**(3), pp. 11–34, 1975.
9. C.S. Wallace and D.L. Dowe. MML estimation of the von Mises concentration parameter. Technical report TR 193, Dept. of Computer Science, Monash University, Clayton, Victoria 3168, Australia, 1993; submitted to Australian J. Statistics.
10. C.S. Wallace and D.L. Dowe. Intrinsic classification by MML – the Snob program. In *Proc. of the 7th Australian Joint Conf. on Artificial Intelligence*, pp. 37–44. World Scientific, 1994.
11. C.S. Wallace and P.R. Freeman. Estimation and inference by compact coding. *Journal of the Royal Statistical Society (Series B)*, **49**, pp. 240–252, 1987.

D.L. DOWE ET AL.

A. Determining the Fisher Information Matrix

A.1. SIMPLIFYING THE NEGATIVE LOG-LIKELIHOOD

We can simplify the negative log-likelihood (Equation 2) by letting

$$x = \sum_{i=1}^{N} \cos(\theta_i) \qquad \text{and} \qquad y = \sum_{i=1}^{N} \sin(\theta_i) \qquad\qquad \text{giving}$$

$$x \cos\mu + y \sin\mu = \cos\mu \sum_{i=1}^{N} \cos\theta_i + \sin\mu \sum_{i=1}^{N} \sin\theta_i = \sum_{i=1}^{N} \cos(\theta_i - \mu) \qquad \text{and}$$

$$-x \sin\mu + y \cos\mu = -\sin\mu \sum_{i=1}^{N} \cos\theta_i + \cos\mu \sum_{i=1}^{N} \sin\theta_i = \sum_{i=1}^{N} \sin(\theta_i - \mu)$$

Thus Equation (2) becomes: $L = N \log(2\pi I_0(\kappa)) - \kappa (x \cos\mu + y \sin\mu)$

A.2. THE FISHER INFORMATION MATRIX

The Fisher matrix has the form:
$$F(\mu, \kappa) = \begin{bmatrix} E(\frac{\partial^2 L}{\partial \mu^2}) & E(\frac{\partial^2 L}{\partial \mu \partial \kappa}) \\ E(\frac{\partial^2 L}{\partial \kappa \partial \mu}) & E(\frac{\partial^2 L}{\partial \kappa^2}) \end{bmatrix}$$

The first partial derivatives of the negative log-likelihood are:

$$\frac{\partial L}{\partial \mu} = -\kappa(-x \sin\mu + y \cos\mu) \qquad \text{and} \qquad \frac{\partial L}{\partial \kappa} = N A(\kappa) - (x \cos\mu + y \sin\mu)$$

where $A(\kappa) = \frac{I_0'(\kappa)}{I_0(\kappa)}$ was defined in Equation (4). We derive each entry of the Fisher in turn. For the first entry, we find the partial derivative:

$$\frac{\partial^2 L}{\partial \mu^2} = \kappa(x \cos\mu + y \sin\mu) = \kappa \sum_{i=1}^{N} \cos(\theta_i - \mu), \qquad \text{and writing } \phi_i = \theta_i - \mu \text{ gives}$$

$$E(\frac{\partial^2 L}{\partial \mu^2}) = \kappa \sum_{i=1}^{N} E(\cos\phi_i) = \kappa N E(\cos\phi_i) \qquad\qquad (5)$$

We find $E(\cos\phi_i)$ by taking the expectation over every possible data value $0 \le \phi_i \le 2\pi$:

$$E(\cos\phi_i) = \frac{1}{2\pi I_0(\kappa)} \int_0^{2\pi} \cos\phi_i e^{\kappa \cos\phi_i} d\phi_i$$

Noting the definitions of the Bessel function (Equation 1) and $A(\kappa)$ (Equation 4):

$$E(\cos\phi_i) = \frac{1}{I_0(\kappa)} \left(\frac{1}{2\pi} \int_0^{2\pi} \cos\phi_i e^{\kappa \cos\phi_i} d\phi_i \right) = \frac{I_1(\kappa)}{I_0(\kappa)} = A(\kappa) \qquad (6)$$

Substituting Equation (6) into (5) gives: $E(\frac{\partial^2 L}{\partial \mu^2}) = \kappa N A(\kappa)$

For the second and third entries, we find the partial derivative

$$\frac{\partial^2 L}{\partial \mu \partial \kappa} = -(-x \sin \mu + y \cos \mu) = -\sum_{i=1}^{N} \sin(\theta_i - \mu) \qquad \text{and hence}$$

$$E(\frac{\partial^2 L}{\partial \mu \partial \kappa}) = -E(\sum_{i=1}^{N} \sin(\theta_i - \mu)) = -NE(\sin(\theta - \mu))$$

We find the expectation, $E(\sin(\theta - \mu))$, integrating over $0 \le \theta \le 2\pi$:

$$E(\sin(\theta - \mu)) = \frac{1}{2\pi I_0(\kappa)} \int_0^{2\pi} \sin(\theta - \mu) e^{\kappa \cos(\theta - \mu)} d\theta = \frac{1}{2\pi I_0(\kappa)} \left[\frac{e^{\kappa \cos(\theta - \mu)}}{\kappa} \right]_0^{2\pi} = 0$$

Hence, $E(\frac{\partial^2 L}{\partial \mu \partial \kappa}) = 0$. Since $\frac{\partial^2 L}{\partial \kappa \partial \mu} = \frac{\partial^2 L}{\partial \mu \partial \kappa}$, $E(\frac{\partial^2 L}{\partial \kappa \partial \mu}) = 0$. For the fourth entry we find the partial derivative

$$\frac{\partial^2 L}{\partial \kappa^2} = NA'(\kappa) \qquad \text{and since } \frac{\partial^2 L}{\partial \kappa^2} \text{ does not depend on the data} \qquad E(\frac{\partial^2 L}{\partial \kappa^2}) = NA'(\kappa)$$

Thus the Fisher matrix is [9]: $\quad F(\mu, \kappa) = \begin{bmatrix} \kappa NA(\kappa) & 0 \\ 0 & NA'(\kappa) \end{bmatrix}$

B. Evaluating the Message Length as $\kappa \to 0$

We overcame this problem using two techniques [9].

B.1. THE H_3 PRIOR

We wish to evaluate the limit of Expression (3) as $\kappa \to 0$ for the h_3 prior. To evaluate this limit, Mardia[5] gave the following expansion for $A(\kappa)$ for small κ:

$$A(\kappa) = \frac{\kappa}{2} \left[1 - \frac{\kappa^2}{8} + \frac{\kappa^4}{48} + \ldots \right] \text{ giving } \lim_{\kappa \to 0} A(\kappa) = 0, \quad \lim_{\kappa \to 0} A'(\kappa) = \frac{1}{2}, \quad \lim_{\kappa \to 0} \frac{A(\kappa)}{\kappa} = \frac{1}{2}$$

and thus
$$\begin{aligned} \lim_{\kappa \to 0} \text{Exp} &= \lim_{\kappa \to 0} p(D|\kappa, \mu) \frac{\kappa}{(1 + \kappa^2)^{3/2} N \sqrt{\kappa A(\kappa) A'(\kappa)}} \\ &= p(D|\kappa = 0, \mu) \frac{1}{N \sqrt{\frac{1}{2} \times \frac{1}{2}}} = p(D|\kappa = 0, \mu) \frac{2}{N} \end{aligned}$$

B.2. THE H_2 PRIOR

For the h_2 prior, the limit of Expression (3) as $\kappa \to 0$ is infinite. However, Wallace and Dowe [9] offered a solution for evaluating the message length as $\kappa \to 0$. They noted that while the Fisher tells us that the uncertainty in μ tends to infinity, we would never actually encode μ to less precision than 2π. They therefore modify the expression to be maximised to:

$$\text{Exp2} = \frac{h_2(\kappa, \mu) \, p(D|\kappa, \mu)}{\sqrt{(\kappa A(\kappa) + \frac{3}{N\pi^2}) A'(\kappa)}} \qquad (7)$$

Peter Cheeseman presenting part one of the tutorial on Bayesian methods.

A CHARACTERIZATION OF THE DIRICHLET DISTRIBUTION WITH APPLICATION TO LEARNING BAYESIAN NETWORKS

DAN GEIGER

Computer Science Department, Technion
Haifa 32000, Israel[‡]

AND

DAVID HECKERMAN

Microsoft Research, 9S
Redmond WA, 98052-6399[§]

Abstract. We provide a new characterization of the Dirichlet distribution. This characterization implies that under assumptions made by several previous authors for learning belief networks, a Dirichlet prior on the parameters is inevitable.

Key words: Bayesian network, Dirichlet, characterization

1. Introduction

In recent years, several researchers have investigated Bayesian methods for learning belief networks [1-4]. These approaches all have the same basic components: a scoring metric and a search procedure. The scoring metric takes data and a network structure and returns a score reflecting the goodness-of-fit of the data to the structure. A search procedure generates networks for evaluation by the scoring metric. These approaches use the two components to identify a network structure or set of structures that can be used to predict future hypotheses or infer causal relationships.

The Bayesian approach can be described as follows. Suppose we have a domain of variables $\{u_1, \ldots, u_n\} = U$, and a set of cases $\{C_1, \ldots, C_m\} = D$ where each case is an instance of some or of all the variables in U. We sometimes refer to D as a database. Let (B_S, B_P) be a belief network, that is, B_S is a directed acyclic graph , each node i of B_s is associated with a random variable u_i and B_P is a set of conditional distributions, $p(u_i|u_{i_1}, \ldots, u_{i_k})$, $1 \leq i \leq n$, where $u_{i_1}, \ldots, u_{i_k}$ are the

---

[‡]Email: dang@cs.technion.ac.il. Part of this work was done while the author visited Microsoft Research Center. An extended version of this work (including all proofs) was presented at the Eleventh Conference on Uncertainty in AI, Montreal, Canada, August, 1995.

[§]Email: heckerma@microsoft.com

K. M. Hanson and R. N. Silver (eds.), Maximum Entropy and Bayesian Methods, 61–68.
© *1996 Kluwer Academic Publishers.*

variables corresponding to the parents of node i in B_S. (For more details, consult [5]). Let B_S^h stand for the hypothesis that cases are drawn independently from a belief network having the structure B_S. Then a Bayesian measure of the goodness-of-fit of a belief network structure B_S is $p(B_S^h|D,\xi)$ given by $p(B_S^h|D,\xi) = c \cdot p(B_S^h|\xi)p(D|B_S^h,\xi)$ where c is a normalizing factor and ξ is the current state of knowledge.

To compute $p(D|B_S^h,\xi)$ in closed form several assumptions were made. First, the database D is a multinomial sample from some belief network (B_S, B_P). Second, for each network structure the parameters associated with one node are independent of the parameters associated with other nodes (global independence [6]) and the parameters associated within a node given one instance of its parents are independent of the parameters of that node given other instances of its parent nodes (local independence [6]). Third, if a node has the same parents in two distinct networks then the distribution of the parameters associated with this node are identical in both networks (parameter modularity [4]). Forth, each case is complete. Fifth, the distribution of the parameters associated with each node is Dirichlet.

The last two assumptions are made so as to create a conjugate sampling situation, namely, after data is seen the distributions of the parameters stay in the same family— the Dirichlet family. A relaxation of the assumption of complete cases was carried out by previous works (e.g., [2]). The contribution of this paper is a characterization of the Dirichlet distribution which enables one to show that the fifth assumption is implied from the first three assumptions and from one additional plausible assumption that if B_1 and B_2 are equivalent belief networks (i.e., they represent the same independence assumptions) then the events B_1^h and B_2^h are equivalent as well (hypothesis equivalence [4]). We make this self-evident assumption explicit because it does not hold for causal networks where two edges with opposing directions correspond to distinct hypotheses.

Our contribution can be described using common statistical terminology as follows. We use this terminology because our result might be found applicable in other statistical uses of the Dirichlet distribution and because it falls under the broad area of characterizations of probability distribution functions. Suppose s and t are two discrete random variables having finite domains, $\{s_i\}_{i=1}^k$ and $\{t_j\}_{j=1}^n$, respectively. We wish to infer the joint probability $p(s,t)$ from a sample of pairs of values (s_i, t_j) of s and t. The standard Bayesian approach to this statistical inference problem is to associate with $p(s_i, t_j)$ a parameter θ_{ij} (sometimes called a multinomial parameter), assign $\{\theta_{ij}|1 \le i \le k, 1 \le j \le n\}$ a prior joint pdf and compute the posterior joint pdf of $\{\theta_{ij}\}$ given the observed set of pairs of values. There are two closely-related variants to this approach which can be described as follows.

Let $\theta_{i\cdot} = \sum_{j=1}^n \theta_{ij}$ stand for the multinomial parameter associated with $p(s = s_i)$ and let $\theta_{j|i} = \theta_{ij} / \sum_j \theta_{ij}$ stand for the multinomial parameter associated with $p(t = t_j | s = s_i)$. Furthermore, let $\theta_{I\cdot} = \{\theta_{i\cdot}\}_{i=1}^{k-1}$ and $\theta_{J|i} = \{\theta_{j|i}\}_{j=1}^{n-1}$. We assume that $\{\theta_{I\cdot}, \theta_{J|1}, \ldots, \theta_{J|k}\}$ are mutually independent and that each has a prior pdf. Now according to Bayesian practice we compute the joint posterior appropriately.

That is, we update the pdf for $\theta_{I\cdot}$ according to the counts of $s = s_i$ in the observed pairs, and update the pdf of $\theta_{J|i}$ according to the counts of $t = t_j$ in all pairs in which $s = s_i$. In a symmetric fashion, let $\theta_{\cdot j} = \sum_{i=1}^{k} \theta_{ij}$, $\theta_{i|j} = \theta_{ij}/\sum_i \theta_{ij}$, $\theta_{\cdot J} = \{\theta_{\cdot j}\}_{j=1}^{n-1}$ and $\theta_{I|j} = \{\theta_{i|j}\}_{i=1}^{k-1}$. Now we assume that $\{\theta_{\cdot J}, \theta_{I|1}, \ldots, \theta_{I|n}\}$ are mutually independent and that each has a prior pdf and we compute the posterior pdf for $\theta_{\cdot J}$ according to the counts of $t = t_j$ and the posterior pdf of $\theta_{I|j}$ according to the counts of $s = s_i$ in all pairs in which $t = t_j$.

To make these techniques operational one must choose a specific prior pdf for the multinomial parameters. The standard choice of a pdf for $\{\theta_{ij}\}$ is a Dirichlet pdf, usually for pragmatic reasons. When such a choice is made, it can be shown that $\{\theta_{I\cdot}, \theta_{J|1}, \ldots, \theta_{J|k}\}$ are indeed mutually independent and that each has a prior Dirichlet pdf. Similarly, $\{\theta_{\cdot J}, \theta_{I|1}, \ldots, \theta_{I|n}\}$ are mutually independent and each has a prior Dirichlet pdf.

The surprising result stated in this article is that if these independence assertions are assumed to hold, and under the assumption of (strictly) positive smooth pdfs, then a prior Dirichlet pdf for $\{\theta_{ij}\}$ is the only possible choice. The assumption of strictly positive pdfs can possibly be dropped without affecting the conclusion but we have not carried out a proof of this claim. Also the smoothness requirement can be dropped as well. The implication of this result to learning Bayesian networks is discussed in Section 3. A preliminary account of analogous results for Gaussian networks is reported in Section 4.

2. Background and Technical Summary

The Dirichlet pdf is defined as follows. Let $\phi_1, \ldots, \phi_l$ be positive random variables that sum to 1. Then $\phi_1, \ldots, \phi_{l-1}$ have a Dirichlet pdf f if

$$f(\phi_1, \ldots, \phi_{l-1}) = \frac{\Gamma(\sum_{i=1}^{l} \alpha_i)}{\prod_{i=1}^{l} \Gamma(\alpha_i)} \prod_{i=1}^{l} \phi_i^{\alpha_i - 1} \tag{1}$$

where $\phi_l = 1 - \sum_{i=1}^{l-1} \phi_i$ and α_i are positive constants (See, e.g., [7]).

We use the following conventions. Suppose $\{\theta_{ij}\}$, $1 \leq i \leq k$, $1 \leq j \leq n$, is a set of positive random variables that sum to 1. Let $\theta_{i\cdot}, \theta_{\cdot j}, \theta_{I\cdot}, \theta_{\cdot J}, \theta_{j|i}, \theta_{i|j}, \theta_{J|i}$, and $\theta_{I|j}$ be defined as in the introduction. Consequently, $\theta_{i\cdot}\theta_{j|i} = \theta_{\cdot j}\theta_{i|j}$ for every i and j. Let f_U be the joint pdf of $\{\theta_{ij}\}$, f_I be the pdf of $\theta_{I\cdot}$, and $f_{J|i}$ be the pdf of $\theta_{J|i}$. Similarly, let f_J be the pdf of $\theta_{\cdot J}$, and $f_{I|j}$ be the pdf of $\theta_{I|j}$. Finally, let f_{IJ} be the joint pdf of $\theta_{I\cdot}, \theta_{J|1}, \ldots, \theta_{J|k}$ and f_{JI} be the joint pdf of $\theta_{\cdot J}, \theta_{I|1}, \ldots, \theta_{I|n}$.

A Dirichlet pdf for $\{\theta_{ij}\}$ is given by

$$f_U(\{\theta_{ij}\}) = c \prod_{i=1}^{k} \prod_{j=1}^{n} \theta_{ij}^{\alpha_{ij} - 1} \tag{2}$$

where $\theta_{kn} = 1 - \sum_A \theta_{ij}$, $A = \{(i,j)|1 \leq i, j \leq n, i \neq k \text{ or } j \neq n\}$, c is the normalization constant and α_{ij} are positive constants.

We observe that f_U and f_{IJ} are related through a change of variables. Since both $\{\theta_{i\cdot}\}_{i=1}^{k}$ and $\{\theta_{j|i}\}_{j=1}^{n}$ are defined in terms of $\{\theta_{ij}\}$ and since $\theta_{ij} = \theta_{i\cdot}\theta_{j|i}$, there exists a one-to-one and onto correspondence between $\{\theta_{ij}\}$ and $\{\theta_{i\cdot}\}\cup\{\theta_{j|i}\}$. The Jacobian $J_{k,n}$ of this transformation is given by

$$J_{kn} = \prod_{i=1}^{k} \theta_{i\cdot}^{n-1} \tag{3}$$

[4].

The following lemma provides a known property of the Dirichlet distribution. A slightly weaker version is stated in [8] (Lemma 7.2).

Lemma 1 *Let $\{\theta_{ij}\}$, $1 \le i \le k$, $1 \le j \le n$, where k and n are integers greater than 1, be a set of positive random variables having a Dirichlet distribution. Then, $f_I(\theta_{I\cdot})$ is Dirichlet, $f_{J|i}(\theta_{J|i})$ is Dirichlet for every i, $1 \le i \le k$, and $\{\theta_{I\cdot}, \theta_{J|1}, \ldots, \theta_{J|k}\}$ are mutually independent.*

Proof: Set $\theta_{ij} = \theta_{i\cdot}\theta_{j|i}$ in Eq. 2, multiply by J_{kn}, and regroup terms. □

The main claim of this article is that, under the assumption of a positive smooth pdf for $\{\theta_{ij}\}$, the converse holds as well. More specifically, we prove the following theorem.

Theorem 2 *Let $\{\theta_{ij}\}$, $1 \le i \le k$, $1 \le j \le n$, $\sum_{ij} \theta_{ij} = 1$, where k and n are integers greater than 1, be positive random variables having a positive smooth pdf $f_U(\{\theta_{ij}\})$. If $\{\theta_{I\cdot}, \theta_{J|1}, \ldots, \theta_{J|k}\}$ are mutually independent and $\{\theta_{\cdot J}, \theta_{I|1}, \ldots, \theta_{I|n}\}$ are mutually independent, then $f_U(\{\theta_{ij}\})$ is Dirichlet.*

Recall that f_U can be written both in terms of f_{IJ} and in terms of f_{JI} by a change of variables and using the Jacobian given by Equation 3. Since both representations must be equal, and using the independence assumptions made by Theorem 2 to factor f_{IJ} and f_{JI}, we get the equality,

$$\left(\prod_{j=1}^{n} \theta_{\cdot j}^{k-1}\right)^{-1} f_J(\theta_{\cdot J}) \prod_{j=1}^{n} f_{I|j}(\theta_{I|j}) = \tag{4}$$

$$\left(\prod_{i=1}^{k} \theta_{i\cdot}^{n-1}\right)^{-1} f_I(\theta_{I\cdot}) \prod_{i=1}^{k} f_{J|i}(\theta_{J|i})$$

This equality, which is in fact a functional equation, summarizes the independence assumptions stated in Theorem 2.

Methods for solving functional equations such as Eq. 4, that is, finding all functions that satisfy them under different regularity assumptions, are discussed in [9]. We have used the following technique. We take repeated derivatives of Eq. 4 and obtain a differential equation, the solution of which after appropriate specialization is the general solution of Eq. 4 (Aczél, 66, Section 4.2, "Reduction to differential equations").

Note that when $n = k = 2$ and by renaming of variable and function names, Eq. 4 can be written as follows:

$$f_0(y)g_1(z)g_2(w) = g_0(x)f_1\left(\frac{yz}{x}\right)f_2\left(\frac{y(1-z)}{1-x}\right) \tag{5}$$

where

$$x = yz + (1 - y)w$$

and where y, z and w replace $\theta_{.j=1}$, $\theta_{i=1|j=1}$, $\theta_{i=1|j=2}$, respectively.

3. Implications For Learning

We now explain how our characterization applies to learning belief networks. We concentrate on belief networks for two discrete variables s and t whose joint distribution is $p(s, t)$. The n-variate case is discussed in [4]. There are three possible belief networks with two nodes. The network that contains no edge between its two nodes s and t, a network $s \to t$ and the network $t \to s$. The first network B_0 corresponds to the assertion that s and t are independent while the second network B_1 and the third one B_2 assert that s and t are dependent. The last two belief networks are equivalent, B_1 represents the factorization $p(s, t) = p(s)p(t|s)$ and B_2 represents the factorization $p(s, t) = p(t)p(s|t)$.

We shall first examine the two complete networks B_1 and B_2. We assume that if two networks B_1 and B_2 are equivalent (as is the case in our example) then the corresponding events B_1^h and B_2^h are equivalent (hypothesis equivalence [4]). Recalling the notations introduced in the introduction, we have that $\theta_{i.} = \sum_{j=1}^{n} \theta_{ij}$ stand for the multinomial parameters associated with $p(s = s_i)$ and $\theta_{j|i} = \theta_{ij} / \sum_j \theta_{ij}$ stand for the multinomial parameters associated with $p(t = t_j|s = s_i)$. Thus,

$$f_{IJ}(\theta_{I.}, \theta_{J|1}, \ldots, \theta_{J|k}|B_1^h) = f_{IJ}(\theta_{I.}, \theta_{J|1}, \ldots, \theta_{J|k}|B_2^h)$$

$$f_{JI}(\theta_{.J}, \theta_{I|1}, \ldots, \theta_{I|k}|B_2^h) = f_{JI}(\theta_{.J}, \theta_{I|1}, \ldots, \theta_{I|k}|B_1^h)$$

Due to these equalities and using local and global independence to factor f_{IJ} and f_{JI}, we immediately obtain Equation 4 (dropping the conditioning events is valid because B_1^h and B_2^h are equivalent). Thus for the two *complete networks* the only possible prior on their parameters is, according to Theorem 2, the Dirichlet distribution.

Note that we only use three assumptions: multinomial sampling, local and global independence, and hypothesis equivalence. Implicit (because we condition on B_i^h) is the assumption that each complete-structure hypothesis has a positive probability.

The prior for any non-complete network follows from the assumption of parameter modularity which says that the pdf associated with a node under the assumption that a specific network generates the data is the same as the pdf of the parameters of that node given another network generates the data provided that the set of parents is identical in the two networks. In our two-variables network, for example, the parameters $\theta_{i.}$ which are associated with node s have the same pdf when conditioned on B_1 and when conditioned on B_0 because in both networks s has the same set of parents (the empty set) and similarly for node t. That is,

$$f_i(\theta_{i.}|B_1^h, \xi) = f_i(\theta_{i.}|B_0^h, \xi)$$

$$f_j(\theta_{\cdot j}|B_2^h,\xi) = f_j(\theta_{\cdot j}|B_0^h,\xi)$$

These equalities imply that the prior for the parameters of B_0 is Dirichlet as well. Thus, parameter modularity is the assumption that extends our result from complete to non-complete networks.

This result of the inevitable choice of a Dirichlet prior for two-variables networks is easily generalized to the n-variate case by induction and without the need to solve any additional functional equations. The inductive proof uses the fact that a cluster of variables each having a Dirichlet distribution is distributed Dirichlet as well. For details consult [4].

Recall that the exponents of θ_{ij} of a Dirichlet distribution can be written as $N\alpha_{ij} - 1$ where N is the "equivalent sample size" (the size of an imaginary database of complete cases—the prior sample—upon which the prior Dirichlet is based) and α_{ij} is the expectation of θ_{ij}. The equivalent sample size reflects the confidence of the user and α_{ij} represents the relative frequency of the pair (i,j) in the prior sample. A joint Dirichlet prior is therefore quite restricting because it allows only one equivalent sample size for the entire domain. That is, there is no way to express different confidence levels regarding the parameters of different parts of the network. Thus the practical ramification of our characterization is that the commonly-made global and local independence assumption is inappropriate whenever a single equivalent sample size is not sufficient to describe prior knowledge. Such a situation occurs, for example, if knowledge about $\theta_I.$ is more precise than knowledge about $\theta_{J|i}$.

One possibility for overcoming this limitation of the Dirichlet prior is to replace the notion of a single equivalent sample size with *equivalent database*. Namely, we ask a user to imagine that she was initially completely ignorant about a domain, having an uninformative prior with an equivalent sample size close to zero. Then, we ask the user to specify a database D_e that would produce a posterior density that reflects her current state of knowledge. This database may contain incomplete cases. Then, to score a real database D, we score the database $D_e \cup D$, using the uninformative prior and a learning algorithm that handles missing data. This way of specifying a prior yields a mixture of Dirichlet distributions which, according to our result, cannot satisfy the local and global independence assumptions.

4. Discussion

The independence assumptions made by Theorem 2 can be divided into two parts: $\{\theta_{J|1}, \ldots, \theta_{J|k}\}$ are mutually independent and $\{\theta_{I|1}, \ldots, \theta_{I|n}\}$ are mutually independent (local independence) and $\theta_I.$ is independent of $\{\theta_{J|1}, \ldots, \theta_{J|k}\}$ and $\theta_{\cdot J}$ is independent of $\{\theta_{I|1}, \ldots, \theta_{I|n}\}$ (global independence). A natural question to ask is whether global independence alone implies a joint Dirichlet pdf for $\{\theta_{ij}\}$.

This question is particularly interesting in light of the analysis of decomposable graphical models given by [8]. Dawid and Lauritzen term a pdf that satisfies global independence *a strong hyper-Markov law* and show the importance of such laws in the analysis of decomposable graphical models. We now show that the class of strong hyper-Markov laws is larger than the Dirichlet class.

When $n = k = 2$, and using the notations of Equation 5 the new functional equation can be written as follows:

$$f_0(y)g(z, w) = g_0(x)f(\frac{yz}{x}, \frac{y(1 - z)}{1 - x}) \tag{6}$$

where $x = yz + (1 - y)w$. Note that Eq. 5 is obtained from this equation by setting $g(z, w) = g_1(z)g_2(w)$ and $f(t_1, t_2) = f_1(t_1)f_2(t_2)$. These equalities correspond to local independence.

Let f_U be a joint pdf of $\{\theta_{ij}\}$ given by

$$f_U(\{\theta_{ij}\}) = K \left[\prod_{i=1}^{2}\prod_{j=1}^{2} \theta_{ij}^{\alpha_{ij}-1}\right] H(\frac{\theta_{11}\theta_{22}}{\theta_{12}\theta_{21}}) \tag{7}$$

where K is the normalization constant, α_{ij} are positive constants and H is an arbitrary positive smooth function. That this pdf satisfies global independence can be easily verified. It can in fact be shown, by solving Eq. 6, that every positive smooth strong Hyper Markov law can be written in this form (when $n = 2$ and $k = 2$). This solution includes the Dirichlet family as a proper subclass.

Since H is a single function that does not depend on a particular network, one can conclude that if local parameter independence is assumed to hold in *one network*, then f_U must still be Dirichlet and therefore, due to Lemma 1, local parameter independence must hold for *all networks*. We have so far proved this claim for two-variables networks but we believe it holds for the n-variate case as well.

As a final comment, we should mention that a functional equation which restricts the possible prior distributions for the parameters of Bayesian networks can be formulated for other sampling situations not necessarily for the multinomial sampling which was assumed in our discussion so far. As another example, consider a two-continuous-variables domain $\{x_1, x_2\}$ having a bivariate-Normal distribution. Constructing a prior for the parameters of such Gaussian networks and performing the prior-to-posterior analysis was carried out in [10,11]. Let $\{m_1, v_1, m_{2|1}, b_{12}, v_{2|1}\}$ and $\{m_2, v_2, m_{1|2}, b_{21}, v_{1|2}\}$ denote the parameters for the network structures $x_1 \rightarrow x_2$ and $x_1 \leftarrow x_2$, respectively. That is, m_1 is the mean of x_1 and v_1 is the variance for x_1. Collectively, these are the parameters associated with node x_1 in the first network. The parameters associated with node x_2 are the conditional mean $m_{2|1}$, the regression coefficient b_{12} of x_2 given x_1 and the conditional variance $v_{2|1}$. Now assuming global parameter independence and hypothesis equivalence and using the Jacobian given in [11] yields the functional equation

$$f_1(m_1, v_1) \ f_{2|1}(m_{2|1}, b_{12}, v_{2|1}) = \frac{v_1^2 v_{2|1}^3}{v_2^2 v_{1|2}^3} \cdot$$

$$f_2(m_2, v_2) \ f_{1|2}(m_{1|2}, b_{21}, v_{1|2}) \tag{8}$$

where f_1, $f_{2|1}$, f_2, and $f_{1|2}$ are arbitrary density functions, and where

$$v_2 = v_{2|1} + v_1 b_{12}^2 \qquad b_{21} = \frac{b_{12} v_1}{v_2} \qquad v_{1|2} = \frac{v_{2|1} v_1}{v_2}$$

$$m_2 = m_{2|1} + b_{12} m_1 \qquad m_{1|2} = m_1 + b_{21} m_2$$

These relationships are well known from path analysis and can be derived from Eq. 4 in [11].

We have solved this functional equation and found that the only smooth solutions are such that $f_1(v_1)$ is an inverse gamma distribution and the conditional distribution $f_{2|1}(b_{12}, v_{2|1})$ is a Normal-gamma distribution times an arbitrary function $H(b_{12}/v_{2|1})$. The arbitrary function is not surprising since the functional equation only encodes global independence and so the solution depends on an arbitrary function just as for multinomial sampling (Equation 7). By adding an appropriate definition of local independence one can further show a bivariate Normal-Wishart distribution is the only possible prior on the joint space parameters (i.e., the inverse covariance matrix and the vector of means) if we assume global parameter independence, local parameter independence for *one* network, and hypothesis equivalence. Indeed this was the prior chosen by [10]. An analogous result appears to hold for the n-variate case as well.

Acknowledgment

We thank J. D. Aczél and M. Ungarish for valuable comments, S. Altschuler, L. Wu, and M. Israeli for their help with some Lemmas, and G. Cooper who helped us define the notion of an equivalent database.

References

1. G. Cooper and E. Herskovits, "A Bayesian method for the induction of probabilistic networks from data," *Machine Learning*, 9, pp. 309–347, 1992.
2. D. Spiegelhalter, A. Dawid, S. Lauritzen, and R. Cowell, "Bayesian analysis in expert systems," *Statistical Science*, 8, pp. 219–282, 1993.
3. W. Buntine, "Operations for learning with graphical models," *Journal of Artificial Intelligence Research*, 2, pp. 159–225, 1994.
4. D. Heckerman, D. Geiger, and D. Chickering, "Learning Bayesian networks: The combination of knowledge and statistical data," *Machine Learning*, to appear, 1995.
5. J. Pearl, *Probabilistic Reasoning in Intelligent Systems: Networks of Plausible Inference*, Morgan Kaufmann, San Mateo, CA, 1988.
6. D. Spiegelhalter and S. Lauritzen, "Sequential updating of conditional probabilities on directed graphical structures," *Networks*, 20, pp. 579–605, 1990.
7. M. DeGroot, *Optimal Statistical Decisions*, McGraw-Hill, New York, 1970.
8. A. Dawid and S. Lauritzen, "Hyper Markov laws in the statistical analysis of decomposable graphical models," *Annals of Statistics*, 21, pp. 1272–1317, 1993.
9. J. Aczel, *Lectures on Functional Equations and Their Applications*, Academic Press, New York, 1966.
10. D. Geiger and D. Heckerman, "Learning Gaussian networks," in *Proceedings of Tenth Conference on Uncertainty in Artificial Intelligence*, Seattle, WA, pp. 235–243, Morgan Kaufmann, July 1994.
11. D. Heckerman and D. Geiger, "Learning Bayesian networks: A unification for dicrete and Gaussian domains," in *Proceedings of Eleventh Conference on Uncertainty in Artificial Intelligence*, Montreal, QU, pp. 274–284, Morgan Kaufmann, August 1995. See also Technical Report TR-95-16, Microsoft, Redmond, WA, February 1995.

THE BOOTSTRAP IS INCONSISTENT WITH
PROBABILITY THEORY

DAVID H. WOLPERT
The Santa Fe Institute
1399 Hyde Park Road
Santa Fe, NM 87501[†]

Abstract. This paper proves that for no prior probability distribution does the bootstrap (BS) distribution equal the predictive distribution, for all Bernoulli trials of some fixed size. It then proves that for no prior will the BS give the same first two moments as the predictive distribution for all size trials. It ends with an investigation of whether the BS can get the variance correct.

Key words: bootstrap, standard distribution, standard error, predictive distribution, Bernoulli trials, error bars

1. Introduction

Say we have N data D that are created by IID (independent and identically distributed) sampling a distribution f, and a statistic $S(D)$ that assigns a number to that data. Our problem is to use D to estimate the probability of getting statistic value s in a subsequent sample of N more data from f.

The vanilla version of the bootstrap procedure (BS) interprets this problem as asking for an estimate of the standard distribution of S, $P(S \mid f, N)$. To create this estimate, first it many times resamples D (with replacement) according to some D-dependent distribution $X(D)$ [1]. In this way it creates many new data sets D'. BS then estimates the standard distribution as the distribution of values of $S(D')$. (In this paper none of the variants of this vanilla version of the BS will be considered.) In particular, the BS can assign an error bar to the observed value of the statistic $S(D)$; its estimate of the standard distribution provides an estimate of the standard error. The power of BS is its wide applicability, the fact that its error bars behave quite "reasonably" (e.g., they often grow as the size of the data shrinks), and the empirical fact that it often gives good estimates of error bars.

This paper investigates whether and when the BS can be exactly correct in its answer to the problem. This question is vacuous if one views the problem of

---

[†] Email: dhw@santafe.edu

K. M. Hanson and R. N. Silver (eds.), Maximum Entropy and Bayesian Methods, 69–76.
© *1996 Kluwer Academic Publishers.*

"estimating the probability of getting ... s in N more samples of f" as estimating $P(S|f, N)$—the BS will be correct if f is the empirical distribution given by D. However from a Bayesian perspective, in the real world we have knowledge of D and none concerning f, so it is more sensible to view the problem as estimating the predictive distribution $P(S$ of N new data points sampled from the unknown $f|D)$ than of estimating $P(S|f, N)$. In fact, *as its results are used in practice*, the BS is usually treated as though it produces the predictive distribution.

Accordingly, this paper analyzes whether the BS's distribution for S can equal the predictive distribution for some prior $P(f)$, and in this sense is "consistent with probability theory". The precise scenario investigated is Bernoulli trials, where S is the number of positive events in the sample. Intuitively, the question is of whether there are scenarios in which the "error bar" assigned by the BS to its estimate of the probability of a positive event p can always agree with the Bayesian "error bar" assigned to the Bayesian estimate of p. In short, can Bayes and the BS always agree in the error bars they assign?

Recursive relationships are used to prove that there is no prior and no $X(D)$ such that the distribution of values of $S(D')$ always equals $P(S|D)$ for all D of fixed size N. Next it is shown that there is no prior and no $X(D)$ such the BS and the predictive distribution will always agree on the first two moments of $P(S|D)$ for all D of any size. Next the question is investigated of whether just the error bars generated by BS can equal those given by a Bayesian calculation (i.e., of whether the BS's variance can be correct). First a preliminary analysis of going from $X(D)$ to a prior giving the same error bar is presented. Then it is shown that for only certain Bernoulli scenarios are there $X(D)$'s that give the same error bar as the uniform prior's predictive distribution, and those $X(D)$'s are derived.

The focus of this paper is the theoretical relationship between a technique having its genesis in one branch of statistics (sampling theory statistics) with techniques from another branch (Bayesian statistics) designed for the same problem. One must be careful not to ascribe too much real world significance to these theoretical conclusions. In particular, none of the results in this paper mean that one should never use the BS in practice, even for Bernoulli trial problems where one is interested in all moments. *Only if one is completely sure of one's prior* should one avoid using the BS for Bernoulli trials. How well the BS performs compared to a Bayesian calculation based on an incorrect prior is the subject of future work.

There has previously been some work on a "Bayesian variant of the BS" [2]. However one can argue that that variant is, in its essentials, equivalent to direct Monte Carlo sampling of the predictive distribution. As such, the question of whether it (or slight modifications of it) can give the predictive distribution is mute. Accordingly, in this paper only the conventional non-Bayesian variant of the BS—by far the more popular variant in the BS community—is addressed.

2. Preliminaries

To make things more precise, say we have N IID Bernoulli trials (e.g., N flips of the same coin), giving n positives. We want the probability of k positives in the next N Bernoulli trials, or moments of that probability. (That number of positives k is the statistic S, and the n out of N positives is the data D.) Let p be the true probability of a positive in any single trial. Then $P(k|n) = \int dp\, P(k|p,n)\, P(p|n)$.

Now $P(k|p,n) = P(k|p) = C_k^N\, p^k\, (1-p)^{N-k}$, where $C_j^i \equiv \frac{i!}{j!\,(i-j)!}$. Furthermore, $P(p|n) = P(n|p)\, P(p)\, /\, P(n)$, which in turn is given by $p^n\, (1-p)^{N-n}\, P(p)\, /\, \int dp\, p^n\, (1-p)^{N-n} P(p)$. Therefore

$$P(k|n) = C_k^N\, \frac{\int dp\, p^{n+k}\, (1-p)^{2N-n-k}\, P(p)}{\int dp\, p^n\, (1-p)^{N-n}\, P(p)}. \qquad (1)$$

As shorthand, define $P_{i,j} \equiv \int dp\, p^i\, (1-p)^j\, P(p)$, so that Eq. (1) becomes

$$P(k|n) = C_k^N\, \frac{P_{n+k,\,2N-n-k}}{P_{n,\,N-n}}. \qquad (2)$$

To simplify the exposition, rather than discuss the BS in terms of the general "$X(.)$", which determines the probability of resampling a particular datum, from now on "$x(.)$" will be used instead, to indicate the probability with which the set of all positives in the data is resampled. With this new notation, the BS consists of making an estimate for p based on n and N, $x(n)$ (the N usually being implicit), and then calculating the probability of k given that $p = x(n)$. So rather than directly calculate $P(k|n)$, in the BS one instead calculates the surrogate $P(k|n,\, p = x(n)) = C_k^N\, [x(n)]^k\, [1 - x(n)]^{N-k}$. (As the BS is usually practiced, this calculation is accomplished by Monte Carlo sampling, but that is not important for current purposes.) The question is whether this surrogate can equal $P(k|n)$.

The crucial difference between the BS's calculation and direct evaluation of the predictive distribution is that the BS is based on a single (estimate of) p, whereas the predictive distribution averages over all p. This is similar to the distinction between ML-II, where one fixes a hyperparameter to a single value, and the full hierarchical Bayesian approach, in which one averages over that hyperparameter. Since ML-II can be a poor approximation to the hierarchical calculation even when the posterior probability of the hyperparameter is sharply peaked [3], one might suspect that the BS has a difficult time agreeing with the predictive distribution.

3. Disagreement for the full distribution, for some k and n

Evidently the BS distribution will agree with the predictive distribution for all $k \in \{0,\dots N\}$ and all $n \in \{0,\dots N\}$ iff the following holds for all such k and n:

$$[x(n)]^k\, [1 - x(n)]^{N-k} = \frac{P_{n+k,\,2N-n-k}}{P_{n,\,N-n}}. \qquad (3)$$

It turns out that the only estimator that (might) meet the constraint of Eq. (3) is the absurd estimator {all $x(i)$ are the same constant, x}. To see this, write

$$\frac{P_{n+k,\,2N-n-k}}{P_{n+k\pm1,\,2N-n-k\pm(-1)}} = \frac{P_{n+k,\,2N-n-k}\,/\,P_{n,\,N-n}}{P_{n+(k\pm1),\,2N-n-(k\pm1)}\,/\,P_{n,\,N-n}}.$$

Now use Eq. (3):

$$\frac{P_{n+k,2N-n-k}}{P_{n+k\pm1,2N-n-k\pm(-1)}} = \frac{[x(n)]^k \, [1-x(n)]^{N-k}}{[x(n)]^{k\pm1} \, [1-x(n)]^{N-k\pm(-1)}}$$

$$= \{\frac{1-x(n)}{x(n)}\}^{\pm1}. \tag{4}$$

For the positive exponent, this equality must hold for all $n \in \{0,\dots N\}, k \in \{0,\dots N-1\}$ ($k \in \{1,\dots N\}$ for the negative exponent). In particular, for the positive exponent it must hold for all $n \in \{1,\dots N\}$ and $k = N-n$ (and similarly for the negative exponent). Therefore for the positive exponent, for all $n \in \{1,\dots N\}$, ($\{0,\dots N-1\}$ for the negative exponent),

$$\frac{P_{N,N}}{P_{N\pm1,N\pm(-1)}} = \{\frac{1-x(n)}{x(n)}\}^{\pm1}.$$

Since the left-hand side is independent of n, by the positive exponents we know that $[1-x(i)] \, / \, x(i) = [1-x(j)] \, / \, x(j)$ for all $i,j \in \{1,\dots N\}$, which means that for fixed $N, x(n)$ is independent of n for $n \in \{1\dots N\}$. Assuming $N > 2$, the negative exponent case then extends this to all $n \in \{0,\dots N\}$. (As an aside, this extension of the exponents means that $P_{N-1,N+1} \, P_{N+1,N-1} = [P_{N,N}]^2$.)

The immediate conclusion is that the distribution calculated by the BS can not be $P(k\,|\,n)$ for any reasonable estimator of p, $x(n)$. Moreover, as is proven in the appendix, the only $P(p)$ which gives rise to a $P(k\,|\,n)$ of the form $C_k^N \, x^k \, [1-x]^{N-k}$ for all $\{k,N\}$ is $P(p) = \delta(p$ - constant$)$, a clearly absurd prior.

Presumably BS probability distributions can, in many regimes of the Bernoulli problem, be good approximators of $P(k\,|\,n)$. However those distributions will never equal $P(k\,|\,n)$ exactly for all n and k for any reasonable prior.

4. Disagreement of the first two moments, for some n and N

This section addresses the issue of whether BS can even get the first two moments correct, for all n and N. First, write

$$E(k\,|\,n,N) \equiv \sum_{k=0}^{N} k \, P(k\,|\,n)$$

$$= \frac{\int dp \, P(p) \, p^n \, (1-p)^{N-n} \sum_{k=0}^{N} k \, C_k^N \, p^k \, (1-p)^{N-k}}{\int dp \, P(p) \, p^n \, (1-p)^{N-n}}.$$

Using the fact that $\sum_{k=0}^{N} k \, C_k^N \, p^k \, (1-p)^{N-k} = Np$, we get

$$E(k\,|\,n,N) = N \, \frac{P_{n+1,N-n}}{P_{n,N-n}}. \tag{5}$$

Using similar reasoning,

$$E(k^2 \,|\, n, N) \;=\; \frac{N\,(N-1)\,P_{n+2,N-n} \;+\; N\,P_{n+1,N-n}}{P_{n,N-n}}. \tag{6}$$

Now by Eq. (5), for BS to get the first moment right, $x(n)$ must be equal to the ratio $P_{n+1,N-n} / P_{n,N-n}$. Then by Eq. (6), for BS to get the second moment right, $x^2(n)\,N\,(N-1) + Nx(n) = \frac{N\,(N-1)\,P_{n+2,N-n} + N\,P_{n+1,N-n}}{P_{n,N-n}}$. Therefore if the first moment is also correct, we have $x^2(n) = P_{n+2,N-n} / P_{n,N-n}$. Combining,

$$\frac{[P_{n+1,N-n}]^2}{P_{n,N-n}} \;=\; P_{n+2,N-n}. \tag{7}$$

We want this to hold for all pairs of values $\{N \geq 1, 0 \leq n \leq N\}$.

Define $D_i \equiv \int dp\, p_i\, P(p)$. Consider the $n = N$ case (so $n \geq 1$). By Eq. (7) $\frac{(D_{n+1})^2}{D_n} = D_{n+2}$, i.e., $\frac{D_{n+2}}{D_{n+1}} = \frac{D_{n+1}}{D_n}$. This must hold for all $n \geq 1$; for all such n, $\frac{D_{n+1}}{D_n} = \frac{D_2}{D_1} \equiv \alpha$. This in turn means that for all such n, $D_n = D_1\,\alpha^{n-1}$.

Now consider the case $N = n+1$ (so $n \geq 0$). We have $\frac{(D_{n+1} - D_{n+2})^2}{(D_n - D_{n+1})} = D_{n+2} - D_{n+3}$. Take $n = 0$, use $D_0 = 1$, and for the other D_n use $D_n = D_1\,\alpha^{n-1}$. This gives $\frac{D_1^2\,(1-\alpha)^2}{1-D_1} = D_1\,[\alpha - \alpha^2]$. Cancelling terms and solving, we get $D_1 = \alpha$. So for all $n \geq 0$, $D_n = \alpha^n$.

This means that $D_2 - (D_1)^2$, the variance of $P(p)$, is 0. The only way this can be is if $P(p)$ is a Dirac delta function about some constant c. This in turn means that $E(k\,|\,n) = c$, which means that $x(n) = c$ for all n; a clear absurdity.

The preceding relied on looking at all n, N. In contrast, the proof concerning the full distribution over k (see the previous section) has N fixed, but varies over all (allowed by N) values of n and k. In addition, the arguments in both sections relied on allowing the $n = N$ and $n = 0$ cases. It is not immediately clear how things are changed if we simply decide to disallow those cases, as in [2].

5. Getting the variance right—going from $x(n)$ to $P(p)$

The results of the previous section notwithstanding, one might wish to use the BS with an $x(n)$ such that the standard deviation of the BS distribution over k, $C_k^N\,[x(n)]^k\,[1-x(n)]^{N-k}$, is also the standard deviation of $P(k\,|\,n)$, for some $P(p)$, for all $N > 0, n \leq N$ (even though the BS distribution can not equal $P(k\,|\,n)$ or even get the first two moments exactly right). In general, the procedure for going from $x(n)$ to a $P(p)$ giving the same variance is the following.

The (squared) variance of k corresponding to $x(n)$ is $N\,x(n)\,(1 - x(n))$. To find the variance given by a $P(p)$, use Eq.'s (5) and (6). Next set the two variances equal and solve for $P_{n+2,N-n}$: for all $n \in \{0, \ldots N\}$,

$$P_{n+2,N-n} \;=\; \frac{x(n)\,(1-x(n))\,P_{n,N-n} + \left(\frac{N\,P_{n+1,N-n}}{P_{n,N-n}} - 1\right)P_{n+1,N-n}}{N - 1}$$

To proceed further one must conduct an analysis similar to that in the appendix; expand the P's in terms of D_j, and solve. As in the appendix, having the requirement on the equalities hold for all possible $\{n, N\}$ might result in there being no $P(p)$ which solves the equations. Unfortunately, time constraints did not allow such an analysis for this paper.

However consider the special case where we use a frequency counts estimator, $x(n) = n/N$, and have $n = 0$ or N, or in any other way allow the estimate of the variance to equal 0. For such a scenario, for that n, $P(k|n)$ must be a delta function. So there are $N - 1$ values of k for which $P(k|n) = 0$. Looking at Eq. (1), we see that since $p^{n+k} (1-p)^{2N-n-k}$ is greater than 0 for all $p \in$ the interval $(0, 1)$, the only way that $P(k|n)$ can equal 0 is if $P(p)$ is 0 for all $p \in (0, 1)$. So $P(p)$ must equal either $\delta(p)$ or $\delta(p-1)$. This means that the variance always equals 0, regardless of n. This rules out the frequency counts estimator, and makes any estimator which can estimate the variance as 0 seem rather absurd.

6. Getting the variance right—going from $P(p)$ to $x(n)$

It is usually easier to go from a $P(p)$ to an $x(n)$ with the same variance than visa-versa. As an example, assume that $P(p)$ is uniform. Then using Eq. (5), $E(k|n, N) = N \frac{n+1}{N+2}$ which of course is just what one would expect from Laplace's law of succession. (In fact, regardless of the prior $P(p)$, $E(k|n, N) = N \int dp\, p\, P(p|n) = N \times$ the "Bayesian" estimate of the average p, given n.)

In a similar manner, we can derive

$$E(k^2|n, N) = \frac{N\,(N-1)\,(n+2)\,(n+1)}{(N+3)\,(N+2)} + \frac{N\,(n+1)}{N+2}.$$

Collecting terms, after a bit of algebra we get

$$\chi(n, N) \equiv \frac{E(k^2|n, N) - [E(k|n, N)]^2}{N}$$

$$= \frac{2(N+1)}{(N+3)\,(N+2)^2} \{(N+2)^2/4 - (n - (N/2))^2\} \tag{8}$$

By inspection, $\chi(N-n, N) = \chi(n, N)$, as it should. We want $N\,x(n)\,[1-x(n)] = N\,\chi(n, N)$ for all $n \in \{0, \dots N\}$. The solution is

$$x(n) = \frac{1 \pm \sqrt{1 - 4\chi(n, N)}}{2}. \tag{9}$$

Now we would like to have $x(n) + x(N-n) = 1$, since intuitively $x(n)$ is the (estimate of) the probability of a positive event, and then by symmetry (redefine what is a "positive" versus a "negative" event), $x(N - n)$ is the probability of a negative event. To obey this equality we can use the negative root for the lower $N/2$ values of n in Eq. (9), and the positive root for the upper $N/2$ values.

So for BS to give the same variance one would have with a uniform $P(p)$, one should do the Monte Carlo sampling according to a distribution in which each of the positive events have probability $x(n)/n$ ($x(n)$ being the probability of the set of all positive events), and all the negative events have probability $\frac{x(N-n)}{N-n} = \frac{1-x(n)}{N-n}$. If no positive events occur ($n = 0$), then one must still assign probability $[1 - x(0)]/N$ to all of the negative events, but one must also assign probability $x(0)$ to a positive event, despite the fact that no such positive event is in the original sample. In other words, one must make up an event and add it to the original sample. (Similarly if no negative events occur.)

Note that $x(n)$ is only real if $\chi(n, N) \leq 1/4$. Therefore, since probabilities must be real, it is necessary that

$$(n - (N/2))^2 \geq (N + 2)^2 [1/4 - \frac{N + 3}{8(N + 1)}]. \tag{10}$$

For $N = 1$, this condition is satisfied by all n. For $N = 2$, it reduces to $|n - 1| \geq \sqrt{2/3}$, which means that n can only equal 0 or 2. For large N, the requirement becomes $|n - (N/2)| \geq N/\sqrt{8}$, which is satisfied by $N(1 - 1/\sqrt{2})$ values of n.

In fact, for N large, we can write down immediately

$$x(n) = \frac{1 \pm \sqrt{-1 + 8(R - 1/2)^2}}{2}, \tag{11}$$

where $R \equiv n/N$. Taking the negative root for $R < 1/2$, and the positive root for $R > 1/2$, this $x(n)$ has the value 0 at $R = 0$, rises to $1/2$ for $R = (1/2) [1 - \sqrt{1/2}]$, is complex up to $(1/2) [1 + \sqrt{1/2}]$, where it again has the value $1/2$, and rises from there up to the value 1 at $R = 1$.

It is interesting to note that for large N, this $x(n)$, the estimator of p that gives the correct variance, agrees more and more with the frequency count estimator of p as one moves towards the limits $R = 0, 1$. In contrast, the Laplace's law of succession estimator of p disagrees more and more with the frequency count estimator as one moves towards those limits. This despite the fact that the Laplace estimator, like the $x(n)$ estimator, is based on a uniform $P(p)$.

7. Future work

The kind of analysis done here can also be done when there are multiple possible events, and when the statistic is a more complicated function than counting the number of events of a given class. Other future work involves seeing how close BS can get to the predictive distribution. It may well be that although it can not given that distribution exactly, it can give a very close approximation to it.

A. Appendix

This appendix solves for the $P(p)$ which satisfies Eq. (3) for all N, n and k. First, since $x(n)$ is independent of n, using the positive exponent and defining $i \equiv n+k$, Eq. (4) tells us that for all $i \in \{0, \ldots 2N - 1\}$, $\frac{P_{i+1,2N-(i+1)}}{P_{i,2N-i}}$ is the constant $x/(1 - x)$ (x being the value shared by all N of the $x(n)$). Define $r \equiv x/(1 - x)$:

$$P_{j,2N-j} = \alpha\, r^j, \tag{12}$$

where α, like r, is an as of yet undetermined constant.[1]

Now recall $D_j \equiv \int dp \, p^j \, P(p)$. Using this, the binomial expansion, Eq. (12), and the definition of $P_{n+k,2N-n-k}$, and defining $m \equiv n+k$, one derives

$$\sum_{i=0}^{2N-m} (-1)^i \, D_{m+i} \, C_i^{2N-m} = \alpha \, r^m \qquad (13)$$

for all $m \in \{0, \ldots 2N\}$.

Start with $m = 2N$, and thereby get $D_{2N} = \alpha \, r^{2N}$. By iteratively decrementing m, we can solve for the other D_{2N-m}. This solution is unique. Therefore any formula for the D_j that solves Eq. (13) for all $2N$ of the m must be the unique solution to Eq. (13). This solution is given by the following:

$$D_j = \alpha \, r^{2N} \, (1 + 1/r)^{2N-j}. \qquad (14)$$

Proof: Plugging Eq. (14) into Eq. (13), we get

$$r^{2N} \sum_{i=0}^{2N-m} (-1)^i \, (1 + 1/r)^{2N-m-i} \, C_i^{2N-m} = r^m$$

as the equality that must be satisfied. This equality in turn implies

$$r^{2N-m} \, (\frac{r+1}{r})^{2N-m} \sum_{i=0}^{2N-m} (\frac{-r}{r+1})^i \, C_i^{2N-m} = 1. \qquad (15)$$

The sum $= [1 - \frac{r}{r+1}]^{2N-m} = (r+1)^{m-2N}$, so we do get 1 as required. QED.

Now D_0 must equal 1, since $P(p)$ is normalized. But by Eq. (14), $D_0 = \alpha \, r^{2N} \, (1 + 1/r)^{2N}$. Therefore $D_j = (1 + 1/r)^{-j}$. This in turn means that $D_2 = (D_1)^2$. An immediate consequence is that the variance of $P(p)$ is 0. The only way that can be is if $P(p)$ is a Dirac delta function. This completes the argument.

Acknowledgements: I would like to thank Bill Macready for helpful discussions. I would also like to thank Ken Hanson for performance well beyond the call of duty in helping me with LaTeXproblems. This work was supported in part by TXN Inc.

References

1. B. Efron and R. Tibshirani, *An Introduction to the Bootstrap*. Chapman-Hall (1993).
2. D. Rubin, *The Bayesian Bootstrap*, The Annals of Statistics, vol. 9, pp. 130-134 (1981).
3. D. Wolpert and C. Strauss, *What Bayes says about the Evidence Procedure*, in *Proceedings of the 1993 Maximum Entropy and Bayesian Methods Conference*, G. Heidbreder (Ed), Kluwer, in press.

---

[1]As an aside, note that by returning to Eq. (3) and setting k to 0, we see that $P_{n,N-n} = \frac{P_{n,2N-n}}{(1-x)^N} = \alpha \, r^n \, (1-x)^{-N}$; so with $\beta \equiv \alpha(1+r)^N$, we can write $P_{i,N-i} = \beta r^i$.

DATA-DRIVEN PRIORS FOR HYPERPARAMETERS
IN REGULARIZATION

DANIEL KEREN
Department of Mathematics, The University of Haifa
Haifa 31905, Israel [‡]

AND

MICHAEL WERMAN
Institute of Computer Science, The Hebrew University
Jerusalem 91904, Israel [§]

Abstract. A popular non-parametric model for interpolating various types of data is based on regularization, which looks for an interpolant that is both close to the data and also "smooth" in some sense. Formally, this interpolant is obtained by minimizing an error functional which is the weighted sum of a "fidelity term" and a "smoothness term". The classical approach is to select weights that should be assigned to these two terms, and minimize the resulting error functional. However, using only these "optimal weights" does not guarantee that the chosen function will be optimal in some sense. For that, we have to consider *all* possible weights. The approach suggested here is to use the full probability distribution on the space of admissible functions, as opposed to the probability induced by using a single combination of weights.

Key words: Bayesian estimation, regularization, Gaussian measures on function spaces.

1. Introduction

In many areas of science and engineering, regularization [1,2] is used to reconstruct functions from partial data. In the field of maximum entropy, a similar idea is used for reconstruction of missing or corrupted data [3-7].

Regularization chooses among the possible functions one which approximates the given data and is also "smooth". A *cost functional* $M(f)$ is defined for every

---

[‡]Email: dkeren@mathcs2.haifa.ac.il

[§]Email: werman@cs.huji.ac.il

K. M. Hanson and R. N. Silver (eds.), Maximum Entropy and Bayesian Methods, 77–84.
© *1996 Kluwer Academic Publishers.*

function f by $M(f) = D(f) + \lambda S(f)$, where $D(f)$ measures the distance of f from the given data, $S(f)$ measures the smoothness of f, and $\lambda > 0$ is a parameter. The f chosen is the one minimizing $M()$.

In the one-dimensional case, one can minimize

$M(f) = \sum_{i=1}^{n} \frac{[f(x_i) - y_i]^2}{2\sigma^2} + \lambda \int_0^1 f_{uu}^2 du$. Due to lack of space only the one-dimensional case will be presented here, however this work was extended and applied to functions of two variables as well.

The Bayesian interpretation of this approach is: we are given the data D and want to find the function f which maximizes $Pr(f/D) \propto Pr(D/f)Pr(f)$. Assuming a Gaussian noise model with variance σ^2, $Pr(f/D) \propto \frac{1}{\sigma^n} \exp(-\frac{1}{2\sigma^2} \sum [f(x_i) - y_i]^2)$. Adopting a physical model, it is common to define $Pr(f) \propto \exp(-\lambda \int f_{uu}^2 du)$. Hence $Pr(f/D) \propto \exp(-M(f))$, and the function minimizing $M()$ maximizes the likelihood. Since the model is Gaussian, the MAP function is also the MSE function.

The question is, how does one choose λ and σ? There are various methods for doing that, and some are mentioned in the following section. However, all regularization schemes we are familiar with choose one combination of weights and use them alone to interpolate the function; but, this approach fails to find the maximum likelihood (MAP) estimate for the interpolant f, as it uses only one set of weights λ and σ to construct f. However, the MAP estimate should maximize the following:

$$\int_w Pr(f/D, w)Pr(w/D)dw$$

where w varies over the set of all possible weights.

If $Pr(w/D)$ has some nice properties – for instance, it is unimodal, symmetric, and concentrated around the pair of weights w_{max} which maximize $Pr(w/D)$ – it may be reasonable to approximate this integral by approximating the integrand with a rectangular function around w_{max}. However, the distribution $Pr(w/D)$ can be complicated and this approximation will then fail [8,9]; see also an example of such a data set and the corresponding probability distribution it induces on the weights, in this paper (Figure 4).

In this paper, it will be shown how to find the function f maximizing $\int_w Pr(f/D, w)Pr(w/D)dw$.

We also address the questions of computing the MSE function, and the pointwise uncertainty associated with it.

These three quantities – the MAP, the MSE, and the uncertainty – are perhaps the three most important estimators for a statistical entity, and it is therefore very important to rigorously compute them.

2. Previous Work

A very popular method for determining the smoothing parameter λ is Generalized Cross Validation, GCV (bootstrapping) [10,2]. In [11], a few methods for choosing the smoothing parameter are analyzed.

A different approach, which also chooses an "optimal" smoothing parameter and uses it, is that of Bayesian model selection which, to the best of our knowledge, was first suggested in the pioneering work of Szeliski [9]. There, the following question is posed: *given the data D, what is the most probable value of the smoothing parameter* λ? More recent work in this direction was done by MacKay [8]. Another method for choosing the smoothing parameter is presented in [12]. In [13], the behavior of the smoothing spline over a range of smoothing parameters is studied, and is then used to construct a confidence interval for the smoothing parameter.

The problem with methods that use a single set of weights is that the choice of the values of λ and σ is sometimes very sensitive to the data. Since these values are crucial to the shape of the fitted curve or surface, it turns out that sometimes a small change in the data drastically changes the shape of the fitted curve or surface (see Figure 1). Another problem is that although it can be proved that GCV has some nice asymptotic properties, the choice of the "optimal" values of λ and σ is heuristic in nature. Nontheless, the algorithm performs well in general and is widely used; there are very sophisticated numerical methods for implementing the GCV algorithm.

Work which proceeds in a direction somewhat similar to the one given here is presented in [3,4]. However, this work is in the realm of entropy and therefore the mathematical framework is rather different from ours; for instance, there is no analog to the calculation of the MSE estimate given here.

Finally, recent work reported in [14,15] concerns the problem of computing the MAP solution, in a Bayesian framework, by integrating over the space of smoothing parameters and noise. For "integrating out" these two parameters, a uniform prior for them is assumed.

3. Computing the MAP Estimate

In order to compute the MAP estimate, we have to maximize $Pr(f/D)$ over all functions f. Using Bayes' rule, $Pr(f/D) \propto Pr(D/f)Pr(f)$. In order to compute this, one needs to integrate over all values of λ, σ, resulting in

$$\int_0^\infty \sqrt{\lambda} \exp(-\lambda \int f_{uu}^2 du) Prior(\lambda) d\lambda \cdot \int_0^\infty \frac{1}{\sigma^n} \exp(-\frac{1}{2\sigma^2} \sum [f(x_i) - y_i]^2) Prior(\sigma) d\sigma$$

where the $\sqrt{\lambda}$ in the first integral normalizes the probability distribution on the function space [16].

The expression above has to be maximized over the space of admissible functions. Let us write it more compactly as $F_1(\int f_{uu}^2 du) F_2(\sum [f(x_i) - y_i]^2)$, where $F_1(\alpha) = \int_0^\infty \sqrt{\lambda} \exp(-\lambda\alpha) Prior(\lambda) d\lambda$ and

$F_2(\beta) = \int_0^\infty \frac{1}{\sigma^n} \exp(-\frac{\beta}{2\sigma^2}) Prior(\sigma) d\sigma.$

Note that, obviously, $F_1()$ and $F_2()$ are monotonically decreasing.

It is possible to turn this optimization problem to a one-dimensional optimization by setting $\int f_{uu}^2 du$ to a constant α, and then minimizing $\sum [f(x_i) - y_i]^2$ over all functions f such that $\int f_{uu}^2 du = \alpha$.

Using Lagrange multipliers, this problem transforms into one resembling "standard" regularization: find a λ such that the function f minimizing $\sum [f(x_i) - y_i]^2 + \lambda \int f_{uu}^2 du$ satisfies $\int f_{uu}^2 du = \alpha$, where λ is the Lagrange multiplier.

We have proved that the f minimizing $\sum [f(x_i) - y_i]^2 + \lambda \int f_{uu}^2 du$ is given by $f(x) = (H_{x_1}(x), ... H_{x_n}(x))(A + \lambda I)^{-1})(y_1 ... y_n)^t$, where

$$H_x(\xi) = \begin{cases} 0 \leq \xi \leq x : & \frac{(x-1)\xi(x^2 - 2x + \xi^2)}{6} \\ x \leq \xi \leq 1 : & \frac{x(\xi-1)(x^2 + \xi^2 - 2\xi)}{6} \end{cases}$$

and $A_{i,j} = H_{x_i}(x_j)$. Let us denote the data vector $(y_1, ... y_n)$ by Y. After some manipulations,

$$\int f_{uu}^2 du = Y^t (A + \lambda I)^{-1}) A (A + \lambda I)^{-1}) Y$$

so, we have to find for which λ this expression equals α. Diagonalizing A by an orthonormal U, $UAU^t = D$, and denoting $Z = UY$, the expression for $\int f_{uu}^2 du$ reduces to

$$\sum \frac{d_i Z_i^2}{(d_i + \lambda)^2}$$

where d_i are the diagonal elements of D. Finding a λ for which this equals α is fast, as this function is monotonically decreasing in λ and we can solve the problem by binary search.

After finding λ, we have to compute $\sum [f(x_i) - y_i]^2$, where f minimizes $\sum [f(x_i) - y_i]^2 + \lambda \int f_{uu}^2 du$. Without going into all the technical details, let us just state that this equals $\beta = \|AU^t(D + \lambda I)^{-1} Z - Y\|^2$, another expression which can be computed fast since it involves inverting a diagonal matrix, and since AU^t needs to be computed only once.

Now, all that's left is to compute $F_1(\alpha)F_2(\beta)$. $F_1()$ and $F_2()$ are one-dimensional integrals with rather simple integrands, and can be computed fast (or perhaps stored in a table).

What remains is to maximize $F_1(\alpha)F_2(\beta)$ over α (recall that β is not a free parameter, as it is determined by α).

The algorithm therefore tries to maximize a function $C(\alpha)$ which is defined as follows:

1) compute $F_1(\alpha)$

2) compute the (single) λ_α which satisfies $\sum \frac{d_i Z_i^2}{(d_i + \lambda_\alpha)^2} = \alpha$. This is fast because, as noted, $\sum \frac{d_i Z_i^2}{(d_i + \lambda)^2}$ is monotonically decreasing in λ (A is positive definite, so $d_i > 0$).

3) define $\beta = \|AU^t (D + \lambda_\alpha I)^{-1} Z - Y\|^2$

4) compute $F_2(\beta)$

5) return $F_1(\alpha) F_2(\beta)$

and we have to maximize $C(\alpha)$ for $0 \le \alpha \le \int (f_{interpolate})^2_{uu} du$, where $f_{interpolate}$ is the interpolant which passes through the data points. This range covers all the relevant functions, because $f_{interpolate}$ is the interpolant of the type we're studying which maximizes $\int f^2_{uu} du$ (it corresponds to $\lambda = 0$).

This is a one-dimensional optimization problem, which we solve numerically. The solution is reasonably fast, taking a few seconds on a workstation.

4. Computing the MSE Estimate

An estimator which for some purposes is more useful than the MAP estimate is the MSE estimate. Its value at x is defined by $E_x = \int f(x) Pr(f/D) \mathcal{D}f$.

In order to compute this integral, the following approach is taken. Let us define a probability structure $M_{\lambda,\sigma}$ on the space of admissible functions. In this space, we assume the measurement noise is σ, and the prior distribution of the function f is $Pr(f) \propto \exp(-\lambda \int f^2_{uu} du)$. Under this probability, which is Gaussian, the MSE function, denoted $(f_{opt})_{\lambda,\sigma}$, is equal to the MAP function and there is a closed-form expression for it (given in the previous section). It can be proved that

$$E_x = \int f(x) Pr(f/D) \mathcal{D}f = \int_\lambda \int_\sigma (f_{opt})_{\lambda,\sigma}(x) Pr(M_{\lambda,\sigma}/D) d\lambda d\sigma$$

After computing $Pr(M_{\lambda,\sigma}/D)$, the following expression for E_x can be derived:

$$\frac{\int \frac{1}{\sqrt{v}} |A + vI|^{-\frac{1}{2}} (H_{x_1}(x)...H_{x_n}(x))(A + vI)^{-1} Y^t [Y(A + vI)^{-1} Y^t]^{\frac{4-n}{2}} dv}{\int \frac{1}{\sqrt{v}} |A + vI|^{-\frac{1}{2}} [Y(A + vI)^{-1} Y^t]^{\frac{4-n}{2}} dv}$$

5. Computing the Uncertainty Associated With the Interpolant

In [17,2,16,8,18,9], the problem of assigning a measure of uncertainty to the regularizing interpolant is addressed. This is very important, because usually one wants not only to know the curve (surface) which is optimal in some sense, but also to know how reliable this curve (surface) is. We chose to extend the method suggested in [16], defining the uncertainty of the interpolant at the point x as $\int [f(x) - E_x]^2 Pr(f/D)\mathcal{D}f$ As was the case with E_x, we obtain a closed-form solution, but its computation is non-trivial.

6. Examples

A simple pattern – one cycle of a sinusoidal function – is contaminated with Gaussian noise, and then the resulting data is interpolated using the GCV algorithm and the methods suggested in the previous sections. The instability of the GCV is demonstrated by noting that changing the value of the data at a single point radically changes the shape of the fitted curve (Figure 1). In Figure 2, The MSE (left) and MAP (right) estimates for these two data sets are presented. In Figure 3, the MSE estimate and confidence intervals for two data sets are given. On the left, the data is a sample of the x-coordinates of a hand-written word. On the right, the interpolant and confidence intervals are given for data unevenly sampled from a sinusoid with noise added to it. One can see that the uncertainty is larger in areas which are far from the sample points. The uncertainty at the endpoints is zero, because we constrain our functions to be zero at the endpoints.

Finally, we give an example which explains why one has to integrate over all possible weights. In Figure 4, two data sets are shown, superimposed. As one can see, they are almost identical. Also, the (scaled) joint probability distribution of the weights λ, σ for one of the data sets is plotted. It has two distinct peaks, which are rather far apart; the location of the peaks correspond to the location of the most probable weights for the two data sets. Therefore, the interpolants for the data sets which use only the most probable weights are drastically different, although the data sets are almost identical.

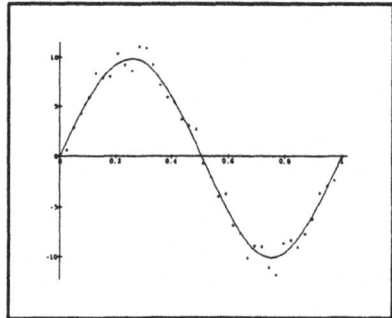 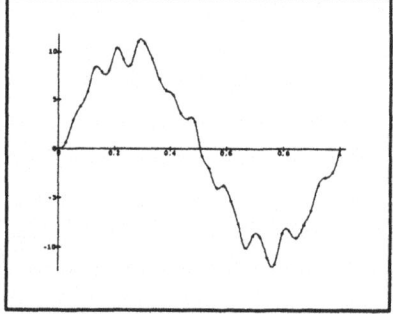

Figure 1: Instability of GCV: for two data sets differing in one point, GCV gives two very different interpolants.

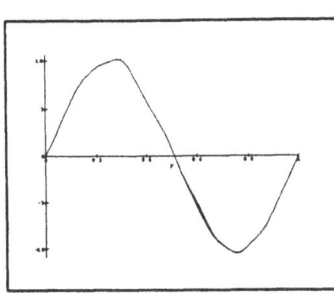
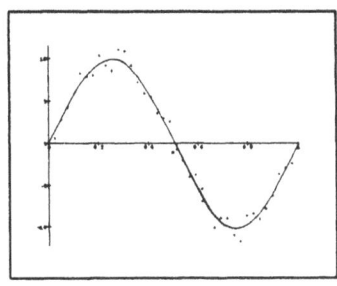

Figure 2: The MSE (left) and MAP (right) estimates for the data sets of Figure 1.

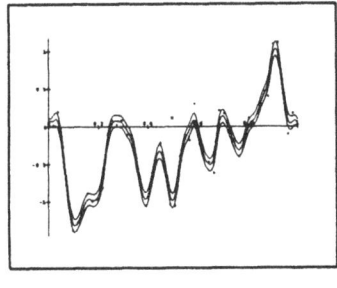
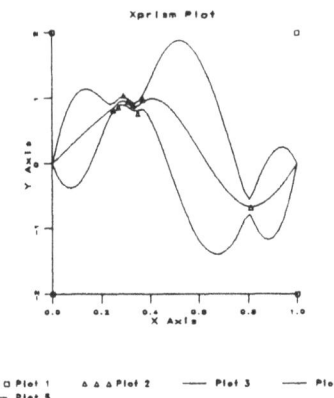

Figure 3: MSE function and confidence intervals for an evenly and unevenly sampled data set.

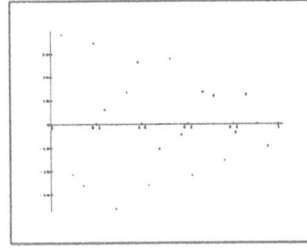
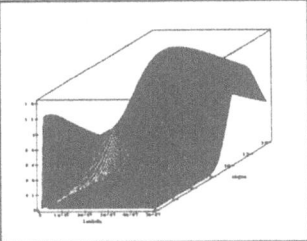

Figure 4: Two nearly identical data sets superimposed, and the (scaled) probability distribution for one of them.

7. Conclusions and Further Research

This work suggests a straightforward approach for solving three basic problems in curve and surface reconstruction, which are very common in many areas: finding the MAP interpolant, finding the MSE interpolant, and computing the uncertainty associated with the interpolant. The solutions are obtained by considering a two-dimensional set of data-driven priors on the function space, and not a single prior as in the standard regularization approach. This leads to a more general solution, which appears to be more stable to perturbations in the data.

References

1. A. Tikhonov and V. Arsenin, "Solution of ill-posed problems," Winston and Sons, 1977.
2. G. Wahba, "Spline models for observational data," Philadelphia: Society for Industrial and Applied Mathematics, 1990.
3. C. Strauss, D. Wolpert, and E. Wolf, "Alpha evidence and the entropic prior," *Maximum Entropy and Bayesian Methods*, Kluwer Academic, 1995.
4. R. Fischer, W. Von Der Linden, and V. Dose, "On the importance of α marginalization in mamimum entropy," *Maximum Entropy and Bayesian Methods*, Kluwer Academic, To Appear.
5. J. Skilling, "Quantified maximum entropy," *Maximum Entropy and Bayesian Methods*, Kluwer Academic, 1990.
6. J. Skilling, "Fundamentals of maxent in data analysis," *Maximum Entropy in Action*, Edited by B. Buck and V.A. Macaulay, Clarendon Press, Oxford, 1991.
7. S. Gull, "Developments in maximum entropy data analysis," *Maximum Entropy and Bayesian Methods*, Kluwer Academic, 1989.
8. D. MacKay, "Bayesian methods for adaptive models," Ph.D thesis, California Institute of Technology, 1992.
9. R. Szeliski, "Bayesian modeling of uncertainty in low-level vision," Kluwer Academic, 1989.
10. P. Craven and G. Wahba, "Optimal smoothing of noisy data with spline functions," *Numerische Mathematik*, **31**, pp. 377–403, 1979.
11. A. Thompson, J. Brown, J. Kay, and D. Titterington, "A study of methods of choosing the smoothing parameter in image restoration by regularization," *IEEE Trans. on Pattern Analysis and Machine Intelligence*, **13**, pp. 326–339, 1991.
12. P. Hall and I. Johnstone, "Emprical functionals and efficient smoothing parameter selection," *Journal of the Royal Statistical Society*, **54**, (1), pp. 475–530, 1992.
13. D. Nychka, "Choosing a range for the amount of smoothing in nonparametric regression," *Journal of the American Statistical Association*, **86**, (415), pp. 653–664, 1991.
14. R. Molina, "On the hierarchical Bayesian approach to image restoration: Applications to astronomical images," *IEEE Trans. on Pattern Analysis and Machine Intelligence*, **16**, (11), pp. 1122–1128, 1994.
15. R. Molina and A. Katsaggelos, "On the hierarchical Bayesian approach to image restoration and the iterative evaluation of the regularization parameter," *Visual Communications and Image Processing'94*, pp. 244–251, Aggelos K. Katsaggelos, Editor, Proc. SPIE 2308, 1994.
16. D. Keren and M. Werman, "Probabilistic analysis of regularization," *IEEE Trans. on Pattern Analysis and Machine Intelligence*, **15**, pp. 982–995, October 1993.
17. D. Nychka, "Bayesian confidence intervals for smoothing splines," *Journal of the American Statistical Association*, **83**, (404), pp. 1134–1143, 1988.
18. F. Larkin, "Gaussian measure in Hilbert space and applications in numerical analysis," *Rocky Mountain Journal of Mathematics*, **2**, pp. 379–421, 1972.

MIXTURE MODELING TO INCORPORATE MEANINGFUL CONSTRAINTS INTO LEARNING

I. TCHOUMATCHENKO AND J.-G. GANASCIA
LAFORIA-CNRS
Université de Pierre et Marie Curie
4 Place Jussieu
75252 Paris Cedex 05
France[†]

Abstract.
This paper addresses the problem of incorporating prior knowledge into learning algorithms. An impressive variety of hand-crafted methods dealing with this difficult problem has been elaborated by both symbolic machine learning and neural networks communities. However, no fairly general methodology has emerged yet. The contribution of this paper is two-folded. First, we propose a Bayesian view of domain knowledge incorporation. In our framework the learning algorithm's designer is proposed to express available prior knowledge in terms of mixture distributions on both parameters and models envolved into a given type of learning. Then the known Bayesian learning machinery is put into work. Second, we fulfill a detailed case study applying our framework to extract logical rules from neural networks. We analyse our approach doing computational experiments on the difficult problem of protein secondary structure prediction. Besides, for this problem, we propose a Dirichlet mixture modeling of training data.

Key words: Bayesian learning, knowledge incorporation, neural networks, protein structure prediction, Gaussian mixtures, Dirichlet mixtures.

1. Introduction

A radical customization of learning algorithms conventionally called *knowledge incorporation* is a hot topic in machine learning research. Numerous modifications of existing symbolic machine learning algorithms have been developed to take into account prior domain knowledge [1][2][3][4]. A rule-based prior knowledge has been successfully incorporated into neural networks [5]. Even more exotically, domain knowledge has been represented as a collection of trained neural networks [6][7]. However, prior knowledge might be and often is available in a form different from

---

[†] Email: irina@laforia.ibp.fr, ganascia@laforia.ibp.fr

K. M. Hanson and R. N. Silver (eds.), Maximum Entropy and Bayesian Methods, 85–92.

logical rules. For example, a large variety of problems is decomposable into sub-problems. This observation leads to a different approach to knowledge incorporation, which consists in constructing a learner per sub-problem and then combining learners in some manner to solve the problem. The most popular amongst the proposed combination methods is the family of hierarchical expert-mixtures models (see, for example, [8]). Furthermore, prior knowledge sometimes has a somewhat technical character as it is the case for regularization methods aimed to improve the generalization ability of neural networks [9] or to clean the data [10]. In this paper we show how different methods of knowledge incorporation can be viewed in a common manner extending the Bayesian approach widely accepted by the neural network community [11]. A Bayesian framework for symbolic learning demonstrated on examples of classification trees and tree-structured Bayesian networks was first proposed in [12]. Moreover, it turns out that enriched with graphical representations, the Bayesian framework can be considered as a specification language for a great deal of learning problems [13]. In this work, we continue to elaborate the Bayesian foundations of learning algorithms.

2. Bayesian incorporation of prior knowledge

As usual in Bayesian considerations, we assume a model M, which corresponds to a knowledge representation formalism of the learning algorithm we study. The model M might correspond to logical rules, classification trees, neural networks, Bayesian networks, etc. The model M has its parameters θ, which might correspond to the number of false positive, false negative, true positive, true negative if we deal with a model representing logical rules. It might be leaf class probabilities in the case of classification trees, network's weights in the case of neural networks or conditional probabilities in the case of Bayesian networks. The data D is assumed to be independently sampled. Prior probability is placed over the model parameters $P(\theta|M)$. The next important component is the data likelihood $P(D|\theta, M)$, which, on the basis of the model assumptions M and given a set of parameters θ says how likely the data is. The model needs to completely determine the data likelihood. Then the posterior probability of parameters is given by Bayes theorem:

$$P(\theta|D, M) = \frac{P(D|\theta, M)P(\theta|M)}{P(D|M)} \tag{1}$$

$P(D|M)$ is the evidence for model M or model likelihood:

$$P(D|M) = \int P(D|\theta, M)P(\theta|M) \, d\theta \tag{2}$$

This integral is often difficult to perform. One can use Markov chain methods, Gaussian or Laplace approximation. Equation (1) is the basis of *parameter learning*. Note that one does not need to calculate evidence in order to learn parameters. In some cases one can escape to maximization procedure to learn parameters. If the prior is a *conjugate prior* for the likelihood distribution, then the posterior is calculated by simple updating.

Another part of the Bayesian learning paradigm is *model learning* which is performed as a mostly heuristic search for a model (models) with high posterior probability:

$$P(M|D) \propto P(D|M)P(M) \tag{3}$$

Now prior knowledge incorporation can be reduced to placing a particular prior distribution on parameters $\mathbf{P}(\theta|\mathbf{M})$ and models $P(M)$. A mixture of distributions gives us a sufficient flexibility to represent a wide spectrum of knowledge. This Bayesian proposition on knowledge incorporation is summarized in table 1.

TABLE 1. Bayesian incorporation of prior knowledge into learning.

Bayesian solution:	Suggest mixture distributions $\mathbf{P}(\theta|\mathbf{M})$ and $P(M)$ on parameters θ and(or) models M respectively corresponding to available prior knowledge
Learning parameters:	$P(\theta|D, M) = \frac{P(D|\theta, M)P(\theta|M)}{P(D|M)}$
Learning model:	$P(M|D) \propto P(D|M)P(M)$ $P(D|M) = \int P(D|\theta, M)P(\theta|M)\, d\theta$

3. A case study: mixture modeling to convert neural networks into logical rules

To prove the operationality of the proposed framework for knowledge incorporation, we study how to convert neural networks into logical rules on its basis. Our a priori knowledge is that a neural network trained to classify and having its weights clustered around $-1, 0, 1$ can be approximately interpreted as *a majority vote rule*.

Let $I_{C_1}, I_{C_2}, ..., I_{C_n}$ be the counters calculating the contribution of input units into the final classification decision of the neural network in favour of class $C_1, C_2, ..., C_n$ respectively.

Let the path p_i^j be a conjunction of two non-zero weighted connections, where the first comes from i-th input unit and the last comes to j-th output unit ($j \in \{1, ..., n\}$). We call p_i^j a positive (negative) path if the product of its weights is positive (negative). We denote a positive (negative) path as p_i^{j+} (p_i^{j-}). The work of the neural network can be approximated as:

1. compute the counters: $I_{C_j} = \sum_{p_i^{j+}} 1 - \sum_{p_i^{j-}} 1 , \quad j \in \{1, ..., n\}$

   This means that the input unit having a positive (negative) path to j-th output unit, $j \in \{1, ..., n\}$ will cause I_{C_j} to be incremented (decremented). We say that such a unit "votes" for (against) the class C_j.

2. compare counters and make a decision:
   if $I_{C_k} = max_j\{I_{C_j}\}$ then class C_k is predicted.

To express this prior knowledge in terms of mixtures of distributions, it is sufficient to use a simple Gaussian mixture. Let the data D be represented as a set of m input-target pairs $D = \{\mathbf{x}^m, \mathbf{t}^m\}$. The parameters θ (weights) are traditionally denoted as $\mathbf{w}$. Then prior probability $P(\mathbf{w}|M)$ corresponding to our knowledge can be written in the form:

$$P(\mathbf{w}|M) \propto \sum_{j=1}^{3} \pi_j p_j(\mathbf{w}) \tag{4}$$

where $p_j(\mathbf{w})$ is a Gaussian density with mean μ_j and standard deviation σ_j, and π_j is a mixing proportion of jth Gaussian ($\sum_j \pi_j = 1$).

Further, we take into account through a scale constant α_i that characteristic sizes of weights are different in different network's layers. Then the cost function corresponding to (1) is[1]:

$$C = \frac{\beta}{2} \sum_m (\mathbf{y}^m - \mathbf{t}^m)^2 - \sum_i \alpha_i \sum_m \log \sum_{j=1; \mathbf{w} \in i}^{3} \pi_j p_j(\mathbf{w}) \tag{5}$$

3.1. LEARNING PARAMETERS

Parameter learning consists in minimizing the cost function (5) complicated by the introduction of *hyperparameters* μ_j, σ_j, π_j, α_i. The method we have chosen to deal with the hyperparameters is based on calculations of their evidence $P(D|\mu_j, \sigma_j, \pi_j, \alpha_i, \beta)$ using Gaussian approximation as it was proposed in [14]. We fulfill Gaussian approximation around each of three maximums of posterior distribution (corresponding to $-1, 0$ and 1 values of network's weights).

3.2. LEARNING MODEL

On one hand, model learning consists in searching the models' space for a model maximizing the posterior probability (3). So we have:

$$P(D|M) = \int P(D|\mu_j, \sigma_j, \pi_j, \alpha_i, \beta, M) P(\mu_j, \sigma_j, \pi_j, \alpha_i, \beta) \, d\mu_j \, d\sigma_j \, d\pi_j \, d\alpha_i \, d\beta \tag{6}$$

where $P(D|\mu_j, \sigma_j, \pi_j, \alpha_i, \beta, M)$ is estimated as explained earlier.

On the other hand, it is known that the ratio between maximal and minimal eigen values of the Hessian matrix of error function indicates convergence problems [16]. So, we have organized our search in such a manner that the Hessian of the model is calculated, its eigen values are analyzed and the model is changed on the basis of this analysis. The organization of the search is inspired by the method of

---

[1] for more detail see [15]

structural risk designed to trade-off between the complexity of a learning system and the amount of available data to garantee the best generalization capacity. In the matter of fact, it is not necessary to have all weights of the neural network clustered around $-1, 0$ and 1 to be able to extract logical rules. Some weaker clusterization of weights is acceptable. For example, the extraction method proposed in [17] clusters weights impinging on *each* unit into five groups, which corresponds naturally to a five-components Gaussian mixture prior. Let us order Gaussian mixture priors from less to more component's ones: $MP_1, MP_2, \ldots, MP_n$. Then our method to search the space of model is summarized in table 2.

TABLE 2. Hessian-based search procedure.

Step1:	Choose a mixture prior ordering $MP_1, MP_2, \ldots, MP_n$.
Step2:	Choose the next mixture prior in the given ordering.
Step3:	Train the neural network under the chosen mixture prior, calculate the posterior of the model.
Step4:	Evaluate the results of training under the chosen mixture prior, if satisfied, stop, if not – evaluate the Hessian using it's diagonal approximation,mark the network's units corresponding to max and min eigen values.
Step5:	Apply the next mixture prior in the given ordering to the parts of the networks around the network's units corresponding to max and min eigen values, rank alternative models by evaluating their posterior, go to Step 3.

3.3. RESULTS ON ONE DIFFICULT PROBLEM

A neural network trained on the secondary structure prediction problem using standard backpropagation technique achieves the accuracy of 62.3%. A neural network $273 - 10 - 3$ trained on the same problem using the proposed above method achieves almost the same accuracy (61.4%) and has its weights strongly clustered. It is interesting to note that our Hessian-based search procedure (see table 2) leads to the model where only 5 from initial 10 hidden units survive. Three of 5 useful hidden units work for coil prediction while only one hidden unit works for each of helix and sheet predictions. It reflects the statistics in the training set. We have extracted from the clustered neural network the majority vote rules for helix, sheet and coil predictions having the prediction accuracy of 61.2%. The helix rule summarizing the positive/negative votes of a given amino acid at a given place in the input window is presented in table 3. We have scored each vote on the test set. The votes that are mostly incorrect have been removed. The resulting slightly refined majority vote rule overperforms the original one only by 0.2%. It means that our method leads to simple rule sets requiring little pruning. Even more interesting results we have obtained applying our extraction technique to predict the secondary structure of α/α proteins. The plain neural networks was at 80.3% level of accuracy while the majority vote extracted after training achieved as high as 78.4%.

TABLE 3. Majority vote rule for helix prediction : the presence of amino acid AA (!AA) at a given position in the input window (P) increments (decrements) helix counter.

P	Amino acids																		
1		!C			E														
2		!C			E														
3	A				E		!G							!P					Y
4	A				E		!G	!H						!P					
5	A				E		!G				L	M		!P			!S		
6	A				E		!G				L	M		!P			!S	!T	
7	A				E	F	!G				L	M	!N	!P			!S	!T	
8	A				E	F	!G			K	L	M		!P	Q	R			
9	A		!D			F			H	K	L	M		!P			!S	!T	
10	A		!D			F	!G		H	K	L	M		!P			!S		
11	A		!D						H	K	L	M		!P					
12	A								H	K	L	M		!P					
13		!C	!D	!E			!G			K				!P			!S	!T	

4. Mixture modeling of the data

While our extraction results are very promising, the prediction accuracy of neural networks for the protein structure prediction is still too low. To increase prediction accuracy we have chosen to apply mixture modeling to the data. In [18], it was proposed to use the information on *multiple alignments* to predict the secondary structure. A neural network was trained on the frequencies of amino acids in the alignment's columns. However, the small sample size of available alignments makes the frequency-based estimates rather poor. To overcome this difficulty we have used Bayesian modeling of the alignments placing Dirichlet mixture priors first developed to align proteins [19]. Namely, let us consider a column $\vec{n}$ of the alignment, such as $\vec{n} = (n_1, \ldots, n_{20})$, where n_i is the number of times the i-th amino acid occurs in the column. We suppose than amino acids are generated according to a probability distribution $\vec{p} = (p_1, \ldots, p_{20})$, and $\vec{p}$ is chosen independently from a Dirichlet mixture density:

where

$$\rho_j(\vec{p}) = \frac{\Gamma(\alpha_0) \prod_{i=1}^{k} p_i^{(\alpha_i - 1)}}{\prod_{i=1}^{k} \Gamma(\alpha_i)} \tag{7}$$

where $\alpha_0 = \sum_{i=1}^{20} \alpha_i$. The 21 × k parameters were estimated on 6472 columns by the research group of David Haussler, UCSF at Santa Cruz. We use these estimates to replace the frequency-based coding by the Dirichlet mixture-based coding at the input of our network predicting the secondary structure as follows:

$$P(j|\vec{n}) = \frac{q_j P(\vec{n}|\rho_j)}{\sum_{l=1}^{k} q_l P(\vec{n}|\rho_l)} \tag{8}$$

where $n = \sum_{i=1}^{20} n_i$ and $\alpha^{(j)} = \sum_{i=1}^{20} \alpha_i^{(j)}$

$$P(j|\vec{n}) = \frac{\Gamma(n+1)\Gamma(\alpha^{(j)})}{\Gamma(n+\alpha^{(j)})} \prod_{n=1}^{20} \frac{\Gamma(n_i + \alpha_i^{(j)})}{\Gamma(n_i + 1)\Gamma(\alpha_i^{(j)})} \tag{9}$$

$$\hat{p}_i = \frac{\Gamma(\alpha^{(j)})(n_i + \sum_{j=1}^{k} P(j|\vec{n})\alpha_i^{(j)})}{\prod_{i=1}^{k} \Gamma(\alpha_i^{(j)})} \tag{10}$$

So, for each column of the alignment we computed $\hat{p}_1, \ldots, \hat{p}_{20}$ and trained the network using these values as input. The prediction accuracy has increased importantly (68.8% against 65.9% for the frequency-based coding).

5. Conclusion

In this paper we have reduced the difficult problem of knowledge incorporation to the better-studied Bayesian learning. Our approach gives very good results for logical rule extraction from neural networks. The success is due to the underlying Bayesian calculations of (hyper)parameters and model's posterior. The Hessian-based heuristics to search the model space also plays important role. We believe our framework for incorporation of prior knowledge in learning algorithms to be useful and general enough. We are working on the learning Bayesian networks in the same framework using Dirichlet mixture priors.

References

1. Ortega J., Fisher D. (1995) Flexibly exploiting prior knowledge in empirical learning, *in Proceedings of the 14th International Joint Conference on Artificial Intelligence*, Montreal, pp. 1041-1047.
2. Clark P., Matwin S. (1993) Using qualitative models to guide inductive learning, *in Proceedings of the Tenth International Conference on Machine Learning*, pp. 49-56, Amherst, MA.
3. Mooney R.J. (1993) Induction over the unexplained: using overly-general theories to aid concept learning, *Machine Learning*, 10:79-110.
4. Pazzani M., Kibler D. (1992) The utility of knowledge in inductive learning, *Machine Learning*, 9(1):57-94.
5. Towell, G.G., Shavlik, J. (1992). Interpretation of artificial neural networks: mapping knowledge-based neural networks into rules, *Advances in Neural Information Processing Systems 4*, pp. 977-984, San Mateo, CA, Morgan Kaufman.
6. Thrun S., Mitchell T. (1993) Integrating inductive neural network learning and explanation-based learning, *in Proceedings of the 13th International Joint Conference on Artificial Intelligence*, Montreal, pp. 930-936.

7. Thrun S., Mitchell T. (1995) Learning one more thing, *in Proceedings of the 14th International Joint Conference on Artificial Intelligence*, pp. 1217-1223.

8. Jordan M., Jacobs R. (1994) Hierarchical mixtures of experts and the EM algorithm, *Neural Computation*, 6:181-214.

9. Hinton G.E., Drew van Camp (1993) Keeping neural networks simple by minimizing the description length of the weights, *in Proceedings of the Conference on Computational Learning Theory*, Santa Cruz.

10. Weigend A.S., Zimmermann H.G. (1995) The observer-oservation dilemma in neuro-forecasting: reliable models from unreliable data through CLEANING, *in Proceedings of the Neural Networks in Capital Markets Conference*, London Business School, October 1995, to appear.

11. Rumelhart D. E., Durbin R., Golden R., Chauvin Y. (1995) Backpropagation: the basic theory, *in Chauvin Y. and Rumelhart D. E., editors, Backpropagation: Theory, Architectures, and Applications*, pp. 1-34, Hillsdale, NJ, Lawrence Erlbaum Associates.

12. Buntine, W. (1991) Classifiers : A theoretical and empirical study, *in Proceedings of the 12th International Joint Conference on Artificial Intelligence*, Sydney, Morgan Kaufmann, pp.638-644.

13. Buntine W. (1994) Operations for learning with graphical models, *Journal of Artificial Intelligence Research*, 2, December 1994.

14. Mackay D.J.C. (1992) A practical Bayesian framework for backpropagation networks, *Neural Computation*, pp. 448-472.

15. Tchoumatchenko I., Ganascia J.-G. (1994) A Bayesian Framework to Integrate Symbolic and Neural Learning, *in Proceedings of the 11th International Conference on Machine Learning*.

16. Le Cun Y., Kanter I., Solla S.A. (1990) Second order properties of error surfaces: learning time and generalization, *Advances in Neural Information Processing Systems 2*, San Mateo, CA, Morgan Kaufman.

17. Craven, M. W., Shavlik, J.W. (1993). Learning symbolic rules using artificial neural networks, *in Proceedings of the 10th International Conference on Machine Learning*, Ahmarest, MA.

18. Rost B., Sander C. (1993) Prediction of protein secondary structure at better than 70% accuracy, *Journal of Molecular Biology* 232.

19. Brown M. et al. (1993) Using Dirichlet mixture priors to derive hidden markov models for protein families, *in Proceedings of the 1st International Conference on Intelligent Systems for Molecular Biology*, pp. 47-55, July 6-9, Bethesda, MD.

MAXIMUM ENTROPY (MAXENT) METHOD
IN EXPERT SYSTEMS AND INTELLIGENT CONTROL:
NEW POSSIBILITIES AND LIMITATIONS

V. KREINOVICH
Department of Computer Science
University of Texas at El Paso
El Paso, TX 79968, USA[§]

AND

H. T. NGUYEN AND E. A. WALKER
Department of Mathematical Sciences
New Mexico State University
Las Cruces, NM 88003, USA[¶] [‖]

Abstract. To describe uncertainty of experts' statements $E_1, ..., E_n$ that form a knowledge base, it is natural to use (subjective) probabilities p_i. Correspondingly, it is natural to use probabilities to describe uncertainty of the system's answer to a given query Q. Since it is impossible to inquire about the expert's probabilities for all possible ($\geq 2^n$) propositional combinations of E_i, a knowledge base is usually *incomplete* in the sense that there are many probability distributions consistent with this knowledge. If we want to return a single probability, we must select *one* of these distributions. MaxEnt is a natural selection rule, but for expert systems, computing the MaxEnt distribution often takes an unrealistically long time.

In this paper, we propose computationally feasible approximate MaxEnt techniques for expert systems and intelligent control.

Key words: MaxEnt, expert systems, uncertainty, commonsense reasoning, intelligent control

[§]Email: vladik@cs.utep.edu. This work was partially supported by NSF Grants No. CDA-9015006 and EEC-9322370, and by NASA Grant No. NAG 9-757.

[¶]Email: {hunguyen,elbert}@nmsu.edu. This work was partially carried out while Hung T. Nguyen was on sabbatical leave at the University of Southern California, Los Angeles, Spring 1995.

[‖]The authors are thankful to Peter C. Cheeseman for his interest and encouragement, and to all the participants of the workshop, especially to Professor Myron Tribus, for helpful discussions.

93

K. M. Hanson and R. N. Silver (eds.), Maximum Entropy and Bayesian Methods, 93–100.
© 1996 *Kluwer Academic Publishers.*

1. Expert Systems

1.1. INTRODUCTION: WHAT ARE EXPERT SYSTEMS, WHY DO WE NEED MAXENT FOR EXPERT SYSTEMS, AND WHY TRADITIONAL MAXENT METHODS ARE NOT ALWAYS SUFFICIENT

The Basic Idea of an Expert System. In many fields (geology, medicine, etc.), there are experts who can make outstanding decisions. Unfortunately, the number of these super-experts is usually small, and we cannot use them for every problem we encounter: e.g., we cannot show all the patients with heart disease to the greatest specialist.

A natural idea is, therefore, to design a computer-based system that would use the knowledge of these experts and make conclusions and decisions based on that knowledge. Such systems need not be completely autonomous. If we can design a system that will prompt one or several different answers, this system will also be of great value to the corresponding community. Such helper systems are called *expert systems* .

To make expert systems successful, we must be able to incorporate the experts' knowledge inside the computer: First of all, we need to describe statements $E_1, ..., E_n$ that comprise the experts' knowledge. Then, if a user asks a query Q, the system must be able to return a "yes" or "no" answer.

To Describe Experts' Knowledge Adequately, We Must Also Describe Uncertainty. Our knowledge is not simply the set of statements that we make. In some of these statements, we are more certain; in some, we are less certain. A natural measure of an expert's certainty in a statement E_i is a (subjective) probability p_i. Therefore, in a knowledge base, we must have not only the experts' statements $E_1, ..., E_n$, but also the probabilities p_i of these statements. This uncertainty changes the desired answer to a query Q: since the experts are not absolutely certain about the statements E_i that comprise the knowledge base, we cannot be absolute certain about the answer to Q deduced from these statements. In other words, due to uncertainty in E_i, we are uncertain about Q. As a result, we want the expert system to return the degree of uncertainty (i.e., the subjective probability) of a queried statement Q.

A Problem: Our Information about Experts' Probabilities is Not Complete. A typical knowledge base, that consists of experts statements $E_1, ..., E_n$ with their probabilities $p_1, ..., p_n$, does not contain the complete information about the expert's probabilities. For example, if the only thing that we know about the two statements E_1 and E_2 is that both have probability 0.5, then:

- it could happen that E_1 and E_2 are equivalent, in which case $E_1 \& E_2$ has probability 0.5, and
- it could also happen that E_1 and E_2 are incompatible, in which case $E_1 \& E_2$ has probability 0.

To get a complete description of subjective probabilities, it is thus not sufficient to know n probabilities p_i of the statements E_i themselves; we also need to know at least 2^n probabilities of the possible propositional combinations of E_i (e.g., probabilities of 2^n statements of the type $E_1 \& \neg E_2 \& ... \& E_n$ that correspond to different combinations of basic statements E_i and their negations). Realistic knowledge bases contain hundreds of statements; it is absolutely impossible to present $2^{100} \approx 10^{30}$ questions to the experts. Hence, no matter how many questions we ask, the probabilistic information contained in the knowledge base will not be complete.

One Possible Solution: Return Intervals of Probabilities As Answers to the Queries. On possible solution to this problem is as follows: There are many probability distributions that are consistent with our knowledge; different distributions may lead to different probabilities $p(Q)$ of the query. Hence, let us provide the user with the *set* of all possible values of $p(Q)$. (Usually, this set is an *interval*.)

In the above pedagogical example, the interval of possible values of $p(E_1 \& E_2)$ is $[0, 0.5]$.

There are several successful expert systems that are based on this idea (see, e.g., Kohout et al. [5] and references therein).

A Problem with Interval Approach. In general, the resulting interval of probability is too wide. In many non-trivial cases, it is so close to the interval $[0, 1]$ (that corresponds to our not knowing anything about $p(Q)$) that this interval becomes absolutely useless to the user who asked this query: e.g., if a doctor is told that the probability of a successful surgery is from the interval $[0.1, 0.9]$, this will not help her decide whether to perform this surgery or not.

MaxEnt Approach. A MaxEnt approach to this problem was proposed by P. Cheeseman (see, e.g., [1]): he proposed to choose, from all probability distributions that are consistent with the given knowledge, the one p with the largest entropy, and for every query Q, return $p(Q)$ (for this chosen p) as a probability that Q is true.

Formally, we can define a *possible world* W as a consistent statement of the type $E_1 \& \neg E_2 \& ... \& E_n$, and choose the *MaxEnt* probability distribution $\{p(W)\}$ so that $S = -\sum p(W) \log p(W) \to \max$ under the following three conditions:

- that the probabilities $p(W)$ are non-negative: $p(W) \geq 0$;
- that the total probability is 1: $\sum p(W) = 1$; and
- that the probability of each statement E_i is equal to the given value p_i, i.e., that $\sum \{p(W) \mid W \vdash E_i\} = p_i$.

Then, for every query Q, we define $p(Q)$ as $p(Q) = \sum \{p(W) \mid W \vdash Q\}$.

Main Problem: MaxEnt Approach Is Often Computationally Intractable. In many cases, MaxEnt approach has lead to successful expert systems (see, e.g., Pearl [11]), but in general, it can be proven to be computationally intractable (see, e.g., Dantsin et al. [2]).

There exist reasonably good techniques that approximate the value $p(Q)$ for the MaxEnt distribution p. One of these techniques is a Monte-Carlo method (see, e.g., [2,7]) that estimates $p(Q)$ by repeatedly asking a query Q to a randomly perturbed knowledge base. To get the probability $p(Q)$ with the desired accuracy $\approx 10\%$, we must repeat this simulation at least 100 times. For a realistic knowledge base, an answer to a query already takes reasonably long (often, around several minutes), so, if we must increase it by a factor of hundred, it becomes unrealistically long.

1.2. METHODS USED IN THE EXISTING EXPERT SYSTEMS

Description of the Existing Methods. MaxEnt approach is not always computationally feasible for expert systems; on the other hand, expert systems do exist, and they are often very successful. How do they do that?

The existing expert systems use a *heuristic* approach. To explain this approach, let us describe how it is applied to the problem with which we started: we know the probabilities p_1 and p_2 of the two statements E_1 and E_2, and we want to know the probability $p(E_1 \& E_2)$. MaxEnt approach is *global* in the sense that to determine this probability, we must know (and take into consideration) the information about all other statements E_i. This necessity of processing all the statements from a knowledge base of size n is what makes the computation time grow exponentially with n. Hence, a natural way to get an easily computable approximation to $p(E_1 \& E_2)$ is to use only the values $p(E_1)$ $(= p_1)$ and $p(E_2)$. In other words, to estimate the probability of $E_1 \& E_2$, a function $f_\& : [0,1] \times [0,1] \to [0,1]$ is chosen, and $p(E_1 \& E_2)$ is estimated as $f_\&(p(E_1), p(E_2))$ (examples of such $\&-operations$ are $f_\& = \min$, $f_\&(a,b) = a \cdot b$, etc.). Similarly, we estimate $p(E_1 \vee E_2)$ as $f_\vee(p(E_1), p(E_2))$ for some $\vee-operation$ $f_\vee$.

If a query Q is more complicated than a disjunction or a conjunction of basic statements, then we express Q in terms of $\&$, $\vee$, and $\neg$, and use this representation to estimate $p(Q)$ step-by-step: e.g., we estimate $p((E_1 \& (E_2 \vee \neg E_1))$ as $f_\&(p(E_1), f_\vee(p(E_2), 1 - p(E_1)))$.

This approach is efficiently used in the existing expert systems, but it has the following two problems:

First Problem with the Existing Approach. The first problem is that for complicated queries, the result of the step-by-step application depends on the representation of the query. For example, a query $E_1 \to E_2$ can be represented either as $\neg E_1 \vee E_2$, or as $\neg E_1 \vee (E_2 \& E_1)$. In classical logic, these two representations are equivalent. However, step-by-step procedure will lead to different results: e.g., for $p_1 = p_2 = 0.5$, $f_\&(a,b) = a \cdot b$, and $f_\vee(a,b) = a + b - a \cdot b$, we get $p(\neg E_1 \vee E_2) = 0.5 + 0.5 - 0.25 = 0.75$, $p(E_2 \& E_1) = 0.25$, and $p(\neg E_1 \vee (E_2 \& E_1)) = 0.5 + 0.25 - 0.125 = 0.625 \neq 0.75$. This problem was recently emphasized by Elkan [3].

Second Problem. In the existing expert systems, , the $\&-$operation $f_\&$ and the $\vee-$operations $f_\vee$ are chosen rather arbitrarily.

1.3. THE PROPOSED APPROACH

The Main Idea. We propose to combine the advantages of MaxEnt (whose formulas are justified and unambiguous) and of the existing approach (whose formulas are computable): Namely, we propose to use the functions $f_\&$ and $f_\vee$, but choose them by using MaxEnt.

Definitions and the Main Result. In precise terms, as $f_\&(a,b)$, we choose $p(A \& B)$ for a distribution p that has the largest entropy among all distributions for which $p(A) = a$ and $p(B) = b$. Similarly, we can define $f_\vee(a,b)$ as $p(A \vee B)$ for the

MaxEnt distribution p. We will call the resulting operations $f_\&$ and $f_\vee$ *MaxEnt operations*.

To find $f_\&(a,b)$ and $f_\vee(a,b)$, we thus have to solve a conditional optimization problem with four unknowns $p(W)$ (for $W = A\&B$, $A\&\neg B$, $\neg A\&B$, and $\neg A\&\neg B$). This problem has the following explicit solution:

Theorem 1. *For* $\&$ *and* $\vee$, *the MaxEnt operations are* $f_\&(a,b) = a \cdot b$ *and* $f_\vee(a,b) = a + b - a \cdot b$.

Comment. For a modified entropy function, a similar problem is solved in Klir et al. [4]; it leads to a different set of $\&-$ and $\vee-$operations. Both pairs of operations are in good accordance with the general group-theoretic approach [6,8,9] for describing operations $f_\&$ and $f_\vee$ which can be optimal w.r.t. different criteria.

The Proposed Approach Is a Solution to the Above Two Problems of Traditional Expert System Methodology. First, we have a *justification* for the chosen $f_\&$ and $f_\vee$ in terms of MaxEnt.

Second, we can repeat the same procedure for any propositional formula $F(A, ..., B)$ (and not only for $A\&B$ and $A \vee B$), and come up with a function $f_F(a, ..., b)$; therefore, we do not need to represent F in terms of $\&$, $\vee$, and $\neg$ any more: we get f_F straight from F. Let us describe this general formula:

Theorem 2. *Let* $F \equiv \vee(A_1^{\varepsilon_1}\&...\&A_n^{\varepsilon_n})$ *be a conjunctive normal form (CNF) for* F, *where* $\varepsilon_i = \pm$, A^+ *means* A, *and* A^- *means* $\neg A$. *Then,* $f_F(a_1, ..., a_n) = \sum a_1^{\varepsilon_1} \cdot ... \cdot a_n^{\varepsilon_n}$, *where* $a^+ = a$ *and* $a^- = 1 - a$.

In particular, we get the following two results:

Theorem 3. *For* $A \to B$, *the corresponding MaxEnt operation is* $f_\to(a,b) = 1 - a + a \cdot b$.

Proof. Indeed, CNF for $a \to b$ is $(a\&b) \vee (\neg a\&b) \vee (\neg a\&\neg b)$; therefore, $f_\to(a,b) = a \cdot b + (1-a) \cdot b + (1-a) \cdot (1-b) = ab + b - ab + 1 - a - b + ab = 1 - a + ab$. Q.E.D.

This result coincides with the result of a step-by-step application of MaxEnt operations $f_\&$ and $f_\vee$ to the formula $B \vee \neg A$ (which is a representation of $A \to B$ in terms of $\&$, $\vee$, and $\neg$).

Theorem 4. *For* $A \equiv B$, *the corresponding MaxEnt operation is* $f_\equiv(a,b) = 1 - a - b + 2a \cdot b$.

Unlike $f_\to$, the resulting expression *does not* cannot be obtained by a step-by-step application of $f_\&$, $f_\vee$, and $f_\neg$ to any propositional formula:

Theorem 5. *The expression* $f_\equiv(a,b) = 1 - a - b + 2a \cdot b$ *cannot be obtained by a step-by-step application of MaxEnt operations* $f_\&(a,b) = a \cdot b$, $f_\vee(a,b) = a + b - a \cdot b$, *and* $f_\neg(a) = 1 - a$ *to a propositional formula* $F(A, B)$.

Idea of the proof. Let us prove this theorem by reduction to a contradiction. Let us assume that there exists a propositional formula F for which the step-by-step procedure leads to $f_\equiv$.

One can easily see that $f_\vee(p,q) = f_\neg(f_\&(f_\neg(p), f_\neg(q)))$ for all p and q, so, we can replace in F every occurrence $P\vee Q$ of $\vee$ by $\neg(\neg P\&\neg Q)$ and still get the same step-by-step result. After doing this, we get a new formula G that only contains $\&$ and $\neg$ and still leads to $f_\equiv$.

If G is of the form $\neg\neg H$, then, since $f_\neg(p) = 1-p$, we have $f_G = 1-(1-f_H) = f_H$, so, we can simply take H as the desired formula.

As a result of this reduction, we get a formula H that represents $f_\equiv$ and that is either of the type $P\&Q$ for some subformulas P and Q, or of the type $\neg(P\&Q)$. Since both operations $f_\&$ and $f_\neg$ are polynomial, the expressions f_P and f_Q that we get after applying these operations step-by-step are also polynomials. Hence:

- In the first case, the function $f_\equiv(a,b)$ is a product of two polynomials $f_P(a,b)$ and $f_Q(a,b)$; however, one can easily show that the polynomial $f_\equiv(a,b)$ is not factorizable.
- In the second case, the function $1 - f_\equiv(a,b)$ is a product of two polynomials f_P and f_Q; however, this polynomial $1 - f_\equiv$ is also not factorizable.

In both cases, we have a contradiction, so, our initial assumption was wrong. Q.E.D.

Comment 1: Relationship to Common Sense Reasoning. Theorem 4 is in good accordance with *common sense reasoning* (that expert systems try to formalize): Indeed, mathematically, "A implies B" means the same as "B or not A", but the resulting formulas of the type "if $2 + 2 = 5$, then I am the King of France" are counterintuitive. From the common sense viewpoint, implication is a separate operation that is different from its mathematical representations in terms of $\&$, $\lor$, and $\neg$. So, we can conclude that *the described MaxEnt approach for choosing "logical" operations with probabilities is closer to common sense reasoning than the traditional expert system approach.*

Comment 2: There Still is a Basis for Propositional Functions. In classical logic, three connectives $\&$, $\lor$, and $\neg$ form a *basis* in the sense that every propositional formula can be expressed in terms of them. The fact that in MaxEnt, we cannot express $\equiv$ in terms of $\&$, $\lor$, and $\neg$, simply means that, unlike classical logic, there connectives do not form a basis for "MaxEnt" probabilistic logic. This does not mean that this new "logic" does not have a finite basis: due to Theorem 2, we can express every function $f_F(a, ..., b)$ in terms of $\&$, $\neg$, and a special partially defined operation $\dot\lor$ ("disjoint union") for which $f_{\dot\lor}(a,b) = a + b$.

Comment 3: MaxEnt Can Also Handle Additional Interval Uncertainty. In practice, for statements E_i from the knowledge base, we often know only approximate values of the probabilities p_i, e.g., *intervals* $[p_i^-, p_i^+]$ of their possible values. In this case, we can first select the probabilities p_i from MaxEnt.

This additional uncertainty is often described by a Dempster-Shafer formalism. For this case, MaxEnt approach is also helpful (see, e.g., Nguyen Walker [10]).

2. Intelligent Control

The Main Problem. If we use an expert system to describe a real-valued variable A, then, for each possible value u of this variable, we can compute the probability $p_A(u)$ that the value A is consistent with the experts' knowledge. Based on this information, a specialist makes a decision (e.g., a doctor chooses whether to operate or not).

If we want an expert system to be used in *control* (e.g., in controlling a space-ship), then, for each possible control value u, we can also get the probability $p(u)$ that u is reasonable, but for control, we often do not have time to ask a specialist. The system must make a decision automatically, all by itself. Such knowledge-based automatic control systems are called *intelligent control* systems. For intelligent control systems, we arrive at the following problem: How can we translate the probabilities $p(u)$ into a single control value $\bar{u}$?

MaxEnt Helps. These probabilities describe the (unknown) set U of control values that are consistent with our knowledge. Therefore, we have a *probability measure* (*random set*) $P(U)$ on the class of all such sets U. Probabilities $p(u)$ are then interpreted as $P(u \in U)$. These probabilities do not determine $P(U)$ uniquely, so, we have to use MaxEnt. MaxEnt leads to the measure in which $u \in U$ and $u' \in U$ are independent events for $u \neq u'$.

Even if the set U of possible values is fixed, we still do not know the probabilities of different values of $u \in U$. To get a reasonable probability distribution $P(u|U)$ on U, we can use MaxEnt again. (E.g., for finite U, we get each $u \in U$ with equal probability.) As a result, we can find the probability $p_{\text{act}}(u)$ of u being actually the best control as $\sum_U P(U)P(u|U)$ (in the finite case, and $E_U(p(u|U))$ in the infinite case). Now, for each value u, we know the probability of this u being the best, so, we can choose $\bar{u}$ as the mathematical expectation of u w.r.t. this distribution. These definitions are easily formalized in a finite case (when u only takes finitely many possible values $u_1, ..., u_m$), and can be extended to $u \in R$ by a limit $m \to \infty$.

How to compute this $\bar{u}$? Due to lack of space, we skip the technical details, and only present the following result:

Result: *For a continuous function $p(u)$, $\bar{u} = \int up(u)\,du / \int p(u)\,du$.*

Idea of the proof: For the finite approximation, when we have N points u_i per unit length, for each choice of U, the average of u is simply $\sum u/\#u$. Since p is continuous, it can be approximated (with arbitrary accuracy) by piecewise-constant functions. On each interval of constancy $I = [\tilde{u} - \Delta u/2, \tilde{u} + \Delta u/2]$, of length Δu, the probability of $u \in I$ to be chosen is the same for all u, and the events of choosing or not choosing two different elements are statistically independent. Totally, we have $\approx N\Delta u$ points u_i on I; therefore, due to large numbers theorem, $\approx p \cdot \Delta u \cdot N$ points will be chosen, and the ratio of the number of chosen points to $\Delta u \cdot N$ tends to p. Since all points $\in I$ are approximately equal to $\tilde{u}$, the total contribution of points from I to $\sum u$ is $\approx \tilde{u} \cdot p(\tilde{u}) \cdot \Delta u \cdot N$, and the total contribution to $\#u$ is $\approx p(\tilde{u}) \cdot \Delta u \cdot N$. Thus, $\sum u \approx N[\sum \tilde{u} \cdot p(\tilde{u}) \cdot \Delta u]$, and $\#u \approx N[\sum p(\tilde{u}) \cdot \Delta u]$. The resulting approximating expressions do not depend on U at all. In the limit, when $N \to \infty$ and $\Delta u \to 0$, both sums (divided by N) turn into the desired integrals. Q.E.D.

The above formula for $\bar{u}$ is actively used in intelligent control (see, e.g., [9]). Thus, *MaxEnt justifies one of the basic formulas used in intelligent control.*

Choice of &– and ∨–Operations Revisited. Since for intelligent control, all we are interested in is u, it makes sense to choose $f_\&$ and $f_\vee$ from the condition that the entropy of the resulting distribution for u is the largest possible. This control-oriented approximating use of MaxEnt leads to different results than the one used for expert systems: namely, the corresponding MaxEnt operations are $f_\&(a, b) = \min(a, b)$ and $f_\lor(a, b) = \min(a + b, 1)$ (for proofs, see [12,13]).

These operations also have a direct control sense: e.g., in many control problems (e.g., in tracking the Space Shuttle), we want the control that leads to the largest *stability* in the sense that if we deviate a little bit from the desired trajectory, we want to guarantee the fastest possible return. It turns out that the above-described MaxEnt operations are exactly the ones that lead to the maximally stable control [9,14]. Thus, *MaxEnt leads to the maximum stability of the controlled system.*

MinEnt? Stability is not the only desired property of control. The most stable system operates very fast. In many control problems, e.g., in docking the Space Shuttle to a space station, such a speed may be dangerous: we can damage the station if we hit it fast. In such cases, we want the *smoothest* possible control. It turns out [9,14] that the smoothest control is attained when we choose the operations $f_\&(a, b) = ab$ and $f_\lor(a, b) = \min(a, b)$. These operations turn out to correspond to the ... *minimal* entropy [12,13].

References

1. P. Cheeseman, "In Defense of Probability", in *Proceedings of the 8-th International Joint Conference on AI*, pp. 1002–1009, Los Angeles, CA, 1985.
2. E. Y. Dantsin and V. Kreinovich, "Probabilistic inference in prediction systems," *Soviet Mathematics Doklady*, 1990, 40(1), pp. 8–12, 1990.
3. C. Elkan, "The paradoxical success of fuzzy logic", in *Proceedings of AAAI-93, American Association for Artificial Intelligence 1993 Conference*, pp. 698–703.
4. G. J. Klir and Bo Yuan, *Fuzzy Sets and Fuzzy Logic*, Prentice Hall, NJ, 1995.
5. L. J. Kohout and W. Bandler, "Fuzzy Interval Inference Utilizing the Checklist Paradigm and BK-Relational Products", pp. 289–334, in R. B. Kearfott and V. Kreinovich, eds., *Applications of Interval Computations*, Kluwer, Boston, MA, 1995 (to appear).
6. V. Kreinovich, "Group-theoretic approach to intractable problems," *Lecture Notes in Computer Science* (Springer-Verlag, Berlin), **417**, pp. 112–121, 1990.
7. V. Kreinovich et al., "Monte-Carlo methods make Dempster-Shafer formalism feasible," in *Advances in the Dempster-Shafer Theory of Evidence*, R. R. Yager et al., eds., pp. 175–191, Wiley, N.Y., 1994.
8. V. Kreinovich and S. Kumar. "Optimal choice of &- and ∨-operations for expert values," in *Proceedings of the 3rd University of New Brunswick Artificial Intelligence Workshop*, pp. 169–178, Fredericton, N.B., Canada, 1990.
9. V. Kreinovich et al., "What non-linearity to choose? Mathematical foundations of fuzzy control", in *Proceedings of the 1992 International Conference on Fuzzy Systems and Intelligent Control*, Louisville, KY, 1992, pp. 349–412.
10. H. T. Nguyen and E. A. Walker, "On decision making using belief functions", in *Advances in the Dempster-Shafer Theory of Evidence*, R. R. Yager, J. Kacprzyk, and M. Pedrizzi, eds., pp. 311–330, Wiley, N.Y., 1994.
11. J. Pearl, *Probabilistic Reasoning in Intelligent Systems*, Morgan Kaufmann, San Mateo, CA, 1988.
12. A. Ramer and V. Kreinovich, "Maximum entropy approach to fuzzy control", *Information Sciences*, **81**(3–4), pp. 235–260, 1994.
13. A. Ramer and V. Kreinovich, "Information complexity and fuzzy control", in *Fuzzy Control Systems*, A. Kandel and G. Langholtz, eds., pp. 75–97, CRC Press, Boca Raton, FL, 1994,
14. M. H. Smith and V. Kreinovich. "Optimal strategy of switching reasoning methods in fuzzy control", in *Theoretical aspects of fuzzy control*, H. T. Nguyen, M. Sugeno, R. Tong, and R. Yager, eds., pp. 117–146, J. Wiley, N.Y., 1995.

THE DE FINETTI TRANSFORM

S. JAMES PRESS
Department of Statistics
University of California, Riverside[†]

Abstract. Bayesian analysis of a problem requires that subjective information be introduced into the model. Sometimes it is more useful or convenient to bring that information to bear through the prior distribution; other times it is preferred to bring it to bear through the predictive distribution. This paper develops procedures for assessing exchangeable, proper, predictive distributions that, when desired, reflect "knowing little" about the model; i.e., exchangeable, maximum entropy, distributions. But what do such distributions imply about the model and the prior distribution? We invoke de Finetti's theorem to define the "de Finetti transform", which permits us, under many frequently occurring conditions (such as natural conjugacy), to find the unique sampling density (conditional on some indexing parameters), and the unique associated prior distribution for those parameters.

1. Introduction

This paper is a condensation of Press, 1995. We begin with de Finetti's theorem (see de Finetti, 1937) in Section 2; we define the de Finetti transform in Section 3, where we also provide a brief table of such transforms; we provide a table of exchangeable maximum entropy distributions in Section 4; and we show how to apply de Finetti transforms for assessing maxent predictive distributions in Section 5.

2. de Finetti's Theorem

We concentrate on the special case where there are densities with respect to Lebesque or counting measures. Let $X_{(1)}, \ldots, X_{(n)} \ldots$ denote an exchangeable sequence of real valued random variables in $\mathcal{R}^p$. Let $X_1, \ldots, X_n$ denote a finite sequence of size n. Once the predictive density h is specified, there exists a unique conditional sampling density $f(x|\omega)$, and a unique cdf $G(\omega)$, a prior (mixing) distribution, such that h may be expressed as in eqn.(2), as the average of the joint

---

[†]
Email: VXPRESS@UCRAC1.UCR.EDU

K. M. Hanson and R. N. Silver (eds.), Maximum Entropy and Bayesian Methods, 101–108.

density of n i.i.d. random variables, conditional on a parameter ω, weighted by a prior (mixing) distribution $G(\omega)$. Symbolically, we have, a.s.,

$$h(x_1, \ldots, x_n) = \int f(x_1 \mid \omega) \cdots f(x_n \mid \omega) dG(\omega), \qquad (1)$$

with

$$G(\omega) = \lim_{n \to \infty} G_n(\omega),$$

and $G_n(\omega)$ denotes the empirical cdf of $(X_1, \ldots, X_n)$. The representation in (2) must hold for all n (this assertion follows from the fact, subsumed within the definition of exchangeability, that there is an infinite sequence of such random variables). Note that x and/or ω may be vector valued.

Conversely, the theorem asserts that if the conditional sampling density f and the prior G are specified, then h is determined uniquely.

If the prior parameter ω is also absolutely continuous with respect to Lebesgue measure, with density $g(\omega)$, we replace $dG(\omega)$ by $g(\omega)d\omega$. From here on, except for counting measure cases, we assume generally that we have the special case of the $f(\omega)$ form in eqn. (3), a.s.,

$$h(x_1, \ldots, x_n) = \int f(x_1 \mid \omega) \cdots f(x_n \mid \omega) g(\omega) d\omega, \qquad (2)$$

and

$$g(\omega) = \frac{d}{d\omega} [\lim_{n \to \infty} G_n(\omega)] .$$

3. The de Finetti Transform

GENERAL

Referring to eqn (3), we think of $h(x)$ as the transform of $g(\omega)$, for preassigned $f(x \mid \omega)$, and correspondingly, $g(\omega)$ as the inverse transform of $h(x)$.

We can relate $g(\omega)$ & $h(x)$ in various ways, depending upon the objective in a given problem. For example, if there is substantive information about $h(x)$ we can assess a predictive distribution for it (see Kadane et al., 1980), and then attempt to find the inverse transform $g(\omega)$ that corresponds (for preassigned $f(x \mid \omega)$).

In some situations, we might know very little about $h(x)$, and might therefore, wish to invoke maxent considerations to specify $h(x)$ as the density of a maxent distribution (see Section 4).

In other situations, $g(\omega)$ or $h(x)$ might be assessed in other ways. There might, for example, be moment-type constraints, tail-type constraints, probability content constraints, algebraic or differential equation-type constraints, or whatever. In any case, we can introduce them through the de Finetti Transform.

In Table 1, we present a brief summary of some readily calculable de Finetti transform pairs for problems in which the model for the data, $f(x|\omega)$, might be binomial, normal, Poisson, exponential, gamma, or Pareto. Prior distributions are often taken to be natural conjugate. All predictive and prior densities given in Table 1 are proper, and relate to natural conjugate families.

TABLE 1: A BRIEF TABLE OF DE FINETTI TRANSFORMS		
MODEL	TRANSFORM	INVERSE TRANSFORM
# Sampling Density $f(x\mid\text{parameter})$	Predictive Density $h(x)$ Kernel	Prior Density Kernel $g(\text{parameter})$
1 binomial $\binom{n}{r}p^r(1-p)^{n-r}$ $r=0,1,\dots,n$ (unknown p)	discrete uniform $h(r)=\frac{1}{n+1}$, $r=\Sigma_1^n x_i,\ \ r=0,1,\dots,n$	uniform $g(p)=1,\ \ 0<p<1$
2 binomial $\binom{n}{r}p^r(1-p)^{n-r}$ $r=0,1,\dots,n$ (unknown p)	beta-binomial $\frac{\binom{n}{r}\Gamma(\alpha+\beta)\Gamma(\alpha+r)\Gamma(n+\beta-r)}{\Gamma(\alpha+\beta+n)\Gamma(\alpha)\Gamma(\beta)}$ $r=\Sigma_1^n x_i,\ \ r=0,1,\dots,n$	beta $p^{\alpha-1}(1-p)^{\beta-1},0<p<1$ $0<\alpha,\ \ 0<\beta$
3 normal $N(0,\sigma^2)$ (unknown σ^2)	multivariate Student t $\frac{1}{\{\nu-2+x'\Sigma^{-1}x\}^{(\nu-2+n)/2}}$ $\Sigma=(\tau/\nu-2)I_n$	inverted gamma $\frac{1}{(\sigma^2)^{\nu/2}}e^{-\tau^2/2\sigma^2}$
4 normal $N(\theta,\tau^2)$ (τ^2-known, θ-unknown)	normal $N(\phi e_n,\Sigma)$ $\Sigma=\sigma^2[(1-\rho)I_n+\rho e_n e_n']$, $\sigma^2=\tau^2+\omega^2$, $\rho=\frac{\omega^2}{\tau^2+\omega^2}$ $e_n=(1,\dots,\dots,1)',e_n:(n\times 1)$	normal $N(\phi,\omega^2)$
5 normal $N(\theta,\sigma^2)$ (θ,σ^2 both unknown)	multivariate Student t $\frac{1}{\{\tau^2+(x-\alpha)'\Sigma^{-1}(x-\alpha)\}^{\frac{n+\nu-2}{2}}}$ $\Sigma=I_n+e_n e_n'$ $e_n=(1,\dots,1)':(n\times 1)$ $\alpha=\frac{\phi}{n+1}\Sigma e_n$	normal-inverted gamma $(\theta\mid\sigma^2)\sim N(\phi,\sigma^2)$ $\sigma^2\sim(\sigma^2)^{-\frac{\nu}{2}}e^{-\tau^2/2\sigma^2}$
6 Poisson $\frac{e^{-\lambda}\lambda^x}{x!}$ (unknown λ)	$\frac{\Gamma[\alpha-1+\Sigma_1^n x_i]}{(n+\beta)^{\alpha+\Sigma_1^N x_i-1}\prod_1^n\Gamma(x_i+1)}$	gamma $\lambda\sim\lambda^{\alpha-1}e^{-\beta\lambda},\ \ 0<\lambda$ $0<\alpha,\ \ 0<\beta$

TABLE 1 (cont.): A BRIEF TABLE OF DE FINETTI TRANSFORMS		
MODEL	TRANSFORM	INVERSE TRANSFORM
# Sampling Density $f(x|\text{parameter})$	Predictive Density $h(\underline{x})$ Kernel	Prior Density Kernel $g(\text{parameter})$
7 exponential $\lambda e^{-\lambda x}$ (unknown λ)	$\dfrac{\beta^{\alpha}\Gamma(\alpha+n)}{\Gamma(\alpha)(\beta+\Sigma_1^n x_i)^{\alpha}}$	gamma $\dfrac{\beta^{\alpha}}{\Gamma(\alpha)}\lambda^{\alpha-1}e^{-\lambda\beta}$
8 gamma $\dfrac{\beta^{\alpha}x^{\alpha-1}e^{-\beta x}}{\Gamma(\alpha)}$ (known - α and unknown - β)	$\dfrac{b^a(\prod_1^n x_i)^{\alpha-1}\Gamma(n\alpha+a)}{\Gamma^n(\alpha)\Gamma(a)(b+\Sigma_1^n x_i)^{n\alpha+a}}$	gamma $\dfrac{b^a\beta^{a-1}e^{-b\beta}}{\Gamma(a)}$
9 Pareto $\dfrac{ax_0^a}{x^{a+1}}, \quad x>x_0$ (unknown shape a) (known threshhold x_0)	$\dfrac{\beta^{\alpha}\Gamma(n+\alpha)}{\Gamma(\alpha)(\prod_i^n x_i)\left[\beta+\log\left(\dfrac{\prod_i^n x_i}{x_0^n}\right)\right]^{n+\alpha}}$ for $\min_i(x_i)>x_0$	gamma $\dfrac{\beta^{\alpha}}{\Gamma(\alpha)}a^{\alpha-1}e^{-\beta a}$

4. Maxent Distributions and Information

SHANNON INFORMATION

There are some statisticians who would not prefer to put distributions on parameters, but would rather put distributions only on observables. Such people may prefer to select a predictive density, $h(x)$, that expresses the idea of knowing little, that is a maximum entropy (maxent) or minimum information distribution. They could accordingly examine a table of de Finetti transforms and select an $h(x)$, and therefore, a prior $g(\theta)$, and a sampling density $f(x|\theta)$, that correspond closely to the situation at hand.

The basic notion of "information" used here was introduced by Shannon (1948), and was explicated, expanded, and applied by many others. The Shannon "entropy" in a distribution with pmf $h(x)$ is defined for discrete random variables (Shannon originally discussed entropy and information for discrete random variables) as:

$$H[h(x)] = -\sum_{i=1}^{n} h(x_i)\log h(x_i) = -E[\log h(x)] . \qquad (3)$$

Once $h(x)$ has been fixed by maxent considerations, for all $n, g(\theta)$ (and also $f(x|\theta)$) are uniquely determined by de Finetti's theorem. But finding the corresponding $g(\theta)$ for a given $h(x)$ may not always be an easy task (very generally, it's a problem in the theory of "integral equations").

CHARACTERIZING $h(x)$ AS A MAXIMUM ENTROPY DISTRIBUTION

It remains to discuss how to specify that a preassigned $h(x)$ is the density of a maxent distribution. This is the opposite problem to the usual one of finding the maxent distribution subject to some prespecified constraints. Our problem here is: given a preassigned density function that we would like to be the density of a maxent distribution, are there constraints that would produce such a result, and if so, what are they? Problems of this type are commonly approached using the Calculus of Variations.

First of all, there is no guarantee that constraints will always exist to make any preassigned distribution a maxent distribution. But it turns out that we can do it for many distributions (including the ones with common names that are usually of interest as natural conjugate distributions; see Table 2). We will adopt "moment-type" constraints in all cases.

For convenience we have collected in Table 2 some of the distributions in the maxent family for n exchangeable variables, along with the moment-type constraints that must be imposed to yield them. In general, we must have the maxent density integrate to unity as the primary constraint (or the pmf must sum to unity, in the discrete case). In addition, however, we usually have one or more secondary constraints that must be imposed. Finally, we must have the distribution invariant with respect to permutations of the subscripts.

In summary, for many situations, we are now able to specify moment-type constraints on a preassigned unconditional distribution with density $h(x)$, such that h corresponds to "knowing little", i.e., a maxent distribution. Then, by taking the inverse de Finetti transform we can determine the proper prior distribution and the sampling distribution that correspond. Conversely, for a given sampling distribution, and a given proper prior distribution (perhaps a natural conjugate distribution) we can often find the maxent distribution that corresponds.

\#	Maxent Distribution pdf/pmf	Secondary Constraints	Domains
	TABLE 2: EXCHANGEABLE MAXIMUM ENTROPY DISTRIBUTIONS		
1	**Uniform** $h(x) = $ constant		$x = (x_i),$ $\alpha_i < x_i < b_i$ $i = 1, \ldots, n$
2	**Exponential** $h(x) = \phi^n e^{-\phi \Sigma_1^n x_i}$	$EX_i = \phi, \quad i = 1, \ldots, n$	$0 < x_i < \infty$
3	**Normal** $h(x) = \frac{1}{(2\pi)^{n/2}|\Sigma|^{1/2}}$ $exp\{-\frac{1}{2}(x-\theta)'\Sigma^{-1}(x-\theta)\}$ for $\theta \equiv \phi e_n, \quad e_n = (1,\ldots,1)',$ $e_n : (n \times 1),$ $\Sigma \equiv \sigma^2[(1-\rho)I_n + \rho e_n e_n']$ $-\frac{1}{n-1} < \rho < 1$	$EX = \theta$ $var(X) = \Sigma$	$x = (x_i),$ $-\infty < x_i < \infty,$ $i = 1, \ldots, n$ $\theta = (\theta_i)$ $-\infty < \theta_i < \infty$ $i = 1, \ldots, n,$ $\Sigma > 0.$
4	**Student t** $h(x) = \frac{C_n |\Sigma|^{-\frac{1}{2}}}{[\nu + (x - \alpha e_n)'\Sigma^{-1}(x-\alpha e_n)]^{\frac{\nu+n}{2}}}$ $e_n : (n \times 1), \quad e_n \equiv (1,\ldots,1)',$ $\Sigma = \sigma^2[(1-\rho)I_n + \rho e_n e_n'],$ $-\frac{1}{n-1} < \rho < 1$	$E\log[\nu + (x-\alpha e_n)'\Sigma^{-1}(x-\alpha e_n)]$ $=$ constant	$x = (x_i),$ $-\infty < x_i < \infty$ $i = 1, \ldots, n$ $\Sigma > 0$
5	**Dirichlet** $h(x|\phi_0, \phi_1) \propto$ $[\prod_{j=1}^n x_j^{\phi_0 - 1}](1 - \Sigma_1^n x_j)^{\phi_1 - 1}$	$E[\log X_j] = g_0,$ $E[\log(1 - \Sigma_1^n X_j)] = g,$ for constants g_0, g	$x = (x_j),$ $j = 1, \ldots, n$ $\Sigma_1^n x_j < 1$ $0 < \phi_0, \quad 0 < \phi_1$
6	**Weibull** $h(x) = a^n \prod_1^n x_i^{a-1} e^{-\Sigma_1^n x_i^a}$	$E[X_i^a] = 1$ $E[\log X_i] = g$	$0 < x_i < \infty$ $x = (x_i)$ $0 < a$
7	**Cauchy** $h(x) = \frac{C_n |\Sigma|^{-\frac{1}{2}}}{[1 + (x - \alpha e_n)'\Sigma^{-1}(x-\alpha e_n)]^{\frac{n+1}{2}}}$ $e_n : (n \times 1), \quad e_n = (1,\ldots,1)',$ $\Sigma = \sigma^2[(1-\rho)I_n + \rho e_n e_n']$ $-\frac{1}{n-1} < \rho < 1, \quad C_n = \frac{\Gamma(\frac{n+1}{2})}{\pi^{\frac{n+1}{2}}}$	$E[\log\{1 + (x - \alpha e_n)'\Sigma^{-1}(x-\alpha e_n)\}]$ $=$ constant	$x = (x_i)$ $-\infty < x_i < \infty$ $i = 1, \ldots, n,$ $\Sigma > 0$

#	Maxent Distribution pdf/pmf	Secondary Constraints	Domains
	TABLE 2 (cont.): EXCHANGEABLE MAXIMUM ENTROPY DISTRIBUTIONS		
8	Wishart x_i are i.i.d. $N(0,\Sigma)$ $V = \Sigma_1^\nu x_i x_i'$, $x_i : (n \times 1)$ $h(V) = h(x_1, \ldots, x_\nu)$ $= \dfrac{C\lvert V\rvert^{(\nu-n-1)/2}e^{-\frac{1}{2}tr\Sigma^{-1}V}}{\lvert\Sigma\rvert^{\nu/2}}$ $C^{-1} = 2^{\frac{\nu n}{2}}\Gamma_n(\nu/2)$, $\Sigma = \sigma^2[(1-\rho)I_n + \rho e_n e_n']$, $-\frac{1}{n-1} < \rho < 1$	$E[\frac{V}{\nu}] = $ constant $E[\log \lvert V\rvert] = $ constant	$V_i(n \times n) > 0$ $\nu \geq n$ $\Sigma : (n \times n) > 0$
9	Gamma $h(x) = C_n(\prod_1^n x_i^{\alpha-1})e^{-\beta\Sigma_1^n x_i}$, $C_n = \frac{\beta^{n\alpha}}{\Gamma^n(\alpha)}$	$EX_i = $ constant $E[\log X_i] = $ constant $i = 1, \ldots, n$	$0 < x_i < \infty$ $x = (x_i)$ $0 < \alpha$ $0 < \beta$
10	Laplace $h(x) = \frac{a^n}{2^n}e^{-a\Sigma_1^n\lvert X_i\rvert}$	$E\{\lvert X_i\rvert\} = $ constant $i = 1, \ldots, n$	$x = (x_i)$ $-\infty < x_i < \alpha$ $a > 0$.
11	Poisson $h(x) = e^{-\lambda}\lambda^{\Sigma_{i=1}^n x_i}/\prod_{i=1}^n(x_i!)$	$E(X_i) = \lambda_1 = $ constant $E\log(X_i!) = \lambda_2 = $ constant	$x_i = 0, 1, 2, \ldots$ $\lambda > 0$.

Note that in Table 2, all distributions in the maxent family must satisfy the primary constraint that they integrate, or sum, to unity; and $\Gamma_n(a)$ denotes the multivariate gamma function, where $\Gamma_n(a) = \pi^{n(n-1)/4}\prod_{j=1}^n \Gamma(a - \frac{i-1}{2})$.

5. Applying de Finetti Transforms

Suppose $(X, \ldots, X_n)$ denotes a finite subset of observable variables from an (infinitely) exchangeable set; this is our data. Suppose further that $f(x\lvert\omega)$ is an appropriate model for the data. The next question is, how should we choose $g(\omega)$ and $h(x)$?

In some situations we might wish to choose $g(\omega)$ and then find the corresponding $h(x)$; in others, we might wish to choose $h(x)$, and then find $g(\omega)$ as the inverse de Finetti transform. We might elect to choose $h(x)$ so that the choice is consistent with "knowing very little" about the predictive distribution of the data, i.e., we want to select a maxent distribution for $h(x)$.

In such a situation we turn to Table I to find the model corresponding to $f(x\lvert\omega)$. Suppose, for example, the model we decide upon is $N(\theta, \sigma^2)$, for both θ and σ^2 unknown. Referring to Table 1, we see that for the normal model with both parameters unknown (Case 5), the de Finetti transform density $h(x)$ is the special multivariate Student t-density shown.

Next, go to Table 2 (Case 4) and note that the multivariate Student t- distribution is maxent for $E\{\log[\nu + (x - \alpha e_n)'\Sigma^{-1}(x - \alpha e_n)]\} = $ constant.

If we believe subjectively that this moment constraint is reasonable in this problem, we can adopt the multivariate Student t-distribution as the maxent distribution for $h(x)$. Then note from Table 1 that the inverse de Finetti transform is

$$g(\omega) = g(\theta, \sigma^2) = g_1(\theta|\sigma^2)g_2(\sigma^2) , \qquad (4)$$

where:

$$\theta|\sigma^2 \sim N(\phi, \sigma^2), \quad g_2(\sigma^2) \propto \frac{1}{(\sigma^2)^{\nu/2}}e^{-\tau^2/2\sigma^2} .$$

Now we must merely assess the hyperparameters (ϕ, ν, τ^2). This might be accomplished using either $g(\theta, \sigma^2)$ or $h(x)$, perhaps with the assistance of the method of Kadane et al., 1980.

For discussion of arbitrary priors not presented in Table 2, see Press, 1995.

References

1. de Finetti. La Prevision: ses lois logique, ses sources subjectives. *Ann. d'Institut Henri Poincare*, Vol. 7, pp. 1-68. Translated in *Studies in Subjective Probability* H. Kyberg and H. Smokler (eds.), New York: John Wiley and Sons, 1964.
2. Kadane J.B., Dickey J.M., Winkler R.L., Smith W.S., and Peters S.C. Interactive Elicitation of Opinion of a Normal Linear Model. *Jour. Amer. Stat. Assn.*, 75, 845-854, 1980.
3. Press S. James The de Finetti Transform. Technical Report #201R, Department of Statistics, University of California, Riverside, 1995.
4. Shannon C.E. The Matahematical Theory of Communication. *Bell System Technical Journal*, July-October 1948, reprinted in C.E. Shannon and W. Weaver, *The Mathematical Theory of Communication*, University of Illinois Press, 3-91, 1949.

CONTINUUM MODELS FOR BAYESIAN IMAGE MATCHING[†]

J. C. GEE
Department of Computer and Information Science
University of Pennsylvania, Philadelphia, PA 19104

AND

P. D. PERALTA
Department of Materials Science and Engineering
University of Pennsylvania, Philadelphia, PA 19104

Abstract. The task of determining the mapping between a pair of images is called image matching and is fundamental in image processing. Prior information is essential to the inference of the mapping because the image features on which matching is based are sparsely distributed and, consequently, underconstrain the problem. In this paper, we describe the Bayesian approach to image matching and introduce suitable priors based on idealized models of continua.

Key words: Bayesian modeling, cerebral anatomy, continuum mechanics, image matching, standard regularization

1. Introduction

Given a pair of images, the object of image matching is to infer the mapping between them. Thus, for each point in one image, this mapping specifies its corresponding point in the second image. Such mappings are required in solutions to a variety of problems in image analysis [1]. For example, to track objects over time–to infer motion or to measure shape change–one must determine the correspondence between successive frames captured with a stationary camera; images of the same scene simultaneously observed by different sensors usually are first fused into the same coordinate space to facilitate their interpretation; and in stereo vision the inference of depth within a scene is accomplished by matching images acquired from different viewpoints. In the latter two examples, the mappings usually take the form of rigid-body motions, sometimes generalized to globally affine or projective transformations to accomodate foreshortening, shear, or scaling effects. The matching problem becomes significantly more difficult when we relax the assump-

[†]This work was supported in part by the U.S.P.H.S. under grant 1-RO1-NS-33662-01A1.

K. M. Hanson and R. N. Silver (eds.), Maximum Entropy and Bayesian Methods, 109–116.

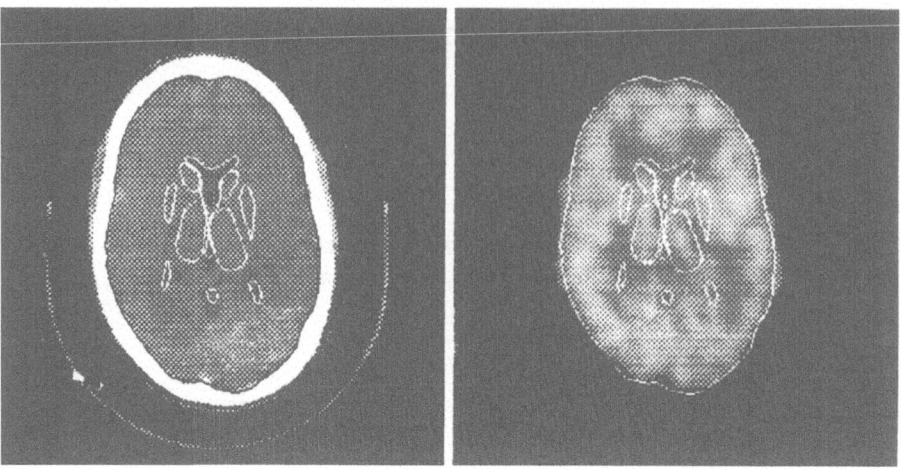

Figure 1.   Corresponding CT and PET brain slices of a patient superimposed with outlines of the cortical surface and several subcortical structures. The outlines were obtained by deforming an anatomical template of the brain to match the CT brain image volume [2].

tion of homogeneous motion and allow the mappings to vary over space and time. In other words, we suppose the objects are deformable in the tracking application.

Another important class of problems for which such flexibility in the mappings is useful involves classification of images by means of pattern recognition. The general idea is to fit a prototypical pattern to objects within a given scene, thus labeling them with attributes associated with the prototype. Note that, unlike the examples above, the images here may not be derived from the same scene. Nonetheless, our intuitive notion of the similarity between instances of a prototype can be formally specified to enable its computer implementation.

Our interest lies with complex prototypes or templates of human cerebral anatomy and their mappings onto the anatomy of individual brains that have been visualized through, say, magnetic resonance imaging (MRI) or X-ray computed tomography (CT). Figure 1 illustrates one such computer interpretation of a pair of CT and positron emission tomography (PET) brain scans. The mapped template, shown as a set of contours in the figure, facilitates quantitative analysis of the images, and makes possible many additional useful applications.

In the next section, the Bayesian approach to image matching is described, in which the unknown mapping is viewed as a stochastic process. We then review a *priori* constraints that have been imposed in related problems in computational vision. Their generalization to continuum models is discussed, and we conclude by briefly commenting on some additional issues. Although our discussion excludes methods for performing global affine registration, they are nearly always used to generate the estimates with which more complex matching algorithms are initialized. Finally, to meet the manuscript length requirement, we have not included numerical results, and have cited some of the relevant work instead.

2. Bayesian image matching

Our Bayesian formulation of the image matching problem follows the general framework Besag eloquently lays out for Bayesian image analysis in [3]. The observations comprise the given pair of pixel images $\{I_O, I_R\}$, where generic $I = \{I_k : k \in T\}$ and T represents, say, a rectangular array in the case of two-dimensional pictures. The two images assume the following relationship:

$$I_O(x) = H(I_R(X)) \oplus N, \tag{1}$$

where H is possibly a blurring transformation, $\oplus N$ represents additive or multiplicative noise, and $x = x(X)$ defines the mapped position of point X. It follows that the point X undergoes displacement $u(X) = x(X) - X$. Under the simplified conditions of zero blur and additive Gaussian noise with covariance Σ and mean zero, the likelihood function is given by the conditional density

$$l(I_O, I_R \mid u) \propto \exp\left\{-\frac{1}{2}(I_O(X+u) - I_R(X))^T \Sigma^{-1}(I_O(X+u) - I_R(X))\right\}. \tag{2}$$

When specialized to pattern matching, I_R in (1) is identified with the template. Thus, each realization I_O of the brain template, for example, combines elements of randomness induced by normal variation in neuroanatomies and by stochastic degradation in the sensing process. As usual, our knowledge about anatomic variability is introduced through the prior distribution $\{p(u)\}$ for the displacements u. Grenander pioneered the probabilistic interpretation of these flexible templates [4,5]. A continuum mechanical analog was developed by Broit in [6].

The importance of the prior in the matching problem is made evident when we consider the calculation of the maximum likelihood estimate of u. Because the image features on which matching is based are sparsely distributed—for example, within the white matter or cerebrospinal fluid spaces—the problem is inherently underconstrained. This difficulty cannot be circumvented by choosing as the point estimate the posterior mean, because the sample displacements from the implied uniform prior violate our expectations about their form, such as local smoothness. In the next section, consideration is given to plausible characteristics that the mappings should exhibit and to how these features can be modeled.

Suppose an appropriate prior $\{p(u)\}$ has been specified, the posterior is the conditional density of u given $\{I_O, I_R\}$:

$$P(u \mid I_O, I_R) \propto l(I_O, I_R \mid u)\, p(u).$$

Following decision theory [7], point estimates of u are inferred by minimizing an expected loss with respect to the posterior. The most common summaries of the posterior are its mode and mean, which correspond to optimal actions under the zero-one and quadratic loss functions, respectively. Estimation of the posterior mode or maximum a *posteriori* solution is complicated in image matching by the nonlinear nature of the (log) likelihood term. We describe in [8] an iterative but deterministic scheme to search for the MAP solution by replacing the likelihood with its linearized or quadratic approximation at successive estimates. To estimate the mean, samples are drawn from a series of Gaussian approximations to the posterior. In a preliminary study [9], the two estimators were found to be equally accurate in recovering an artificial mapping under various degradation conditions; the uncertainty in their values was explored in [10]. The computation of these kinds of Bayesian estimates through Monte Carlo methods is the subject of active research interest [11].

3. Prior distributions for image mappings

We began by enumerating several examples in which the matching problem figures importantly. The optical flow and stereo problems are representative of a wider class of ill-posed problems in early vision that have been approached with good results through standard regularization theory and its generalizations [12,13]. In the next section, we briefly review the method and note its relation to Bayesian estimation. As would be required in standard regularization, it is not unreasonable that our image mappings should in general be smooth. Indeed, for the special case of template matching, expected deviations from the prototype are typically idealized to assume the form of continuous, one-to-one mappings. The application of such ideas to the modeling of biological variation originated with Thompson [14]—see also Murray [15], especially chp. 18, and Rohlf and Bookstein [16] for contemporary views of morphogenesis and morphometrics, respectively. Once this idealization is accepted, it is natural to consider the images as continua and to analyze their mappings in terms of deformation kinematics. Such an approach was first studied by Broit [6], who applied it to the brain template matching problem of immediate interest here. Broit represented the template as another image volume but endowed it with elasticity. We describe a simple example of a continuum model in Section 3.2 and then consider the treatment of large deformations in Section 3.3.

3.1. STANDARD REGULARIZATION

The general idea of the most widely applied regularization method is to reformulate the given problem in terms of a variational principle [17]. The associated functional comprises a term which enforces fidelity of the solution to the data and a stabilizer through which a *priori* constraints on the solution are imposed. The influence of each term on the final solution is balanced against the other by varying a weighting or regularization parameter. In the original development of standard Tikhonov regularization, the stabilizing functional is given by the following Sobolev norm:

$$\|C\,u\|^2 = \sum_{m=0}^{p} \int_{\Re} w_m(x) \left(\frac{\partial^m u(x)}{\partial x^m} \right)^2 dx,$$

where the weighting functions w_m are strictly positive and continuous. Similar constraints on the smoothness of solutions for early vision problems have been motivated by physical considerations. For example, to reconstruct visual surfaces, a surface consistency constraint is introduced in [18] and then carefully shown to imply minimization of the quadratic variation or thin-plate spline functional.

There is a ready Bayesian interpretation of standard regularization: the norm measuring the compatibility of the estimate with the observations corresponds to the likelihood (consider, for example, $-\ln l(I_O, I_R | u)$ in (2)); and the stabilizer can be viewed as the energy of a Gibbs prior which assigns small probability to "rough" solutions. It is evident that the regularized and MAP solutions are equivalent.

Gibbs modeling of physical constraints, first advocated by Grenander [4] and Geman and Geman [19], has become an indispensable technique in practical Bayesian image analysis. It enables an arbitrary stabilizing functional to be associated with a probability distribution. For example, the prior models defined by the membrane [20] and thin-plate energies are correlated Gaussians [21]—see also their relation to the pairwise difference priors described in [3,11].

3.2. CONTINUUM MODELS

A body experiences strain when the relative position of its points are changed. The change in position of the points or deformation, effected by some motion $x = x(X, t)$, results in a new configuration of the body at time t, where $X = x(X, t_0)$ describes the original undeformed configuration. Points within the body may also be displaced without deformation; they are said to undergo rigid motion.

Consider the displacement of the material point P to position $x = X + u(X, t)$ in Figure 2. Continuum theory requires every mapping $x = x(X, t)$ to be one-to-one and its Jacobian determinant everywhere greater than zero. These conditions ensure that the continuum is both indestructible and impenetrable—the topology of our templates are therefore always preserved after deformation. Suppose the displacements also take a neighboring point Q at $X + dX$ to $x + dx = X + dX + u(X + dX, t)$. The relative position of P and Q changes to

$$dx = dX + u(X + dX, t) - u(X, t) = dX + (\nabla u)dX,$$

where ∇u is the displacement gradient. Note that $\nabla u = 0$ implies $dx = dX$ or a rigid-body translation. ∇u must therefore embody both the rotational component of the rigid transformation and the pure deformation in the motion. To characterize the latter, consider two infinitesimal fibers emanating from the same point P such that $dx^1 = dX^1 + (\nabla u)dX^1$ and $dx^2 = dX^2 + (\nabla u)dX^2$. Their dot product provides the deformation measure we are after:

$$dx^1 \cdot dx^2 = dX^1 \cdot dX^2 + 2\, dX^1 \cdot E\, dX^2,$$

where $E = 1/2 \{(\nabla u) + (\nabla u)^T + (\nabla u)^T(\nabla u)\}$ is the Lagrangian strain tensor and represents pure deformation; in the absence of strain, $dx^1 \cdot dx^1$ is equal to $dX^1 \cdot dX^2$, and the motion is rigid. In the linearized theory, the displacement gradients are assumed small. We drop the product $(\nabla u)^T(\nabla u)$ in E and obtain the infinitesimal strain tensor ϵ, whose components are given by

$$\epsilon_{ij} = \frac{1}{2}\left(\frac{\partial u_i}{\partial X_j} + \frac{\partial u_j}{\partial X_i}\right).$$

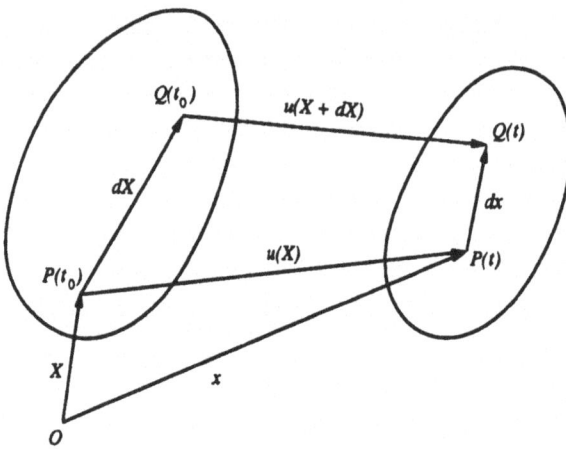

Figure 2.   Deformation and motion of a continuum.

Given the above measures of deformation, the choice of an appropriate norm is naturally made by considering the internal work done by the conjugate stress σ or internal force that arises in response to the strain. To illustrate, for isotropic isothermal elastic solids under small strain, Hooke's law reduces to the following linear relation in indicial notation:

$$\sigma_{ij} = \lambda\epsilon_{kk}\delta_{ij} + 2\mu\epsilon_{ij},$$

where λ and μ are known as the Lamé constants [22]. The potential of the corresponding Gibbs prior is then given by the internal strain energy

$$\int_V \sigma_{ij}\epsilon_{ij}\,dV = \int_V \left\{\lambda(\epsilon_{kk})^2 + 2\mu\epsilon_{ij}\epsilon_{ij}\right\}dV, \tag{3}$$

where the integral is taken over the volume of the body and can be numerically evaluated with, for example, the finite element method [23]. Note that in the current example the norm compares with the first-order quadratic stabilizers used in standard regularization. Moreover, the elastic constants are related to the regularization parameter: by varying their values, we can modulate the stiffness of the body and thus the degree of smoothness in the deformations. It is clear that each set of constitutive equations defining the particular idealized solid yields a different internal energy or regularizing behavior.

3.3. LARGE DEFORMATIONS

When the variations we expect to encounter in template matching are large, the theory of more general continua may be applied [22]. Plastic behavior is advantageous, as large deformations are made more probable than in the linear theory; the penalty is quadratic up to the yield strain, then slopes more gently in the plastic regime, as the stress increases more slowly with strain. The effect is similar to that of using a more robust version of the quadratic estimator, in which large strains are less severely penalized. The main difficulty with the analysis of large deformations is its computational complexity because of nonlinearities introduced through both the kinematic variables and constitutive relations. Current solution methods are sufficiently costly that an investigator should determine whether physically more accurate models are worthwhile or even appropriate in his or her particular application of image matching.

A more fundamental problem with large motions is that they give rise to "false matches" [24], which cannot be avoided simply by increasing the likelihood of large deformations in the prior model. A strategy is needed to search for the solution, and one widely used approach involves solving the matching problem at progressively finer spatial scales [25]. Larger scale motions are inferred by matching the lower spatial frequencies of the images, and then used to remove their effect in motion estimation of the higher frequency content. Priors can be imposed that are scale specific [21] and the uncertainty in the estimates propagated from one scale to the next [26]. The coarse-to-fine "warping" algorithm has been successfully applied to brain template matching using linear elastic [27] and viscous fluid [28] priors.

4. Discussion

The construction of prior distributions based on physical considerations is appealing because the resultant priors possess an intuitive interpretation and can be analyzed with continuum mechanical theory. Also, it becomes natural to represent the underlying displacement field by a finite element approximation, affording additional flexibility and efficiency. It is important to emphasize, however, that the use of continuum priors in our template matching problem is without rigorous physical justification. Our objective mainly concerns the imposition of certain constraints on the mappings, which is conveniently done within the framework of continuum mechanics.

Now, suppose, contrary to the prerequisites of continuum theory, discontinuous mappings actually exist and are expected to appear as, for example, in pathological anatomy or scans acquired before and after surgery. The usual solution is to introduce auxiliary processes to model the discontinuities, and is a form of robust estimation [29].

Intuitively, the likelihood of a mapping should reflect the degree to which one image is made similar to the other through the mapping. One such measure was provided in (2), but the similarity measurements may be based on image attributes other than the actual intensity values, such as the intensity gradient at a point or even its underlying tissue content [9]. When observations of more than one type of image feature are available, their influence on the final solution must again be moderated, perhaps, through a logarithmic opinion pool [7].

References

1. L. Brown, "A survey of image registration techniques," *ACM Comput. Surveys*, **24**, pp. 325–376, 1992.

2. R. Bajcsy, M. Reivich, and S. Kovačič, *Evaluation of registration of PET images with CT images*. Technical Report MS-CIS-88-89, Department of Computer and Information Science, University of Pennsylvania, Philadelphia, 1988.

3. J. Besag, "Towards bayesian image analysis," *J. Appl. Stat.*, **16**, pp. 395–407, 1989.

4. U. Grenander, *Tutorial in Pattern Theory*. Division of Applied Mathematics, Brown University, 1983.

5. U. Grenander, Y. Chow, and D. M. Keenan, *Hands: A Pattern Theoretic Study of Biological Shapes*. Springer-Verlag, New York, 1991.

6. C. Broit, *Optimal registration of deformed images*. Ph.D. Thesis, Department of Computer and Information Science, University of Pennsylvania, Philadelphia, 1981.

7. J. O. Berger, *Statistical Decision Theory and Bayesian Analysis*. Springer-Verlag, New York, 1985.

8. J. C. Gee, D. R. Haynor, M. Reivich, and R. Bajcsy, "Finite element approach to warping of brain images,"in *Medical Imaging 1994: Image Processing*, M. H. Loew, ed., SPIE, Be'lingham, 1994.

9. J. C. Gee, L. Le Briquer, C. Barillot, D. R. Haynor, and R. Bajcsy, "Bayesian approach to the brain image matching problem,"in *Medical Imaging 1995: Image Processing*, M. H. Loew, ed., SPIE, Bellingham, 1995.

10. J. C. Gee, L. Le Briquer, C. Barillot, and D. R. Haynor, "Probabilistic matching of brain images,"in *Information Processing in Medical Imaging*, Y. Bizais, C. Barillot, and R. D. Paola, eds., pp. 113–125, Kluwer Academic, Dordrecht, 1995.

11. J. Besag, P. Green, D. Higdon, and K. Mengersen, "Bayesian computation and stochastic systems," *Stat. Sci.*, **10**, (1), pp. 3–66, 1995.

12. T. Poggio, V. Torre, and C. Koch, "Computational vision and regularization theory," *Nature*, **317**, (26), pp. 314–319, 1985.

13. M. Bertero, T. A. Poggio, and V. Torre, "Ill-posed problems in early vision," *Proc. IEEE*, **76**, (8), pp. 869–889, 1988.

14. D. W. Thompson, *On Growth and Form*. Cambridge University Press, Cambridge, 1917.

15. J. D. Murray, *Mathematical Biology*. Springer-Verlag, Berlin, 1993.

16. F. J. Rohlf and F. L. Bookstein, *Proc. Michigan Morphometrics Workshop*. The University of Michigan Museum of Zoology, Ann Arbor, 1990.

17. A. N. Tikhonov and V. A. Arsenin, *Solutions of Ill-Posed Problems*. Winston, Washington, DC, 1977.

18. W. E. L. Grimson, *From images to surfaces: a computational study of the human early visual system*. MIT Press, Cambridge, 1981.

19. S. Geman and D. Geman, "Stochastic relaxation, gibbs distribution, and the bayesian restoration of images," *IEEE Trans. Pattern Anal. Machine Intell.*, **6**, pp. 721–741, 1984.

20. J. Marroquin, S. Mitter, and T. Poggio, "Probabilistic solution of ill-posed problems in computational vision," *J. Amer. Stat. Assoc.*, **82**, (397), pp. 76–89, 1987.

21. R. Szeliski, *Bayesian Modeling of Uncertainty in Low-level Vision*. Kluwer Academic, Norwell, 1989.

22. L. E. Malvern, *Introduction to the Mechanics of a Continuous Medium*. Prentice-Hall, Englewood Cliffs, 1969.

23. K.-J. Bathe, *Finite Element Procedures in Engineering Analysis*. Prentice-Hall, Englewood Cliffs, 1982.

24. D. Marr, *Vision: A Computational Investigation into the Human Representation and Processing of Visual Information*. W. H. Freeman, San Francisco, 1982.

25. D. Terzopoulos, "Image analysis using multigrid relaxation methods," *IEEE Trans. Pattern Anal. Machine Intell.*, **8**, pp. 129–139, 1986.

26. E. P. Simoncelli, *Distributed Representation and Analysis of Visual Motion*. Vision and Modeling Group Technical Report 209, MIT Media Laboratory, 1993.

27. R. Bajcsy and S. Kovačič, "Multiresolution elastic matching," *Comput. Vision, Graphics, Image Process.*, **46**, pp. 1–21, 1989.

28. G. E. Christensen, *Deformable Shape Models for Anatomy*. Ph.D. Thesis, Department of Electrical Engineering, Washington University, St. Louis, 1994.

29. M. J. Black and A. Rangarajan, "On the unification of line processes, outlier rejection, and robust statistics with applications in early vision," *Int. J. Comput. Vision*, to appear.

MECHANICAL MODELS AS PRIORS IN
BAYESIAN TOMOGRAPHIC RECONSTRUCTION

ANAND RANGARAJAN
Departments of Diagnostic Radiology and Computer Science
Yale University
New Haven, CT 06520-8042[§]

AND

SOO-JIN LEE AND GENE GINDI
Departments of Radiology and Electrical Engineering
SUNY at Stony Brook
Stony Brook, NY 11794-8460[¶] [‖]

Abstract. We introduce a new prior—the *weak plate*—to Bayesian tomographic reconstruction. The weak plate captures the piecewise ramplike spatial structure evident in primate autoradiograph source distributions. The weak plate is a part of a family of "mechanical" models—weak membrane (1st order), weak plate (2nd order), and weak quadric (3rd order)—in which a class of smoothness constraints derived from properties of ideal physical materials are used as models in the associated reconstruction problem. Since "weak" priors generate local minima in MAP estimation, we have designed novel Generalized Expectation–Maximization deterministic annealing algorithms to alleviate this problem. Our simulation studies qualitatively demonstrate the improvements over the weak membrane and maximum likelihood reconstructions.

Key words: tomographic reconstruction, weak plate, piecewise smooth, deterministic annealing, expectation–maximization, free energy, saddle point.

1. Introduction

Maximum likelihood (ML) tomographic reconstruction [1] is a very illustrative example of the "overfitting" problem in statistics. Given emission projection data—which can roughly be visualized as line integrals of a 2-D source distribution taken at different angles—the problem is to reconstruct the underlying source distribution (image). Unfortunately, the desired image resolution dictates that we esti-

[§]Email: anand@noodle.med.yale.edu
[¶]Email: sjlee@clio.rad.sunysb.edu, gindi@clio.rad.sunysb.edu
[‖]This work was supported by NIH R01-NS32879.

K. M. Hanson and R. N. Silver (eds.), Maximum Entropy and Bayesian Methods, 117–124.
© *1996 Kluwer Academic Publishers.*

mate N parameters from approximately the same amount of data leading to the aforementioned overfitting problem [2]. Bayesian tomographic reconstruction is a natural way of specifying an *effective* number of parameters that is much less than the original number. From the Bayesian viewpoint, this leads quite naturally to the question: what is a good prior?

Priors can be seen as means of incorporating actual, known information regarding the local spatial character of the source. Several such prior distributions have been proposed in the literature [3–5]. In these proposals, the source was assumed to be piecewise smooth. Our own early work incorporated one such prior, a "weak membrane" (WM) model, to capture notions of piecewise smoothness [5,6]. Most of the simulation experiments were done with the mathematical phantoms which were themselves piecewise constant. Reconstruction algorithms, including our own, that use this kind of piecewise constant assumption as prior perform better, according to most reasonable metrics, than those that make no such assumption, when tested on such phantoms. However, the self-consistent loop of piecewise constant prior and piecewise constant phantom leads one to question whether these good results generalize to a realistic clinical setting where the underlying (patient) source distribution may not be piecewise constant.

Here we argue for more expressive priors able to model more complicated forms than a piecewise constant source. Our work is motivated by the observations of piecewise ramplike spatial structure in primate autoradiographs [7]. The essential feature of our new prior model, the *weak plate* (WP) is that it favors piecewise *linear* ramplike regions. The weak plate is a part of a family of "mechanical" models—weak membrane (1st order), weak plate (2nd order), and weak quadric (3rd order)—in which a class of smoothness constraints derived from properties of ideal physical materials are used as models in the associated reconstruction problem [8]. Smoothness priors often correspond to a notion of fitting a member of a class of smooth functions to the data; "weak" constraints refer to the allowed inclusion of spatial discontinuities in the fit. Thus a weak membrane is a low order (1st derivative) spline allowed to have breaks (0th-order discontinuities) and a weak plate is the familiar (2nd-order) thin plate spline that is allowed to have breaks and creases (discontinuities in 0th and 1st derivatives).

2. Bayesian Tomographic Reconstruction

We model our problem on a 2-D discrete lattice indexed by i, j. Lowercase bold quantities denote 2-D vector fields and the corresponding lowercase, italicized quantities denote the elements of the vector field. Similarly, uppercase bold quantities denote 2-D random fields. Thus, $\Pr(\mathbf{F} = \mathbf{f})$ denotes the probability that the random field $\mathbf{F}$ takes the value $\mathbf{f}$ and $f_{i,j}$ denotes an element of $\mathbf{f}$ at location (i, j).

To enforce the notion of *piecewise* smoothness, Geman and Geman [9] introduced an unobservable line process l into the image model to preserve the discontinuities in the reconstructions, and versions of the line processes have been proposed for medical imaging in [4–6]. We begin by formulating our reconstruction problem from Bayes' Theorem with the aid of the line processes:

$$\Pr(\mathbf{F} = \mathbf{f}, \mathbf{L} = \mathbf{l}|\mathbf{G} = \mathbf{g}) = \frac{\Pr(\mathbf{G} = \mathbf{g}|\mathbf{F} = \mathbf{f}, \mathbf{L} = \mathbf{l})\Pr(\mathbf{F} = \mathbf{f}, \mathbf{L} = \mathbf{l})}{\Pr(\mathbf{G} = \mathbf{g})},$$

where $\mathbf{f}$, $\mathbf{l}$, and $\mathbf{g}$ are the source intensities, line processes, and projection data, respectively, and $\mathbf{F}$, $\mathbf{L}$, $\mathbf{G}$ are the associated 2-D random fields. We require the maximum *a posteriori* (MAP) estimate of both the source field and the discontinuity field $(\hat{\mathbf{f}}, \hat{\mathbf{l}})$: $(\hat{\mathbf{f}}, \hat{\mathbf{l}}) = \arg\max_{(\mathbf{f},\mathbf{l})} \Pr(\mathbf{G} = \mathbf{g}|\mathbf{F} = \mathbf{f})\Pr(\mathbf{F} = \mathbf{f}, \mathbf{L} = \mathbf{l})$.

We model both the likelihood and the prior as Gibbs distributions [9]. A Gibbs distribution is specified by an energy function E from which the partition function Z can be obtained. Since the number of detected counts in emission tomography (ET) is independently Poisson distributed, we model the likelihood as

$$\Pr(\mathbf{G} = \mathbf{g}|\mathbf{F} = \mathbf{f}) = \prod_{t\theta} \frac{(\sum_{ij} \mathcal{H}_{t\theta,ij} f_{i,j})^{g_{t\theta}} \exp(-\sum_{ij} \mathcal{H}_{t\theta,ij} f_{i,j})}{g_{t\theta}!}. \tag{1}$$

In (1), $g_{t\theta}$ is the number of detected counts in the detector bin indexed by t at angle θ, and $\mathcal{H}_{t\theta,ij}$ is the probability that a photon emitted from source location (i, j) hits detector bin t at angle θ. The major physical factors can be adequately modeled as linear effects and summarized by the system matrix $\mathcal{H}_{t\theta,ij}$.

The likelihood energy in (1) is

$$E_D(\mathbf{f}) = -\log\Pr(\mathbf{G} = \mathbf{g}|\mathbf{F} = \mathbf{f}) = \sum_{t\theta} \left[-g_{t\theta} \log \sum_{ij} \mathcal{H}_{t\theta,ij} f_{i,j} \right] + \sum_{t\theta,ij} \mathcal{H}_{t\theta,ij} f_{i,j},$$

where the term $\log g_{t\theta}!$ was dropped since it does not involve $\mathbf{f}$. The prior is defined over the source intensities and line processes. The corresponding prior energy is $E_P(\mathbf{f},\mathbf{l}) = -\log\Pr(\mathbf{F} = \mathbf{f}, \mathbf{L} = \mathbf{l})$.

2.1. THE WEAK PLATE AND THE WEAK MEMBRANE PRIORS

The energy function corresponding to the WM prior is a function of the source $\mathbf{f}$, and the line processes $\mathbf{l}$:

$$E_P^M(\mathbf{f},\mathbf{l}) = \lambda \sum_{ij} [f_v^2(i,j)(1 - l_{i,j}^h) + f_h^2(i,j)(1 - l_{i,j}^v)] + \alpha \sum_{ij} (l_{i,j}^h + l_{i,j}^v). \tag{2}$$

In (2), f_v and f_h are the first order derivatives along the vertical and the horizontal directions, respectively, and are defined as $f_v(i,j) \overset{\text{def}}{=} f_{i+1,j} - f_{i,j}$ and $f_h(i,j) \overset{\text{def}}{=} f_{i,j+1} - f_{i,j}$. The binary variables l^v and l^h are vertical and horizontal line processes, respectively, and λ and α are positive hyperparameters. Note that if $\mathbf{l}$ is set to zero everywhere, the prior reduces to a familiar smoothness regularizing term penalizing the first squared derivatives. The last term penalizes the creation of the discontinuities, charging an amount α at each such site. Therefore, the two terms in (2) encourage smoothness except where discontinuities ($l_{i,j} = 1$) occur.

However, due to its nature in favoring piecewise constant reconstructions, the WM prior has the unfortunate effect of turning a ramp into a single step or stepped terraces depending on the parameter settings. This is the "gradient limit" effect [10], which is a fundamental limitation of the WM (and of many first order Markov models [11] as well). If the gradient of a ramp exceeds the limit, a discontinuity appears in its reconstruction. If the gradient is much greater than the limit, a solution with multiple breaks might have the smallest energy. The gradient limit is inversely proportional to the scale λ. This disadvantage of the WM leads to significant errors in tomographic reconstructions as shown in our simulations.

While the WM prior favors the piecewise constant reconstructions, the WP prior favors piecewise linear reconstructions. Our corresponding energy function for the WP is [10]:

$$E_P^P(\mathbf{f}, \mathbf{l}) = \lambda \sum_{ij} \left\{ \left[f_{vv}^2(i,j) + 2f_{hv}^2(i,j) + f_{hh}^2(i,j) \right] (1 - l_{i,j}) \right\} + \alpha \sum_{ij} l_{i,j}, \quad (3)$$

where our discrete approximations of the partial second derivatives are $f_{vv}(i,j) \overset{\text{def}}{=} f_{i+1,j} - 2f_{i,j} + f_{i-1,j}$, $f_{hh}(i,j) \overset{\text{def}}{=} f_{i,j+1} - 2f_{i,j} + f_{i,j-1}$, $f_{hv}(i,j) \overset{\text{def}}{=} f_{i,j} - f_{i+1,j} - f_{i,j+1} + f_{i+1,j+1}$. Note that this form has the line processes $l_{i,j}$, associated with the six nodes, $f_{i-1,j}, f_{i,j-1}, f_{i,j}, f_{i,j+1}, f_{i+1,j}$, and $f_{i+1,j+1}$. (An advantage of mechanical models written in a functional form as in (2) and (3) is that Markov models with appropriate neighborhoods are automatically generated upon discretization.)

If $\mathbf{l}$ is set to zero everywhere, the prior reduces to the thin plate spline smoothing functional However, unlike the WM which encourages only the formation of piecewise constant regions, the WP encourages smoothness even in ramplike regions without incurring a penalty. That is, a "crease", a discontinuity in the first derivative, will turn on the line process. Thus the discontinuities in the WP correspond to discontinuities in the source gradient in addition to those in the source itself.

The posterior energy functions can be written as

$$E^P(\mathbf{f}, \mathbf{l}) = E_D(\mathbf{f}) + E_P^P(\mathbf{f}, \mathbf{l}), \quad E^M(\mathbf{f}, \mathbf{l}) = E_D(\mathbf{f}) + E_P^M(\mathbf{f}, \mathbf{l}) \quad (4)$$

where the WM and WP energy functions are defined in (2) and in (3) respectively.

2.2. MAP ESTIMATION VIA DETERMINISTIC ANNEALING

Minimization of the WP and WM energy functions is difficult due to the presence of the binary valued line processes $\mathbf{l}$. In this section, we present a deterministic annealing method to minimize the above energy functions. While deterministic annealing is not guaranteed to find the global minimum, it tends to avoid poor local minima.

Deterministic annealing methods begin by minimizing the *free energy* at high temperature $(\frac{1}{\beta})$ and then tracking the minimum through the variation of the temperature parameter. At high temperatures, the free energy is nearly convex and easily minimized. The free energy $F(\beta)$ is equal to $-\frac{1}{\beta} \log Z(\beta)$ where Z is the partition function. Since the partition function integral is intractable, we employ the well known *saddle-point* approximation from statistical physics. This is briefly summarized here—for more details, see [12].

$$
\begin{aligned}
Z(\beta) &= \sum_{\{\mathbf{f},\mathbf{l}\}} \exp[-\beta E(\mathbf{f},\mathbf{l})] = \sum_{\{\mathbf{f},\mathbf{l}\}} \prod_{ij} \int_R dz_{i,j}\delta(l_{i,j} - z_{i,j})\exp[-\beta E(\mathbf{f},\mathbf{z})] \\
&= \sum_{\{\mathbf{f},\mathbf{l}\}} \int_R dz_{i,j} \int_I du_{i,j} \exp\sum_{ij} u_{i,j}(l_{i,j} - z_{i,j})\exp[-\beta E(\mathbf{f},\mathbf{z})] \\
&= \sum_{\{\mathbf{f}\}} \int_R dz_{i,j} \int_I du_{i,j} \exp\sum_{ij}[-u_{i,j}z_{i,j} + \log(1 + \exp u_{i,j})]\exp[-\beta E(\mathbf{f},\mathbf{z})]
\end{aligned}
$$

where $\mathbf{z}$ is the *analog* line process; $\mathbf{z} \in [0,1]^N$ [13]. The original MAP combinatorial optimization problem on the binary valued line processes $\mathbf{l}$ has been converted into a nonlinear optimization problem on the continuous valued analog line processes $\mathbf{z}$ through the application of deterministic annealing [14]. Approximating the integral by the value of the integrand at its saddle points, we get

$$
\begin{aligned}
F(\beta) &\approx -\frac{1}{\beta}\log Z(\beta,\hat{\mathbf{f}},\hat{\mathbf{z}},\hat{\mathbf{u}}) = \min_{\mathbf{f},\mathbf{z}}\max_{\mathbf{u}} E(\mathbf{f},\mathbf{z}) + \frac{1}{\beta}\sum_{ij}[u_{i,j}z_{i,j} - \log(1 + \exp u_{i,j})] \\
&= \min_{\mathbf{f},\mathbf{z}} E(\mathbf{f},\mathbf{z}) + \frac{1}{\beta}\sum_{ij}[z_{ij}\log z_{ij} + (1 - z_{ij})\log(1 - z_{ij})]. \quad (5)
\end{aligned}
$$

From (5), it can be seen that at high temperature (low β), the entropy term ($\sum_{ij}[z_{ij}\log z_{ij} + (1 - z_{ij})\log(1 - z_{ij})]$) convexifies the free energy. At very low temperatures (high β), the entropy term drops out leaving us with the original MAP energy function.

One last detail needs to be resolved before we can summarize the deterministic annealing (DA) algorithm. The global connectivity of the likelihood term precludes the use of standard descent methods. The Expectation–Maximization (EM) algorithm (with the familiar complete/incomplete data formulation) is usually pressed into service to alleviate this problem. This approach can be extended to the MAP estimation problem as well. Below, we take a somewhat idiosyncratic approach in presenting the EM algorithm as coordinate descent on a new likelihood objective function [15]:

$$
\begin{aligned}
E_D(\mathbf{c},\mu,\mathbf{f}) &= \sum_{t\theta,ij}[\mathcal{H}_{t\theta,ij}f_{i,j} - c_{t\theta,ij}\log\mathcal{H}_{t\theta,ij}f_{i,j}] \\
&+ \sum_{t\theta}\mu_{t\theta}(\sum_{ij}c_{t\theta,ij} - g_{t\theta}) + \sum_{t\theta,ij}c_{t\theta,ij}(\log c_{t\theta,ij} - 1) \quad (6)
\end{aligned}
$$

where $\mathbf{c}$ is associated with the complete data and μ is a Lagrange parameter satisfying the constraint $\sum_{ij} c_{t\theta,ij} = g_{t\theta}$. The EM algorithm results when (6) is minimized in two stages—first w.r.t. $(\mathbf{c},\mu)$ and then w.r.t. $\mathbf{f}$. The same approach can be extended to cover the MAP case with either the WP or the WM energy functions:

$$
E(\mathbf{c},\mu,\mathbf{f},\mathbf{z}) = E_D(\mathbf{c},\mu,\mathbf{f}) + E_P(\mathbf{f},\mathbf{z}) + \frac{1}{\beta}\sum_{ij}[z_{ij}\log z_{ij} + (1 - z_{ij})\log(1 - z_{ij})].
$$

We apply a coordinate descent method to minimize the above energy function at each temperature setting. Step sizes necessary for gradient descent methods are eschewed. We first minimize the energy function w.r.t. $c_{t\theta,ij}$ (while satisfying the complete/incomplete data constraint using a simple normalization) followed by succesive minimization w.r.t. $\mathbf{f}$ and $\mathbf{z}$. After suitable convergence criteria are met, β is increased and the procedure repeated.

The update equations for $\mathbf{c}$, $\mathbf{f}$ and $\mathbf{z}$ in the WP case are:

$$\frac{\partial E}{\partial \mathbf{c}} = 0 \Rightarrow c_{t\theta,ij} = g_{t\theta} \frac{\mathcal{H}_{t\theta,ij} f_{i,j}}{\sum_{kl} \mathcal{H}_{t\theta,kl} f_{k,l}} \tag{7}$$

$$\frac{\partial E}{\partial f_{i,j}} = 0 \Rightarrow f_{i,j} = \frac{-(\sum_{t\theta} \mathcal{H}_{t\theta;ij} - 2\lambda X_3) + \sqrt{(\sum_{t\theta} \mathcal{H}_{t\theta;ij} - 2\lambda X_3)^2 + 8\lambda X_2 X_1}}{4\lambda X_2}$$

$$\frac{\partial E}{\partial z_{i,j}} = 0 \Rightarrow z_{i,j} \stackrel{\text{def}}{=} \frac{1}{1 + \exp\{-\beta(\lambda [f_{vv}^2(i,j) + 2f_{hv}^2(i,j) + f_{hh}^2(i,j)] - \alpha)\}} \tag{8}$$

where $X_1 \stackrel{\text{def}}{=} \sum_{t\theta} c_{t\theta,ij}$, and X_2 and X_3 are functions of the $\mathbf{f}$ and $\mathbf{z}$ that are in the neighborhood of $f_{i,j}$. More details on this derivation (and for the WM case as well) can be found in [16,6]. Below we summarize the DA algorithm:

Pseudo-code for the MAP-DA algorithm
Initialize β to β_0, $\mathbf{f}$ and $\mathbf{z}$ to random values.
Begin A: Do A until $z_{i,j} \leq \tau$ or $z_{i,j} \geq (1 - \tau)$.

    Begin B: Do B until $\frac{E_n - E_{n-1}}{E_n - E_0} < \epsilon$.

    Update complete data $\mathbf{c}$ using (7) (E-step).

    Update $\mathbf{f}$ and $\mathbf{z}$ using (8) (M-step).

    End B

$\beta \leftarrow (\beta \times \gamma)$ where γ governs the annealing schedule.
End A

3. Experiments and Results

We performed 2-D simulation studies with projection data from a 128×128 "ground truth" autoradiograph phantom [7], with 128 projection angles over 360° and 192 detector bins at each projection. To generate projection data, we simply add Poisson noise to each of the attenuated projections of the phantom, using the noiseless projection value as the mean in a Poisson random number generator.

We tested the four reconstruction algorithms—MAP-DA with the WM prior, MAP-DA with the WP prior, and ML-EM with two different stopping rules on the phantom shown in Fig. 1. For convenience, we will refer to these simply as the WM, WP, EM-1, and EM-2 algorithms.

For the ML-EM reconstructions, we used two different stopping rules. Our EM-1 algorithm adhered to a stopping rule based on minimum RMS error in the reconstruction. With our simulations, the average number of iterations for the minimum RMS error averaged 16. For our noise trials, we fixed the number of iterations at these average values. Our EM-2 algorithm adhered to a stopping rule related to the χ^2 stopping rule reported in [17].

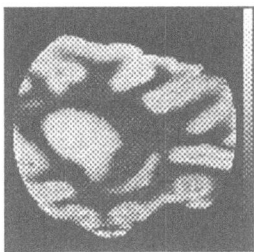

Figure 1.   "Ground-truth" autoradiograph phantom used in experiments

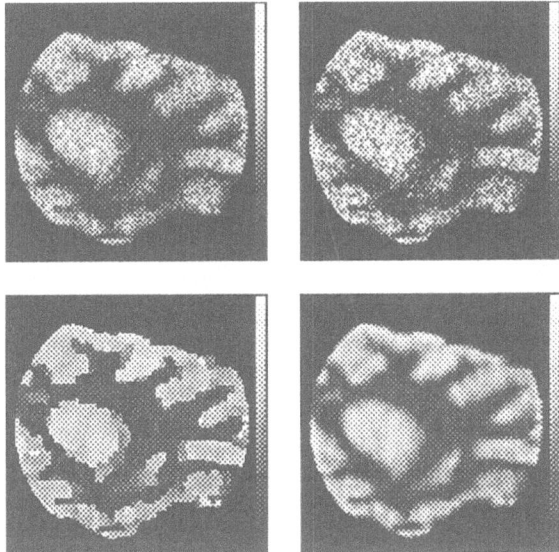

Figure 2.   Anecdotal reconstructions. Top left: EM-1 reconstruction. Top right: EM-2 reconstruction. Bottom left: WM reconstruction. Bottom right: WP reconstruction.

Figure 2 shows anecdotal reconstructed images from the four reconstruction algorithms. Several qualitative observations may be noted in Fig. 2. The WM reconstruction, Figs. 2(c), looks artificially patchy, a result not unexpected since WM tends to favor piecewise constant reconstructions. The gradient limit effects ("patchiness") may be reduced by increasing λ, but then a tradeoff results and the reconstructed images become excessively smooth. The WM attempts to create step edges even in the ramp regions, resulting in spurious discontinuities. By extension to a higher order model, the WP reconstructions avoid the artifacts of WM reconstructions as seen in Figs. 2(d). Comparisons of WP to the EM reconstructions in Fig. 2 show that WP reduces noise without introducing artifacts. Close inspection reveals that the WP reconstruction captures subtle aspects of the phantom in Fig. 1 that are missed by the EM algorithms. More quantitative analyses using bias/variance metrics can be found in [16].

4. Conclusions

We have introduced the weak plate prior—a higher order mechanical model—to emission tomography. The new prior is capable of representing inhomogeneous source distributions frequently assumed to be uniform. A novel Generalized EM deterministic annealing algorithm is used to obtain MAP estimates. Our initial efforts [18] in the difficult problem of estimating the hyperparameters λ and α, utilize maximum pseudo-likelihood methods with training data, and recently other efforts have been reported for determining λ [19].

References

1. L. Shepp and Y. Vardi, "Maximum likelihood reconstruction for emission tomography," *IEEE Trans. Med. Imag.*, **1**, pp. 113–122, 1982.
2. S. Geman, E. Bienenstock, and R. Doursat, "Neural networks and the bias/variance dilemma," *Neural Computation*, **4**, pp. 1–58, 1992.
3. S. Geman and D. E. McClure, "Statistical methods for tomographic image reconstruction," *Bulletin of the International Statistical Institute*, **LII-4**, pp. 5–21, 1987.
4. T. Hebert and R. Leahy, "A generalized EM algorithm for 3-D Bayesian reconstruction from Poisson data using Gibbs priors," *IEEE Trans. Med. Imag.*, **8**, pp. 194–202, June 1989.
5. G. Gindi, M. Lee, A. Rangarajan, and G. Zubal, "Bayesian reconstruction of functional images using anatomical information as priors," *IEEE Trans. Med. Imag.*, **12**, pp. 670–680, Dec. 1993.
6. G. Gindi, A. Rangarajan, M. Lee, P. J. Hong, and G. Zubal, "Bayesian reconstruction for emission tomography via deterministic annealing," in *Info. Proc. Med. Imag.*, H. Barrett and A. Gmitro, eds., pp. 322–338, Springer–Verlag, 1993.
7. G. Gindi and A. Rangarajan, "What can SPECT learn from autoradiography?," in *Proc. Nucl. Sci. Symp./Med. Imag. Conf.*, vol. 4, pp. 1715–1719, IEEE Press, 1994.
8. D. Geman and G. Reynolds, "Constrained restoration and the recovery of discontinuities," *IEEE Trans. Patt. Anal. Mach. Intell.*, **14**, pp. 367–383, March 1992.
9. S. Geman and D. Geman, "Stochastic relaxation, Gibbs distributions and the Bayesian restoration of images," *IEEE Trans. Patt. Anal. Mach. Intell.*, **6**, pp. 721–741, Nov. 1984.
10. A. Blake and A. Zisserman, *Visual Reconstruction*, MIT Press, Cambridge, MA, 1987.
11. A. Rangarajan and R. Chellappa, "Markov random field models in image processing," in *The Handbook of Brain Theory and Neural Networks*, M. A. Arbib, ed., pp. 564–567, MIT Press, 1995.
12. I. Elfadel, "Convex potentials and their conjugates in analog mean-field optimization," *Neural Computation*, **7**, pp. 1079–1104, Sept. 1995.
13. A. Rangarajan and R. Chellappa, "A continuation method for image estimation using the adiabatic approximation," in *Markov Random Fields: Theory and Application*, R. Chellappa and A. K. Jain, eds., pp. 69–91, Academic Press, 1993.
14. D. Geiger and F. Girosi, "Parallel and deterministic algorithms from MRFs: Surface reconstruction," *IEEE Trans. Patt. Anal. Mach. Intell.*, **13**, pp. 401–412, May 1991.
15. R. Hathaway, "Another interpretation of the EM algorithm for mixture distributions," *Stat. and Prob. Lett.*, **4**, pp. 53–56, 1986.
16. S.-J. Lee, A. Rangarajan, and G. Gindi, "Bayesian image reconstruction in SPECT using higher-order mechanical models as priors," *IEEE Trans. Med. Imag.*, **14**, pp. 669–680, Dec. 1995.
17. J. Llacer and E. Veklerov, "Feasible images and practical stopping rules for iterative algorithms in emission tomography," *IEEE Trans. Med. Imag.*, **8**, pp. 186–193, 1989.
18. S.-J. Lee, G. Gindi, G. Zubal, and A. Rangarajan, "Using ground-truth data to design priors in Bayesian SPECT reconstruction," in *Info. Proc. Med. Imag.*, Y. Bizais, C. Barillot, and R. D. Paola, eds., pp. 27–38, Kluwer Acad. Pub., 1995.
19. Z. Zhou, R. M. Leahy, and E. U. Mumcuoglu, "Maximum likelihood hyperparameter estimation for Gibbs priors with applications to PET," in *Info. Proc. Med. Imag.*, Y. Bizais, C. Barillot, and R. D. Paola, eds., pp. 39–51, Kluwer Acad. Pub., 1995.

THE BAYES INFERENCE ENGINE

K.M. HANSON AND G.S. CUNNINGHAM
Los Alamos National Laboratory, MS P940
Los Alamos, New Mexico 87545, USA[†]

Abstract. We are developing a computer application, called the Bayes Inference Engine, to provide the means to make inferences about models of physical reality within a Bayesian framework. The construction of complex nonlinear models is achieved by a fully object-oriented design. The models are represented by a data-flow diagram that may be manipulated by the analyst through a graphical-programming environment. Maximum *a posteriori* solutions are achieved using a general, gradient-based optimization algorithm. The application incorporates a new technique of estimating and visualizing the uncertainties in specific aspects of the model.

Key words: Bayesian analysis, MAP estimator, uncertainty estimation, object-oriented programming, adjoint differentiation, optimization

1. Introduction

As scientists, we use models to describe and understand physical reality. In building our models from data, we must be able to answer such questions as: Which models are appropriate? What are the values of the model parameters? How certain can we be in our interpretations drawn from measurements that are subject to uncertainty? The methodology of Bayesian analysis provides the framework to address these questions. In the Bayesian approach our degree of certainty is represented as a probability density function. Appropriate calculation and use of the posterior, the probablity that summarizes all available information concerning a particular physical situation, permits us to quantitatively answer the above questions regarding our models of the physical world.

We are developing a computer application, which we call the Bayes Inference Engine (BIE), to provide a tool with which to easily perform Bayesian analysis of physical models. The BIE represents a computational approach to Bayesian inference, as opposed to the traditional analytical approach [1]. The computational approach affords great flexibility in modeling, which facilitates the construction

---

[†] Email: kmh@lanl.gov, cunning@lanl.gov

K. M. Hanson and R. N. Silver (eds.), Maximum Entropy and Bayesian Methods, 125–134.
© *1996 Kluwer Academic Publishers.*

of complex models for the objects under study. For instance, the BIE easily deals
with data that are nonlinearly dependent on the model parameters. For example,
radiographic data are not linearly related to material densities [2]. One of our
first goals is to reconstruct objects from several radiographs taken of them. Furthermore, the computational approach allows one to use nonGaussian probability
distributions, such as entropic priors, which have been used with great success in
certain kinds of applications [3,4].

Given the context of these Proceedings, it should not be necessary to provide a
detailed description of Bayesian analysis. To learn the basics of Bayesian analysis,
the reader is referred to the collected works of the Proceedings of this workshop
series, in particular to those by Gull and Skilling and their colleagues [3–5], to
textbooks that explain the modern view of Bayesian methodology [6,7], or to
several introductory articles related to image analysis written by one of the authors
[8,9]. Lack of space also precludes us from showing examples produced by the BIE.
The reader is encouraged to look at some of the many references cited, many of
which are available on the World Wide Web (http://planck.lanl.gov/~kmh).

2. The Bayes Inference Engine

Our goals for the BIE are that it should be easy to learn and to use and that
it should provide a high degree of interactivity with good visualization of the
inference process and the models. Additionally, we are building an application
that provides the user with a great deal of flexibility in configuring object models
and measurement models. We deem these features essential to the usefulness of
the BIE.

In Bayesian analysis, the state of knowledge about the parameters $\mathbf{x}$ associated
with a model that describes the physical object being studied is summarized by
the posterior, which is the probability density function $p(\mathbf{x}|\mathbf{d})$ of the parameters
given the observed data $\mathbf{d}$. Bayes law gives the posterior as

$$p(\mathbf{x}|\mathbf{d}) \propto p(\mathbf{d}|\mathbf{x})\, p(\mathbf{x}) \ . \tag{1}$$

The probability $p(\mathbf{d}|\mathbf{x})$, called the likelihood, comes from a comparison of the
actual data to the data predicted on the basis of the model of the object. The predicted data are generated using a model for how the measurements are related to
the object, which we call the measurement model. The prior $p(\mathbf{x})$ expresses what is
known about the object, exclusive of the present measurements, and may represent
knowledge acquired from previous measurements, specific information regarding
the object itself, or simply general knowledge about the parameters, e.g. that they
are nonnegative. Bayes law says that for a given object model the posterior can be
evaluated by combining the likelihood, which requires the data values predicted
for that object model, and with the numerical value of the prior. This calculation
is usually straightforward. It involves calculating the predicted measurements for
the given object model, which we refer to as the forward measurement calculation.

A typical nonBayesian approach to estimating model parameters from a given
set of data is to attempt to apply the inverse of the forward measurement process

to the data. Such an approach is plagued by problems, particularly when there are insufficient data to uniquely determine all aspects of the actual object or when the measurements are degraded by excessive noise. Common remedies for overcoming such problems is to invoke some sort of regularization to permit the inverse solution or to reduce in the model the number of parameters to be determined.

We avoid the calculational difficulties of direct inversion in the BIE by basing the estimation procedure on the forward measurement calculation, which results in an evaluation of $\varphi = -\log[p(\mathbf{x}|\mathbf{d})]$ (neglecting a constant normalization term). The parameters for the object model are found using an algorithm to minimize φ with respect to those parameters, which results in the well known maximum *a posteriori* (MAP) solution because it maximizes the posterior. This optimization process is facilitated by making use of the derivatives of φ with respect to the object parameters, which are calculated using the adjoint differentiation technique described in Sect. 4. The use of priors provides the means of regularization in a probabilistic way that has a quantifiable basis, which may potentially be verified experimentally. We note that this numerical approach has many benefits, which were outlined in the Introduction. Thus with the computer we can obtain accurate Bayesian solutions to fairly complex problems that are intractable using analytic approaches. The computer also allows us to explore complex situations employing data visualization to enhance understanding, fulfilling the promise of using the full posterior provided by the Bayesian approach.

The BIE incorporates many innovative features, including:
1) a graphical programming tool programmed in an object-oriented language, which greatly enhances the flexibility of modeling objects and measurements,
2) adjoint differentiation to calculate the gradient of φ, with respect to all object parameters,
3) new approaches to solving the constrained optimization problem, which is required to find the MAP solution,
4) geometrical representations of physical objects, and
5) a new method to explore the reliability of the Bayesian solution.

We will describe each of these new developments in the following sections.

3. Data Flow Diagram

The analyst interacts with the BIE through a graphical programming environment [10], as shown in Fig. 1. We believe that this mode of interaction provides a very intuitive interface for building models because most scientists and engineers have had some experience with data-flow diagrams. The square icons represent transforms, which are connected by lines drawn between them to describe flow of data.

We are programming the BIE using the object-oriented (OO) language Smalltalk in the version supplied by ParcPlace Systems[1], which includes a complete class library, including classes for easy development of a graphical user-interface. In our description of object classes in the BIE, we capitalize the class names in accor-

---

[1] ParcPlace Systems, Inc., 999 East Arques Ave., Sunnyvale, CA 94086, tel: 408-481-9090

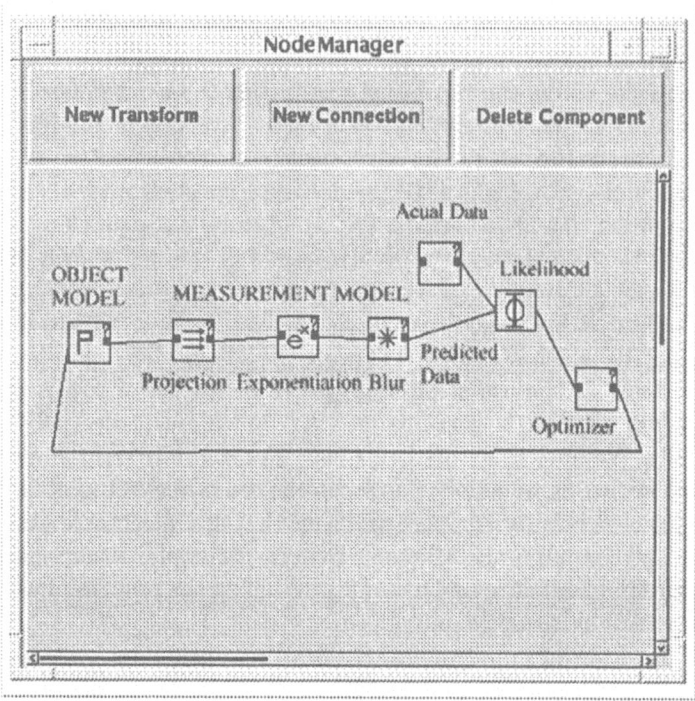

Figure 1. The canvas of the Bayes Inference Engine permits one to specify a data-flow diagram by connecting together Transforms, represented visually by squares on the canvas.

dance with the naming convention of this object-oriented language. A distinctive feature of the BIE is that the Transforms represented in Fig. 1 are 'living' objects with which one can interact. By clicking on the icon representing the Transform with the middle mouse button, a menu pops up that allows one to specify a request of the Transform. One can see a description of a Transform and change the parameters that define it. One can have the Transform display its output data structure.

Referring to the data-flow diagram in Fig. 1, the Parameters of the object model (the lefthand icon) provide input to the measurement model. The radiographic measurement model shown consists of the next three icons, which sequentially take the projection of the object, exponentiate the result, and perform a convolution with a point-spread function kernel to account for radiographic blur. The output of the measurement model represent predicted Data, which is fed into a (minus) LogLikelihood function, designated by Φ, along with the actual data, the uppermost icon. A LogPrior, which operates on the model parameters, can also be specified. The output of the LogLikelihood is fed into the Optimizer, the lower right-hand icon, whose task it is to find the values of the object-model parameters that result in a minimum value for Φ. One specifies the Parameters of the object model that are to be optimized by connecting their icons to the Optimizer.

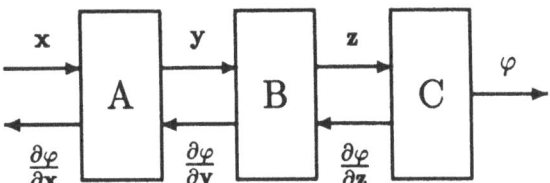

Figure 2. A generic data-flow diagram showing a sequence of transformations, represented by the boxes A, B, and C, starting with the data structure **x** and resulting in the scalar φ. The flow of data for the adjoint derivatives is in the reverse direction, from right to left.

After optimization, the object model and its Parameter values represent the MAP solution.

We have come to realize that the OO approach has provided much more than just a productive programming environment; it has aided in the design of the overall application, as well as the numerical algorithms at a basic level. For example, the following aspects of the BIE have been elucidated by OO design: the adjoint differentiation technique, the accommodation of constraints in optimization, the automatic connection of the Optimizer to any parameter, and the appropriate role of Connectors and Transforms in the data-flow diagram. An interesting aspect of the OO design of the data-flow diagram is that no supervisor of the sequence of calculations is needed; the Transforms simply do what they are asked to do, when they are asked to do it.

4. Adjoint Differentiation

To obtain the MAP estimate in the BIE, we need to minimize the scalar function φ by varying the variables that comprise the parameters of the object model. This optimization problem would be intractable without knowing the gradient of φ, or sensitivities, with respect to the many parameters on which it depends. We have uncovered a technique to calculate these crucial sensitivities, called adjoint differentiation [11], that is apparently little known. Using the adjoint differentiation technique the calculation of all these derivatives can be done in a computational time that is comparable to the forward calculation through the data-flow diagram. The adjoint sensitivity technique is crucial to the efficient operation of the BIE. We believe that it could be beneficially employed in many types of forward modeling codes.

Consider a calculation such as the sequence of transformations depicted in Fig. 2. In the context of the BIE, the transformations are implemented by the Transform class. The independent variables in the data structures designated by the vector **x** are transformed by block A to produce the dependent variables **y**. This is transformed by block B to produce the dependent data structure **z**, and finally by C to produce the scalar φ, which would correspond to the minus logarithm of the posterior in the BIE.

We make no assumptions about the transformations except that they are differentiable. The transforms can be nonlinear, as is the case for projections of objects defined in terms of their geometry (see Sect. 6) and for exponentiation, which is needed to model radiographic measurements.

The flow diagram indicates that φ depends on z, which depends on y, which in turn depends on x, the independent variables. The chain rule for differentiation gives the derivatives of φ with respect to the ith component of x,

$$\frac{\partial \varphi}{\partial x_i} = \sum_{jk} \frac{\partial \varphi}{\partial z_k} \frac{\partial z_k}{\partial y_j} \frac{\partial y_j}{\partial x_i} . \tag{2}$$

Even if the transformations are nonlinear, this expression amounts to a product of matrices, each element of which specifies the differential response of an output variable with respect to a differential change of an input variable.

The essence of adjoint differentiation is to perform the sum on the indices in the reverse order from that used in the forward calculation. If the sum on k is done before that on j, the sequence of calculations is

$$I \xrightarrow{C'^t} \frac{\partial \varphi^\dagger}{\partial z} \xrightarrow{B'^t} \frac{\partial \varphi^\dagger}{\partial y} \xrightarrow{A'^t} \frac{\partial \varphi^\dagger}{\partial x} ,$$

where I represents the indentity vector and, for example, B'^t implements the adjoint of the matrix $\frac{\partial z}{\partial y}$. This sequence implies intermediate data structures (e.g. $\frac{\partial \varphi}{\partial y}$) that resemble the normal data structures (e.g. y). Thus the requirement for storing these data structures is merely double that required to store the structures for the forward calculation, which may be required for the sensitivity calculation if the transformations are nonlinear. The backward flow of the adjoint derivatives is depicted in Fig. 2.

If the sums over the indices were done in the opposite order, mimicking the forward calculation, the intermediate data structures would be matrices containing a number of elements equal to the product of the sizes of the forward data structures. Since we are considering very large forward data structures, this way of accumulating the sensitivities could be untenable.

Our use of objects to represent transformations greatly helps implement this adjoint calculation [12]. In accordance with the OO approach, each transformation is self-contained; it requires only its input variables to calculate its output variables, e.g. module B uses only its input y to calculate its output z. Therefore, each transformation should require nothing more than its input to implement the derivative of its output variables with respect to its input variables. We emphasize that in the OO approach we are using, each Transform has the responsibility to propagate the adjoint derivative from its output side to its input side. The Transform "knows how" to do this because it knows how to accomplish the forward calculation.

In reality, what this means is that when a new Transform is created by a programmer, the code for propogating the adjoint derivative should be developed using the logic of the forward calculation to determine the derivatives of the output variables with respect to the input variables. We stress that the derivative matrix need not be explicitly calculated and stored. The adjoint code has the responsibility of calculating the *effect* of multiplying derivatives on its output side by the derivative matrix, which can often be achieved using computer code that is very similar to the code for the forward calculation. To illustrate, for linear transformations, which can be characterized in terms of a matrix multiplication, the implementation of the adjoint differentiation calculation simply amounts to multiplication by the transpose of that matrix. In this case the transpose operation can often be realized by trivially altering the computer code for the forward transformation.

5. Optimization

In the BIE the MAP solution is found by minimizing φ with respect to all the model parameters. Our optimization procedure is based on knowing the gradient of φ with respect to the parameters, which is calculated as described in the previous section. Our use of OO programming imposes certain restrictions on how the optimization can be implemented. For example, many of the models in the BIE impose constraints on parameters. Some constraints involve fixed limits on individual parameters, e.g. nonnegativity. Constraints can also exist between parameters. The approach to optimization must include these in an OO way. That is, the Optimizer should only request that the Parameters act on themselves. Examples of possible actions include a) add a specified vector to the present values of the parameters and b) satisfy constraints on the parameters.

The general method that we employ to guarantee that the constraints on the parameters are met is by projection onto convex sets (POCS) [13]. Each Parameter checks whether constraints are violated. If they are, the Parameters minimally change themselves to meet the constraints [14].

6. Geometric Representation of Objects

Deformable models have been developed in a number of fields to describe objects geometrically, particularly in computer vision where the aim is to decompose a scene (image) in terms of geometrical objects [15–17].

We are pioneering the use of deformable geometric models to improve tomographic reconstructions of objects from just a few views [18–20]. This tack is quite different from the normal one of representing a 2D object in terms of its density, typically described by square pixels on an ordered grid. The reconstruction process amounts to deforming an initial object geometry in a minimal way to match the data. In the Bayesian approach, one controls the geometric deformation by placing a prior on it. The net effect is to add to φ a deformation energy that penalizes larger deformations. This approach has proven to be a valuable means to achieve good reconstructions in situations where all other methods fail, for example when only two radiographs are available [20]. However, it must be emphasized that this approach can only be successful when the objects being reconstructed have a fairly simple morphology that is approximately known beforehand.

7. Exploration of the uncertainty in a Bayesian solution

One of the most important features of the Bayesian approach is that the posterior characterizes of the degree of certainty in the models used in an analysis. Although many articles have been written about Bayesian image analysis, surprisingly few of these have fully exploited the full posterior as a measure of uncertainty. The reason probably lies in the fact that it is computationally difficult to cope with the posterior in a large dimensional space.

One way to visualize the reliability of an inferred model is to display a sequence of solutions that are randomly chosen from the posterior probability distribution. This approach, proposed by Skilling et al. [21], provides a stochastic look at the range of possible solutions. The sequence of images, typically calculated off line, is presented as a video loop. By showing a representative range of alternative solutions, the degree of variability of this presentation provides the viewer with a visual impression of the degree of uncertainty in the inferred model. One would expect that the present emphasis in Bayesian research on Markov chain Monte Carlo methods [22] to generate random samples of the posterior will be useful for this type of visualization.

We have proposed a new approach [23,24], which is based on drawing an analogy between φ and a physical potential. Then the gradient of φ is analogous to a force. From this viewpoint an unconstrained MAP solution can be interpreted as the situation in which the forces on all the variables in the problem balance so that the net force on each variable is zero. Further, when a variable is perturbed from the MAP solution, the derivative of φ with respect to that variable is the force that drives it back towards the MAP solution. The phrase "force of the data" takes on real meaning in this context.

To explore the reliability of a particular feature of a MAP solution, the user specifies it by directly perturbing the selected combination of parameters that characterize the feature of interest. The posterior is incremented on the basis of this perturbation to effectively apply a constant force to the parameters in question. Then, all the parameters are readjusted to minimize the new φ. The uncertainty in the parameters is indicated by the amount that they move away from their MAP values for a given applied external force. The correlations between parameters experiencing the external force and the others is demonstrated by how much and in what direction the parameters change. We have shown that this approach leads to a quantitative estimate for an appropriate part of the covariance matrix for problems in which the parameters are unconstrained [23,24]. Ideally, these correlations could be seen through direct interaction with a rapidly-responding dynamical Bayesian system. Alternatively, they may be demonstrated as a video loop produced off line.

8. Future Directions

The BIE provides us with an ideal basic tool with which to make Bayesian inferences regarding physical models. We aim to extend its existing capabilities in a number of directions.

In developing the BIE we have concentrated on two-dimensional models, in order to impact our programmatic goals. We expect to develop both 1D and 3D models. One-dimensional models will enable us to develop demonstrations of how the Bayesian approach addresses many familiar problems such as deconvolution, or restoration, of blurred 1D signals, spectral estimation, interpolation, and line fitting. Calculations in 1D can be done very quickly, so that the concepts discussed in Sect. 7 can be demonstrated in a real-time interactive environment. We also expect to develop three-dimensional modeling capabilities before long so that we can address the problem of tomographic reconstruction of 3D objects.

The interactivity of the BIE allows the analyst to fully diagnose the models he creates. Feedback about what is needed from the model to match the data is provided by displaying the gradient of φ, which shows the force of the data. The full interactivity with the object models makes it easy to augment the models to achieve a better fit to the data. The Bayesian methodology allows one to make inferences about the choice of models appropriate to describe reality. Our preference for simpler models over more complex ones can be incorporated through a prior on model complexity [5]. An interesting example of model selection is supplied in the context of an object defined by its boundary, which is smooth by default. The boundary might be allowed to develop a kink, i.e. an abrupt change in slope, thereby negating the smoothness constraint at a particular place, if the data provide enough evidence for such a departure from the default model [25]. We anticipate that the gradient of φ will play a fundamental role in making such such decisions about when the complexity of a model needs to be increased.

We will implement the means to generate random samples from the posterior [22]. This capability could be used to estimate the posterior mean (as an alternative to the posterior mode) and variance, which is one way to summarize the uncertainty in solutions. This technique permits marginalization with respect to any nuisance parameters. It also can provide a visualization of the uncertainty in solutions by displaying as a video loop the sequence of random samples [21].

Acknowledgements

This work was supported by the United States Department of Energy under contract number W-7405-ENG-36. We would like to acknowledge the helpful conversations we have had with the following people: John Skilling, Stephen F. Gull, Kyle J. Myers, and Robert F. Wagner.

References

1. K. M. Hanson and G. S. Cunningham, "A computational approach to Bayesian inference," in *Proc. Interface Conf.*, M. Meyer, ed., 1995 (to be published).

2. K. M. Hanson, "A Bayesian approach to nonlinear inversion: Abel inversion from x-ray data," in Transport Theory, Invariant Imbedding, and Integral Equations, *Lect. Notes in Pure and Appl. Math.* **115**, P. Nelson, V. Fabor, D. L. Seth, and A. B. White, Jr., eds., pp. 363–368, Marcel Dekker, New York, 1989.

3. J. Skilling and R. K. Bryan, "Maximum entropy image reconstruction: general algorithm," *Mon. Not. R. Ast. Soc.*, **211**, pp. 111–124, 1984.

4. S. F. Gull, "Developments in maximum-entropy data analysis," in *Maximum Entropy and Bayesian Methods*, J. Skilling, ed., pp. 53–71, Kluwer Academic, Dordrecht, 1989.

5. S. F. Gull, "Bayesian inductive inference and maximum entropy," in *Maximum Entropy and Bayesian Methods in Science and Engineering (Vol. 1)*, G. J. Erickson and C. R. Smith, eds., pp. 53–74, Kluwer Academic, Dordrecht, 1989.

6. S. J. Press, *Bayesian Statistics: Principles, Models, and Applications*, Wiley, New York, 1989.

7. A. Gelman, J. B. Carlin, H. S. Stern, and D. B. Rubin, *Bayesian Data Analysis*, Chapman & Hall, London, 1995.

8. K. M. Hanson, "Bayesian and related methods in image reconstruction from incomplete data," in *Image Recovery: Theory and Application*, H. Stark, ed., pp. 79–125, Academic, 1987.

9. K. M. Hanson, "Introduction to Bayesian image analysis," in *Image Processing*, M. H. Loew, ed., *Proc. SPIE*, **1898**, pp. 716–731, 1993.

10. G. S. Cunningham, K. M. Hanson, G. R. Jennings, Jr., and D. R. Wolf, "An object-oriented implementation of a graphical-programming system," in *Image Processing*, M. H. Loew, ed., *Proc. SPIE*, **2167**, pp. 914–923, 1994.

11. W. C. Thacker, "Automatic differentiation from an oceanographer's perspective," in *Automatic Differentiation of Algorithms: Theory, Implementation, and Application*, A. Griewank and G. F. Corliss, eds., pp. 191–201, SIAM, Philadelphia, 1991.

12. G. S. Cunningham, K. M. Hanson, G. R. Jennings, Jr., and D. R. Wolf, "An interactive tool for Bayesian inference," in *Review of Progress in Quantitative Nondestructive Evaluation*, D. O. Thompson and D. E. Chimenti, eds., vol. 14A, pp. 747–754, Plenum, New York, 1995.

13. D. C. Youla, "Mathematical theory of image restoration by the method of convex projections," in *Image Recovery: Theory and Application*, H. Stark, ed., pp. 29–77, Academic, 1987.

14. G. S. Cunningham, K. M. Hanson, G. R. Jennings, Jr., and D. R. Wolf, "An object-oriented optimization system," in *Proc. IEEE Int. Conf. Image Processing, vol. III*, pp. 826–830, IEEE, 1994.

15. M. Kass, A. Witkin, and D. Terzopoulos, "Snakes: active contour models," *Inter. J. Comp. Vision*, **1**, pp. 321–331, 1988.

16. R. Szeliski, "Probabilistic modeling of surfaces," *Proc. SPIE*, **1570**, pp. 154–165, 1991.

17. U. Grenander and M. Miller, "Representations of knowledge in complex systems," *Jour. Roy. Stat. Soc. B*, **56**, pp. 549–603, 1994.

18. K. M. Hanson, "Flexible prior models in Bayesian image analysis," in *Maximum Entropy and Bayesian Methods*, A. Mohammad-Djafari and G. Demoment, eds., pp. 399–406, Kluwer Academic, Dordrecht, 1993.

19. K. M. Hanson, "Bayesian reconstruction based on flexible prior models," *J. Opt. Soc. Amer. A*, **10**, pp. 997–1004, 1993.

20. K. M. Hanson, G. S. Cunningham, G. R. Jennings, Jr., and D. R. Wolf, "Tomographic reconstruction based on flexible geometric models," in *Proc. IEEE Int. Conf. Image Processing, vol. II*, pp. 145–147, IEEE, 1994.

21. J. Skilling, D. R. T. Robinson, and S. F. Gull, "Probabilistic displays," in *Maximum Entropy and Bayesian Methods*, W. T. Grandy, Jr. and L. H. Shick, eds., pp. 365–368, Kluwer Academic, Dordrecht, 1991.

22. J. Besag, P. Green, D. Higdon, and K. Mengersen, "Bayesian computation and stochastic systems," *Stat. Sci*, **10**, pp. 3–66, 1995.

23. K. M. Hanson and G. S. Cunningham, "The hard truth," in *Maximum Entropy and Bayesian Methods*, J. Skilling, ed., Kluwer Academic, Dordrecht, 1994 (to be published).

24. K. M. Hanson and G. S. Cunningham, "Exploring the reliability of Bayesian reconstructions," in *Image Processing*, M. H. Loew, ed., *Proc. SPIE*, **2434**, pp. 416–423, 1995.

25. K. M. Hanson, R. L. Bilisoly, and G. S. Cunningham, "Kinky tomographic reconstruction," to be published in *Image Processing*, M. H. Loew and K. M. Hanson, eds., *Proc. SPIE*, **2710**, 1996.

A FULL BAYESIAN APPROACH FOR INVERSE PROBLEMS

ALI MOHAMMAD-DJAFARI

Laboratoire des Signaux et Systèmes (CNRS-ESE-UPS),
Supélec, Plateau de Moulon, 91192 Gif-sur-Yvette, France [†]

Abstract. The main object of this paper is to present some general concepts of Bayesian inference and more specifically the estimation of the hyperparameters in inverse problems. We consider a general linear situation where we are given some data y related to the unknown parameters x by $y = Ax + n$ and where we can assign the probability laws $p(x|\theta)$, $p(y|x,\beta)$, $p(\beta)$ and $p(\theta)$. The main discussion is then how to infer x, θ and β either individually or any combinations of them. Different situations are considered and discussed. As an important example, we consider the case where θ and β are the precision parameters of the Gaussian laws to whom we assign Gamma priors and we propose some new and practical algorithms to estimate them simultaneously. Comparisons and links with other classical methods such as maximum likelihood are presented.

Key words: Bayesian inference, Hyperparameter estimation, Inverse problems, Maximum likelihood

1. Introduction

In a general Bayesian inference, we have the data y, a known relation between the unknown parameters x and y and finally the hyperparameters β and θ. The Bayesian estimation technique is now well established [1-7] and has been used since many years to resolve the inverse problems in signal and image reconstruction and restoration [10-14,17,18,20,21].

The first step before applying the Bayes' rule is to assign the prior probability laws $p(x|\theta)$, $p(y|x,\beta)$, $p(\theta)$ and $p(\beta)$. The next step is to determine the posterior laws and then to infer the unknowns. In this paper we are focusing more on the second step than on the first step. So we assume that all the direct probability laws are known.

The main object of this paper is to show how can we infer simultaneously the unknown parameters x and the hyperparameters β and θ from the data y.

Before going more in details let us give one example. This will permit us to fix the situations. Consider the case where the unknown parameters x represent

---

[†]Email: djafari@lss.supelec.fr

K. M. Hanson and R. N. Silver (eds.), Maximum Entropy and Bayesian Methods, 135–144.
© *1996 Kluwer Academic Publishers.*

the pixel values of an unobserved image and the data y are the pixel values of an observed image which is assumed to be a degraded version of it. If we consider a linear degradation we have

$$y = Ax + n, \tag{1}$$

where A is a $(m \times n)$ matrix representing the degradation process and n represents the measurement uncertainty (noise) which is assumed to be additive, centered, white, Gaussian and independent of x. This hypothesis leads us to

$$p(y|x, \beta) = \frac{1}{Z_1(\beta)} \exp\left\{ -\frac{1}{2}\beta(y - Ax)^t(y - Ax) \right\}. \tag{2}$$

In this case β is a positive parameter which is related to the noise variance σ_b^2 by $\beta = 1/\sigma_b^2$ and $Z_1(\beta) = (2\beta/\pi)^{m/2}$ is the normalizing factor.

Consider also, for this example, a Gaussian prior law for x :

$$p(x|\theta) = \frac{1}{Z_2(\theta)} \exp\left\{ -\frac{1}{2}\theta\phi(x) \right\} \quad \text{with} \quad \phi(x) = x^t P_0^{-1} x, \tag{3}$$

where $\theta = 1/\sigma_x^2$ is a positive parameter, P_0 is the *a priori* covariance matrix of x and $Z_2(\theta) = (2\theta/\pi)^{n/2}|P_0|^{1/2}$.

A well known case is the situation where θ, β and P_0 are known and we only want to estimate x. In fact, in this special case, the joint law $p(y, x|\theta, \beta)$ and the posterior law $p(x|y, \theta, \beta)$ are both Gaussian and we have

$$p(x|y, \theta, \beta) \propto \exp\left\{ -\frac{1}{2}\beta(y - Ax)^t(y - Ax) - \frac{1}{2}\theta x^t P_0^{-1} x \right\}, \tag{4}$$

and, if we note by

$$\hat{x} = \arg\min_x \left\{ J(x) = (y - Ax)^t(y - Ax) - \lambda x^t P_0^{-1} x \right\} \quad \text{with} \quad \lambda = \theta/\beta, \tag{5}$$

then, it is easy to show that

$$x|y \sim \mathcal{N}\left(\hat{x}, \hat{P}\right) \quad \text{with} \quad \begin{cases} \hat{x} = \beta\hat{P}A^t y \\ \hat{P} = \beta^{-1}\left(A^t A + \lambda P_0^{-1}\right)^{-1}. \end{cases} \tag{6}$$

One can make a comparison with the classical regularization techniques for inverse problems with smoothness hypothesis, where $P_0^{-1} = D^t D$ with D a matrix approximating a differentiation operator and λ is called the regularization parameter [14].

What we address here is the generalization of the problem of the determination of the regularization parameter λ which has been studied for a long time [22-28,15,18,30,31] and is still an open problem.

What is proposed here is to consider the general case where θ and β are considered to be unknown and we are facing to make inference as well about x as about them. What we propose is to consider the hyperparameters θ and β in the same manner than x, *i.e*; translate our prior knowledge about them by the probability laws $p(\theta)$ and $p(\beta)$, then determine the posterior laws and finally infer about them from these posterior laws.

2. General Bayesian inference approach

Assume now that we know the expressions of all the prior laws. We can then calculate the joint probability law:

$$p(y, x, \theta, \beta) = p(y|x, \beta)\, p(x|\theta)\, p(\theta)\, p(\beta). \tag{7}$$

In an ideal case where we are given A, y, β and θ, to infer x we can calculate the posterior law $p(x|y, \theta, \beta)$ and if we choose as the solution to our problem the Maximum *a posteriori* (MAP) estimate, we have:

$$\hat{x} = \arg\max_{x} \{p(x|y, \theta, \beta)\} = \arg\max_{x} \{p(y|x, \beta)\, p(x|\theta)\}. \tag{8}$$

But, unfortunately, in practical situations we are not given β and θ and the main problem is how to infer them. We consider the following situations:

1. The first is to estimate the three quantities simultaneously. We call this method Joint Maximum *a posteriori* (JMAP) and the estimates are defined as

$$(\hat{x}, \hat{\theta}, \hat{\beta}) = \arg\max_{(x, \theta, \beta)} \{p(y|x, \beta)\, p(x|\theta)\, p(\beta)\, p(\theta)\}. \tag{9}$$

   One practical way to do this joint optimization is to use the following algorithm

$$\begin{cases} \hat{x}^{k+1} = \arg\max_{x} \left\{ p(y|x, \hat{\beta}^{k})\, p(x|\hat{\theta}^{k}) \right\} \\ \hat{\theta}^{k+1} = \arg\max_{\theta} \left\{ p(\hat{x}^{k}|\theta)\, p(\theta) \right\} \\ \hat{\beta}^{k+1} = \arg\max_{\beta} \left\{ p(y|\hat{x}^{k}, \beta)\, p(\beta) \right\} \end{cases} \tag{10}$$

2. In the second case θ and β are considered as the nuisance parameters and are integrated out of the problem and x is estimated by

$$\begin{aligned} \hat{x} &= \arg\max_{x} \{p(x|y)\} = \arg\max_{x} \left\{ \int p(y, x, \theta, \beta)\, d\theta\, d\beta \right\} \\ &= \arg\max_{x} \left\{ \int p(y|x, \beta)\, p(\beta)\, d\beta \int p(x|\theta)\, p(\theta)\, d\theta \right\}. \end{aligned} \tag{11}$$

   We call this method Marginalized MAP type one (MMAP^1).
3. In the third case only θ is considered as the nuisance parameter and is integrated out of the problem and x and β are estimated by

$$\begin{aligned} (\hat{x}, \hat{\beta}) &= \arg\max_{(x, \beta)} \{p(x, \beta|y, \theta)\} = \arg\max_{(x, \beta)} \left\{ \int p(y, x, \theta, \beta)\, d\theta \right\} \\ &= \arg\max_{(x, \beta)} \left\{ p(y|x, \beta)\, p(\beta) \int p(x|\theta)\, p(\theta)\, d\theta \right\}. \end{aligned} \tag{12}$$

   We call this method Marginalized MAP type two (MMAP^2).

4. Finally, in the last case we may first estimate $\widehat{\theta}$ and $\widehat{\beta}$ by

$$
\begin{aligned}
(\widehat{\theta}, \widehat{\beta}) &= \arg\max_{(\theta, \beta)} \{p(\theta, \beta | y)\} = \arg\max_{(\theta, \beta)} \left\{ \int p(y, x, \theta, \beta) \, dx \right\} \\
&= \arg\max_{(\theta, \beta)} \left\{ p(\beta) \, p(\theta) \int p(y | x, \beta) \, p(x | \theta) \, dx \right\} \\
&= \arg\max_{(\theta, \beta)} \{p(\beta) \, p(\theta) l(\theta, \beta | y)\} \,.
\end{aligned}
\tag{13}
$$

and then used them for the estimation of x by

$$
\widehat{x} = \arg\max_{x} \left\{ p(x | y, \widehat{\theta}, \widehat{\beta}) \right\} \,.
\tag{14}
$$

We call this method Marginalized MAP type three (MMAP³).

Note that if $p(\theta)$ and $p(\beta)$ are uniform functions of θ and β, then $\widehat{\theta}$ and $\widehat{\beta}$ correspond to the classical maximum likelihood (ML) estimates because $l(\theta, \beta | y)$ is, for a given y, the likelihood function of θ and β.

The calculus of $l(\theta, \beta | y)$ is not easy and so is its optimization. Many works have been done on the subject. We distinguish three kind of methods:

– The first is to use the Expectation-Maximization (EM) algorithm which has been developed exactly in the context of ML parameter estimation [32,4,33].

– The second is to estimate the integral using a Monte Carlo simulation method (Stochastic EM: SEM).

– The third is to make some approximations. For example, at each iteration during the optimization, one may obtain an analytical expression for that integral by approximating the expression inside it by a second order polynomial (Gaussian quadrature approximation).

We will consider this last method.

3. A case study

Let us consider the following simple linear inverse problem $y = Ax + n$ and make the following hypothesis:

– The noise n is considered to be white, centered and Gaussian with precision β, so that we have

$$
y | x, \beta \sim \mathcal{N}(Ax, \beta^{-1}I) \longrightarrow p(y | x, \beta) = \frac{1}{Z_1(\beta)} \exp\left\{ -\frac{1}{2}\beta \|y - Ax\|^2 \right\} \,.
\tag{15}
$$

where $Z_1(\beta) \propto \beta^{m/2}$.

– Our prior prior knowledge about x can be translated by

$$
p(x | \theta) = \frac{1}{Z_2(\theta)} \exp\left\{ -\frac{1}{2}\theta\phi(x) \right\} \,.
\tag{16}
$$

where we will consider the following special cases for $\phi(x)$:

- Gaussian priors:

$$\phi_G(\boldsymbol{x}) = \boldsymbol{x}^t \boldsymbol{P}_0^{-1} \boldsymbol{x} = \|\boldsymbol{D}\boldsymbol{x}\|^2 \longrightarrow \boldsymbol{x}|\theta \sim \mathcal{N}(0, \theta^{-1} \boldsymbol{P}_0^{-1}),$$

which can also be written $\phi_G(\boldsymbol{x}) = \sum_j \sum_i p_{ij} x_i x_j$ with some special cases:

$$\phi_G(\boldsymbol{x}) = \sum_j x_j^2, \quad \text{or} \quad \phi_G(\boldsymbol{x}) = \sum_j |x_j - x_{j-1}|^2.$$

- Generalized Gaussian priors:

$$\phi_{GG}(\boldsymbol{x}) = \sum_j |x_j - x_{j-1}|^p, \quad 1 < p \leq 2.$$

- Entropic priors:

$$\phi_E(\boldsymbol{x}) = \sum_{j=1}^{n} S(x_j) \text{ where } S(x_j) = \left\{ x_j^2, \ x_j \ln x_j - x_j, \ \ln x_j - x_j \right\}.$$

- Markovian priors:

$$\phi_M(\boldsymbol{x}) = \sum_j \sum_{i \in N_j} V(x_j, x_i), \quad \text{where} \quad V(x_j, x_i) \text{ is a potential function}$$

and where N_j is a set of sites considered to be neighbors of site j, for example $N_j = \{j-1, j+1\}$, or $N_j = \{j-2, j-1, j+1, j+2\}$.

Note that, in all cases θ is generally a positive parameter. Note also that in the first case we have $Z_2(\theta) \propto \theta^{n/2}$. Unfortunately we have not an analytic expression for $Z_2(\theta)$ in the other cases. However, in the situations we are concerned with, $Z_2(\theta)$ can either be calculated numerically or approximated by

$$Z_2(\theta) \propto \theta^{\alpha n/2}. \tag{17}$$

— θ and β are both positive parameters. We choose Gamma prior laws for them:

$$\theta \sim \mathcal{G}(a, \zeta) \longrightarrow p(\theta) \propto \theta^{(a-1)} \exp\{-\zeta\theta\} \longrightarrow \mathrm{E}\{\theta\} = a/\zeta, \quad \mathrm{Var}\{\theta\} = a/\zeta^2$$

$$\beta \sim \mathcal{G}(b, \zeta) \longrightarrow p(\beta) \propto \beta^{(b-1)} \exp\{-\zeta\beta\} \longrightarrow \mathrm{E}\{\beta\} = b/\zeta, \quad \mathrm{Var}\{\beta\} = b/\zeta^2$$

Now, using the following notations

$$Q(\boldsymbol{x}) = \|\boldsymbol{y} - \boldsymbol{A}\boldsymbol{x}\|^2, \qquad J_0(\boldsymbol{x}) = \beta Q(\boldsymbol{x}) + \theta \phi(\boldsymbol{x}),$$

$$\nabla Q(\boldsymbol{x}) = -2\boldsymbol{A}^t(\boldsymbol{y} - \boldsymbol{A}\boldsymbol{x}), \quad \text{and} \quad \nabla J_0(\boldsymbol{x}) = \beta \nabla Q(\boldsymbol{x}) + \theta \nabla \phi(\boldsymbol{x}),$$

we can calculate the expression of the joint pdf $p(\boldsymbol{y}, \boldsymbol{x}, \theta, \beta) = p(\boldsymbol{y}|\boldsymbol{x}, \beta) p(\boldsymbol{x}|\theta) p(\theta) p(\beta)$, which can be written

$$p(\boldsymbol{y}, \boldsymbol{x}, \theta, \beta) \propto \theta^{-(\alpha n/2 - a + 1)} \beta^{-(m/2 - b + 1)} \exp\left\{ -\frac{1}{2} J_1(\boldsymbol{x}) \right\}, \tag{18}$$

with $\quad J_1(x) = \beta[Q(x) + 2\zeta] + \theta[\phi(x) + 2\zeta] = J_0(x) + 2\zeta(\theta + \beta).$ $\quad$ (19)

This will let us to go further in details of some of the above mentioned cases. For example in the Gaussian case we have:

$$x|y, \theta, \beta \sim \mathcal{N}(\hat{x}, \hat{P}) \text{ with } \hat{x} = \beta\left(\beta A^t A + \theta P_0^{-1}\right)^{-1} A^t y \text{ and } \hat{P} = \left(\beta A^t A + \theta P_0^{-1}\right)^{-1}$$

$$\theta|y, x, \beta \sim \mathcal{G}(a - \alpha n/2, \frac{1}{2}[\phi(x) + 2\zeta]) \longrightarrow \mathrm{E}\{\theta|y, x, \beta\} = \frac{2a - \alpha n}{[\phi(x) + 2\zeta]},$$

$$\beta|y, x, \theta \sim \mathcal{G}(b - m/2, \frac{1}{2}[Q(x) + 2\zeta]) \longrightarrow \mathrm{E}\{\beta|y, x, \theta\} = \frac{2b - m}{[Q(x) + 2\zeta]}.$$

Now, let us consider the four aforementioned methods a little more in details.

3.1. JOINT MAXIMUM A POSTERIORI (JMAP)

Using the expressions and the notations of the last paragraph in (11) we have to deal with the following algorithm:

$$\hat{x}^{k+1} = \arg\max_{x}\left\{p(y|x, \hat{\beta}^k)\, p(x|\hat{\theta}^k)\right\} = \arg\min_{x}\left\{J_0(x, \hat{\beta}^k, \hat{\theta}^k)\right\},$$

$$\hat{\theta}^{k+1} = \arg\max_{\theta}\left\{p(\hat{x}^k|\theta)\, p(\theta)\right\} = \arg\min_{\theta}\left\{[\phi(\hat{x}^k) + 2\zeta]\theta - (2a - \alpha n - 2)\ln\theta\right\},$$

$$\hat{\beta}^{k+1} = \arg\max_{\beta}\left\{p(y|\hat{x}^k, \beta)\, p(\beta)\right\} = \arg\min_{\beta}\left\{[Q(\hat{x}^k) + 2\zeta]\beta - (2b - m - 2)\ln\beta\right\}.$$

The two last equations have explicit solutions. In the case of Gaussian priors, the first equation has also an explicit solution. However, in general, we propose the following gradient based algorithm:

Algorithm 1: $\quad \hat{x}^{k+1} = (1 - \mu)\hat{x}^k - \mu\nabla J_0(\hat{x}^k, \hat{\beta}^k, \hat{\theta}^k)$

$$= (1 - \mu)\hat{x}^k - \mu[\hat{\beta}^k\nabla Q(\hat{x}^k) + \hat{\theta}^k\nabla\phi(\hat{x}^k)], \quad 0 < \mu < 1,$$

$$\hat{\theta}^{k+1} = \frac{(2a - \alpha n - 2)}{[\phi(\hat{x}^k) + 2\zeta]}, \quad a > (\alpha n + 2)/2,$$

$$\hat{\beta}^{k+1} = \frac{(2b - m - 2)}{[Q(\hat{x}^k) + 2\zeta]}, \quad b > (m + 2)/2.$$

The conditions $a > (\alpha n + 2)/2$ and $b > (m + 2)/2$ are added to satisfy, when necessary, the positivity constraint of $\hat{\theta}$ and $\hat{\beta}$.

3.2. MARGINALIZED MAXIMUM A POSTERIORI MMAP[1]

Considering θ and β as the nuisance parameters and integrating out them from $p(y, x, \theta, \beta)$ we obtain

$$p(y, x) = \iint p(y, x, \theta, \beta)\, d\beta\, d\theta \propto [Q(x) + 2\zeta]^{-(m - 2b)/2}\,[\phi(x) + 2\zeta]^{-(\alpha n - 2a)/2} \quad (20)$$

Now, defining $\hat{x}_{\text{MMAP}} = \arg\max_{x} \{p(x|y)\} = \arg\min_{x} \left\{ \frac{1}{2} J_2(x) \right\},$

with $J_2(x) = (2a - \alpha n)\ln[Q(x) + 2\zeta] + (2b - m)\ln[\phi(x) + 2\zeta],$ (21)

and trying to calculate this solution by an iterative gradient based algorithm, we have to calculate

$$\nabla J_2(x) = \frac{(2a - \alpha n)}{[Q(x) + 2\zeta]}\nabla Q(x) + \frac{(2b - m)}{[\phi(x) + 2\zeta]}\nabla\phi(x).$$

We propose then the following iterative algorithm:

Algorithm 2: $\begin{aligned} \hat{x}^{k+1} &= (1 - \mu)\hat{x}^k - \mu\nabla J_2(\hat{x}^k) \\ &= (1 - \mu)\hat{x}^k - \mu[\hat{\beta}^k\nabla Q(\hat{x}^k) + \hat{\theta}^k\nabla\phi(\hat{x}^k)], \quad 0 < \mu < 1, \\ \hat{\theta}^k &= \frac{(2a - \alpha n)}{[\phi(\hat{x}^k) + 2\zeta]}, \quad a > \alpha n/2, \\ \hat{\beta}^k &= \frac{(2b - m)}{[Q(\hat{x}^k) + 2\zeta]}, \quad b > m/2. \end{aligned}$

3.3. MARGINALIZED MAXIMUM A POSTERIORI MMAP^2

In this case, θ only is considered as a nuisance parameter and is integrated out:

$$\begin{aligned} p(y, x, \beta) &= \int p(y, x, \theta, \beta)\,d\theta \\ &\propto \beta^{-m/2 + b - 1}[\phi(x) + 2\zeta]^{-(\alpha n - 2a)/2}\exp\left\{ -\frac{1}{2}\beta[Q(x) + 2\zeta] \right\}.\end{aligned} \quad (22)$$

Then, x and β are estimated by

$$(\hat{x}, \hat{\beta}) = \arg\max_{x, \beta} \{p(y, x, \beta)\}. \quad (23)$$

Noting

$$-2\ln p(y, x, \beta) = -(2b - m - 2)\ln\beta + (2a - \alpha n)\ln[\phi(x) + 2\zeta] + \beta[Q(x) + 2\zeta]$$

and differentiating it with respect to β gives

$$\hat{\beta} = \frac{2b - m - 2}{[Q(x) + 2\zeta]}.$$

So, noting

$$J_3(x, \beta) = (2a - \alpha n)\ln[\phi(x) + 2\zeta] + \beta[Q(x) + 2\zeta] \quad (24)$$

and $\nabla J_3(x, \beta) = \frac{2a - \alpha n}{[\phi(x) + 2\zeta]}\nabla\phi(x) + \beta\nabla Q(x),$

and using a gradient based algorithm for minimizing J_3 with respect to x we propose the following:

Algorithm 3:
$$\widehat{x}^{k+1} = (1-\mu)\widehat{x}^{k+1} - \widehat{\beta}^k \nabla Q(x) - \widehat{\theta}^k \nabla \phi(x), \quad 0 < \mu < 1,$$
$$\widehat{\theta}^k = \frac{2a - \alpha n}{[\phi(\widehat{x}^k) + 2\zeta]}, \quad a > (\alpha n + 2)/2$$
$$\widehat{\beta}^k = \frac{2b - m - 2}{[Q(\widehat{x}^k) + 2\zeta]}, \quad b > (m+2)/2.$$

3.4. MAXIMUM LIKELIHOOD OR MMAP[3]

In this case first x integrated out x from $p(y, x, \theta, \beta)$ to obtain:

$$p(y, \theta, \beta) = \int p(y, x, \theta, \beta)\, dx = \frac{\beta^{(b-1)}}{Z_2(\beta)} \frac{\theta^{(a-1)}}{Z_1(\theta)} \int \exp\left\{-\frac{1}{2} J_1(x, \beta, \theta)\right\} dx \quad (25)$$

with
$$J_1(x, \beta, \theta) = \beta[Q(x) + 2\zeta] + \theta[\phi(x) + 2\zeta]. \quad (26)$$

Excepted the Gaussian case where J_1 is a quadratic function of x, in general, it is not easy to obtain an analytical expression for this integral. One can then try to make a Gaussian approximation which means to develop J_1 around its minimum $\widehat{x}_{\text{MAP}} = \arg\min_x \{J_1(x, \beta, \theta)\}$ by

$$J_1(x, \beta, \theta) \simeq \frac{1}{2}(x - \widehat{x}_{\text{MAP}})^t M (x - \widehat{x}_{\text{MAP}}) + g^t(x - \widehat{x}_{\text{MAP}}) + c, \quad (27)$$

where $g = \beta \nabla Q(x) + \theta \nabla \phi(x)$ is the gradient of J_1 and M is its Hessian, both calculated for $\widehat{x}_{\text{MAP}}$. With this approximation we obtain

$$p(y, \theta, \beta) = \beta^{-m/2 + b - 1} \theta^{-\alpha n/2 + a - 1} |M(\beta, \theta)|^{-\frac{1}{2}} \exp\left\{-\frac{1}{2} J_1(\widehat{x}_{\text{MAP}}, \beta, \theta)\right\}. \quad (28)$$

Differentiating $l(\theta, \beta|y) = \ln p(y, \theta, \beta)$ with respect to β and θ gives

$$\widehat{\beta} = \frac{2b - m - 2}{[Q(\widehat{x}^k) + 2\zeta] + \text{trace}[M^{-1} A^t A]}, \quad \widehat{\theta} = \frac{2a - \alpha n - 2}{[\phi(\widehat{x}^k) + 2\zeta] + \text{trace}[M^{-1} P_0^{-1}]}.$$

where P_0^{-1} is the Hessian of $\phi(x)$, $A^t A$ is the Hessian of $Q(x)$ and M is the Hessian of $J_1(x)$:

$$M(\beta, \theta) = \beta A^t A + \theta P_0^{-1}.$$

Using these expressions we propose the following algorithm:

Algorithm 4:
$$\widehat{x}^k = \arg\min_x \left\{J_1(x, \widehat{\beta}^k, \widehat{\theta}^k)\right\} = M(\widehat{\beta}^k, \widehat{\theta}^k)^{-1} A^t y,$$
$$\widehat{\theta}^{k+1} = \frac{2a - \alpha n - 2}{[\phi(\widehat{x}^k) + 2\zeta] + \text{trace}[M^{-1} P_0^{-1}]},$$
$$\widehat{\beta}^{k+1} = \frac{2b - m - 2}{[Q(\widehat{x}^k) + 2\zeta] + \text{trace}[M^{-1} A^t A]}.$$

This algorithm needs the inversion of the matrix M which is very costly in practice.

4. Comparison and the main structure of the proposed algorithms

Comparing the **Algorithms 1 to 4**, one can see that they all have the same structure:

- for fixed θ and β optimize locally a criterion $J(x, \beta, \theta)$, and
- update θ and β using the solution $\hat{x}$ just obtained and iterate until convergence.

Note also that only in **Algorithm 4**, the updating step takes account of the measurement system operator A and the covariance structure P_0 of the input x.

5. Conclusions and perspectives

We considered the inverse problem of infering the unknowns x from the data y in a special case of linear inverse problems $y = Ax + n$ using a full bayesian approach and presented four algorithms to estimate simultanously the hyperparameters θ and β and the unknowns x. The main structure of all of these algorithms are the same even if the procedure to deduce them have been different. However, we have not yet really tested them to give any conclusion about their relative performances. Note however that one of them distinguishes itself from the others by taking account of the measurement system operator A and the covariance structure P_0 of x in the hyperparameters updating step and, by the same way, by its calculation cost. We hope to be able to give some measure of their relative performances in simulation and in real applications in near future.

References

1. G. Box and G.C. Tiao, *Bayesian inference in statistical analysis*. Addison-Wesley publishing, 1972.
2. H. Sorenson, *Parameter estimation*. Marcel Dekker, Inc., 1980.
3. J. Besag, "Digital image processing : Towards Bayesian image analysis," *Journal of Applied Statistics*, vol. 16, no. 3, pp. 395–407, 1989.
4. P. J. Green, "Bayesian reconstructions from emission tomography data using a modified EM algorithm," *IEEE Transactions on Medical Imaging*, vol. 9, pp. 84–93, Mar. 1990.
5. D. Malec and J. Sedransk, "Bayesian methodology for combining the results from different experiments when the specifications for pooling are uncertain," *Biometrika*, vol. 79, no. 3, pp. 593–601, 1992.
6. G. Gindi, M. Lee, A. Rangarajan, and Z. I., "Bayesian reconstruction of functional images using anatomical information as priors," *IEEE Transactions on Medical Imaging*, vol. MI-12, no. 4, pp. 670–680, 1993.
7. J. Bernardo and A. Smith, *Bayesian Theory*. Chichester, England: John Wiley, 1994.
8. Barndorff-Nielsen, *Information and Exponential Model in Statistics*. New-York: John Wiley, 1978.
9. H. Derin, H. Elliott, R. Cristi, and D. Geman, "Bayes smoothing algorithms for segmentation of binary images modeled by markov random fields," *IEEE Transactions on Pattern Analysis and Machine Intelligence*, vol. PAMI-6, p. 4, 1984.
10. S. Geman and D. Geman, "Stochastic relaxation, Gibbs distributions, and the Bayesian restoration of images," *IEEE Transactions on Pattern Analysis and Machine Intelligence*, vol. PAMI-6, p. 2, 1984.

11. A. Tarantola, *Inverse problem theory : Methods for data fitting and model parameter estimation.* Amsterdam: Elsevier Science Publishers, 1987.

12. J. Skilling, *Maximum-Entropy and Bayesian Methods.* Dordrecht, The Netherlands: Kluwer Academic Publisher, 1988.

13. Titterington and Rossi, "Another look at a Bayesian direct deconvolution method," *Signal Processing,* vol. 9, pp. 101–106, 1985.

14. G. Demoment, "Image reconstruction and restoration : Overview of common estimation structure and problems," *IEEE Transactions on Acoustics Speech and Signal Processing,* vol. 37, pp. 2024–2036, Dec. 1989.

15. K.-Y. Liang and D. Tsou, "Empirical Bayes and conditional inference with many nuisance parameters," *Biometrika,* vol. 79, no. 2, pp. 261–270, 1992.

16. R. E. McCulloch and P. E. Rossi, "Bayes factors for nonlinear hypotheses and likelihood distributions," *Biometrika,* vol. 79, no. 4, pp. 663–676, 1992.

17. J. Idier and Y. Goussard, "Markov modeling for Bayesian restoration of two-dimensional layered structures," *IEEE Transactions on Information Theory,* vol. 39, pp. 1356–1373, July 1993.

18. A. Mohammad-Djafari, "On the estimation of hyperparameters in Bayesian approach of solving inverse problems," in *Proceedings of IEEE ICASSP,* pp. 567–571, 1993.

19. A. Nallanathan and W. J. Fitzgerald, "Bayesian model selection applied to spatial signal processing," *Proceedings of the IEE,* vol. 141, pp. 76–80, Feb. 1994.

20. J. Diebolt and C. P. Robert, "Estimation of finite mixture distributions through Bayesian sampling," *Journal of Royal Statistical Society B,* vol. 56, no. 2, pp. 363–375, 1994.

21. H. Carfantan and A. Mohammad-Djafari, "A Bayesian approach for nonlinear inverse scattering tomographic imaging," in *Proceedings of IEEE ICASSP,* vol. IV, pp. 2311–2314, May 1995.

22. J. Cullum, "The effective choice of the smoothing norm in regularization," *Math. Comp.,* vol. 33, pp. 149–170, 1979.

23. Titterington, "General structure of regularization procedures in image reconstruction," *Astrononmy and Astrophysics,* vol. 144, pp. 381–387, 1985.

24. L. Younès, "Estimation and annealing for Gibbsian fields," *Annales de l'institut Henri Poincaré,* vol. 24, pp. 269–294, Feb. 1988.

25. S. Lakshmanan and H. Derin, "Simultaneous parameter estimation and segmentation of Gibbs random fields using simulated annealing," *IEEE Transactions on Pattern Analysis and Machine Intelligence,* vol. PAMI-11, no. 8, pp. 799–813, 1989.

26. A. Mohammad-Djafari and J. Idier, "Maximum likelihood estimation of the lagrange parameters of the maximum entropy distributions," in *Maximum Entropy and Bayesian Methods in Science and Engineering* (C. Smith, G. Erikson, and P. Neudorfer, eds.), pp. 131–140, Kluwer Academic Publishers, 1991.

27. Thompson, Brown, Kay, and Titterington, "A study of methods of choosing the smoothing parameter in image restoration by regularization," *IEEE Transactions on Pattern Analysis and Machine Intelligence,* vol. 13, Apr. 1991.

28. E. Gassiat, F. Monfront, and Y. Goussard, "On simultaneous signal estimation and parameter identification using a generalized likelihood approach," *IEEE Transactions on Information Theory,* vol. IT-38, pp. 157–162, 1992.

29. T. J. Hebert and R. Leahy, "Statistic-based map image reconstruction from poisson data using Gibbs prior," *IEEE trans. on Signal Processing,* vol. 40, pp. 2290–2303, Sept. 1992.

30. C. Bouman and K. Sauer, "Maximum likelihood scale estimation for a class of markov random fields penalty for image regularization," in *Proceedings of IEEE ICASSP,* vol. V, pp. 537–540, 1994.

31. A. N. Iusem and B. F. Svaiter, "A new smoothing-regularization approach for a maximum-likelihood problem," *Applied Mathematics and Optimization,* vol. 29, pp. 225–241, 1994.

32. A. Dempster, N. Laird, and D. Rubin, "Maximum likelihood from incomplete data via the EM algorithm," *Journal of Royal Statistical Society B,* vol. 39, pp. 1–38, 1977.

33. Vardi and Lee, "From image deblurring to optimal investments maximum likelihood solutions for positive linear inverse problems," *Journal of Royal Statistical Society B,* vol. 55, no. 3, pp. 569–612, 1993.

PIXON-BASED MULTIRESOLUTION IMAGE RECONSTRUCTION AND QUANTIFICATION OF IMAGE INFORMATION CONTENT

R.C. PUETTER

Center for Astrophysics and Space Sciences
University of California, San Diego
9500 Gilman Drive
La Jolla, CA 92093-0111[†]

Abstract. This paper describes the theory of pixon-based image reconstruction. After a brief introduction of the basic concepts of the pixon, the paper concentrates primarily on our current implementation of the techniques along with the approximations and shortcuts that seem to provide practical software algorithms. We then present an example of application of the method to astrophysical data, i.e. imaging of the Einstein ring in the gravitational lens of FSC 10214+4724.

Key words: pixons, image reconstruction, Bayesian estimation, multiresolution algorithmic complexity, maximum entropy, maximum likelihood

1. Image Reconstruction and the Inverse Problem

This paper presents a practical tutorial on how to implement pixon-based image reconstruction. Pixons (or *informatons*—their extension to generalized data sets), represent an optimized, information theory-based coordinate system for performing parameter estimation. Pixon-based methods are now becoming recognized as one of the most successful methods for practical image reconstruction, with performance exceeding those of other commonly used techniques such as Maximum Likelihood methods and Maximum Entropy methods. Like many of these methods, pixon-based reconstruction can be viewed within the Bayesian estimation framework. Hence before going into the particulars of pixon-based image reconstruction, we shall describe Bayesian methods briefly and the interpretation of pixon-based concepts within this formalism. Since we intend this paper to provide a practical tutorial for the implemetation of pixons, we shall gloss over much of the theoretical discussion presented in other papers on pixons. For readers more interested in these topics, we refer you to the papers [1-4], and especially my review article [5].

[†]Email: rpuetter@ucsd.edu

K. M. Hanson and R. N. Silver (eds.), Maximum Entropy and Bayesian Methods, 145–152.
© 1996 *Kluwer Academic Publishers.*

2. Bayesian Methods

For the general inverse problem, the method of Bayesian estimation seeks to infer the most probable set of "parameters-of-interest" from a set of measured or otherwise given parameters. In the case of image restoration, this normally means estimating a higher resolution (or undistorted) image from a blurry (or distorted) image. In order to use probabilitistic methods, one typically examines the joint probability distribution of the data, D, the underlying, unblurred image which is to be estimated, I, and the model, M, which describes the detailed relationship of the data to the image. This joint probability distribution, $p(D, I, M)$, can be factored using conditional probabilites in the following manner:

$$
\begin{align}
p(D, I, M) &= p(D|I, M)p(I, M) \tag{1} \\
&= p(D|I, M)p(I|M)p(M) \tag{2} \\
&= p(I|D, M)p(D, M) \tag{3} \\
&= p(I|D, M)p(D|M)p(M) \tag{4} \\
&= p(MI|D, I)p(D, I) \tag{5} \\
&= p(MI|D, I)p(D|I)p(I) \quad , \tag{6}
\end{align}
$$

where $p(x|y)$ is the conditional probability of x given y, i.e. the probability of x given that y has a particular value. Equations (1) through (6) can be rearranged to give:

$$
\begin{align}
p(I|D, M) &= \frac{p(D|I, M)p(I|M)p(M)}{p(D|M)p(M)} \tag{7} \\
&= \frac{p(D|I, M)p(I|M)}{p(D|M)} \tag{8} \\
&\propto p(D|I, M)p(I|M) \quad , \tag{9}
\end{align}
$$

or

$$
\begin{align}
p(I, M|D) &= \frac{p(D|I, M)p(I, M)}{p(D)} \tag{10} \\
&\propto p(D|I, M)p(I|M) \quad . \tag{11}
\end{align}
$$

The formulae given in equations (7)-(9) are the basic starting point for Bayesian image reconstruction. These equations essentially assume that the appropriate model, M, is known and will not be varied during the reconstruction process. This assumption is made more explicit in equation (9) where the term $p(D|M)$ is dropped (it is assumed to be constant—M and D are not varied during the reconstruction). Our preferred formulation, however, is given in equations (10)-(11). Here, we do not assume that all of the parameters associated with the model are "nuisance" parameters. Indeed, in pixon-based reconstruction the model parameters associated with the local smoothness scale (see below) are extremely important. Hence we have chosen to keep some of the "model" parameters on an equal footing with the image, and to seek the most likely estimate of their values simultaneously with the image.

3. Choice of Image Representation and Pixon-Based Methods

Notable successes of previous Bayesian image reconstruction methods, e.g. Maximum Likelihood (ML) and Maximum Entropy (ME), are greatly reduced noise propagation errors relative to linear methods (e.g. Fourier convolution inversion) and consequently greatly enhanced resolution and spurious source rejection. While these methods represent a significant advance, they still display significant problems. ML methods typically over-fit the data and ME and ML methods usually under-estimate source brightnesses. Both methods suffer from spurious source production although at lower levels than linear methods.

One of the fundamental sources of the problems with these methods is their selection of the coordinate system in which to represent the image. For most modern electronic imaging systems, the seemingly natural choice for the image coordinate system is the same rectangular pixel grid that is used for the data. However this is actually a very arbitrary coordinate system. This is easily seen if the data is very noisy or the underlying image is intrinsically very smooth. In either case the collected data typically is inadequate to constrain as many independent numbers as are contained in the pixel coordinate grid. Such a situation will give rise to numerical artifacts since the inversion algorithm cannot distinguish between solutions contained in such a large solution space. Furthermore, since the typical member of the solution space has many spurious sources, the selection of such a solution is virtually guaranteed. The only way to prevent these problems is to constraint the solution in some manner.

Both Bayesian methods and concepts from information science suggest parts of the solution to this problem. Modern Bayesian interpretations of Occam's razor indicate that all that is required to solve the spurious source problem is to use the simplest hypothesis to explain the data consistent with its noise properties. In other words we need to optimize the prior while keeping the Goodness-of-Fit criterion, $p(D|I, M)$, optimized. The information science concepts of Algorithmic Information Content (AIC) further tells us that the terseness of a description (and hence its simplicity) is a strong function of the complexity (descriptiveness) of the language in which it is expressed [6-9]. Thus in order to achieve the Bayesian ideal of Occam's razor we need to select a descriptive language in which the image can be described in the most concise manner. Since the language for describing the image is normally associated with the model, i.e. part of the nuisance parameters, in order to achieve this goal we must be willing to vary the "image description language", which is part of M, as part of the problem.

We are now left with the problem of how to select a highly descriptive language for generic images. Again, work in other disciplines that deal routinely with images suggest at least one practical solution. In a variety of endeavors (e.g. image classification, morphology, and segmentation; computational methods; image compression; image restoration; video encoding and high definition TV; wavelet theory and applications; medical imaging; neural networks, and motion detection—see [5] and references therein) multiresolution image languages are receiving considerable attention. The success of multiresolution languages is not surprising. In fact, these ideas are familiar to most scientists and are fundamental to simple, well known concepts, e.g. the use of fine grain photographic film to capture fine detail—the grain size is smaller than the finest scale required to be captured. It is quite clear, for example, that one approach to concisely describing the information in an image is to use fewer degrees-of-freedom (DOFs) per unit area in portions of the image which are smooth and a greater density of degrees-of-freedom where there is greater detail. Indeed, each of these degrees-of-freedom might be likened to a single photographic grain or a generalized pixel. The value of this pixel represents the average brightness in a given region. This, in fact, is the origin of the name "pixon". Each pixon is a single DOF used to describe the image in a particular region. The "pix" part of the name recognizes its pixel heritage, while the "on" suffix recognizes its more fundamental nature. The pixon is fundamental to the image, not to the instrument that took the picture. In the section below, we shall outline in more detail our specific implementation of a multiresolution, pixon language for image description.

4. A Multiresolution Pixon Language for Image Description

In the section above, we argued that a multiresolution image description language would be suitable (i.e. concise) for describing generic images. In fact, what we discovered is that in order to optimize the Bayesian prior we need the simplest image/model combination possible. While detailed prior knowledge of the properties of the image would allow even greater optimization of the prior, our knowledge is usually incomplete. That is why we are talking about *generic* images and *generic* languages to describe images. Use of detailed prior knowledge will always allow us to construct better image models, but if we are clever about our construction of models for generic images, we should not do too badly for most types of images we shall encounter.

Again, since our goal is to optimize the image/model prior, $p(I, M)$, we see that this can be accomplished by model simplification, i.e. using the fewest degrees-of-freedom necessary to fit the data within the accuracy demanded by the noise [remember the product of the terms $p(D|I, M)$ and $p(I, M)$, i.e. $p(I, M|D)$ is the quantity to be optimized]. Thus any scheme which controls the local density of DOFs used to describe the local image information is suitable. Thus rather than using image signal contained in cells with hard boundaries, we have chosen to use a local correlation scale formulation to control the DOF density. We call these "fuzzy pixons". To formalize the definition, at each point in the image, $\vec{x}$, we write the image, I, as

$$I(\vec{x}) \;=\; (K \otimes I_{pseudo})(\vec{x}) \tag{12}$$

$$(K \otimes I_{pseudo})(\vec{x}) \;=\; \int dV_{\vec{y}}\; K(\vec{x}, \vec{y}, \delta(\vec{x})) I_{pseudo}(\vec{y}) \tag{13}$$

$$K(\vec{x}, \vec{y}, \delta(\vec{x})) \;=\; K\left(\frac{\|\vec{x} - \vec{y}\|}{\delta(\vec{x})}\right) \quad ; \text{ for radially symmetric pixons , } \tag{14}$$

where equations (12) through (14) show that the image is a local convolution of a "pseudo-image" with a blurring function with a given local scale, $\delta(\vec{x})$. Note that the local scale varies with position $\vec{x}$ in the image and the integration in equation (13) is carried out over volume in pseudo-image space. We have also indicated in equation (14) that one suitable functional form for the pixon kernel function, K, might be a radially symmetric function that depends only on the distance between the kernel center and the image position ratioed to the local scale. We have found this functional form to work quite well for centrally peaked kernel functions with a finite foot-print. We normally use truncated paraboloids, i.e.

$$K\left(\vec{x}, \vec{y}, \delta(\vec{x})\right) = \begin{cases} \left(1 - \frac{\|\vec{x}-\vec{y}\|^2}{\delta(\vec{x})^2}\right) / \int dV_{\vec{y}} \left(1 - \frac{\|\vec{x}-\vec{y}\|^2}{\delta(\vec{x})^2}\right) & ; \|\vec{x} - \vec{y}\| \leq \delta(\vec{x}) \\ 0 & ; \|\vec{x} - \vec{y}\| > \delta(\vec{x}) \end{cases} \quad (15)$$

Of course AIC concepts suggest that a richer language, e.g. elliptical pixons, would yield a more concise image description which Occam's razor then says should have a more optimized prior. In fact, we have used elliptical pixons recently to perform some image restorations—see below. Nonetheless, for generic images it seems clear that we have reached a point of diminishing returns and it is unlikely that pixon kernels more complicated than ellipses are warranted for generic images.

5. Solving for the M.A.P. Image/Model Pair

Now that we have described a suitable language in which to solve the problem, we shall move on to the details of finding the M.A.P. (Maximum *A Posteriori*) image/model pair, i.e. the image/model pair that maximizes $p(I, M|D)$. As is clear from equation (11), we need to maximize the product of $p(D|I, M)$ and $p(I, M)$. The appropriate choice for $p(D|I, M)$ in the case of pixelized data with independent Gaussian noise is $p(D|I, M) = (2\pi\sigma^2)^{1/2n} \exp(-\chi^2/2)$ where χ^2 is the standard chi-squared value, [i.e. $\chi^2 = \sum_1^n (x_i - \langle x_i \rangle)^2/\sigma^2$], σ is the standard deviation of the noise, and n is the number of independent measure parameters (here the number of pixels). A choice for $p(I, M)$ is less obvious. The choice of Maximum Entropy enthusiasts is something like

$$p(I|M) = \frac{N!}{n^N \prod_{i=1}^n N_i!} = e^S \quad (16)$$

$$p(M) = constant \quad , \quad (17)$$

where n is the number of pixels, N is the total number of counts in the image, N_i is the number of counts in pixel i, and S is the entropy. This choice is made on the basis of counting arguments. The arguments are basically sound with the only exception being that the pixel cells are not the most appropriate choice for an image's basis. A more appropriate choice would be the multiresolution pixon basis discussed above. While the ME prior of equations (16)-(17) states that the most likely *a priori* image is flat in the pixel coordinate system, evaluating equation (16) using fuzzy pixon cells rather than pixels, would state that once the structure of the image is known (i.e. all the local smoothness scales are determined) we would guess *a priori* that each of the cells would have equal brightness.

In essence then, we could use the same prior as used by ME workers with the pixons substituted for the pixels. The *a priori* probability arguments for the prior of equations (16)-(17) remain valid. It is just that we now recognize that the pixons are the appropriate coordinate system and that we will probably use vastly fewer pixons (i.e. DOFs) to describe the image than there are pixels in the data. To make the formulae explicit, we would define the pseudo-image on a psuedo-grid which is as least as fine as the pixel grid (we use this finer grid for the image too) and then use the following substitutions:

$$p(I|M) = \frac{N!}{n^N \prod_{i=1}^{n_{pixons}} N_i!} \tag{18}$$

$$N = \sum_{i=1}^{n_{pixons}} N_i = \sum_{j=1}^{n_{pixels}} N_j \tag{19}$$

$$N_i = \int dV_{\vec{y}} \, k(\vec{x}, \vec{y}, \delta(\vec{x})) I_{pseudo}(\vec{y}) \tag{20}$$

$$\prod_{i=1}^{n_{pixons}} N_i! = \prod_{j=1}^{n_{pixels}} (N_j!)^{p_j} \tag{21}$$

$$p(\vec{x}) = 1 / \int dV_{\vec{y}} \, k(\vec{x}, \vec{y}, \delta(\vec{x})) \tag{22}$$

$$p(M) = constant \tag{23}$$

where N_i is the number of counts in pixon i, N_j is the number of counts in pixel j, p is the pixon density, i.e. the number of pixons per pseudo-pixel (in image space), and $k(\vec{x}, \vec{y}, \delta(\vec{x}))$ is the pixon shape kernel normalized to unity at $\vec{y} = \vec{x}$.

To obtain the M.A.P. image/model pair, one could now proceed directly by minimizing the product of $(2\pi\sigma^2)^{1/2n_{pixels}} \exp(-\chi^2/2)$ and equation (18) with respect to the local scales, $\{\delta(\vec{x_j})\}$—the pixon map, and the pseudo-image values, $\{I_{pseudo,j}\}$. However, this is not what we have done in practice. Instead, we have divided the problem into a sequential chain of two repeated steps: (1) optimization of the pseudo-image with the local scales held fixed, foloowed by (2) optimization of the local scales with the pseudo-image held fixed. The sequence is then repeated until convergence. To formally carry out this proceedure, in step (1) we should find the M.A.P. pseudo-image, i.e. the pseudo-image that maximizes $p(I|D, M)$—note that we are using the notation here that the local scales belong to M, while the pseudo-image values are associated with I. In step (2) we would then find the M.A.P. model, i.e. the scales that maximize $p(M|D, I)$.

While the above proceedure is quite simple, we have made still further simplifications. In neither step do we evaluate the prior. We simply evaluate the GOF term $p(D|I, M)$. So in step (1) we essentially find the Maximum Likelihood pseudo-image with a given pixon map. In step (2) we must take into account some knowledge of the pixon prior, but we simply use the fact the the pixon prior of equation

Raw, Flat-Fielded Data

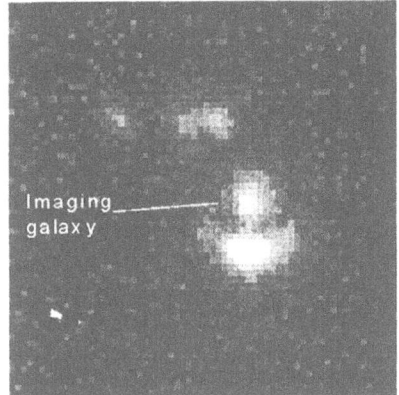

Imaging galaxy

The Einstein Ring of FSC 10214+4724

Keck K-band imaging data of Graham and Liu (1995)

Elliptical Pixon Reconst.

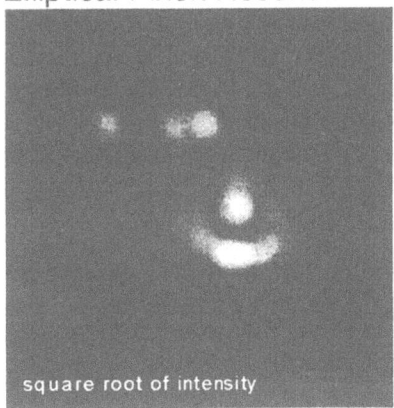

square root of intensity

Max Entropy Reconst.

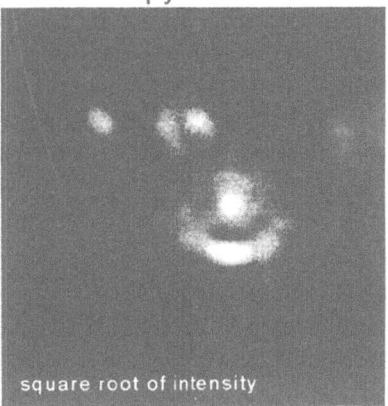

square root of intensity

Figure 1. Comparison of image reconstruction techniques for the Einstein ring of FSC 10214+4724. The elliptical pixon reconstruction used 196 different elliptical pixon kernel functions. Relative to radially symmetric reconstructions, the elliptical pixon reconstruction used roughly 50% fewer DOFs in the image model.

(18) increases rapidly as the number of pixons are decreased and the remaining pixons are packed as full of signal as possible. So what we do is, at each pseudo-grid point, j, we attempt to increase the local scale until it is no longer possible to fit the data within the noise. In other words at each pseudo-grid point we progressively smooth the current pseudo-image until the GOF criterion is "violated". We then select the largest scale at each point which was acceptable and use these values in the next iteration of step (1).

There is one more practical matter to consider. As with any interative method, there can be convergence problems. With the approach outlined above, we have noticed that if small scales are allowed early in the solution, then these scales become "frozen-in", even if they would have later proved inappropriately small. To solve this problem, we start out the pseudo-image calculation with initially large pixons scales. We then use our pixon-calculator [the code that performs step (2) by selecting the largest scales that still provide an adequate fit to the data] to determine new scales. The pixon-calculator reports that over some of the image the initial scales are fine, but over other parts of the image smaller scales are required. At this point, however, we do not use the smallest scales requested. Instead, we let the scales get somewhat smaller over the portion of the image which was determined to require smaller scales and proceed to step (1) of the next iteration. We repeat this process letting the smallest scales allowed get smaller and smaller until the method converges. The process has proven to be very robust.

6. A Sample Restoration: The Einstein Ring in FSC 10214+4724

Figure 1 presents an elliptical pixon-based reconstruction of the Keck Telescope K-band imaging data of [10]. Also shown is an ME reconstruction of the same data from [10]. As can be seen, the pixon-based reconstruction reveals considerably more of the Einstein ring. This improved sensitivity is possible because of the ability of pixon-based methods to develop a critical image model. Also note the rejection by the pixon-based reconstruction of the many extended fine structure features in the ME reconstruction. The apparently finer resolution in the ME reconstruction is unwarranted since the pixon-based solution has an equally good statistical fit.

Acknowledgements

The author would like to acknowledge the many valuable contributions from his collaborators and colleagues, epecially R. Piña and N. Weir. The author also thanks J. Graham for making his data and ME reconstruction available. This research was supported by the NSF, NASA, the California Association for Research in Astronomy, and Cal Space.

References

1. Piña, R.K., and Puetter, R.C. (1993) Bayesian Image Reconstruction: The Pixon and Optimal Image Modeling, *P.A.S.P.*, **105**, pp. 630–637
2. Puetter, R.C., and Piña, R.K. (1993a) The Pixon and Bayesian Image Reconstruction, *Proc. S.P.I.E.*, **1946**, pp. 405–416
3. Puetter, R.C., and Piña, R.K. (1993b) Pixon-Based Image Reconstruction, *Proc. MaxEnt '93*, in press
4. Puetter, R.C. (1994) Pixons and Bayesian Image Reconstruction, *Proc. S.P.I.E.*, **2302**, pp. 112-131
5. Puetter, R.C. (1995) Pixon-Based Multiresolution Image Reconstruction and the Quantification of Picture Information Content, *Int. J. Image Sys. & Tech.*, in press
6. Solomonoff, R. (1964) *Inf. Control*, **7**, p. 1
7. Kolmogorov, A.N. (1965) *Inf. Transmission*, **1**, p. 3
8. Chaitin, G.J. (1966) *J. Ass. Comput. Mach.*, **13**, p. 547
9. Zurek, W.H. (1989) Thermodynamic Cost of Computation, Algorithmic Complexity and the Information Metric, *Nature*, **341**, p. 119
10. Graham, J.R., and Liu, M.C. (1995) High Resolution Infrared Imaging of FSC 10214+4724: Evidence for Gravitational Lensing, *Ap.J.(Letters)*, **449**, p. L29

BAYESIAN MULTIMODAL EVIDENCE COMPUTATION BY ADAPTI TEMPERING MCMC

MIAO-DAN WU AND WILLIAM J. FITZGERALD

Signal Processing & Communications Laboratory
Cambridge University Engineering Department
Trumpington Street, Cambridge CB2 1PZ, U.K. [†]

Abstract. The key to the numerical implementation of Bayesian model selection and Bayesian model averaging is the computation of the model evidences. Defined as the probability of the observed data given a model, the model evidence can be shown to be equal to the normalising constant of the underlying posterior distribution. When posterior unimodality is a valid assumption, it can be adequately evaluated using methods such as the Laplace method and standard Markov chain Monte Carlo (MCMC) simulation techniques. However, in cases where posterior distributions are significantly multimodal, which typically arise in ill-conditioned problems, these methods tend to get trapped in local modes and consequently do not yield robust results. To facilitate movement around the multimodal parameter space, we introduce the reciprocal temperature parameter β and define an auxiliary distribution over both the parameters of interest and β. The Gibbs sampler is then employed to simulate this joint distribution and an importance reweighting procedure is incorporated for adaptively updating the ratios of the partition functions. The thermodynamic integration approach is adopted to evaluate the partition function at the highest temperature, at which point departure from unimodality is expected to be insignificant. The partition function at $\beta = 1$, which in the new formulation corresponds to the desired evidence value, can thus be obtained by multiplying the partition function evaluated at the highest temperature with the ratio of these two partition functions. The algorithm is illustrated by applying it to a simulated trimodal distribution.

Key words: Evidence, MCMC, Simulated tempering, Adaptive tempering, Thermodynamic integration

---

† Email: mdw@cam.ac.uk

K. M. Hanson and R. N. Silver (eds.), Maximum Entropy and Bayesian Methods, 153–161.
© 1996 *Kluwer Academic Publishers.*

1. Introduction

There are numerous statistical inference problems that require model selection with respect to a set of competing models. The Bayesian solution to the model selection problem is to choose one that fits the prior knowledge and data the best, i.e., one that has the highest posterior probability. There are also circumstances in which model selection is not an end in itself and the purpose of some inference process concerns estimation or prediction. In those cases, employing just one model ignores model uncertainty. To alleviate the undesirable effects of model uncertainty, one can adopt the Bayesian model averaging (BMA) approach [1][2] , in which the final estimation or prediction is a result of averaging across a variety of plausible models.

To illustrate the idea of BMA, consider the case in which one has K parametric probability models $\mathcal{M}_\infty, \ldots, \mathcal{M}_\mathcal{K}$ and assume that

$$P(\mathbf{D}) = \sum_{j=1}^{K} \mathbf{P}(\mathcal{M}_|, \mathbf{D}) = \sum_{j=1}^{K} \mathbf{P}(\mathbf{D} \mid \mathcal{M}_|) \mathcal{P}(\mathcal{M}_|) \tag{1}$$

where $\mathbf{D}$ is the observed data vector. The posterior probability of model $\mathcal{M}_\|$, $\forall \| \in \{\infty, \ldots, \mathcal{K}\}$, is given by

$$P(\mathcal{M}_\| \mid \mathbf{D}) = \frac{\mathbf{P}(\mathbf{D} \mid \mathcal{M}_\|) \mathcal{P}(\mathcal{M}_\|)}{\mathbf{P}(\mathbf{D})} = \frac{\mathbf{P}(\mathbf{D} \mid \mathcal{M}_\|) \mathcal{P}(\mathcal{M}_\|)}{\sum_{j=1}^{K} \mathbf{P}(\mathbf{D} \mid \mathcal{M}_|) \mathcal{P}(\mathcal{M}_|)} \tag{2}$$

Inference about a quantity of interest $\mathbf{D}^*$ in BMA is a weighted sum of the contributions from all the individual models, i.e.

$$P(\mathbf{D}^* \mid \mathbf{D}) = \sum_{k} \mathbf{P}(\mathbf{D}^* \mid \mathbf{D}, \mathcal{M}_\|) \mathcal{P}(\mathcal{M}_\| \mid \mathbf{D}) \tag{3}$$

Since $P(\mathcal{M}_\| \mid \mathbf{D}) \propto \mathbf{P}(\mathbf{D} \mid \mathcal{M}_\|) \mathcal{P}(\mathcal{M}_\|)$, the key to the successful numerical implementation of both model selection and BMA can be readily seen to lie in the computation of the term $P(\mathbf{D} \mid \mathcal{M}_\|)$, which is frequently referred to as the evidence for model $\mathcal{M}_\|$. Denoting $\theta_\mathbf{k}$ to be the parameter vector of model $\mathcal{M}_\|$, one can express the evidence as

$$P(\mathbf{D} \mid \mathcal{M}_\|) = \int \mathcal{P}(\mathbf{D} \mid \theta_\mathbf{k}, \mathcal{M}_\|) \mathcal{P}(\theta_\mathbf{k} \mid \mathcal{M}_\|) \lceil \theta_\mathbf{k} \tag{4}$$

which indicates that the evidence is the normalising constant of the underlying posterior distribution of the parameter vector.

Reliable evidence evaluation, which involves high-dimensional integration, can be prohibitively difficult. Most methods can work reasonably well provided the posterior Normality assumption is not severely violated. This is true not only of those using deterministic approximations such as that adopted by Mackay [8], but also of various MCMC based methods [2][3][4][5][6][7]. Typically MCMC algorithms rely on local moves to explore a posterior probability surface. When the surface exhibits multimodality , this can lead to excruciatingly slow convergence (multimodality has been generally considered as one of the most pressing issues confronting MCMC). This paper presents an MCMC algorithm aimed at yielding robust evidence estimation for models to which the assumption of posterior unimodality does not apply well. The remainder of the paper is organised as follows: Section 2 developes the adaptive tempering algorithm, which is an elaboration of the simulated tempering scheme devised independently by Geyer [9] and Marinari and Parisi [10] and addresses relevant implementation issues. Section 3 applys the algorithm to a two-dimensional trimodal distribution and compares the result with that of the unmodified Gibbs sampler. Sections 4 concludes the paper.

2. Adaptive Tempering Algorithm

2.1. SIMULATED TEMPERING ALGORITHM

MCMC methods originated in statistical physics. Owing to this link, interesting insights can be gained by associating a given distribution of interest with a corresponding energy function and a temperature parameter, namely, to express it in the form of a Gibbs distribution

$$P(\theta \mid \mathbf{D}, \mathcal{M}) = \frac{\exp\left[-\beta \mathcal{E}(\theta)\right]}{\mathcal{Z}(\beta)} \tag{5}$$

where the energy function is defined as

$$E(\theta) \stackrel{\text{def}}{=} -\log P(\theta \mid \mathcal{M}) - \log \mathcal{P}(\mathbf{D} \mid \theta, \mathcal{M}) \tag{6}$$

and β is the reciprocal temperature ($\beta = \frac{1}{k_B T}$, k_B being the Boltzmann constant), which has to be set at 1 to attain exact correspondence. The evidence can be shown to be equivalent to the partition function $Z(\beta)$ when $\beta = 1$.

A significant benefit of adopting such a canonical formulation arises from the introduction of the temperature parameter. As shall be demonstrated in the ensuing analysis, it can play an important role in combatting multimodality . For a multimodal distribution with a high degree of correlation amongst the individual components, high probability (low energy) regions of the state space are typically considerably separated by regions of low probability (high energy). Single-site Metropolis-Hastings or Gibbs sampler algorithms will accordingly be very slow to leave a region of high probability and to make substantial changes to nearly all the components to evolve into another distinctly different one. Indeed, it is often noticed that in those cases ordinary MCMC algorithms are no less prone than their deterministic iterative counterparts to getting trapped in spurious local modes. A logical solution is then to resort to algorithms of even greater stochasticity. It follows from the Metropolis updating procedure that for a given energy function, the Metropolis acceptance probability for a random proposal grows exponentially with temperature. By greatly increasing acceptance rates, raising the temperature effectively compensates for the disadvantages resulting from the local updating intrinsic to MCMC and enhances the global propagation of the Markov chain around the state space.

One systematic way of exploiting the *global characteristics* induced by heating up the system is to introduce an auxiliary probability distribution $P(\beta, \theta)$ over the state space $S = \{\beta_i\} \times \wp_\theta$, where $\{\beta_i\}$ is a sequence of temperatures in which $\beta_i < \beta_{i+1}$, $\beta_N = 1$, $i = 1, \ldots, N-1$, and $\wp_\theta$ is the parameter space. The fact that there is a range of temperatures to choose from offers a temperature detour mechanism whereby instead of being rejected outright at the desired temperature, a high energy state can move to a higher temperature subspace where it is more sustainable; the chain can subsequently move back to the low temperature subspace when it has found a state that is supportable by the corresponding Gibbs distribution. The connection between this joint distribution and a given distribution of interest is that the conditional distribution $P(\theta \mid \beta)$ is made equal to the corresponding Gibbs distribution.

Sampling from $P(\theta \mid \beta)$ can be achieved using a Metropolis-Hastings or Gibbs sampler updating procedure. To sample from $P(\beta \mid \theta)$, one can run a Metropolis Markov chain on the discrete space $\{\beta_i\}$. Given θ, the Metropolis acceptance probability for a move from β_j to β_i is

$$
\min\left[1, \frac{P(\beta_i \mid \theta)}{P(\beta_j \mid \theta)}\right] \;=\; \min\left[1, \frac{P(\beta_i)P(\theta \mid \beta_i)}{P(\beta_j)P(\theta \mid \beta_j)}\right]
$$
$$
\;=\; \min\left[1, \frac{P(\beta_i)}{P(\beta_j)} \exp\left[(\beta_j - \beta_i)E(\theta)\right] \frac{Z(\beta_j)}{Z(\beta_i)}\right] \quad (7)
$$

Here the marginal $\{P(\beta_i)\}$ can be viewed as a *pseudoprior* on $\{\beta_i\}$, which has to be specified prior to the tempering run. The acceptance probability can also be expressed in the form $\frac{P(\beta_i)\exp[-U(\beta_i|\theta)]}{P(\beta_j)\exp[-U(\beta_j|\theta)]}$ where

$$
U(\beta \mid \theta) = \beta * [E(\theta) - F(\beta)] \quad (8)
$$

and $F(\beta)$ is the Helmholtz free energy at temperature β.

2.2. DETERMINING THE RATIOS OF THE PARTITION FUNCTIONS

What remains to be specified to construct the Gibbs Markov chain is the values for $\{Z(\beta_i)\}$. Geyer [9] suggests using preliminary runs and other trial and error methods to estimate them, which can be time-consuming and inaccurate in the context of a multimodal distribution indexmultimodal distribution since $\{Z(\beta_i)\}$ may vary over orders of magnitude. The approach adopted here is not to evaluate $\{Z(\beta_i)\}$ explicitly, but to embed an importance reweighting element to adaptively evaluate the ratio $\frac{Z(\beta_i)}{Z(\beta_1)}$ (hence the name, Adaptive Tempering). As a consequence, with $Z(\beta_1)$ assumed to be readily estimable, the evidence which equals $Z(\beta_N)$ (note $\beta_N = 1$) can be obtained by $Z(\beta_N) = Z(\beta_1)\prod_{i=1}^{N-1} \frac{Z(\beta_{i+1})}{Z(\beta_i)}$.

If β_i and β_{i+1} are sufficiently close, either of the two Gibbs distributions at these two temperatures can be a good importance sampling distribution for the other. The following reweighting step can therefore be expected to work well,

$$
\frac{Z(\beta_{i+1})}{Z(\beta_i)} \;=\; \int \frac{\exp\left[-\beta_{i+1}E(\theta)\right]}{Z(\beta_i)} \, d\theta
$$

$$= \int \frac{\exp\left[-\beta_i E(\theta)\right]}{Z(\beta_i)} \exp\left[(\beta_i - \beta_{i+1})E(\theta)\right] d\theta$$

$$= \langle \exp\left[(\beta_i - \beta_{i+1})E(\theta)\right] \rangle_i \tag{9}$$

where $\langle \cdot \rangle_i$ denotes ensemble averaging with respect to the corresponding Gibbs distribution at temperature β_i.

The adaptive tempering algorithm can thus be implemented as follows:

1. At iteration t, draw $\theta^{(t)} \sim \exp\left[-\beta^{(t-1)}E(\theta)\right]$.
2. Use a uniform distribution to propose $\beta^{(t)}$; Accept it with probability

$$\min\left(1, \frac{P(\beta^{(t)})}{P(\beta^{(t-1)})} \exp\left[(\beta^{(t-1)} - \beta^{(t)})E(\theta^{(t)})\right] \frac{Z(\beta^{(t-1)})}{Z(\beta^{(t)})}\right)$$

otherwise let $\beta^{(t)} = \beta^{(t-1)}$.
3. At every N_Zth iteration, where N_Z is a prespecified constant, update $\frac{Z(\beta_{i+1})}{Z(\beta_i)}$, $\forall i$, by

$$\frac{Z(\beta_{i+1})}{Z(\beta_i)} = \frac{\sum_{t_n^{(i)}} \exp\left[(\beta_i - \beta_{i+1})E(\theta^{t_n^{(i)}})\right]}{N^{(i)}}$$

where $\{t_n^{(i)}\}$ are the times at which $\beta = \beta_i$ and $N^{(i)}$ is the total number of them up to the point of updating.

The resultant non-stationarity of the Markov chain should not be an issue of concern. Suffice it to say that the fluctuations in the ratios of $\{\frac{Z(\beta_{i+1})}{Z(\beta_i)}\}$ prior to convergence merely have the effect of preventing the simulation from operating in optimal conditions. Whatever the values taken on by the partition functions, step 1 leaves the conditional distribution $P(\theta \mid \beta)$ invariant, which is what really counts.

2.3.   ESTIMATING $Z(\beta_1)$ BY THERMODYNAMIC INTEGRATION

Due to the complicated dynamics of the adaptive tempering algorithm, it is difficult to give definitive and generic criteria on how to specify the temperature sequence. The spacings between the temperatures should be chosen as wide as possible, while still maintaining the condition that neighbouring distributions can serve as good importance sampling distributions for each other. The value of β_1 should be chosen to be sufficiently low so as to allow virtually uninhibited movement around the parameter space. With β_1 thus chosen, $Z(\beta_1)$ ought to be significantly less difficult to estimate than its unheated counterpart $Z(\beta_N)$.

The thermodynamic integration algorithm [15] is adopted here as a general approach to estimating $Z(\beta_1)$. Consider two Gibbs distributions , $\frac{\exp[-E'(\theta)]}{Z'}$ and $\frac{\exp[-E_{ref}(\theta)]}{Z_{ref}}$, where Z' is the quantity of interest and Z_{ref} is assumed known. In the same state space define an intermediate system $P_\alpha(\theta)$ with the energy function $E_\alpha(\theta) = \alpha E'(\theta) + (1 - \alpha)E_{ref}(\theta)$, $0 < \alpha < 1$, and the partition function

$Z_\alpha = \int \exp\left[-(\alpha E'(\theta) + (1 - \alpha)E_{ref}(\theta))\right] d\theta$. Differentiating $\log Z_\alpha$ with respect to α yields

$$\frac{\partial \log Z_\alpha}{\partial \alpha} = \frac{\int \exp\left[-E_\alpha(\theta)\right]\left[E_{ref}(\theta) - E'(\theta)\right] d\theta}{Z_\alpha} = \langle E_{ref}(\theta) - E'(\theta)\rangle_\alpha \qquad (10)$$

where $\langle \cdot \rangle_\alpha$ denotes statistical expectation with respect to $P_\alpha(\theta)$. Consequently, Z' can be calculated by the approximation

$$\log Z' - \log Z_{ref} = \int \log Z_\alpha \, d\alpha \approx \sum_{i=2}^{m} (\alpha_i - \alpha_{i-1})\langle E_{ref}(\theta) - E'(\theta)\rangle_{\alpha_i} \qquad (11)$$

where $\{\alpha_i\}$ is a sequence of numbers in ascending order ranging from $\alpha_1 = 0$ to $\alpha_m = 1$. The equivalent energy function for the estimation of $Z(\beta_1)$ is $\beta_1 \times E(\theta)$ and a Gaussian can be used as a reference system. Amongst the samples generated from a tempering run, use can be made of those associated with β_1 to determine the parameter values of the reference Gaussian.

2.4. IMPLEMENTATION ISSUES

To have a good intitial value setting for $\{Z(\beta_i)\}$, one can use the following Taylor approximation,

$$\log Z(\beta) = \log Z(\beta_0) + \sum_{n=1}^{+\infty} \frac{(\beta - \beta_0)^n}{n!} \frac{d^n}{d\beta^n} \log Z(\beta) \big|_{\beta=\beta_0} \qquad (12)$$

where

$$\frac{d^n}{d\beta^n} \log Z(\beta) = (-1)^n \langle E^n(\beta)\rangle \qquad (13)$$

The values of $\langle E^n(\beta)\rangle$ can be obtained by performing a pilot MCMC run at a suitably high temperature β_0 prior to the tempering run.

Ideally, the pseudoprior $P(\beta_i)$ ought to be chosen to be proportional to the integrated autocorrelation time $\tau_{int,f}(\beta_i)$ associated with the observable function $f(\theta) = \exp\left[(\beta_i - \beta_{i+1})E(\theta)\right]$ Unfortunately, the very complexity of the problem which led to consideration of MCMC in the first place makes it impossible to reliably calculate the integrated autocorrelation times at different temperatures. As a general guideline, however, qualitative measures can still be taken to put relatively large pseudo-priors on those temperatures which are expected to give rise to slow-mixing chains. What's more, when dealing with complicated problems, one may find it useful to adopt an *adaptive* procedure to adjust the pseudoprior values as the simulation proceeds.

Aside from combatting multimodality, the adaptive tempering method has an additional advantage of providing a concrete means for assessing *convergence*. One can know precisely when to stop the simulation by comparing the relative frequency with which the temperatures are visited with their respective pseudopriors, since under the joint distribution, the frequency with which a temperature is visited is proportional to its marginal density after convergence.

3. Performance of the Adaptive Tempering algorithm on a Trimodal Distribution

A bivariate distribution, which is plotted in Figure 1, is studied to test the performance of the proposed algorithm. The analytical expression for it is

$$P(x,y) = \frac{\exp\left[-E(x,y)\right]}{Z(1)} \tag{14}$$

where

$$
\begin{aligned}
E(x,y) = -\log\{ \quad & 9 \quad \frac{1}{\sqrt{2\pi}0.5}\frac{1}{\sqrt{2\pi}0.5}\exp[-\frac{1}{2}\frac{(x+5)^2}{0.5^2}-\frac{1}{2}\frac{(y-8.6605)^2}{0.5^2}] \\
+ \quad & 6 \quad \frac{1}{\sqrt{2\pi}0.5}\frac{1}{\sqrt{2\pi}0.5}\exp[-\frac{1}{2}\frac{(x+5)^2}{0.5^2}-\frac{1}{2}\frac{(y+8.6605)^2}{0.5^2}] \\
+ \quad & 6 \quad \frac{1}{\sqrt{2\pi}0.5}\frac{1}{\sqrt{2\pi}0.5}\exp[-\frac{1}{2}\frac{(x-10)^2}{0.5^2}-\frac{1}{2}\frac{y^2}{0.5^2}] \quad \} \tag{15}
\end{aligned}
$$

20 temperatures are used to implement the tempering algorithm, geometrically spaced from $\beta_1 = 0.01$ to $\beta_{20} = 1$. Identical pseudopriors are adopted in this simple problem. Metropolis updates for (x,y) are alternated with those for β. At each Gibbs iteration, 20 base Metropolis updates are performed to sample from $P(x,y \mid \beta)$, using a Gaussian proposal distribution; and for ease, β is updated just once with a bi-state proposal distribution under which the candidate is equally likely to be either of the two adjacent temperatures of the current one. 10,000 Gibbs iterations are carried out for this tempering run, starting at $(x,y) = (0,0)$ and $\beta = 1$. Figure 2 shows the trajectory of the resultant 656 samples corresponding to $\beta = 1$ and Figure 3 shows that of the 505 samples corresponding to $\beta = 0.01$. Figure 4 plots the unnormalised partition function $\frac{Z(\beta_i)}{Z(0.01)}$ versus β, from which the evidence $Z(1.0)$ is calculated to be $0.0454 \times Z(0.01)$. The fact that out of the 10,000 samples 656 stayed at $\beta = 1$ implies that convergence had yet to be attained, for if it had, the chain would, according to the assigned pseudoprior, have visited the subspace for which $\beta = 1$ close to $\frac{10000}{20} = 500$ times.

The normal Gibbs sampler algorithm is also applied at the unheated temperature $\beta = 1$ and run for 100,000 Gibbs iterations. Figure 5 shows the evolution of the Markov chain with time (note only every 10th iteration is plotted). Whereas in the adaptive tempering run, with just 10,000 Gibbs iterations, all the three modes have been visited, the Gibbs sampler did not leave the mode centred at $(-5, -8.6605)$ for more than 100,000 iterations. Running the tempering algorithm with 100,000 Gibbs iterations gives even better results. Figure 6 plots the trajectory of the samples corresponding to $\beta = 1$. In this case, unlike what is depicted in Figure 2, the 3 modes have been visted in the correct proportion, indicative of convergence to the true distribution.

$Z(0.01)$ is estimated by the thermodynamic integration method. Based on the samples generated from the shorter run and using 100 intermediate systems each of which is sampled 10 times, $Z(0.01)$ is estimated to be 545.3454. $Z(1)$ is thus equal to 545.3454 $\times$ 0.0454 = 24.7587, which is slightly larger than the true value 21.

The difference can be chiefly accounted for by the figure 0.0454, which was derived when the chain had not quite reached the equilibrium. Applying thermodynamic integration to samples generated by the unmodified Gibbs sampler yields $Z(1) = 5.9986$. As a consequence of failing to explore the state space in totality, this result merely corresponds to the probability mass of one mode and significantly underestimates the true evidence value.

4. Conclusions

The adaptive tempering algorithm is directly applicable to the problem of sampling from a multimodal distribution . It can also conceivably be an alternative to Simulated Annealing . The deterministic problem of minimizing $E(\theta)$ is equivalent to the stochastic one of generating random samples from $\exp[-\beta E(\theta)]$ when $\beta \to +\infty$. Therefore to optimise a multimodal energy function , one can employ the adaptive tempering scheme to sample the corresponding Gibbs distribution at a very low temperature.

References

1. Chatfield, C. (1995) Model Uncertainty, Data Mining and Statistical Inference. *J. R. Statist. Soc. A*, **158**, 1-26.
2. Raftery, A. E. (1994) Hypothesis Testing and Model Selection via Posterior Simulation. Technical Report, Department of Statistics, University of Washington.
3. Geyer, C. J. (1994) Estimating Normalising Constants and Reweighting Mixtures in Markov Chain Monte Carlo. Technical Report No. 568, School of Statistics, University of Minnesota.
4. Carlin, B. P. and Chib, S. (1994) Bayesian Model Choice via Markov Chain Monte Carlo. Research Report 93-006, Division of Biostatistics, University of Minnesota.
5. Green, P. J. (1994) Reversible Jump MCMC Computation and Bayesian Model Determination. Presented at the *Workshop on Model Criticism in Highly Structured Stochastic Systems*, Wiesbaden, September 1994.
6. Grenander, U. and Miller, M. (1994) Representation of knowledge in complex systems (with Discussion). *J. R. Statist. Soc. B*, **56**, No. 4, 549-603.
7. Madigan, D. and York, J. (1993) Bayesian Graphical Models for Discrete Data. Technical Report No. 259, Department of Statistics, University of Washington.
8. Mackay, D. J. C. (1992) Bayesian Methods for Adaptive Models. Ph.D thesis, California Institute of Technology.
9. Geyer, C. J. and Thompson, E. A. (1994) Annealing Markov Chain Monte Carlo with Applications to Ancestral Inference. Technical Report No. 589, School of Statistics, University of Minnesota.
10. Marinari, E. and Parisi G. (1992) Simulated Tempering: A New Monte Carlo Scheme. *Europhysics Letters*, **19**, 451-458.
11. Goodman, J. and Sokal, A. D. (1989) Multigrid Monte Carlo method: conceptual foundations. *Phys. Rev. D.*, **40**, 2035-2071.
12. Besag, J. and Green, P. J. (1993) Spatial Statistics and Bayesian Computation. *J. R. Statist. Soc. B*, **55**, No. 1, 25-37.
13. (1993) Discussion on the Meeting on the Gibbs Sampler and other Markov Chain Monte Carlo methods. *J. R. Statist. Soc. B* **55**, No. 1, 53-102.
14. Neal R. (1994) Sampling from Multimodal Distributions using Tempered Transitions. Technical Report No. 9421, Department of Statistics, University of Toronto.
15. Ó Ruanaidh, J. J. K. and Fitzgerald, W. J. (1995). *Numerical Bayesian Methods applied to Signal Processing*. Berlin: Springer-Verlag.

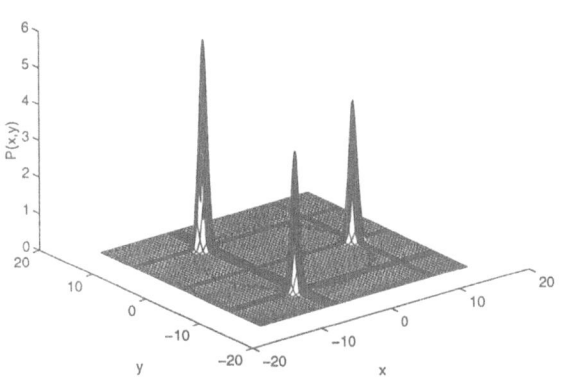

Figure 1. Plot of the trimodal distribution.

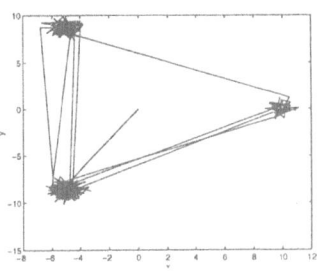

Figure 2. Trajectory of the samples corresponding to $\beta = 1$ using a tempering run of 10,000 iterations.

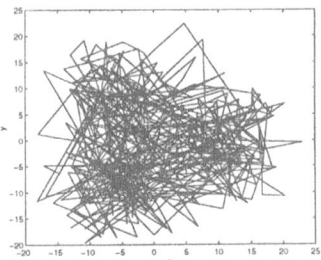

Figure 3. Trajectory of the samples corresponding to $\beta = 0.01$ using a tempering run of 10,000 iterations.

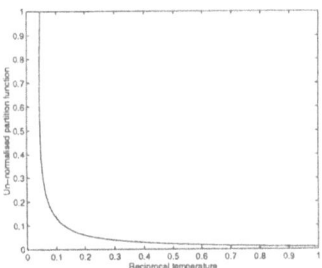

Figure 4. Plot of the un-normalised partition function versus the reciprocal temperature using a tempering run of 10,000 iterations.

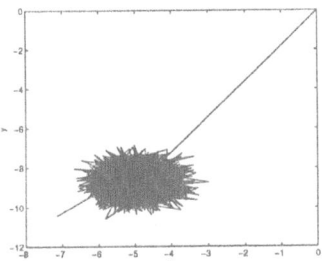

Figure 5. Trajectory of the samples using a Gibbs sampler run of 100,000 iterations.

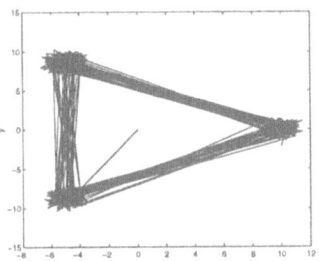

Figure 6. Trajectory of the samples corresponding to $\beta = 1$ using a tempering run of 100,000 iterations.

Wray Buntine presenting his part of the tutorial on Bayesian methods.

BAYESIAN INFERENCE AND THE ANALYTIC CONTINUATION OF IMAGINARY-TIME QUANTUM MONTE CARLO DATA

J. E. GUBERNATIS AND J. BONČA

Theoretical Division, Los Alamos National Laboratory
Los Alamos, NM 87545, U.S.A.

AND

MARK JARRELL

Department of Physics, University of Cincinnati
Cincinnati, OH 45221, U.S.A.

Abstract. We present a brief description of how methods of Bayesian inference are used to obtain real frequency information by the analytic continuation of imaginary-time quantum Monte Carlo data. We present the procedure we used, which is due to R. K. Bryan, and summarize several bottleneck issues.

Key words: entropic prior, maximum entropy, analytic continuation

1. Introduction

Most often, discussions about the application of methods of Bayesian inference focus on the choice of prior probabilities. The choice of the likelihood function is usually viewed as less problematic. Typically, the natural choice is that of Gaussian distribution. In this paper, we will discuss a successful application of the methods of Bayesian inference to the problem of the analytic continuation imaginary-time data obtained from quantum Monte Carlo simulations. Here, a "natural" choice of the prior probability was the entropic prior. The choice of the likelihood function as a Gaussian distribution was an unnatural one but the only workable one.

The analytic continuation problem is a very ill-posed one. It amounts to numerically performing an inverse, two-sided Laplace transformation of noisy and incomplete data. Bayesian methods are used to develop a procedure to regularize the problem. A major difficulty in achieving success was overcoming the naturally non-Gaussingly-distributed data produced by the simulation. To achieve such a Gaussian distribution, large amounts of data were produced to force the central limit theorem to hold approximately. Experience shows that unless the data used as input to the analytic continuation problem consistent with the procedures used to perform the continuation, then unsatisfying, and often incorrect, results follow.

K. M. Hanson and R. N. Silver (eds.), Maximum Entropy and Bayesian Methods, 163–170.
© 1996 *Kluwer Academic Publishers.*

While the need for such consistency in retrospect seems obvious, it was overlooked in several of the initial numerical approaches to the problem.

In what follows, we first will summarize and define the problem of the analytic continuation of imaginary-time quantum Monte Carlo data. Next, we will state the Bayesian approach we used. We simply borrowed a procedure proposed by Bryan [1]. In this context, we will give several illustrations showing why both the generation and statistical quality of the data can be bottleneck issues. We conclude with a summation. We comment that recently two of us wrote a review article that details our methods and procedures [2].

2. The Analytic Continuation Problem

For finite-temperature, quantum Monte Carlo simulations, the analytic continuation problem is "Given noisy and incomplete estimates for the imaginary-time correlation function $G(\tau)$, what is the spectral density $A(\omega)$?" The imaginary-time correlation function (or many-body Green's function) arises naturally in the field-theoretic approach to quantum many-body problems [3]. The analytic properties of this approach leads to the association of a spectral density to each type of many-body Green's function. The spectral density often exhibits in its structure (its peaks and their location) information about the subtle microscopic processes taking place in physical systems. Often these functions can be measured experimentally. Theoretical physicists and chemists find the computation of these function important and useful for the physics they reveal about the physical models being studied and about the relevance of the models to real physical systems.

Many-body quantum theory states [3] that $G(\tau)$ and $A(\omega)$ are related by

$$G(\tau) = \int_{-\infty}^{+\infty} d\omega \, \frac{e^{-\tau\omega} A(\omega)}{1 \pm e^{-\beta\omega}}$$

The $\pm$ refers to Fermi and Bose statistics, $\tau0$, and β is the inverse of the temperature. A precise definition of $A(\omega)$ can be given in terms of the eigenvalues and vectors of the Hamiltonian specifying the physical system, quantities which are very hard to obtain in part because the number of eigenstates generally increases exponentially with the size of the system. This situation points to utility for a Monte Carlo simulation of the problem. Depending on the definition of $G(\tau)$, $A(\omega)$ is related to such experimentally measurable quantities as the photoemission spectra, optical conductivity, dynamic magnetic susceptibility, etc. These quantities, not the real-time correlations, is usually all that is measured.

2.1. THE PROBLEM

The main difficulty in the analytic continuation problem is that at large positive and negative frequencies the kernel

$$K(\tau,\omega) = \frac{e^{-\tau\omega}}{1 \pm e^{-\beta\omega}}$$

is exponential. This condition makes $G(\tau)$ insensitive to the high-frequency details of $A(\omega)$. In terms of obtaining $A(\omega)$ from $G(\tau)$, this insensitivity leads to an ill-posed problem, and the ill-posedness implies an infinite-number of solutions exist. The task is to select from this infinity of solutions one that is "best" by some criterion.

2.2. GENERAL FEATURES

The spectral density satisfies

$$A(\omega) \geq 0$$
$$\int d\omega \, A(\omega) < \infty$$

These features allows us to interpret $A(\omega)$ as a probability function. Often, the bound on the integral of the spectral density is precisely known or is computable by the simulation. In these cases, called *sum rules*, the bound also reveals physical information about the problem.

The correlation functions $G(\tau)$ satisfy

$$G(\tau + \beta) = \pm G(\tau)$$

which allows us to restrict $0 \leq \tau < \beta$. Within this range

$$G(\beta - \tau) = G(\tau)$$

$G(\tau)$ is usually bounded from above and below.

2.3. IMAGINARY-TIME QUANTUM MONTE-CARLO

Quantum Monte Carlo simulations at finite temperatures are done in imaginary time τ. The Wick rotation, $it \rightarrow \tau$, transforms Schroedinger's equation for a Hamiltonian operator H

$$i\frac{\partial \psi}{\partial t} = H\psi$$

and its formal solution

$$\psi(t) = e^{-iHt}\psi(t = 0)$$

into

$$\frac{\partial \psi}{\partial \tau} = -H\psi$$

and

$$\psi(\tau) = e^{-\tau H}\psi(\tau = 0)$$

Other operators evolve and transform as

$$A(t) = e^{itH}Ae^{-itH} \quad \rightarrow \quad A(\tau) = e^{\tau H}Ae^{-\tau H}$$

The Wick rotation thus transforms oscillatory exponentials e^{-itH} into diffusive ones, $e^{-\tau H}$. The importance of such exponential lies with the dynamics of quantum mechanical systems.

Why perform the Wick rotation? From the point of view of performing a Monte Carlo simulation, at long times real-time such simulations are inefficient: sampling on smaller and smaller times scales becomes necessary as time increases just to achieve proper self-cancelations. Quantum simulations, however, are readily (and naturally) performed in imaginary-time.

3. A Solution Path

The functional form of the analytic continuation problem clearly reveals its ill-posed nature. In general, the solution of such a problem requires a regularizer. In the present situation, we have the added difficulties of the estimates of $G(\tau)$ obtained from the simulations being incomplete and noisy. They are incomplete because we only determine $G(\tau)$ at discrete values of τ. They are noisy because of the nature of a Monte Carlo simulation.

To proceed, we first convert the integral equation to a linear system of equations

$$G_i = \sum_j K_{ij} A_j$$

where $G_i = G(\tau_i)$, $K_{ij} = K(\tau_i, \omega_j)$, and $A_i = A(\omega_i)\Delta\omega_i$. Next, we pretend to consider a constrained least-squares approach, i.e., we consider seeking A_i that maximizes

$$Q = \alpha S - \frac{1}{2}\chi^2$$

$$\begin{aligned}
\chi^2 &= \sum_{i,j}(\bar{G}_i - G_i)[C^{-1}]_{ij}(\bar{G}_j - G_j) \\
&= \sum_{i,j}(\bar{G}_i - \sum_k K_{ik}A_k)[C^{-1}]_{ij}(\bar{G}_j - \sum_k K_{jk}A_k)
\end{aligned}$$

and

$$\bar{G}_i = \frac{1}{M}\sum_{m=1}^{M} \bar{G}_i^{(m)}$$

C is the covariance matrix

$$C_{ik} = \frac{1}{M(M-1)}\sum_{m=1}^{M}(\bar{G}_i - \bar{G}_i^{(m)})(\bar{G}_k - \bar{G}_k^{(m)})$$

The $\bar{G}_i^{(m)}$ are statistically independent estimates of G_i S is the information theory entropy

$$S = \sum_i [A_i - m_i - A_i \ln(A_i/m_i)]$$

with m_i being the default model.

There are several things being assumed. One is the entropic prior. The rationale for this choice follows from the observation that the possible interpretation of the spectral density as a probability function and the utility of the principle of maximum entropy for assigning probabilities. In practice, the choice is very convenient because of the ease of insuring the non-negativity of the computed spectral density. Another assumption is the choice of a multivariate Gaussian for the likelihood function. We know of no other analytic form that fits or can be made to match the distribution of the data, short of mindless function fitting that most likely would be needed for each simulation.

What does one chose for α? We opt for the following [1]

$$\begin{aligned} \langle A \rangle &= \int \mathcal{D}A \, d\alpha A(\alpha) \Pr[A, \alpha | \bar{G}] \\ &= \int \mathcal{D}A \, d\alpha A(\alpha) \Pr[A | \bar{G}, \alpha] \Pr[\alpha | \bar{G}] \\ &\approx \int d\alpha \, \hat{A}_\alpha \Pr[\alpha | \bar{G}] \end{aligned}$$

In the above,

$$\begin{aligned} \Pr[A | \bar{G}, \alpha] &= \Pr[\bar{G} | A] \Pr[A | \alpha] \Pr[\alpha] \\ &= \frac{e^{\frac{-1}{2} \chi^2}}{Z_L} \frac{e^{\alpha S}}{Z_S(\alpha)} \Pr[\alpha] \\ &= \frac{e^Q}{Z_L Z_S(\alpha)} \Pr[\alpha] \end{aligned}$$

where Z_L and $Z_S(\alpha)$ are the normalization constants for the likelihood function and the prior. Also,

$$\begin{aligned} \Pr[\alpha | \bar{G}] &= \int \mathcal{D}A \Pr[\bar{G} | A] \Pr[A | \alpha] \Pr[\alpha] \\ &= \Pr[\alpha] \int \mathcal{D}A \frac{e^Q}{Z_L Z_S} \end{aligned}$$

To proceed, for a fixed value of α, we solve

$$\left. \frac{\delta Q}{\delta A} \right|_{A = \hat{A}_\alpha} = 0$$

to find $\hat{A}_\alpha$. (For a given value of α, Q is a convex function of A_i.) Then, we numerically perform

$$\bar{A} = \int d\alpha \, \hat{A}_\alpha \Pr[\alpha | \bar{G}]$$

The functional integration over A and the integral over α are done numerically.

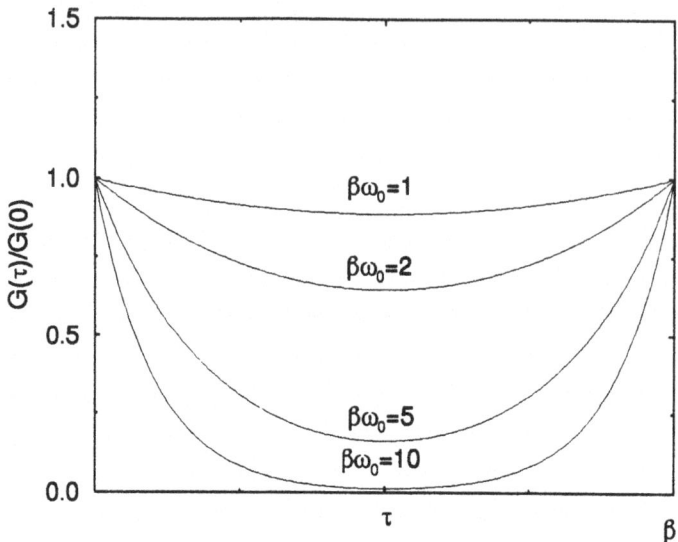

Figure 1. The exact Green's function as a function of τ for a simple harmonic oscillator. ω_0 is the oscillator frequency and β is the inverse of the temperature.

4. Bottlenecks

Only for a very large number of statistically independent measurements does the distribution of the Monte Carlo data approach the assumed multivariate Gaussian. The results are very sensitive to the extent to which the central limit theorem can "save the day." There are several other issues regarding the data. Compared to Monte Carlo simulations where the analytic continuation is not an objective, the data in the continuation problem require a more careful statistical characterization, particularly with respect to statistical independence. (The assumed likelihood function assumes statistical independence.) Also the variance of the Monte Carlo results needs a greater reduction. (This need is a consequence of the ill-posedness of the problem.) The techniques we used to statistically characterize and qualify the data are discussed in [2].

In Fig. 2, we show the exact Green's function for a single quantum harmonic oscillator (or well). Those obtained from almost any simulation look like these. In general, some "characteristic energy scale" ω_0 exists. When it is smaller than the the temperature ($\beta\omega_0 < 1$), the Green's function is flat. If one envisions error bars on the measured values, then a number of measurements are "within the error" of each other, and hence not all the measured data make independent contributions. When it is larger then the temperature ($\beta\omega_0 1$), only in a small region near $\tau = 0$ and by symmetry near $\tau = \beta$ is the relative error small. Additionally, the smoothness of the curve implies that the different values of $G(\tau)$ are correlated with one another. The correlation is the reason why the covariance matrix appears in the likelihood function.

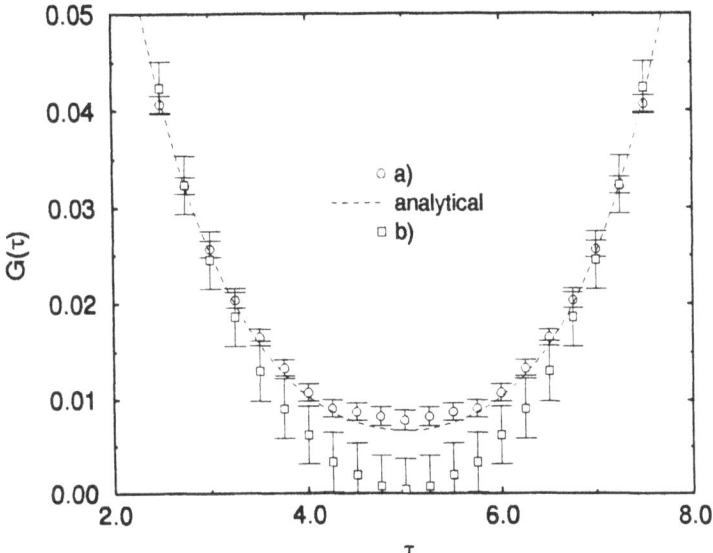

Figure 2.   Near $\tau = \beta/2$, the Green's function (dashed line) from Fig. 2 for the $\beta\omega_0 = 10$ case compared with the results obtained by two different quantum Monte Carlo algorithms. The results from the non-ergodic algorithm lie below the exact result.

In Fig. 3, we shown a portion of Fig. 2 (the dashed line) near $\tau = \beta/2$ together with the values of the Green's functions computed for the same problem from two slightly different quantum Monte Carlo algorithms. The results which show the greatest deviation from the exact results were obtained with an algorithms that was slightly non-ergodic. We point out the reduced y-axis scale compared to Fig. 2. If we used in Fig. 3 the same scale as in Fig. 2, the differences between the results from the two simulations would be barely noticeable. Figure 4 illustrates that these small differences can make a big difference in the results [4]: the exact $A(\omega)$ should be a δ-function positioned at 1 with a weight (sum rule) of 0.5. The result from the ergodic algorithm gives a peak at the right location with the right weight. The width of the peak will narrow as the statistical error associated with data narrows (i.e., if more data is used). The result obtained with the other algorithm is broader and incorrectly placed and weighted. The breadth can be reduced by reducing the statistical error of the data, but this will not overcome the systematic error caused by the algorithm.

5. Remarks

The methods [1] briefly discussed provide a framework to approach the analytic continuation problem in which the assumptions and approximations in the approach can be clearly defined. They opened a completely new set of opportunities for the applications of quantum simulations. Several other important points are: (1) There is nothing about the methods that is intrinsic to quantum Monte Carlo

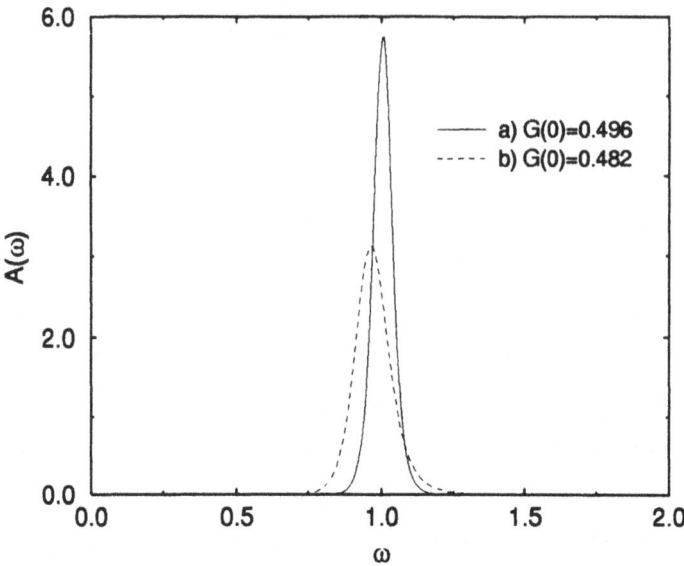

Figure 3.   The spectral densities obtained from the simulation data partially represented in Fig. 3. The result (dashed line) obtained from the non-ergodic algorithm is improperly positioned and weighted. For the particular physical model, the value of $G(0)$ gives the value of the sum rule, i.e., the expected value of the integral of $A(\omega)$.

or analytic continuation problems. (2) The Bayesian methods lead to a procedure with no adjustable parameters. This removes one of the difficulties often associated with other regularization methods. We also remark that the requirement for Gaussian-distributed data is a result of a particular choice of the likelihood function. Making the data as consistent as possible with this assumption is the hardest part of the problem and is the principal source of the increased computational cost associated with this problem. Recently, we have successfully obtained the actual real-time Green's function from the imaginary-time one [4].

This work was supported in part by the U. S. Department of Energy. We also gratefully acknowledge the influence and past collaborations with Devinder Sivia and Richard Silver.

References

1.   R. K. Bryan, "Maximum entropy analysis of oversampled data problems," *Eur. Biophys. J.,* 18, pp. 165-174, 1990.
2.   M. Jarrell and J. E. Gubernatis, "Bayesian inference and the analytic continuation of imaginary-time quantum Monte Carlo data," Phys. Reports, to appear.
3.   See for example, A. Abrikosov, L. Gorkov, and I. Dzyaloshinskii, *Quantum Field-Theoretical Methods in Statistical Physics*, Pergamon, Oxford, 1965.
4.   J. Bonča and J. E. Gubernatis, "Real-time dynamics from imaginary-time quantum Monte Carlo simulations: test on oscillator chains," Phys. Rev E, to appear.

SPECTRAL PROPERTIES FROM QUANTUM MONTE CARLO DATA: A CONSISTENT APPROACH

R. PREUSS, W. VON DER LINDEN* AND W. HANKE

Institut für Theoretische Physik, Universität Würzburg
Am Hubland, D-97074 Würzburg, Germany ‡
*) Max-Planck-Institut für Plasmaphysik, EURATOM Association
D-85740 Garching b. München, Germany §

Abstract.
   Bayesian statistics in the frame of the maximum entropy concept has widely been used for inferential problems, particularly, to infer dynamic properties of strongly correlated fermion systems from Quantum-Monte-Carlo (QMC) imaginary time data. In current applications, however, a consistent treatment of the error-covariance of the QMC data is missing. Here we present a closed Bayesian approach to account consistently for the QMC-data.

Key words: Quantum Monte Carlo, Analytic Continuation, Spectral Properties, Covariance.

1. Introduction

Since the discovery of the high temperature superconductors (HTSC) in 1986 [1] these materials are a major topic in condensed matter physics. Not only the unique properties of the HTSC's attract attention, but also their most promising use in various fields of industrial application. But despite a tremendous effort in this area a consistent theory is still lacking.

   The crystal structures of all HT_c-superconductors have one thing in common: planes consisting of copper and oxygen atoms, where the Cu-atoms span a square lattice, and the O-atoms are located between the copper sites. Shown by experiment the major physics happens to take place in these CuO-planes. Therefore in our theoretical approach only these planes are considered.

   The CuO-planes can be mapped onto a so-called 3 band Hubbard Hamiltonian [2], describing the motion of the electrons traveling around in the plane, undergoing a Coulomb repulsion for two electrons on the same site and taking into account chemical properties like the energy difference in the copper and oxygen

‡Email: roland@physik.uni-wuerzburg.de
§Email: wvl@ibmop5.ipp-garching.mpg.de

K. M. Hanson and R. N. Silver (eds.), Maximum Entropy and Bayesian Methods, 171–178.
© *1996 Kluwer Academic Publishers.*

orbitals. Being further simplified to only one kinetic and one potential term (giving the so-called 1 band Hubbard model) and describing now merely a plane of copper atoms, this model still cannot be treated neither analytically nor by using perturbation theory methods (both terms are of same order of magnitude). Only numerical methods are capable of revealing the unusual properties arising from the correlation effects of the interacting electrons. Among these are the exact diagonalization (ED) of a small sized system (up to 25 sites), which will be used in this paper for reasons of comparison, and the Quantum-Monte-Carlo (QMC) simulation of somewhat larger systems (up to 300 sites). We apply the later method to extract results on one-particle spectra, which can then be compared with data from photoemission and inverse photoemission experiments on the HTSC-materials.

The problem we have to deal with is that the QMC method provides information of dynamical quantities on the imaginary time axis only [3,4]. The necessary analytic continuation to the real frequency spectral weight function is done by inverting the spectral theorem being a Laplace transformation, which is an ill-posed problem. The difficulties inverting such a function are enforced by the availability of only a finite number of the QMC data, additionally blurred by statistical errors. A direct inversion of the Laplace transform would tremendously overfit the noise and the desired signal would be buried underneath it.

Here MaxEnt has become a standard and successful technique to perform this inversion (for a review see [5]). Originally been introduced to infer celestial images from incomplete and noisy radio-astronomic data [6,7], it has been applied successfully to various other data-analysis problems [3,8-10].

A complication arises, however, in the present application as the errors of the QMC-data are correlated. It has been proposed [3] to include the error-covariance matrix as exact data-constraints. This approach has been heavily debated as it leads to the following dilemma. For a standard QMC sample size, the off-diagonal elements of the covariance matrix are not negligible and ought to be taken into account. However, the errors of the covariance matrix are huge and the information provided by the QMC covariance matrix is useless and in many applications even disadvantageous. The dilemma obviously arises due to the neglect of the errors of the covariance matrix. Here we present a closed Bayesian approach to account fully for the noisy QMC data plus covariance matrix. This scheme allows to infer reliable results whenever only a small sample size is available.

2. Quantified Maximum Entropy

Bayes' theorem is used to determine the posterior probability $P(A|DCH)$ (*Posterior*) of the sought-for quantity A, given hypotheses H and QMC data D, which are related to A via $D_l^{ex} = D_l^{ex}(\{A\})$. The experimental data D_l deviate from the exact values by the statistical QMC errors η_l. Due to the QMC-algorithm, the errors are correlated and the information about the error-covariance matrix will be denoted by C. Bayes' theorem, $P(A|DCH) = P(D|ACH)P(A|CH)/P(D|CH)$, relates the *Posterior* to the likelihood function $P(D|ACH)$, which contains the error statistic of the QMC data, and the prior probability $P(A|CH)$. Our prior knowledge is part of the hypotheses H.

In the case of a positive, additive distribution function (PAD) – e.g. the spectral density – the most ignorant *Prior* is the entropic *Prior*[11]

$$P(A|CH) = \frac{1}{Z_S} \exp\left(\alpha \underbrace{\int [A(\omega) - m(\omega) - A(\omega)\ln(\frac{A(\omega)}{m(\omega)})]d\omega}_{S}\right) \qquad . \qquad (1)$$

Z_S is the normalization constant guarantying $\int P(A|CH)\mathcal{D}A = 1$. In the following we will assume, as indicated in Eq. 1 that $A = A(\omega)$ is a function of the frequency ω. The entropy S is measured relative to a default-model $m(\omega)$ which contains "weak prior assumptions" which can still be overruled by the data constraints.

Next we will turn to the determination of the *Likelihood*, the central topic of this paper. The *Likelihood* quantifies the probability for the realization of the specific data-values D measured in the experiment, supposing the exact function A were known. The *Likelihood* describes therefore the error statistics of the QMC-data

$$P(D|ACH) =: \rho(D^{ex} - D) = \rho(\eta) \quad \text{given C and H} \quad . \qquad (2)$$

In the following, C stands for the error-covariance matrix C_{ij}

$$C_{ij} = \frac{1}{N-1}\langle \eta_i \eta_j \rangle = \frac{1}{N-1}\langle (D_i^{ex} - D_i)(D_j^{ex} - D_j) \rangle \qquad , \qquad (3)$$

measured by QMC. N is the number of data-values D_i.

Assuming that the exact values of the error-covariance are known one obtains for the *Likelihood* the following result, which is the ubiquitous normal distribution

$$\rho(\eta) = \frac{1}{\sqrt{\det(2\pi C)}} e^{-\frac{1}{2}\sum_{ij} \eta_i C_{ij}^{-1} \eta_j} \qquad (4)$$

and simplifies to a Gaussian if the errors are uncorrelated $C_{ij} = \delta_{ij}\sigma_i^2$,

$$\rho(\eta) = \frac{1}{\sqrt{\prod(2\pi\sigma_i^2)}} e^{-\frac{1}{2}\sum_i \frac{\eta_i^2}{\sigma_i^2}} \qquad (5)$$

Unfortunately, this handy result is only valid if the QMC error-covariance were known exactly, which is not the case. Setting the off-diagonal elements of C to zero, which means to assume no correlations in the data, may result in over-fitting the data [5]. But the errors of the covariance matrix are significant: in our data we had to discover an average of 30% relative error for the diagonal elements, and up to 100% for the off-diagonal elements in the small bin case, decreasing only to about an average of 20% for the diagonal elements in the large bin case.

Therefore, the errors of the covariance matrix have to be treated on the same footing as the errors of the QMC data D in the first place – by quantified MaxEnt [6].

3. Consideration of the error of the covariance

Again, the *Posterior* for $\rho(\eta)$, given the QMC error-covariance C_{ij}, and the statistical errors σ_{ij} of C_{ij} and all our hypotheses can be determined via Bayes' theorem

$$P(\rho|C\sigma H) = \frac{P(C|\rho\sigma H)P(\rho|H)}{P(C|H)} \quad . \tag{6}$$

Superfluous conditions have been discarded. Again the entropic *Prior* is invoked. QMC simulations provide the statistical error σ_{ij} of the covariance matrix. Further information is not available by present QMC simulations. We therefore assume that the error σ_{ij} are uncorrelated and known, this is part of our hypotheses H. A generalization beyond this assumption is straightforward. The *Likelihood* reads

$$P(C|\rho\sigma H) = \exp\left(-\frac{1}{2}\underbrace{\sum_{ij}\frac{(C_{ij} - \int \rho(\eta)\eta_i\eta_j d\eta)^2}{\sigma_{ij}^2}}_{\chi^2}\right) \quad . \tag{7}$$

The MaxEnt result is hence obtained upon maximizing the *Posterior*, or rather

$$\mathcal{L}(\rho, C) = \alpha S - \frac{1}{2}\chi^2 - \lambda_0 \sum_i (\int \rho(\eta)d\eta - 1) \quad . \tag{8}$$

Introducing the Legendre transform $\tilde{\mathcal{L}}$ [12]

$$\alpha\lambda_{ij} := \frac{\partial \mathcal{L}}{\partial C_{ij}} = \frac{C_{ij}^{\mathrm{QMC}} - C_{ij}}{\sigma_{ij}^2} \tag{9}$$

$$\tilde{\mathcal{L}}(\rho, \lambda) = \mathcal{L}(\rho, C) - \alpha \sum_{ij} \lambda_{ij} C_{ij} \quad , \tag{10}$$

we obtain for the PAD

$$\rho(\eta) = \frac{1}{Z}\exp \eta^T \lambda \eta \quad , \quad Z = \int e^{-\eta^T \lambda \eta} d^N \eta = \frac{\pi^{N/2}}{\sqrt{\det(\lambda)}} \quad . \tag{11}$$

The covariance is now

$$C_{ij}^{\mathrm{QMC}} = \int \rho(\eta)\eta_i\eta_j d\eta = -\frac{\partial}{\partial \lambda_{ij}}\ln(Z) = \frac{1}{2}[\lambda^{-1}]_{ij} \quad . \tag{12}$$

The solution can be cast into the same functional form (Eq. 4) as in the case of exact data-constraints [10,12], merely the determination of the Lagrange parameters is modified due to the presence of noise to

$$\lambda = \frac{1}{2}[C + \alpha M]^{-1} \quad , \quad M_{ij} = \sigma_{ij}^2 \lambda_{ij} \quad . \tag{13}$$

Hence the *Likelihood* $P(D|ACH)$ remains a normal distribution, merely the covariance is not the QMC error-covariance, it rather has to be determined via Eq. 13. The regularization parameter α entering Eq. 13 can be determined either self-consistently upon maximizing the marginal *Posterior* $P(\alpha|C\sigma H)$ [6,7], or via the historic condition $\chi^2 = N$. We employ the historic approach since we know that the number of good degrees of freedom is small and both stopping criteria will yield essentially the same result [7].

4. Application to the spectral properties of strongly correlated electrons

As a typical and topical problem we study the dynamic properties of the Hubbard model which is presently the subject of intense analytical and numerical studies. The detailed understanding of the dynamic properties of strongly correlated electrons is essential for the theoretical description of the high temperature superconductors. The Hubbard model reads

$$H = -t \sum_{<i,j>,\sigma} \left(c_{i,\sigma}^\dagger c_{j,\sigma} + h.c. \right) + U \sum_i n_{i\uparrow} n_{i\downarrow} \qquad (14)$$

with the hopping matrix element t between two adjacent sites. $c_{i,\sigma}^{(\dagger)}$ destroys (creates) an electron of spin σ on site i, $< i,j >$ denotes nearest neighbors, U is a Coulomb repulsion for two electrons of opposite spin on the same site and $n_{i,\sigma} = c_{i,\sigma}^\dagger c_{i,\sigma}$.

Unfortunately, dynamic properties cannot be measured directly by QMC simulations. Dynamical information is provided by the one-particle Matsubara Greens function $D_l = - < T_\tau c(\eta) c^\dagger(0) >$ for discrete η-values on the imaginary time axis, where $l = n * \beta/L$, $n = 0, ..., L$ and L is the number of time slices. In order to determine the spectral density $A(\omega)$ for real frequencies ω the spectral theorem is applied:

$$D_l = - \int A(\omega) \frac{e^{-\tau_l \omega}}{1 + e^{-\beta\omega}} d\omega \qquad (15)$$

which is, as already stated above, an inverse Laplace transformation problem and pathologically ill-posed.

Further information on the spectrum is provided by making use of the lowest order moments of $A(\omega)$, $\mu_m = \int \omega^m A(\omega) d\omega$, which are given by commutation relations, $\mu_m = < [c, H]_m, c^\dagger >$, and are of simple shape for m=1,2 [13].

5. Results

In order to compare the QMC/MaxEnt-data with exact results, we consider a chain of $N = 12$ sites, which is still accessible by exact diagonalization (ED) techniques. The QMC-simulations were done for an inverse temperature of $\beta t = 20$ ($T = 0.05t$), where ground state behavior is achieved for this system size and a comparison with the $T = 0$-ED results is possible. After obtaining thermal equilibrium, up to 640000 sweeps through the space-time Ising-fields were performed.

The ED-spectra were artificially broadened with a Gaussian distribution to get the shape normally observed by experiment, as is standard procedure on handling these results. Therefore the comparison persists on the overall distribution of spectral weight rather than looking for the redisplaying of each (infinitely high) peak of the ED-data.

Data from consecutive measurements are highly correlated, even for the same statistical variable D_l. A study of the skewness (third moment) and the kurtosis (fourth moment) of the data showed, that to get Gaussian behavior at least 200 measurements, each separated by 4 sweeps, have to be accumulated to form one bin. Then the results of a certain number of bins are used for the inversion process.

But binning the data does not suffice to get rid of all the correlations. Still one has to consider the correlations in imaginary time τ(i. e. between D_l and $D_{l'}$). In Fig. 1 the QMC-result for one single bin is compared to the final shape of the Greens function for $N^{bin} = 800$. Instead of being distributed 'at random' around the average the data for this bin are systematically lower than the average for $\tau < \beta/2$ and systematically higher for $\tau > \beta/2$. These correlations may be reduced by forming larger and larger bins (i. e. using more and more computation time). But it is the aim of this paper to show that this is not a sensible thing to do and that one can do better by taking into account the correlations [5] and particularly by accounting for the statistical errors of the covariance matrix.

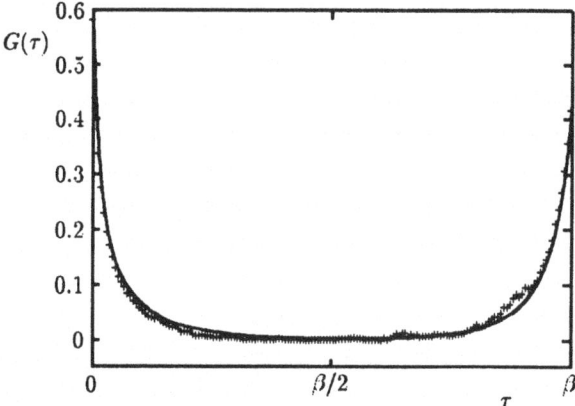

Figure 1. The data from one bin (points) compared to the final average over 800 bins (line).

In the following we will discuss the results of the MaxEnt-procedure considering three cases: (1) Neglecting any information of the covariance matrix, (2) using the covariance matrix only, and (3) taking into account both the covariance and its errors. To determine the dependence on the number of bins and to give an quantitative argument for the amount of the computational effort, which has to be taken, we examined spectra resulting from QMC-data for 200, 400 and 800 bins (160000, 320000 and 640000 sweeps, respectively). The number of time-slices, which corresponds to the dimension of the covariance matrix, is 160 in the present study.

Starting with 200 bins, which is slightly larger than the number of timeslices, one can see in Fig. 2 that neither the MaxEnt reconstruction of the plain data (Fig. 2a) nor the additional use of the covariance matrix (Fig. 2b) gives a reliable result. In both spectra the structures are too pronounced and at the wrong position. It appears that the results are generally better if the covariance matrix is not included, since the additional information is treated as exact data-constraints although it suffers from pronounced statistical noise. If, however, the statistical errors of the covariance matrix are taken into account (Fig. 2c) the result reproduces the ED-result very well. There is a small overestimation of the spectral weight at $\omega \approx -3$ only.

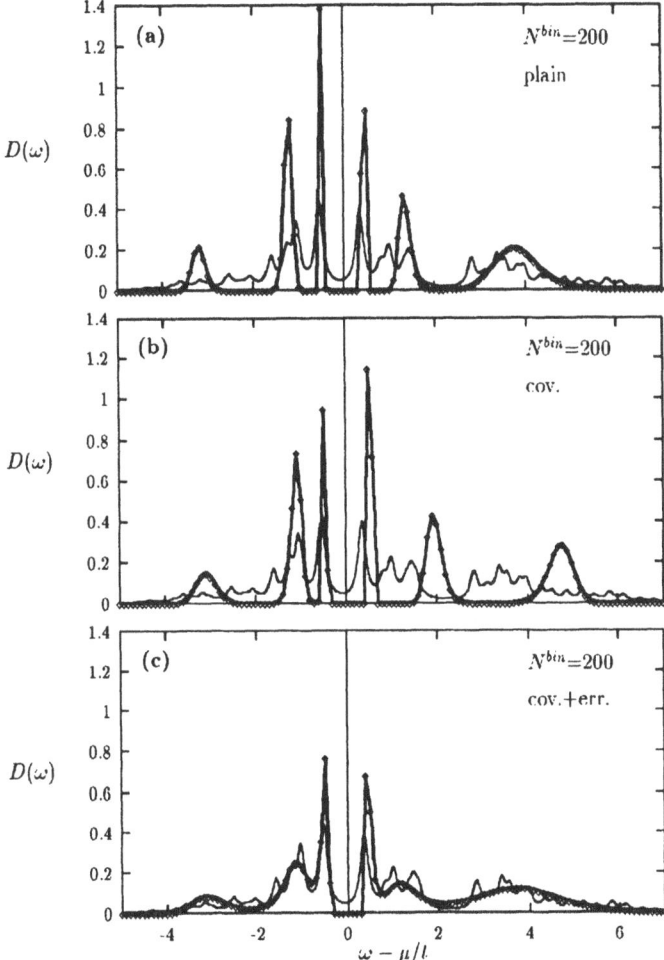

Figure 2. Comparison of the MaxEnt-spectra (thick line) with the ED-result (thin line) for 200 bins: (a) without use of the covariance (first case, see text), (b) with covariance, (c) with the error of the covariance matrix.

Increasing the number of bins to 400 (not shown, see [14]) gives still an overfitted result for the first case. Taking the covariance matrix into account the spectrum is slightly improved for $\omega > 0$, but for $\omega \approx -2$ the spectral weight is suppressed completely. Again the best spectrum is obtained if the errors of the covariance matrix are properly accounted for. Eventually for the large number of 800 bins all three spectra show satisfactory results.

The convergency of the various approaches is reasonable, since with increasing number of bins, the correlation of the QMC-errors for different imaginary times vanishes and the covariance matrix becomes diagonal and the covariance of the errors can be ignored. At the same time, the errors of the covariance matrix decrease and assuming exact data-constraints becomes also exact.

Further investigations of the effect of the error of the covariance matrix revealed that the procedure can be simplified by assuming a sufficiently large constant relative error (in our case 20%). The results show no significant deviation from the results obtained by taking the correctly generated errors.

6. Conclusion

Imaginary time data, obtained by standard QMC simulations, suffer from strongly correlated statistical errors if only a small sample is used. It appears reasonable to include the additional knowledge provided by the covariance matrix. But it has been generally observed, the QMC data of the covariance matrix are useless if the sample size is smaller than at least two times the number of time-slices. We have shown that treating consistently the covariance matrix and its errors in the Bayesian frame reliable results can be obtained almost regardless of the number bins. At least our approach represents a way of acquiring reliable data if only a small sample is at hand.

7. Acknowledgments

We would like to thank R. Silver, S. R. White and J. Gubernatisfor stimulating discussions. R. P. is grateful for support by the DFG-project Ha 1537/12-1. The calculations were performed at HLRZ Jülich and at LRZ München.

References

1. G. Bednorz and K. A. Müller *Z. Phys. B*, **64**, p. 188, 1986.
2. V. J. Emery *Phys. Rev. Lett.* , **58**, p. 2794, 1987.
3. R. N. Silver, D. S. Sivia, and J. E. Gubernatis *Phys. Rev. B*, **41**, p. 2380, 1990. See references therein.
4. R. N. Silver et al. *Phys. Rev. Lett.* , **65**, p. 496, 1990.
5. M. Jarrell and J. E. Gubernatis, "Bayesian inference and the analytic continuation of imaginary-time quantum monte carlo data," *Phys. Rep.* , 1995. To be published.
6. J. Skilling in *Maximum Entropy and Bayesian Methods*, J. Skilling, ed. , p. 45, Kluwer Academic, Dordrecht, 1989.
7. S. F. Skilling in *Maximum Entropy and Bayesian Methods*, J. Skilling, ed. , p. 53, Kluwer Academic, Dordrecht, 1989.
8. B. Buck and V. A. Macaulay, eds. , *Maximum Entropy in Action*, Oxford Science, Oxford, 1990.
9. W. von der Linden, M. Donath, and V. Dose *Phys. Rev. Lett.* , **71**, p. 899, 1993.
10. W. von der Linden *Appl. Phys.* , **A60**, p. 155, 1994.
11. J. Skilling in *Maximum Entropy in Action*, B. Buck and V. A. Macaulay, eds. , p. 19, Oxford Science, Oxford, 1990.
12. R. Silver and H. Martz in *Maximum Entropy and Bayesian Methods*, G. Heidbreder, ed. , Kluwer Academic, Dordrecht, 1993. To be published.
13. S. R. White *Phys. Rev. B*, **44**, p. 4670, 1991.
14. W. von der Linden, R. Preuss, and W. Hanke, "Consistent application of maximum entropy to quantum monte carlo data," submitted to Phys. Rev. B (1995).

AN APPLICATION OF MAXIMUM ENTROPY METHOD
TO DYNAMICAL CORRELATION FUNCTIONS
AT ZERO TEMPERATURE

HANBIN PANG, H. AKHLAGHPOUR AND M. JARRELL
Department of Physics, University of Cincinnati,
Cincinnati, OH 45221

Abstract. The ground state dynamical correlation function of the spin 1/2 Heisenberg chain is obtained with the maximum entropy method. Using sum rules, moments of a dynamical correlation function are expressed in terms of the equal-time correlation functions which be calculated with density matrix renormalization group method. With finite number of moments, we apply the maximum entropy method to obtain the dynamical correlation function.

Key words: dynamical correlation function, Heisenberg

1. Introduction

Dynamical correlation functions are important physical quantities, because they can provide a deeper understanding of a physical system. So far there are only a few general ways to obtain dynamical correlation functions for strong correlated systems, and they all have some limitations. Bosonization or conformal field theory [1] can only provide asymptotic behavior of correlation functions for one-dimensional models in the quantum critical regime. Quantum Monte Carlo simulations with the maximum entropy method [2,3] may give the dynamical correlation functions at finite temperature, but it is hard to obtain them at zero temperature. One method to calculate the ground-state dynamical properties [4] of a finite size system is the so called continue fraction method, which is based on Lanczos method. This method is limited to very small systems.

In this paper, we briefly describe a new numerical method for calculating ground-state dynamical correlation functions. For more details about this method one may find in the reference [5]. The idea is that one can always express the moments of a dynamical correlation function as a summation of some equal-time correlation functions, and if these equal-time correlation functions can be calculated, the dynamical correlation function may be obtained by doing an inverse transformation. We calculate equal-time correlation functions using the Density

K. M. Hanson and R. N. Silver (eds.), Maximum Entropy and Bayesian Methods, 179–186.
© *1996 Kluwer Academic Publishers.*

Matrix Renormalization Group method (DMRG)[6]. Since we cannot obtain infinite number of moments, we use the maximum entropy method (MEM) to obtain
an approximate solution of the inverse problem. We apply this method to the one-
dimensional spinless fermion system with nearest neighbor interaction. This model
is equivalent to the spin-1/2 XXZ chain. We consider two special cases of this
model, corresponding to the XY model and the Heisenberg model. The dynamical
density-density correlation is obtained.

2. Description of the method

2.1. SPINLESS FERMION MODEL

The Hamiltonian of the one-dimensional spinless fermion model is

$$H = -t \sum_i (c_i^\dagger c_{i+1} + h.c.) + V \sum_i n_i n_{i+1}, \tag{1}$$

where $c_i^{(\dagger)}$ are annihilation (creation) operators for a fermion at site i, and $n_i = c_i^\dagger c_i - \frac{1}{2}$. This model may be mapped to the XXZ model by the Jordan-Wigner
transformation. At $V = 0$ this model is equivalent to the XY model, while at
$V = 2t$ it is equivalent to the Heisenberg model.

2.2. MOMENT AND SUM RULE

The first step of our method is to express the moments of a dynamical correlation
function as some static correlation functions with sum rules. The density-density
correlation functions are defined as the following:

$$\begin{aligned}
\chi_c(q,t) &= \frac{1}{2}\langle\{n(q,t), n(-q,0)\}\rangle - \langle n(q,t)\rangle\langle n(-q,0)\rangle, \\
\chi''(q,t) &= \frac{1}{2}\langle[n(q,t), n(-q,0)]\rangle,
\end{aligned} \tag{2}$$

where $n(q) = N^{-1/2} \sum n_l e^{iql}$, and N is the system size. The curly bracket indicates
an anticommutator, and $\langle n(q)\rangle = \text{Tr}(n(q)e^{-\beta H})$. From the fluctuation-dissipation
theorem, we have $\chi_c(q,\omega) = \coth(\omega/2k_B T)\chi''(q,\omega)$. Due to the parity and time
reversal symmetry in our model, $\chi''(q,\omega)$ and $\chi_c(q,\omega)$ have following properties:
$\chi''(q,-\omega) = -\chi''(q,\omega)$ and $\chi_c(q,-\omega) = \chi_c(q,\omega)$. At zero temperature $\chi_c(q,\omega) = \chi''(q,\omega)$ for $\omega > 0$, therefore the moment is defined as

$$m_l(q) = \int_0^\infty \frac{d\omega}{\pi} \omega^{l-1} \frac{\chi''(q,\omega)}{\omega}. \tag{3}$$

Using sum rules, the moments are equal to some static or equal-time correlation
functions.

$$m_1(q) = \frac{1}{2}\chi(q,\omega = 0)$$

$$m_2(q) = \chi_c(q, t = 0)$$

$$m_3(q) = -\frac{1}{2}\langle[[H, n(q)], n(-q)]\rangle = 2\langle c_i^\dagger c_{i+1}\rangle(1 - \cos(q)) \qquad (4)$$

where $\chi(q, \omega = 0)$ is the static susceptibility. The general form of the sum rules for higher moments is in the Appendix A. Apart from the first moment which is given by the static susceptibility, all the other moments can be expressed as equal-time correlation functions. In real calculations, it is tedious to calculate the commutators for higher moments. However it is still reasonable to obtain the expression for the first several moments using a symbolic manipulator, such as Mathematica, to calculate the commutators. In this work we have calculated the expressions for the first five moments.

2.3. EQUAL TIME CORRELATION FUNCTION

The second step is to obtain the moments by calculating the corresponding equal-time correlation functions with DMRG. The infinite lattice method (see Ref. 6 for details) is used in our calculations for open ended chains. $t = 1$ is chosen, and states kept at each iteration varies from 52 to 64. Usually equal-time correlations, for example $\langle n_i n_j\rangle$, are calculated by putting n_i and n_j on separate blocks. In contrast to the conventional method, we choose n_i and n_j on the same block and site j at the edge of the block connecting to the other block. Since the site j is one of the two sites which are just added into system, the Hilbert space for site j is complete. Therefore we can calculate the matrix elements of the combined operator $n_i n_j$ accurately. In Appendix B, we prove that $\langle n_i n_j\rangle$ calculated in this way has much higher precision than that calculated by putting n_i and n_j on separate blocks. For a system which has parity and translational symmetries, $\langle n_j n_i\rangle$ only depends on $|i - j|$. Therefore $\langle n_q n_{-q}\rangle = \sum_l \langle n_j n_{j+l}\rangle e^{iql}$ is independent of j. We calculate the moments for system sizes varying from 100 sites to 200 sites, and obtain their values for infinite system by extrapolation. Since the first moment is not an equal-time correlation but the static susceptibility, it is calculated with the finite lattice method.

2.4. MAXIMUM ENTROPY METHOD

The next step is to use MEM to obtain the dynamical correlation functions. This method has been applied to the analytic continuation of the Quantum Monte Carlo data [9], and in this paper we apply a similar method to extract the dynamic susceptibility $\chi''(q, \omega)$ from the finite number of moments m_l with the corresponding errors σ_l. We define $f(\omega) = \chi''(\omega)/\omega$, which is positive definite, as the distribution function, and the entropy or the information function $S = \sum_\omega f(\omega) - f(\omega)\log f(\omega)$. By maximizing the entropy under the constrains $m_l - \int_0^\infty \frac{d\omega}{\pi}\omega^{l-1}f(\omega) = 0$, $f(\omega)$ has the following form

$$f(\omega) = e^{-\sum_{l=1}^n (\lambda_l \omega^{l-1})}, \qquad (5)$$

where n is the number of moments and λ_l are the Lagrange multipliers. At this point one may try to find λ_l by requiring the $f(\omega)$ to satisfy the constrains without considering the error bars of the moments. However, in general, the error bars cannot be neglected. The kernel of the transformation is singular, so small errors in the moments may produce large errors in $f(\omega)$. By maximizing the posterior probability $e^{\alpha S - L}$ where $L \equiv \sum_l (m_l - \int_0^\infty \frac{d\omega}{\pi} \omega^{l-1} f(\omega))^2 / \sigma_l^2$, one can find the most probable $f(\omega)$, which gives us the moments within the range of error bars.

3. Discussion of results

3.1. EXTRAPOLATION

Let us first discuss the extrapolation and the error bar of our DMRG results. There are two major contributions to the error: that from finite size effects and that from basis set truncation in the DMRG calculations. The error bar of DMRG calculation for any finite size is obtained by varying the number of states kept at each iteration, whereas the finite size error is obtained by varying the system size. The asymptotic behavior of correlation functions is known for this model [1], which decay as a power of system size. In Fig. 1, we plot the second and third moments at $q = \pi/2$ for $V = 2t$ as a function of $1/N$, where N is the number of sites of the system. The error from basis set truncation produces the error in the extrapolated values. We use this resultant error to estimate the error bar of the moments. Extrapolating to $1/N \to 0$ gives $m_2 = 0.1700$, and the error bar is estimated as 10^{-4}. For the third moment we have $m_3 = 0.59085$ and the estimated error bar is 2×10^{-5}. Actually the third moment is known exactly: $m_3 = -\frac{2}{3} E(1 - \cos(q))$ with the ground state energy per site $E = -2(\ln 2 - 1/4)$. The exact value at $\pi/2$ is 0.590863.

3.2. NON-INTERACTING CASE (*XY* MODEL)

We test our method for the non-interacting case. In this case, $\chi''(q, \omega)$ is known exactly [10]:

$$\chi''(q, \omega) = \frac{\theta(\omega - 2t|\sin(q)|)\theta(4t|\sin(q/2)| - \omega)}{[16t^2 \sin^2(q/2) - \omega^2]^{1/2}}, \tag{6}$$

where $\theta(x)$ is the step function. The moments can also be calculated analytically. We compare the moments calculated by DMRG with the exact results, the error bars obtained by the extrapolation provide a good estimate. Apart from the five moments, there are two more pieces of information in this case: the energy boundaries $2t|\sin(q)| < \omega < 4t|\sin(q/2)|$ for $\chi''(q, \omega)$. We use these information in our MEM calculations. In Fig. 2, we plot $\chi''(q, \omega)$ at $q = 2\pi/3$ obtained by MEM with different number of moments and the exact one from Eq. (5). It shows that the $\chi''(q, \omega)$ obtained by MEM converge toward the exact one when the number of moments is increased, and $\chi''(q, \omega)$ calculated with five moments is a good approximation for the exact result.

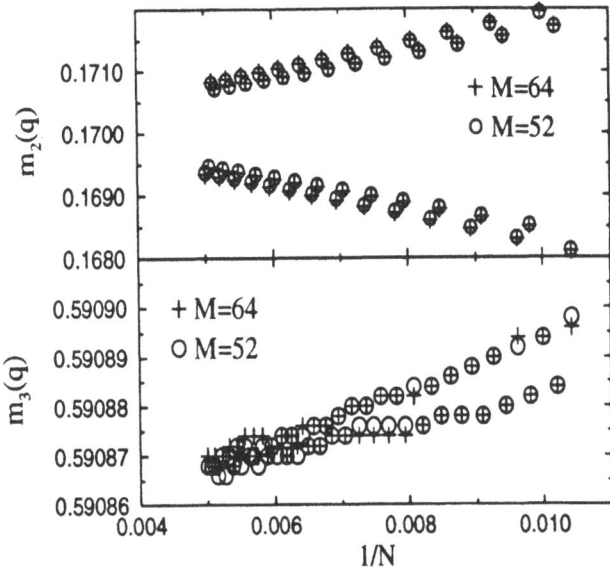

Figure 1. Moments versus the inverse system size $1/N \to 0$ for $V = 2t$ and $q = \pi/2$, where M is the number of states kept at each iteration in DMRG calculations. The extrapolation is to $1/N \to 0$, and the error is estimated by the different extrapolations caused by the errors in slope.

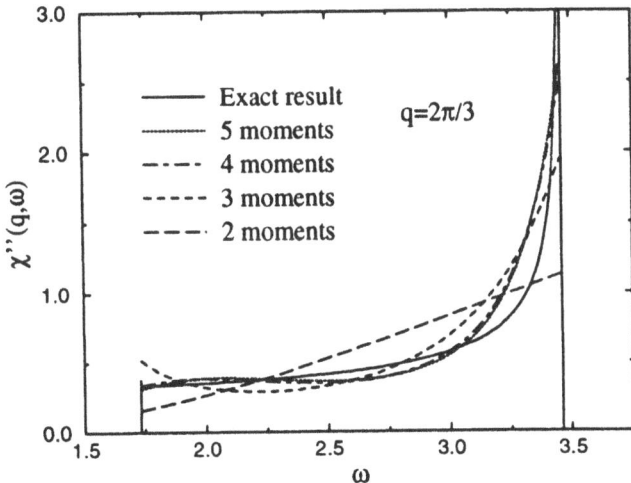

Figure 2. The dynamical structure function $\chi''(q,\omega)$ for $V = 0$ (the XY model) and $q = 2\pi/3$. We plot the results obtained by MEM with different number of moments. Two solid vertical lines are the energy boundaries.

3.3. INTERACTING CASE (HEISENBERG MODEL)

For the interacting case with $V = 2t$, which corresponding to the Heisenberg model, the elementary excitations are known as $S = 1/2$ objects [11] (spinons). The dispersion relation is $\epsilon(q) = \pi t |\sin(q)|$ [12], which provides the lower bound

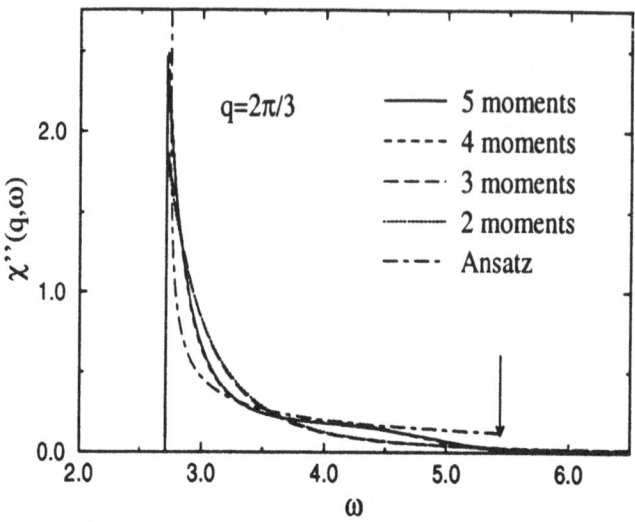

Figure 3.   The dynamical structure function $\chi''(q,\omega)$ for $V = 2t$ (the Heisenberg model) and $q = 2\pi/3$. The results obtained by MEM with different number of moments are plotted. The analytic *ansatz* which satisfies the third sum rule is also plotted for comparison. The solid vertical line is the lower boundary. The arrow marks the position of the upper boundary for the two-spinon excited states.

of excitation energies for each momentum q. The spectral weight is dominated by the continuum of the two-spinon excited states [13], and the energy range for the continuum is $t\pi|\sin(q)| < \omega < 2t\pi|\sin(q/2)|$. Based on this fact an analytic ansatz for $\chi''(q,\omega)$ was proposed as [13]

$$\chi''(q,\omega) = A\frac{\theta(\omega - \pi t|\sin(q)|)\theta(2\pi t|\sin(q/2)| - \omega)}{[\omega^2 - \pi^2 t^2 \sin^2(q)]^{1/2}}, \qquad (7)$$

where $\theta(x)$ is the step function. This is an approximate solution, because it cannot satisfy the sum rules with a fixed paramenter A. Since the contributions from the excited states of more than two spinons are finite, we only have the low energy bound. We use it as a requirement on $\chi''(q,\omega)$ in our MEM calculations. In Fig. 3, the $\chi''(q,\omega)$ at $q = 2\pi/3$ obtained by MEM with different number of moments are plotted. The analytic *ansatz* which satisfies the third sum rule is also plotted for comparison. One can see the tendency of the curves as the number of moments increase. $\chi''(q,\omega)$ tends to diverge at lower-bound. We also mark the position of upper-bound for two-spinon excited states. It is obvious that the contributions from the two-spinon continuum is dominate.

4. Summary

In conclusion, we have developed a numerical method for calculating the ground-state dynamical correlation functions in one-dimensional quantum systems based on the Density Matrix Renormalization Group Method and the maximum entropy method. We demonstrate this method on the dynamical density-density correlation function $\chi''(q,\omega)$ of the spinless fermion system with nearest neighbor interaction. For the non-interacting case, the dynamical density-density correlation function obtained by our method shows a very good agreement with the exact result. For the interacting case with $V = 2t$, we obtain the $\chi''(q,\omega)$, which was not known before.

5. Acknowledgement

We would like to acknowledge useful discussions with J.E. Gubernatis, Shoudan Liang, and R.N. Silver. This work was supported by the National Science Foundation grant No. DMR-9107563. In addition MJ would like to acknowledge the support of the NSF NYI program.

A. Sum Rules

In general, the high order of moments ($l > 2$) are

$$
\begin{aligned}
m_l &= \int_{-\infty}^{\infty} \frac{d\omega}{2\pi} \omega^{l-2} \int_{-\infty}^{\infty} dt \chi''(t) e^{i\omega t} \\
&= \frac{1}{2} \int_{-\infty}^{\infty} \frac{d\omega\, dt}{2\pi} e^{i\omega t} i^{l-2} i^{l-2} [[H,[H,...,[H,n(q,t)]...],n(-q,0)] \\
&= -\frac{1}{2} \langle [[H,[H,...,[H,n(q,0)]...],n(-q,0)] \rangle
\end{aligned}
$$

for l odd, and for l even, the moments are

$$
m_l = \int_{-\infty}^{\infty} \frac{d\omega}{2\pi} \omega^{l-2} \int_{-\infty}^{\infty} dt \chi_c(t) e^{i\omega t} = \frac{1}{2} \langle \{ [H,[H,...,[H,n(q,0)]...],n(-q,0) \} \rangle.
$$

Here we have used the facts that $\chi''(q,-\omega) = -\chi''(q,\omega)$, $\chi_c(q,-\omega) = \chi(q,\omega)$, $\chi_c(q,\omega) = \chi''(q,\omega)$ for $\omega > 0$, and $\frac{dn}{dt} = i[H,n]$.

B. Equal-time correlatin function

In the DMRG calculations, the system is separated into two blocks: left block and right block. In this Appendix, we prove that equal-time correlation function calculated by putting operators in one block has high precision. One of equal-time correlation functions we calculate is $\langle \Psi_0 | n_l n_m | \Psi_0 \rangle$ with the ground state

$$
|\Psi_0\rangle = \sum_{ij} \Psi_{ij} |i\rangle |j\rangle.
$$

Here i and j run through the Hilbert spaces of the left block and the right block respectively, and the wavefunction is real for simplicity. If the operators n_l and n_m are in the same block, we have

$$\langle \Psi_0 | n_l n_m | \Psi_0 \rangle = \sum_{ii'} \sum_j \Psi_{ij} \Psi_{i'j} \langle i | n_l n_m | i' \rangle$$

$$= \sum_{ii'} \rho_{ii'} \langle i | n_l n_m | i' \rangle = \sum_\alpha w_\alpha \langle \alpha | n_l n_m | \alpha \rangle,$$

where $\rho_{ii'}$ is the density matrix, and w_α and $|\alpha\rangle$ are its eigenvalues and eigenstates. If the correlation function is calculated in a reduced Hilbert space, the discarded weight $w_d = \sum_\alpha w_\alpha$, with α running through the states in the rest of Hilbert space. The error caused by this approximation is less than $0.25 w_d$ because $\langle \alpha | n_l n_m | \alpha \rangle \leq 0.25$ for every state $|\alpha\rangle$). Now it comes to the question whether we can calculate $\langle i | n_l n_m | i' \rangle$ in the reduced Hilbert space. If m is the site just added into the left block, then $\langle i | n_l n_m | i' \rangle$ can be calculated accurately, because we have $|i\rangle = |k\rangle |s\rangle$, where the states $|s\rangle$ for site m are complete, and $|k\rangle$ are the states in the reduced Hilbert space for the rest of sites in the left block. The matrix elements

$$\langle i | n_l n_m | i' \rangle = \langle s | \langle k | n_l n_m | k' \rangle | s' \rangle = \sum_{p\sigma} \langle s | \langle k | n_l | p \rangle | \sigma \rangle \langle \sigma | \langle p | n_m | k' \rangle | s' \rangle$$

$$= \sum_{p\sigma} \langle k | n_l | p \rangle \delta_{s\sigma} \delta_{p,k'} \langle \sigma | n_m | s' \rangle = \langle k | n_l | k' \rangle \langle s | n_m | s' \rangle$$

where $|p\rangle |\sigma\rangle$ is a complete basis, but only states in the reduced Hilbert space are involved in calculating $\langle i | n_l n_m | i' \rangle$.

References

1. Luther A. and Peschel I. (1975) *Phys. Rev. B*, **Vol. 12**, pp. 3908; Affleck I. (1990) *Fields, Strings and Critical Phenomena*, Ed. Brezin E. and Zinn-Justin J. (North Holland, Amsterdam, 1990).
2. Deisz J., Jarrell M. and Cox D. L. (1990) *Phys. Rev. B*, **Vol. 42**, pp. 4869
3. Preuss R., Muramatsu A., Linden W. von der, Dieterich P., Assaad F. F., and Hanke W. (1994) *Phys. Rev. Lett.*, **Vol. 73**, pp. 732.
4. Gagliano E. R. and Balseiro C. A. (1987) *Phys. Rev. Lett.*, **Vol. 59**, pp. 2999.
5. Pang Hanbin, Akhlaghpour H., and Jarrell M., to be published in *Phys. Rev. B*.
6. White S. R. (1992) *Phys. Rev. Lett.*, **Vol. 69**, pp. 2863; White S. R. (1993) *Phys Rev. B*, **Vol. 48**, pp. 10345.
7. Brandt T. (1983) *Statistical and Computational Methods in Data Analysis*, (North-Holland, Amsterdam, 1983), chap. 13; Gull S.F. and Skilling J. (1984) *IEE Proceedings*, **Vol. 131**, pp. 646; Bryan R.K. (1990) *Eur. Biophys. J.*, **Vol. 18**, pp. 165.
8. Hohenberg P. C. and Brinkman W. F. (1974) *Phys. Rev. B*, **Vol. 10**, pp. 128.
9. Silver R.N., Sivia D.S., Gubernatis J.E., and Jarrell M. (1990) *Phys. Rev. Lett.*, **Vol. 65**, pp. 496; Gubernatis J.E., Jarrell M., Silver R.N., and Sivia D.S. (1991) *Phys. Rev. B*, **Vol. 44**, pp. 6011.
10. Katsura S., Horiguchi T., and Suzuki M. (1970) *Physica*, **Vol. 46**, pp. 67.
11. Faddeev L. D. and Taktajan L. A. (1981) *Phys. Lett. A*, **Vol. 85**, pp. 375.
12. Des Cloizeaux J. and Pearson J. J. (1962) *Phys. Rev.*, **Vol. 128**, pp. 2131.
13. Müller G., Beck H., and Bonner J. (1979) *Phys. Rev. Lett.*, **Vol. 43**, pp. 75.

CHEBYSHEV MOMENT PROBLEMS:
MAXIMUM ENTROPY AND KERNEL POLYNOMIAL METHODS

R. N. SILVER, H. ROEDER, A. F. VOTER AND J. D. KRESS
Theoretical Division
MS B262 Los Alamos National Laboratory
Los Alamos, NM 87545 [†]

Abstract. Two Chebyshev recursion methods are presented for calculations with very large sparse Hamiltonians, the kernel polynomial method (KPM) and the maximum entropy method (MEM). They are applicable to physical properties involving large numbers of eigenstates such as densities of states, spectral functions, thermodynamics, total energies for Monte Carlo simulations and forces for tight binding molecular dynamics. This paper emphasizes efficient algorithms.

Key words: computational physics, convex optimization, density of states, electronic structure, kernel polynomial method, large sparse matrices, moment problems, maximum entropy.

1. Introduction

Many computational physics problems involve calculations with very large sparse Hamiltonian matrices. Finding all eigenvectors and eigenvalues requires cpu time scaling as $O(N^3)$ and memory scaling as $O(N^2)$, which is impractical. For ground or isolated eigenstates the preferred method is Lanczos diagonalization, which uses only matrix-vector multiply operations and requires cpu and memory scaling as $O(N)$. But new $O(N)$ methods are needed for properties involving many eigenstates such as the density of states (DOS) and spectral functions, and for quantities that can be derived from DOS such as thermodynamics, total energies for electronic structure and forces for molecular dynamics and Monte Carlo simulations. In such applications, limited energy resolution and statistical accuracy are often acceptable provided the uncertainties can be quantified. Maximum entropy (MEM) [1,2] has been a popular approach to such problems, usually fitting power moments of a DOS or spectral function. However, the non-linear convex optimiza-

---

[†]to appear in Maximum Entropy and Bayesian Methods 1995, eds. K. Hanson and R. N. Silver, Kluwer Academic Press; e-mail: rns@loke.lanl.gov

K. M. Hanson and R. N. Silver (eds.), Maximum Entropy and Bayesian Methods, 187–194.

tion algorithms required to find MEM solutions may be difficult to implement for large numbers of power moments and for singular structures in DOS.

Calculation of Chebyshev moments is much less sensitive than power moments to the limitations of machine precision. Chebyshev series are also preferred because of their isomorphism to Fourier series which enables use of advanced methods of Fourier analysis such as FFT's, Gibbs damping, etc. This paper discusses the generation of Chebyshev moment data, describes a simple linear Chebyshev approximation termed the *kernel polynomial method* (KPM) [3,4], and then it presents an efficient MEM algorithm [5].

Consider the DOS as representative of the properties of interest. The first step is to scale the Hamiltonian, $\mathbf{H} = a\mathbf{X} + b$ such that all eigenvalues x_n of $\mathbf{X}$ lie between -1 and $+1$. The DOS is then

$$D(x) = \frac{1}{N} \sum_{n=1}^{N} \delta(x - x_n) \quad . \tag{1}$$

The data about $D(x)$ consists of Chebyshev moments,

$$\mu_m = Tr\{T_m(\mathbf{X})\} = \int_{-1}^{1} T_m(x)D(x)dx \quad . \tag{2}$$

Calculation of moments uses Chebyshev recursion,

$$T_{m+1}(\mathbf{X}) = 2\mathbf{X}T_m(\mathbf{X}) - T_{m-1}(\mathbf{X}) \quad , \tag{3}$$

requiring the same optimized matrix-vector-multiply algorithm in Lanczos methods. Unlike Lanczos recursion, Chebyshev recursions are numerically stable to arbitrarily large numbers of moments without any need for expensive reorthogonalization. Exact evaluation of M moments uses cpu time $\propto O(N^2 M)$. A stochastic method [3], scaling as $O(NMN_r)$, uses estimators

$$\hat{\mu}_m \approx \frac{1}{N_r} \sum_r < r|T_m(\mathbf{X})|r > \quad , \tag{4}$$

where $|r>$ are N_r Gaussian random vectors. Such data have calculable statistical variance proportional to $(NN_r)^{-1}$. If the Hamiltonian has only local off-diagonal elements, as in tight-binding Hamiltonians, a non-stochastic locally truncated approximation to the Hamiltonian $\mathbf{H}_i$ may be adequate [6]. The estimator,

$$\hat{\mu}_m \approx \sum_i < i|T_m(\mathbf{X}_i)|i > \quad , \tag{5}$$

generates data with a systematic error determined by the truncation range. Cpu scales as $O(NMJ)$, where J is the average number of states in the truncation range. Exact moment derivatives (which are related to forces) can also be calculated.

2. The Kernel Polynomial Method

Cpu time and memory limit the number of moments M and their statistical and systematic errors. Given such limited data, KPM and MEM are two ways to estimate DOS. KPM [3,4] provides a linear Chebyshev approximation to a DOS with a uniform resolution in $\phi = \cos^{-1}(x)$. It is based on an exact moment expansion,

$$D(x) = \frac{1}{\pi\sqrt{1-x^2}} \left[\mu_0 + 2 \sum_{m=1}^{\infty} \mu_m T_m(x) \right] \quad . \tag{6}$$

The KPM truncates this expansion at M moments and introduces a factor g_m^M to damp the Gibbs phenomenon,

$$D_K(x) = \frac{1}{\pi\sqrt{1-x^2}} \left[\mu_0 + 2 \sum_{m=1}^{M} \mu_m g_m^M T_m(x) \right] \quad . \tag{7}$$

Let $D(\phi) \equiv \sin(\phi)D(X)$ and $T_m(x) = \cos(m\phi)$. Then $D_K(\phi)$ is both a simple convolution and a truncated Fourier series,

$$D_K(\phi) = \int_0^{2\pi} \delta_K(\phi - \phi_o)D(\phi_o)d\phi_o \quad ; \quad \delta_K(\phi) = \frac{1}{2\pi} \left[g_0 + 2 \sum_{m=1}^{M} g_m^M \cos(m\phi) \right] \quad . \tag{8}$$

The "kernel" $\delta_K(\phi)$ is a 2π-periodic polynomial approximation to a Dirac delta function, analogous to the resolution function of a spectrometer. Resolution is uniform in ϕ with width $\Delta\phi \propto M^{-1}$. If $g_m^M = 1$, at large $|\phi|$ the kernel is oscillatory with period $\Delta\phi = \pi/M$ within an envelope function decreasing slowly as $1/\phi^2$. The result is the Gibbs phenomenon of a lack of uniform convergence at singular structures in DOS. An optimal g_m^M can be determined variationally by requiring the kernel to be a polynomial of degree M, strictly positive, normalized and have minimal variance in ϕ [7]. Specifically, by the Fejer-Riesz theorem

$$\delta_K(\phi) = \frac{1}{2\pi} \left| \sum_{\nu=0}^{M} a_\nu e^{i\nu\phi} \right|^2 \quad ; \quad g_m^M = \sum_{\nu=0}^{M-m} a_\nu a_{\nu+m} \quad . \tag{9}$$

To obtain the best energy resolution minimize the variance,

$$\Delta\phi^2 \equiv \int_{-\pi}^{\pi} \phi^2 \delta_K(\phi)d\phi \simeq \int_{-\pi}^{\pi} (2 - 2\cos(\phi))\delta_K(\phi)d\phi = 2g_0 - 2g_1 \quad , \tag{10}$$

subject to a normalization constraint, $\int_{-\pi}^{\pi} \delta_K(\phi)d\phi = 1$, or equivalently $g_0^M = 1$. That is, the variational problem,

$$Q = g_1 - \lambda g_0 = \sum_{\nu=0}^{M-1} a_\nu a_{\nu+1} - \lambda \sum_{\nu=0}^{M} a_\nu a_\nu \quad , \tag{11}$$

results in

$$\frac{\delta Q}{\delta a_\nu} = 0 \quad \Rightarrow a_{\nu+2} - 2\lambda a_{\nu+1} + a_\nu = 0 \quad . \tag{12}$$

The solution to Eq. (12) is

$$a_\nu = \frac{U_\nu(\lambda)}{\sqrt{\sum_{\nu=0}^{M} U_\nu^2(\lambda)}} \quad ; \quad U_\nu(\lambda) = \frac{\sin((\nu+1)\phi_\lambda)}{\sin(\phi_\lambda)} \quad ; \quad \cos(\phi_\lambda) = \lambda \quad , \tag{13}$$

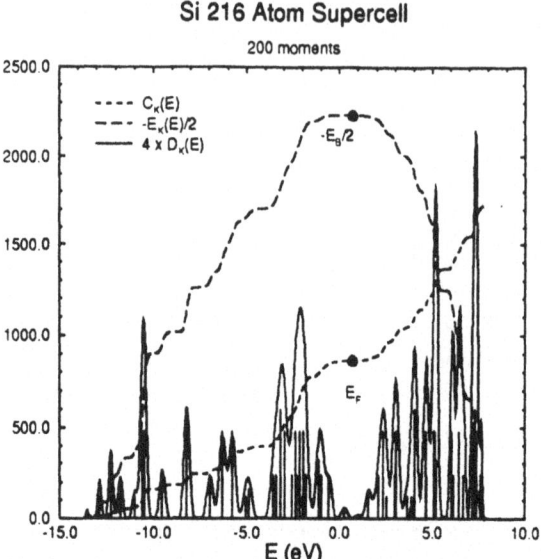

Figure 1. KPM DOS and band energy calculation for Si 216 atom supercell.

where the U_ν are Chebyshev polynomials of the second kind. The same kernel is also obtained by minimizing the uniform norm [8]. Its envelope function decreases exponentially at large $|\phi|$.

Figure 1 illustrates the application of KPM to the electronic structure of a 216 atom Si supercell using a tight binding Hamiltonian [7] based on the parameterization of Goodwin, et al. [9]. This system is small enough to be exactly diagonalized. Vertical lines are at the energies of the exact eigenstates and their height is proportional to their degeneracy. The solid line is the KPM approximation to the DOS obtained for 200 Chebyshev moments. A Fermi energy E_F is the energy at which the cumulative DOS $C_K(E)$ equals the number of electrons. The total band energy E_B is then the cumulative energy $E_K(E)$ at E_F. For band energies KPM converges $\propto M^{-2}$ reaching 10^{-5} relative accuracy at about $M \approx 150$.

KPM can be applied to other properties such as spectral functions [7],

$$A(\omega) = \lim_{\eta \to 0+} \frac{1}{\pi} Im \left\{ < \Psi_0 | O^\dagger \frac{1}{\omega - \mathbf{H} - i\eta} O | \Psi_0 > \right\} \quad , \tag{14}$$

where $\mathbf{O}$ is an operator. KPM approximations use moments $\mu_m^\mathbf{O} = < \Psi_0 | O^\dagger T_m(\mathbf{X}) O | \Psi_0 >$.

Applications to thermodynamics use a rapidly converging Fourier-Bessel expansion of the partition function [3],

$$Z = e^{-\beta b} \left[I_0(\beta a) + 2 \sum_{m=1}^{\infty} I_m(\beta a) \mu_m \right] \quad . \tag{15}$$

The $I_m(\beta a)$ are modified Bessel functions. The partition function involves integral rather than pointwise convergence, so the optimal choice is $g_m^M = 1$.

3. The Maximum Entropy Method

MEM uses the same Chebyshev moment data as KPM. The entropy,

$$S \equiv \int_0^\pi \left[D(\phi) - D_o(\phi) - D(\phi) \ln \left(\frac{D(\phi)}{D_o(\phi)} \right) \right] d\phi \quad . \tag{16}$$

Here $D_o(\phi)$ is a default model for the DOS in the absence of data. Consider first the case where the data are subject to Gaussian independent statistical noise,

$$\hat{\mu}_m = \mu_m + \eta_m \quad ; \quad \mathbf{E}\eta_m = 0 \quad ; \quad \mathbf{E}\eta_m\eta_{m'} = \sigma_m^2 \delta_{mm'} \quad . \tag{17}$$

($\mathbf{E}$ denotes the statistical expectation value of the random variable that follows it.) In case the data are exact, σ_m represents the numerical precision required of the MEM fit to the data. The *primal optimization problem* is to maximize entropy as a function of $D(\phi)$ constrained by the known moments. That is, maximize

$$Q_p \equiv S - \frac{\chi^2}{2\alpha} \quad ; \quad \chi^2 = \sum_{m=0}^M \left(\frac{\hat{\mu}_m - \mu_m}{\sigma_m} \right)^2 \quad . \tag{18}$$

The statistical regularization parameter α sets a balance between the fit, measured by χ^2, and an information measure, $-S$, of distance between the inferred $D(\phi)$ and the default model $D_o(\phi)$. (Alternatively, $1/\alpha$ is a Lagrange multiplier.) The $m = 0$ term is included to constrain normalization, $\hat{\mu}_0 = 1$. Taking the limit $\sigma_0 \to 0$ strictly enforces normalization.

Our MEM algorithm consists of three nested loops: iterations in α, until a stopping criterion is reached; at each α, Newton-Raphson iterations of a dual optimization problem to solve for the MEM $D(\phi)$; at each α and MEM $D(\phi)$ conjugate gradient iterations to apply the Hessian onto a vector.

Popular stopping criteria for α are $\chi^2 = M$ and $\chi^2 - 2\alpha S = M$, although many other criteria are discussed in the literature. However, the algorithm for finding the MEM $D(\phi)$ tends to be unstable if initiated at such small α. Instead, start at large $\alpha^1 \approx \chi_o^2$, and use $D_o(\phi)$ to initiate the optimization of $D^1(\phi)$. Progress down in α such that $\alpha^{k+1} = \alpha^k/2$. If this is unstable, halve the step down in α repeatedly until stability is reached. At each α, use $D^k(\phi)$ as the starting point for the optimization of $D^{k+1}(\phi)$. Once the stopping criterion is passed, perform a golden search for the optimal α.

In the case of exact moment data, set σ_m to the numerical precision required, which can be very small. In our applications to electronic structure, errors of one part in 10^5 or smaller were used. Iterate $\alpha \to 0$ until the entropy S saturates at an α-independent value.

Given an α, a variety of algorithms have been developed to find MEM solutions [10,11]. The *primal optimization problem* maximizes Q_p as a function of $D(\phi)$,

$$\frac{\delta Q_p}{\delta D(\phi)} = -\ln \left(\frac{D(\phi)}{D_o(\phi)} \right) + \sum_{m=0}^M \frac{\hat{\mu}_m - \mu_m}{\alpha \sigma_m^2} \cos(m\phi) = 0 \quad , \tag{19}$$

which has a unique solution. Define parameters $\vec{\lambda}$ by

$$\hat{\mu}_m - \mu_m + \alpha\sigma_m^2\lambda_m = 0 \quad . \tag{20}$$

Then the $D(\phi)$ satisfying Eq. (19) is

$$D(\phi) = D_o(\phi)\exp\left(-\sum_{m=0}^{M}\lambda_m\cos(m\phi)\right) \quad . \tag{21}$$

This form is also obtained by maximizing entropy subject to Lagrange contraints on moments with Lagrange multipliers $\vec{\lambda}$.

However, a *dual optimization problem* as a function of the M Lagrange multipliers [12] solves the same problem, and it is more stable numerically than the primal problem. The $\vec{\lambda}$ of the dual problem vary more slowly than the $D(\phi)$, and they are a finite rather than a continuous set of variables. The quantity,

$$Q_d \equiv \ln\left(\int_0^{\pi} D(\phi)d\phi\right) + \sum_{m=0}^{M}\left[\hat{\mu}_m\lambda_m + \frac{\alpha\sigma_m^2\lambda_m^2}{2}\right] \quad , \tag{22}$$

is maximized as a function of the $\vec{\lambda}$ when Eq. (20) is satisfied. Away from the maximum, define

$$\frac{\delta Q_d}{\delta\lambda_m} \equiv \xi_m \equiv \hat{\mu}_m - \mu_m + \alpha\sigma_m^2\lambda_m \quad . \tag{23}$$

Then,

$$Q_d = Q_p + \sum_{m=0}^{M}\frac{\xi_m^2}{2\alpha\sigma_m^2} \quad . \tag{24}$$

The Hessian of the dual problem is a positive definite $M \times M$ matrix and a simple function of the moments,

$$H_{mm'} \equiv \frac{\partial^2 Q_d}{\partial\lambda_m\partial\lambda'_m} = \frac{\mu_{m+m'} + \mu_{|m-m'|}}{2} + \alpha\sigma_m^2\delta_{mm'} \quad . \tag{25}$$

A solution to Eq. (20) may be found by Newton-Raphson iteration. Beginning with some starting $\vec{\lambda}^0$, the $n+1$'th step is

$$\vec{\lambda}^{n+1} = \vec{\lambda}^n - \mathbf{H}_n^{-1}\vec{\xi}^n \quad . \tag{26}$$

The quantity, $\mathbf{H}^{-1}\vec{\xi}$, may be calculated, e.g., by conjugate gradients. In view of Eq. (24), converging bounds at the n'th iteration are $Q_d^n \geq Q^\infty \geq Q_p^n$ where $Q^\infty \equiv \lim_{n\to\infty}\{Q_d^n, Q_p^n\}$. This provides stopping criteria for the iteration.

For electronic structure applications, high numerical precision (e.g. $\approx 10^{-6}$) is needed for accurate energy derivatives. Careful attention to how the MEM algorithms are discretized then becomes very important. Practical fast Fourier transform (FFT) algorithms calculate the $\mu_m = \int_0^{\pi}\cos(m\phi)D(\phi)d\phi$ by sampling the domain $0 \leq \phi \leq \pi$ at a discrete set of N_p equally spaced points. The Shannon

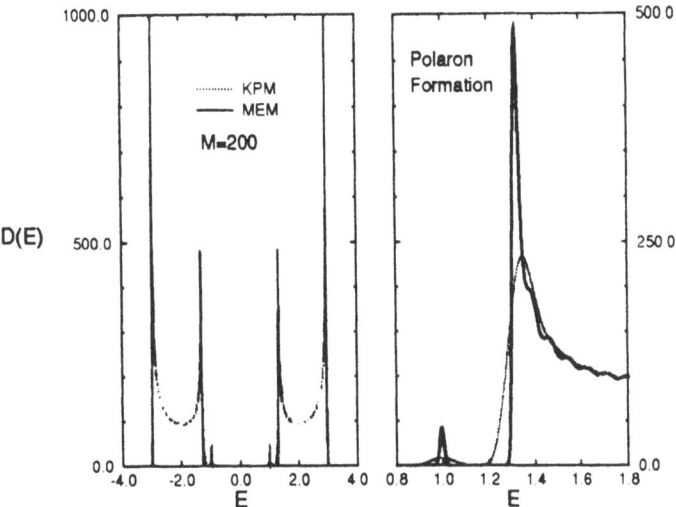

Figure 2.   Comparison of KPM and MEM for the DOS of a polaron formation problem using 200 moments.

sampling theorem says that such naive discretization corresponds to representing a DOS in a N_p-order truncated Chebyshev series. In effect, MEM is used to infer an N_p-order Chebyshev approximation from knowledge of M true moments.

But typical DOS contain singular structures such as δ-functions, van Hove singularities, band edges, etc. These structures are properly described by an infinite order Chebyshev expansion. As discussed previously for the KPM, abrupt truncation of a Chebyshev expansion at N_p terms results in the Gibbs phenomenon; i.e. singular structures in the true DOS at ϕ_o induce oscillations in Chebyshev approximated DOS of period $\Delta\phi = \pi/N_p$ with an envelope function decreasing slowly as $1/(\phi - \phi_o)^2$. While the moment data satisfy the Hausdorff conditions for the existence of a positive solution [1], the added requirement that the solution be an N_p-order Chebyshev series is stronger. For the exact moment problem, the α-iteration may have difficulty forcing $\chi^2 \to 0$ and saturating the entropy S.

Fortunately, the kernel polynomial method (KPM) provides a solution to this discretization problem. In the MEM problem, replace the M Chebyshev moment data $\hat{\mu}_m$ by modified moments $\hat{\mu}_m g_m^{N_p}$, where the $g_m^{N_p}$ are the Gibbs damping factors in Eq. (9). In other words, change the goal of the MEM algorithm to the inference of a N_p-order KPM approximation to the DOS. Iteration toward $\chi^2 \to 0$ and saturation of S becomes easy.

By choosing $N_p \gg M$, MEM can achieve significant resolution enhancements over KPM from the same M moments. In tests with tight binding Hamiltonians for the electronic structure of Si [5], band energies converge approximately 4 times faster with MEM than with KPM. For the example in Fig. 1, MEM reaches 10^{-5}

accuracy at $M \approx 35$. Setting $N_p \geq 4M$ is adequate to achieve this gain. The cpu time required by MEM scales as $O(M^2)$, and it is negligible compared to the cpu time required to generate the moment data. Use of MEM cuts the overall cpu requirements by at least a factor of 4 over KPM. Isolated features in DOS, such as individual states and band edges, may converge even faster.

Figure 2 compares MEM and KPM for the DOS of a 1D polaron formation problem [5]. The Hamiltonian consists of an electron placed into a 10,000 atom chain with a Peierls distortion, which is then allowed to relax resulting in the polaron state at $E = 1.0$. MEM achieves dramatically better energy resolution than KPM for isolated states and band edges, but it tends to "ring" (or oscillate) in smooth positive regions of a DOS when singular structures, such as Van Hove singularities, are nearby. For such regions of a spectrum, MEM may converge more slowly than KPM. A solution to the ubiquitous MEM ringing problem most likely will require a modification to the entropy functional to include local smoothness constraints [13].

References

1. L. R. Mead and N. Papanicolaou, "Maximum entropy in the problem of moments," *J. Math. Phys.*, **25**, p. 2404, 1984.
2. D. A. Drabold and O. F. Sankey, "Maximum entropy approach to linear scaling in the electronic structure problem," *Phys. Rev. Letts.*, **70**, p. 3631, 1993.
3. R. N. Silver and H. Roeder, "Density of states of megadimensional hamiltonian matrices," *Int. J. Mod. Phys. C*, **5**, pp. 735–753, 1994.
4. L. W. Wang, "Calculating the density of states and optical-absorption spectra for large quantum systems by the plane wave moments method," *Physical Review B*, **49**, pp. 10154–10158, 1994.
5. R. N. Silver and H. Roeder, "Maximum entropy algorithms for chebyshev moment problems," *J. Comp. Phys.*, to appear 1995.
6. A. F. Voter, J. D. Kress, and R. N. Silver, "Linear scaling tight binding from a truncated moments approach," *submitted to Physical Review B*, 1995.
7. R. N. Silver, H. Roeder, A. F. Voter, and J. D. Kress, "Kernel polynomial approximations for densities of states and spectral functions," *J. Comp. Phys.*, to appear 1995.
8. J. D. Jackson, *Theory of Approximation*, Cambridge University Press, 1930.
9. L. Goodwin, A. J. Skinner, and D. G. Pettifor, "Generating transferable tight binding parameters: Application to silicon," *Europhysics Letters*, **9**, pp. 701–706, 1989.
10. I. Turek, "A maximum entropy approach to the density of states within the recursion method," *J. Phys. C: Solid State Physics*, **21**, pp. 3251–3260, 1988.
11. J. Skilling, "Bayesian numerical analysis," in *Physics & Probability: Essays in honor of Edwin T. Jaynes*, J. W. T. Grandy and P. W. Milonni, eds., pp. 207–221, Cambridge University Press, 1993.
12. C. Auyeng and R. M. Mersereau, "A dual approach to signal restoration," in *Digital Image Restoration*, A. K. Katsaggelos, ed., pp. 21–55, Springer, 1991.
13. R. N. Silver and H. F. Martz, "Applications of quantum entropy to statistics," in *1994 Proceedings of the Statistical Computing Section*, pp. 61–70, American Statistical Association, 1994.

CLUSTER EXPANSIONS AND ITERATIVE SCALING
FOR MAXIMUM ENTROPY LANGUAGE MODELS

JOHN D. LAFFERTY AND BERNHARD SUHM
School of Computer Science
Carnegie Mellon University
5000 Forbes Avenue
Pittsburgh, PA 15217 USA[†]

Abstract. The maximum entropy method has recently been successfully introduced to a variety of natural language applications. In each of these applications, however, the power of the maximum entropy method is achieved at the cost of a considerable increase in computational requirements. In this paper we present a technique, closely related to the classical cluster expansion from statistical mechanics, for reducing the computational demands necessary to calculate conditional maximum entropy language models.

1. Introduction

In this paper we present a computational technique that can enable faster calculation of maximum entropy models. The starting point for our method is an algorithm [1] for constructing maximum entropy distributions that is an extension of the generalized iterative scaling algorithm of Darroch and Ratcliff [2,3]. The extended algorithm relaxes the assumption of [2,3] that the constraint functions sum to a constant, and results in a set of decoupled polynomial equations, one for each feature, that must be solved to obtain the scaling terms. For each iteration, the distribution must be normalized (that is, the partition function must be calculated), and the coefficients of the polynomials must be determined; these steps have roughly the same computational cost.

For language modeling applications the partition function and coefficient calculations entail summing over the target vocabulary, typically on the order of 10,000–100,000 words, and determining those features that apply to each possible word for each context that appears in the training data. When this calculation is implemented directly by carrying out the summation while hashing to determine features and feature weights, it can be exceedingly slow. We address this problem

---

[†]Research supported in part by NSF and ARPA under grant IRI-9314969 and the ATR Interpreting Telecommunications Research Laboratories.

K. M. Hanson and R. N. Silver (eds.), Maximum Entropy and Bayesian Methods, 195–202.
© *1996 Kluwer Academic Publishers.*

by use of a technique that we call the *cluster expansion*, due to its resemblance to series expansion methods in statistical physics, that carries out both the partition function and coefficient calculations efficiently. Our basic idea is to avoid hashing and an explicit summation over the entire target vocabulary for each context by calculating the partition function (or coefficients) for all contexts simultaneously as a telescoping sum of polynomials in the feature weights. By choosing the data structures in the implementation appropriately, the cluster expansion can be easily implemented for a class of language models that includes n-gram constraints in addition to state constraints from an underlying automaton, or other long-distance constraints.

In this paper we present a description of the basic technique as well as its application to the construction of a simple language model for use in a speech recognition system.

2. Language Modeling

2.1. LANGUAGE MODELS AS PRIORS FOR BAYESIAN DECODING

Language modeling attempts to identify regularities in natural language and capture them in a statistical model. Language models are crucial ingredients in automatic speech recognition [4] and statistical machine translation [5] systems, where their use is naturally viewed in terms of the *noisy channel* model from information theory. In this framework an information source emits messages X from a distribution $P(X)$ which then enter into a noisy channel and emerge transformed into observables Y according to a conditional probability distribution $P(Y \mid X)$. The problem of decoding is to determine the message $\widehat{X}$ having the largest posterior probability given the observation:

$$\widehat{X} = \arg\max_{X \in \mathcal{H}} P(X \mid Y) = \arg\max_{X \in \mathcal{H}} P(Y \mid X)\, P(X)\,.$$

Thus, Bayesian decoding is carried out using a prior distribution $P(X)$ on messages, a channel model $P(Y \mid X)$, and a decoder $\arg\max_{X \in \mathcal{H}}$. For speech recognition and machine translation, the prior distribution is called a *language model*, and it must assign a probability to every string of symbols that can be hypothesized by the decoder. The most common language models used in today's speech systems are the n-gram models, constructed in terms of simple word frequencies.

2.2. CONDITIONAL MAXIMUM ENTROPY LANGUAGE MODELS

In the usual application of the maximum entropy principle [6], prior information, typically in the form of frequencies, is represented as a set of constraints which collectively determine a unique maximum entropy distribution. For example, if we observe certain bigram word frequencies $c_{ij} = \tilde{p}(w_i\, w_j)$ and we constrain a language model to agree with these observations, the maximum entropy distribution assigns a probability $p_\lambda(W)$ to a word string W according to a Gibbs distribution of the form

$$p_\lambda(W) = \frac{1}{Z_\lambda} \exp\left(\sum_{ij} \lambda_{ij} f_{ij}(W)\right)$$

where the *feature* $f_{ij}(W)$ counts the number of times the bigram $w_i w_j$ occurs in the string W, and where the partition function Z_λ is obtained by summing over all possible word strings W.

In contrast to this use of the joint distribution, recent applications of the maximum entropy method in language modeling [7,8] have employed *conditional* models. Such models employ features to represent various frequencies in the training text, such as the bigram features just mentioned, but they use this information to constrain a family of conditional exponential models. Factoring a word string $W = w_0 w_1 \cdots w_N$ into conditional probabilities we can write

$$p(W) = p(w_0) \prod_{i=1}^{N} p(w_i \mid w_0 w_1 \cdots w_{i-1}) = p(w_0) \prod_{i=1}^{N} p(w_i \mid h_i)$$

where h_i is the *history at time i*. In terms of conditional models, the constraints are presented as

$$\sum_h \tilde{p}(h) \sum_w p(w \mid h) f_\alpha(h, w) = \sum_{h,w} \tilde{p}(h, w) f_\alpha(h, w)$$

where h is a history, and the maximum entropy model subject to these constraints is given by

$$p_\lambda(w \mid h) = \frac{1}{Z_\lambda(h)} \exp\left(\sum_\alpha \lambda_\alpha f_\alpha(h, w)\right) \tag{1}$$

The partition function $Z_\lambda(h)$ is now obtained from summing over the target word vocabulary, rather than over all word strings. Constraining a family of conditional models in this manner is typically much more manageable computationally than working with a single constrained joint distribution. In addition, the use of conditional models is desirable for applications which process the input in a left-to-right fashion.

3. Iterative Scaling

The generalized iterative scaling algorithm of Darroch and Ratcliff [2] is one method for calculating the maximum entropy distribution (1). This algorithm assumes that the features $f_\alpha(h, w)$ are non-negative and sum to a constant, independent of h and w:

$$M(h, w) \equiv \sum_\alpha f_\alpha(h, w) = M, \quad \text{for all } h, w. \tag{2}$$

Given these restrictions, the Darroch-Ratcliff algorithm begins with an initial model, typically the uniform distribution obtained by setting $\lambda_\alpha = 0$. In the iterative step, when the current model is $p_\lambda(w \mid h)$, the algorithm increments each parameter λ_α by an amount $\Delta\lambda_\alpha$ determined by

$$\Delta\lambda_\alpha = \frac{1}{M} \log \left(\frac{\sum_{h,w} \tilde{p}(h,w) f_\alpha(h,w)}{\sum_{h,w} \tilde{p}(h) p_\lambda(w \mid h) f_\alpha(h,w)} \right)$$

Letting $\Delta\beta_\alpha = e^{\Delta\lambda_\alpha}$, we can express this update as choosing $\Delta\beta_\alpha$ to be the unique solution of the equation

$$\tilde{p}_\lambda[f_\alpha \Delta\beta_\alpha^M] = \tilde{p}[f_\alpha] \tag{3}$$

where $q[\,\cdot\,]$ denotes expectation with respect to q and we use $\tilde{p}_\lambda$ to denote the distribution $\tilde{p}_\lambda(h,w) = \tilde{p}(h) p_\lambda(w \mid h)$.

While the restriction (2) on M can always be enforced by introducing a "slack variable," it can be inconvenient to do so for conditional maximum entropy language models that typically have hundreds of thousands of features. In [1] an algorithm was introduced that extends the Darroch-Ratcliff procedure by relaxing the assumption that $M(h,w)$ is a constant. The updates for the improved algorithm are again given by equation (3), but with M now interpreted as a random variable. When (2) holds, the algorithms are identical. In general, the algorithm which allows M to vary is more natural and easier to implement. It also converges more quickly, by effectively increasing the step size taken toward the maximum entropy solution at each iteration.

4. Cluster Expansions

4.1. THE MAYER EXPANSION FOR A CLASSICAL GAS

If the Hamiltonian for a classical N-particle system is given by $H = \frac{1}{2}\sum_i p_i^2 + \sum_{i<j} v_{ij}$ and the system occupies a volume V, then the classical partition function of the system at temperature T is given by

$$Q_N(V,T) = \frac{1}{h^{3N} N!} \int_{\mathbf{R}^{3N}} \int_V dp\,dq\,\exp\left(-\tfrac{1}{2}\beta \sum_i p_i^2 - \beta \sum_{i<j} v_{ij}\right)$$

where $\beta = 1/kT$ and h is a constant introduced to make Q_N dimensionless. Computing the integral over the momenta reduces this to

$$Q_N(V,T) = \frac{1}{\lambda^{3N} N!} \int_V dq\,\exp\left(-\beta \sum_{i<j} v_{ij}\right) \equiv \frac{1}{\lambda^{3N} N!} Z_N(V,T)$$

where $\lambda = \sqrt{2\pi\hbar^2/kT}$. The idea of the cluster expansion is to make a change of variables

$$\phi_{ij} = e^{-\beta v_{ij}} - 1$$

and expand Z_N as a sum of products of ϕ_{ij}:

$$Z_N(V,T) = \int_V dq \prod_{i<j}(1 + \phi_{ij}) = \int_V dq \left(1 + \sum_{i<j} \phi_{ij} + \sum_{i<j}\sum_{k<l} \phi_{ij}\phi_{kl} + \cdots\right)$$

A convenient way to think about the integrals that need to be computed comes from expressing the various terms as graphs. If $N = 3$, for example, the integrands are represented as graphs as follows:

In terms of this correspondence, $Z_N = \sum_G S(G)$, where the sum is over all N-particle graphs and $S(G)$ is the appropriate integral; for example,

$$S\left(\begin{array}{c} {\scriptstyle 3} \\ \diagup \\ {\scriptstyle 1 \quad 2} \end{array} \right) = \int_V dq\, \phi_{12}\, \phi_{13}\,.$$

If a graph G is disconnected, then $S(G)$ factors into a product of terms, and each connected component is referred to as a *cluster*. The *Mayer cluster integral* b_l is given by $b_l = 1/l! \sum_{l\text{-clusters } G_l} S(G_l)$. Thus,

$$b_3 = \frac{1}{3!} S\left(\begin{array}{c} {\scriptstyle 3} \\ \diagup \\ {\scriptstyle 1\ \ 2} \end{array} + \begin{array}{c} {\scriptstyle 3} \\ \diagup \\ {\scriptstyle 1\ \ 2} \end{array} + \begin{array}{c} {\scriptstyle 3} \\ \wedge \\ {\scriptstyle 1\ \ 2} \end{array} + \begin{array}{c} {\scriptstyle 3} \\ \triangle \\ {\scriptstyle 1\ \ 2} \end{array} \right).$$

Simple combinatorial arguments lead to an expression for Z_N in terms of the integrals b_l. While this is then carried further to obtain a series expansion for the grand partition function, our use of the method will simply make use of the discrete analogues of the integrals b_l for conditional models. For more details on the statistical physics calculations we refer to [9].

4.2. CLUSTER EXPANSIONS FOR CONDITIONAL MAXENT MODELS

The computation necessary to carry out the iterative scaling algorithm described in Section 3 is naturally divided into two parts. First, for a conditional maximum entropy model of the form (1), it is necessary to compute the partition functions $Z_\lambda(h)$ for each history h such that $\tilde{p}(h) > 0$. Using the notation from statistical physics, we make the change of variables $\phi_\alpha(h, w) = e^{\lambda_\alpha f_\alpha(h,w)} - 1$ so that $Z_\lambda(h)$ can be expressed as

$$Z_\lambda(h) = \sum_w \prod_\alpha (1 + \phi_\alpha) = \sum_w \left(1 + \sum_\alpha \phi_\alpha + \sum_{\alpha,\alpha'} \phi_\alpha\, \phi_{\alpha'} + \cdots \right).$$

In analogy with the classical expansion, this expresses the normalization $Z_\lambda(h)$ as a sum of *cluster integrals*, where $\sum_w \sum_\alpha \phi_\alpha(h, w)$ is the order one cluster, $\sum_w \sum_{\alpha,\alpha'} \phi_\alpha\, \phi_{\alpha'}$ is the order two cluster, and the highest order cluster that needs to be computed is the order-M cluster where M is the largest value of $\sum_\alpha f_\alpha(h, w)$.

This gives an *exact* expression for $Z_\lambda(h)$ as a telescoping sum. The point of using this technique, as we will explain further in the following section, is that computation of the individual clusters can be significantly more efficient than computing $Z_\lambda(h)$ directly. Furthermore, the computation of the clusters can be shared across different histories. The use of Cheeseman's method [10,11] of reordering summations within a cluster can provide further savings.

The second computation that is necessary is the calculation of the coefficients of $\Delta\beta_\alpha$ in the expectation $\tilde{p}_\lambda[f_\alpha\Delta\beta_\alpha^M]$ that appears in the scaling equation (3). In a manner similar to that described above, we expand in terms of ϕ_γ to obtain

$$\tilde{p}_\lambda[f_\alpha\Delta\beta_\alpha^M] = \sum_h \frac{\tilde{p}(h)}{Z_\lambda(h)} \sum_w \left(1 + \sum_\gamma \phi_\gamma + \sum_{\gamma,\gamma'} \phi_\gamma\,\phi_{\gamma'} + \cdots\right) f_\alpha(h,w)\,\Delta\beta_\alpha^{M(h,w)}.$$

Here again, indirect computation of the coefficients through the calculation of the individual cluster terms can be significantly more efficient than direct computation.

The primary savings that this technique affords results from its avoidance of an explicit summation over the entire target vocabulary for each history. In addition, it can make hashing for feature lookup unnecessary. While we do not generally obtain better theoretical computational complexity, this simple trick can result in substantial savings in the computation necessary for carrying out generalized iterative scaling. We will now give further details of these calculations for a simple topic-dependent bigram model developed for use in a speech recognition system.

5. Example: A Topic-Dependent Language Model

In this section we describe the application of the cluster expansion to the training of a topic-dependent bigram model of the *Switchboard* corpus [12] for use in a speech recognition system. This corpus comprises approximately three million words of text, transcribed from more than 150 hours of speech collected from telephone conversations. An important aspect of the Switchboard corpus is that the conversations are restricted to 70 different topics. To take advantage of this structure, we trained a maximum entropy language model whose constraints were of three types. In addition to unigram and bigram constraints, we introduced topic-dependent unigram constraints for those words having the greatest mutual information with the topic.

More precisely, the model that we constructed was specified as follows. Conditioning on a word history h which ends in a word w', the probability of predicting w is given by

$$p(w\,|\,h) = \sum_t p(\text{topic} = t\,|\,h)\,p(w\,|\,h,t) = \sum_t p(\text{topic} = t\,|\,h)\,p(w\,|\,w',t).$$

This model has two components: a *topic prediction* model $p(\text{topic} = t\,|\,h)$ and a *word prediction* model $p(w_j\,|\,t,w_i)$. (The topic prediction model is not discussed here.) The word prediction model is constructed as a conditional maximum entropy distribution of the form

$$p_\lambda(w_j\,|\,t,w_i) = \frac{1}{Z_\lambda(i,t)}\exp\left(\lambda_{ij} + \lambda_i + \lambda_{tj}\right).$$

We thus place constraints on the model so that it agrees with the bigram and unigram frequencies as they appear in the data. In addition, we constrain the topic-dependent unigrams, corresponding to the parameters λ_{tj}, for those words w_j that appear with sufficiently high mutual information with topic t. For example, the topic-independent bigram constraint equations take the form

$$\sum_t \tilde{p}(w_i, t)\, p(w_j \mid w_i, t) = \tilde{p}(w_i, w_j) \equiv c_{ij}$$

where $\tilde{p}$ is the empirical distribution, and the corresponding scaling equations update λ_{ij} by an amount $\Delta\lambda_{ij} = \log \Delta\beta_{ij}$, where $\Delta\beta_{ij}$ is the unique positive solution to the equation

$$\sum_t \tilde{p}(w_i, t)\, p_\lambda(w_j \mid w_i, t)\, \Delta\beta_{ij}^{M(i,t,j)} = c_{ij}\,.$$

The constraint and scaling equations for the parameters λ_i and λ_{tj} are similar.

To apply the cluster expansion technique to this model we express the partition functions $Z_\lambda(i, t)$ in terms of the variables $\phi_\alpha = e_\alpha^\lambda - 1$ and expand $Z_\lambda(i, t) = \sum_j (1 + \phi_j)(1 + \phi_{tj})(1 + \phi_{ij})$ into a sum of four cluster "integrals"

$$Z_\lambda(i, t) = b_0 + b_1(i, t) + b_2(i, t) + b_3(i, t)\,.$$

Using a variant of the physicists' graph notation that is appropriate for conditional models, we can express these terms as a sum over all configurations of a set of graphs; for example,

In these figures the unlabeled vertex is summed over, and an edge connecting the vertex labeled ε denotes a unigram term ϕ_j. Thus,

We use the fact that $\phi_\alpha = 0$ unless λ_α is a parameter that is being estimated. This is what allows the above telescoping summation to be carried out efficiently; for example, the summation $\sum_j \phi_j \phi_{ij}$ is carried out only over those indices j for which the bigram (w_i, w_j) is constrained. The largest cluster, $b_3(i, t)$, involves a summation over all those indices j for which the bigram (w_i, w_j) is constrained *and* w_j is a topic word for topic t. The cluster integrals for the various values of (i, t) with $\tilde{p}(w_i, t) > 0$ can be calculated simultaneously by a single pass through appropriately constructed data structures, and require no expensive hashing of the bigram parameters. A very similar analysis is applied to the task of computing the coefficients of the iterative scaling equations for all of the parameters. When we implemented this technique for the topic-dependent model, the resulting calculation was more than 200 times faster than the direct implementation of the iterative scaling algorithm.

6. Summary

Our use of the cluster expansion for the language model presented in Section 5 demonstrates that this technique can be an important tool for reducing the computational burden of computing maximum entropy language models. The method also applies to higher order models such as "trigger models" [8], where occurrences of words far back in the history can influence predictions by the use of long-distance bigram parameters. As a general technique, however, the method is limited in its usefulness. As in statistical mechanics, when the number of interacting constraints is large (*i.e.*, when the gas is dense), the cluster expansion is of little use in computing the exact maximum entropy solution. For such cases the use of approximation techniques should be investigated.

References

1. S. D. Pietra, V. D. Pietra, and J. Lafferty, "Inducing features of random fields," tech. rep., CMU-CS-95-144, Department of Computer Science, Carnegie Mellon University, 1995.
2. J. Darroch and D. Ratcliff, "Generalized iterative scaling for log-linear models," *Ann. Math. Statistics*, **43**, pp. 1470–1480, 1972.
3. I. Csiszár, "A geometric interpretation of Darroch and Ratcliff's generalized iterative scaling," *The Annals of Statistics*, **17**, (3), pp. 1409–1413, 1989.
4. L. R. Bahl, F. Jelinek, and R. L. Mercer, "A maximum likelihood approach to continuous speech recognition," *IEEE Trans. on Pattern Analysis and Machine Intelligence*, **PAMI-5**, (2), pp. 179–190, 1983.
5. P. Brown, J. Cocke, S. D. Pietra, V. D. Pietra, F. Jelinek, J. Lafferty, R. Mercer, and P. Roosin, "A statistical approach to machine translation," *Computational Linguistics*, **16**, pp. 79–85, 1990.
6. E. T. Jaynes, *Papers on Probability, Statistics, and Statistical Physics*, D. Reidel Publishing, Dordrecht–Holland, 1983.
7. A. Berger, S. D. Pietra, and V. D. Pietra, "A maximum entropy approach to natural language processing," *Computational Linguistics*, to appear, 1995.
8. R. Lau, R. Rosenfeld, and S. Roukos, "Adaptive language modeling using the maximum entropy principle," in *Proceedings of the ARPA Human Language Technology Workshop*, pp. 108–113, Morgan Kaufman Publishers, 1993.
9. R. P. Feynman, *Statistical Mechanics: A Set of Lectures*, W. A. Benjamin, Reading, MA, 1972.
10. P. C. Cheeseman, "A method for computing generalized Bayesian probability values for expert systems," in *Proc. Eighth International Conference on Artificial Intelligence*, pp. 198–202, 1983.
11. S. A. Goldman, "Efficient methods for calculating maximum entropy distributions," tech. rep., MIT Department of Electrical Engineering and Computer Science (Masters thesis), 1987.
12. J. Godfrey, E. Holliman, and M. McDaniel, "Switchboard: Telephone speech corpus for research development," in *Proc. ICASSP-92*, pp. I–517–520, 1992.

A MAXENT TOMOGRAPHY METHOD FOR ESTIMATING FISH DENSITIES IN A COMMERCIAL FISHERY

S LIZAMORE
Institute of Statistics and Operations Research
Victoria University, Wellington, NZ [§]

M VIGNAUX
National Institute of Water and Atmospheric Research Ltd.
Wellington, NZ [¶]

AND

G A VIGNAUX
Institute of Statistics and Operations Research
Victoria University, Wellington, NZ [‖]

Abstract. A Maxent tomography method is used to estimate densities of fish in the Hoki fishery off the West coast of New Zealand using data from commercial trawling. The individual trawls are the tomography lines with information on the start and end positions of the tow and the weight of fish caught. The Maximum Entropy algorithm is used to find the best estimate of the fish density distribution. Contour and density plots of relative fish density have been obtained for sequences of weekly data in several fishing seasons.

Key words: Tomography, Fish, New Zealand, Maximum Entropy

1. Introduction

Hoki, *Macruronus novaezelandiae*, is a fish species widely distributed in New Zealand waters. The largest stock is thought to have feeding grounds in the Sub-Antarctic region south of New Zealand but their spawning occurs each July and August off the West Coast of the South Island (See Figure 1). The fishery on the spawning grounds is the largest commercial fishery in New Zealand. About 60 trawlers, each up to 100 metres long, tow nets up to 100 metres square for up to 50 km at one time. One catch can contain 100 tonnes of hoki.

[§] Email: suzette@isor.vuw.ac.nz
[¶] Email: vignauxm@frc.niwa.cri.nz
[‖] Email: vignaux@isor.vuw.ac.nz

K. M. Hanson and R. N. Silver (eds.), Maximum Entropy and Bayesian Methods, 203–210.

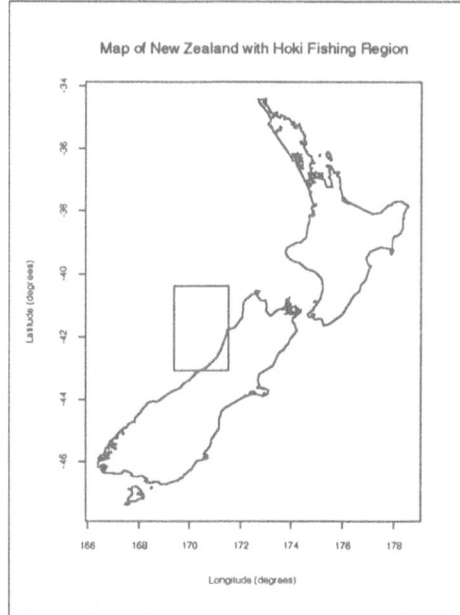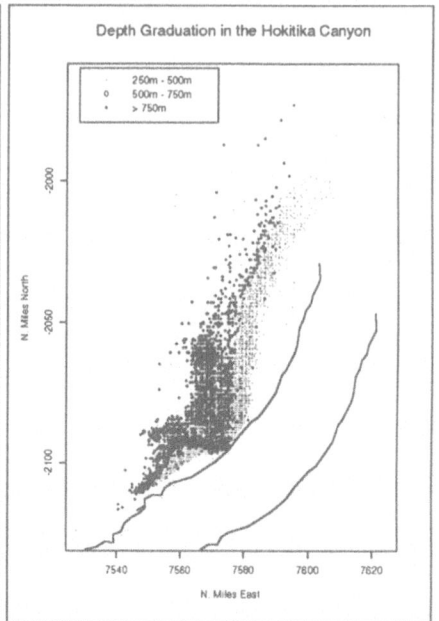

Figure 1. The location of the New Zealand hoki spawning fishery is shown in the left figure. The right figure shows the start positions of tows for 1993, the coast, and the the 25 n. mile fishing limit. The depth indication shows the deep underwater Hokitika Canyon where the hoki congregate at the start of the season

The fishery is controlled by the New Zealand Government which is required to set sustainable quotas. Large vessels are restricted from fishing within 25 nautical miles (n. miles) from the coast.

The Government monitors the hoki resource by a variety of fisher-independent and fisher-dependent methods. Information obtained by these methods is combined with population models to provide estimates of biomass. Fisher-independent methods include acoustic surveys on the spawning grounds and random trawl surveys on the feeding grounds. These observations are too sparse to give accurate spatial information about the fishery.

Fisher-dependent data from the commercial trawling activities are used for Catch Per Unit Effort (CPUE) calculations [2]. Here catch per nautical mile is assumed to be related to fish density. The true relationship depends on the spatial dynamics of the fishery, but these are poorly understood. Better information about local density fluctuations would be useful for managing the fishery and predicting the effect of new management measures. Attempts have been made by Vignaux [1] to determine more detailed structure using CPUE data.

Maximum Entropy (MaxEnt) imaging methods appear to be a promising technique for using the trawl data to get this more detailed information about spatial and temporal density variations. The individual trawls (called tows) form segments of tomographic scans from which densities might be estimated. Maxent methods such as those by Gull and Daniell [3] have been used to develop images from tomographic data in astronomy. Here they are applied to a more down to earth situation.

2. Data and Method

The data for our analysis comes from standard forms filled in for each tow undertaken by the commercial trawling fleet during the last six seasons. Each report, made aboard the vessel, includes the date and time of the tow, the start and end positions, and the size of catch. The size, net and other characteristics of the vessel are also known.

Tows tend to be irregularly spaced and are neither orthogonal nor randomly distributed. The reason for this is that fishers aim to trawl high concentrations of fish, thus the majority of the fleet will be found fishing in the areas which have recently had high catches. The data source is therefore not well designed for estimating fish densities.

The data collected may not be reliable as recording is done under uncontrolled conditions. Error checking is particulary difficult as it is done some time after the observation. Obvious errors are removed, but it is probable that many remain. It is therefore important to check that results obtained are not too sensitive to their effects.

The weight of fish caught in a tow not only depends on the density of fish but also on the size of the net, characteristics of the ship towing it, and other factors. A linear multiplicative model was fitted to standardise the catches so that data from tows would be more comparable:

$$\frac{C_i}{l_i} = k_1 k_2 k_3 \ldots \exp(\epsilon_i).$$

Here C_i is the observed catch in tow i, l_i is the length of that tow, k_1, k_2, k_3, etc. are fitted regression terms reflecting the effect of other factors on the catch. ϵ_i is the residual from the fitted model and represents fluctuations in catch rate that cannot be explained by other factors. Some of this residual may be explained by spatial variation in fish density on a local scale. The expression is made additive by a logarithmic transformation.

The model was fitted using stepwise regression. The variables included were time within the season, vessel nationality (a surrogate for many behavioural factors), time of day, year, vessel length, latitude and headline height. The most significant variable is a seasonal effect, modelled as a cubic continuous variable. The second most important variable is nationality, then come the time of the day and the year effect (a surrogate for changes in total biomass from year to year). Vessel size, net size and tow latitude are minor variables. These variables explain only about a quarter of the total variance. Apart from any variation due to spatial fluctuations in fish density there will also be a great deal of noise.

A standardised catch is defined as:

$$C_i' = \frac{C_i}{k_1 k_2 k_3 \ldots} = l_i \exp(\epsilon_i) = \int_l \rho(l) dl,$$

where $\rho(l)$ is the relative linear density at a point in the tow. This model is approximated by dividing the area into a number of squares, j, setting ρ_j (the image) as the mean density in square j and l_{ij} (the transition matrix) as the distance traversed by tow i across square j. It is assumed that the tows are linear from start to finish. The model becomes

$$C_i' = \sum_j l_{ij} \rho_j$$

and provides the image to data tranformation.

V. A. Macauley's [4] programs, based on the algorithm by Skilling and Bryan [5], were modified for use in this study. These programs are written in C and were originally used for MaxEnt spectral analysis. The method was tested out on dummy data and it proved successful in reproducing images even with high noise levels in the simulated tow data.

In an earlier study Vignaux [1] found evidence for spatial correlations of fish density up to 11 n. miles and temporal effects lasting up to 14 days. Three different block sizes (4×4, 8×8 and 12×12 n. miles) were therefore tested in this study. Similarly data was aggregated into periods of 3, 7, and 14 days.

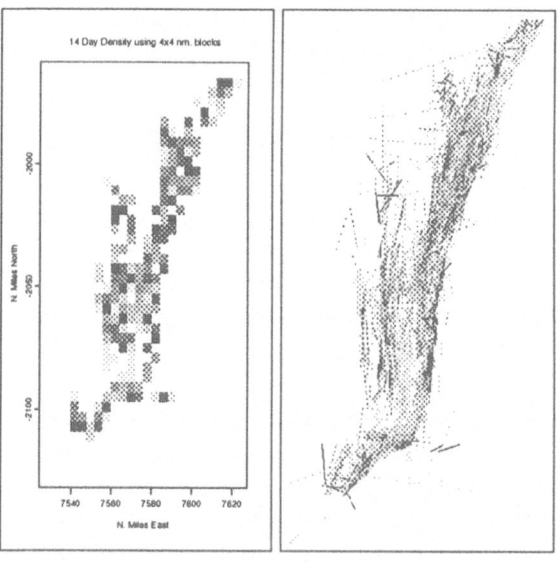

Figure 2.   MaxEnt estimated densities for 4×4 squares and a 14-day period compared with catch rate. The darker the line for a tow, the more fish per n. mile.

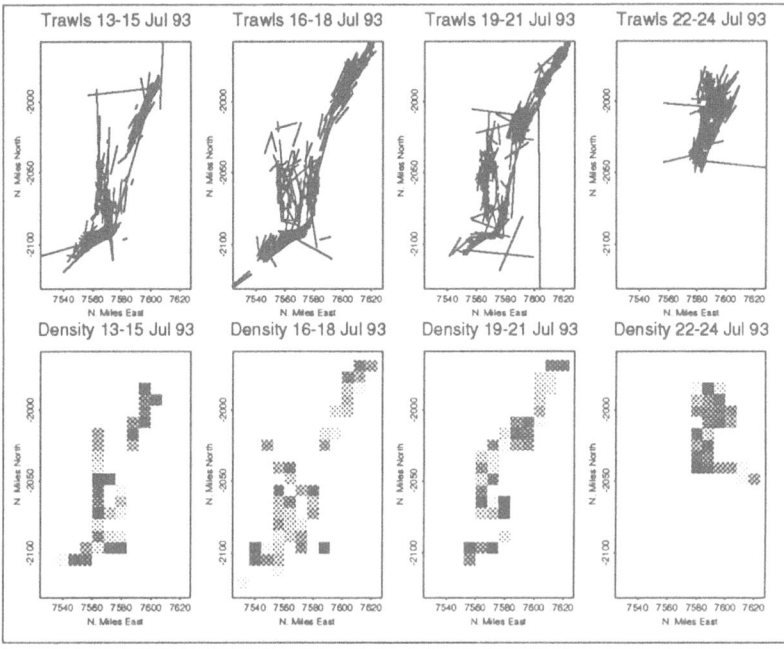

Figure 3.   Four sequential 3-day periods in the 1993 season. The upper diagram show the tows, the lower the corresponding MaxEnt density estimates in 8 × 8 squares.

3. Results

The left side of Figure 2 shows that the MaxEnt method was successful in producing images of fish density at these scales. The data set used for that figure covers a period of 14 days and uses 4 × 4 n. mile squares. The error parameter had to be set quite high for the method to converge with this noisy data.

To compare the estimated densities in relation to the raw data, a catch intensity was calculated for each tow as the standardised weight of fish caught per n. mile. This is plotted on the right side of Figure 2 with darker lines indicating the stronger intensities. Comparing this plot with the MaxEnt plot on the left it appears that the MaxEnt results are consistent with the catch intensities on the tows. In this image there appear to be two major regions that had a lower catch rates, the first in the lower centre region of the plot and the other in the upper right region of the plot. Both methods highlight these features well.

Figure 3 shows the tows and corresponding density plots for a set of sequential 3-day periods from the peak of the 1993 season. 240 8 × 8 n. mile squares are used to cover the 15 thousand square n. mile fishing zone.

The density plots provide information about the fishery that cannot be determined by the tow information alone. Fishers in the northern region on 13-15 July had good catch rates and were followed by others so that from 16-21 July there were many tows in the northern fishing area, though densities do not seem to have been particularly high. By the end of the period all the fishers were concentrated in the northern region and we have no information about the density in the southern region.

Clearly not all of the anomalous tows have been removed from the data. For example the North-South tow in the third period which is over 120 n. mile long is likely to be the result of a recording or transcription error in the finish position. It is therefore necessary to be careful in interpreting the density when the data comes from only a few tows. The method was tested for sensitivity to the presence of such outliers. When an anomolous tow is removed from the data the image contrast generally increases without any change in the overall image.

3.1. DIFFERENT SIZE BLOCKS

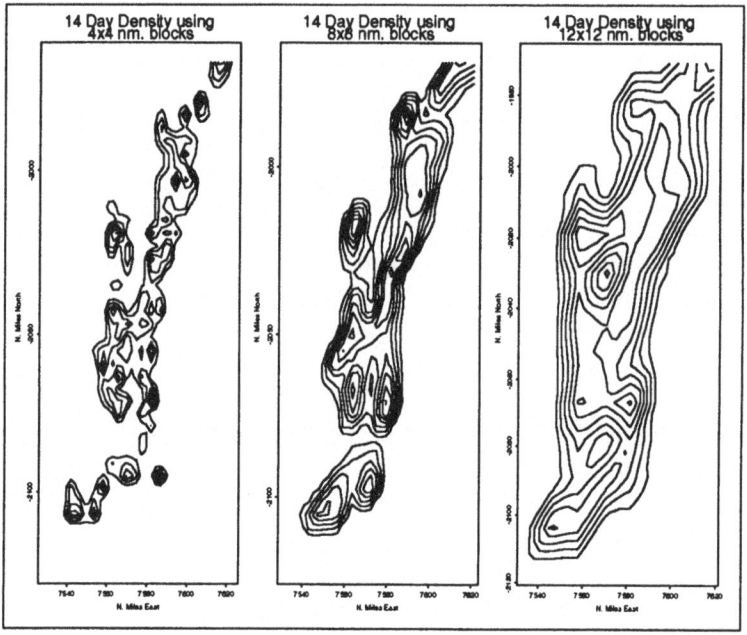

Figure 4.   Comparison of contours produced by 4 × 4, 8 × 8 and 12 × 12 n. mile blocks

The data can also be presented in contour form as shown in Figure 4 for the same 14-day period using 4, 8 and 12 n. mile square blocks.

When the 4 × 4 blocks are used, the target sea area is divided into 960 squares. The image produced is reasonably fine but very variable in density making it difficult to pick up any overall patterns. The 8 × 8 blocks make it easier to spot these patterns, while still retaining the basic shapes and features of the data. Each of the 12 × 12 blocks represent the same sea area as nine 4 × 4 block and produce a very coarse image which lacks many of the interesting features of the data. The 8 × 8 n. mile blocks were selected for further analyses.

3.2. THE 1991 SEASON

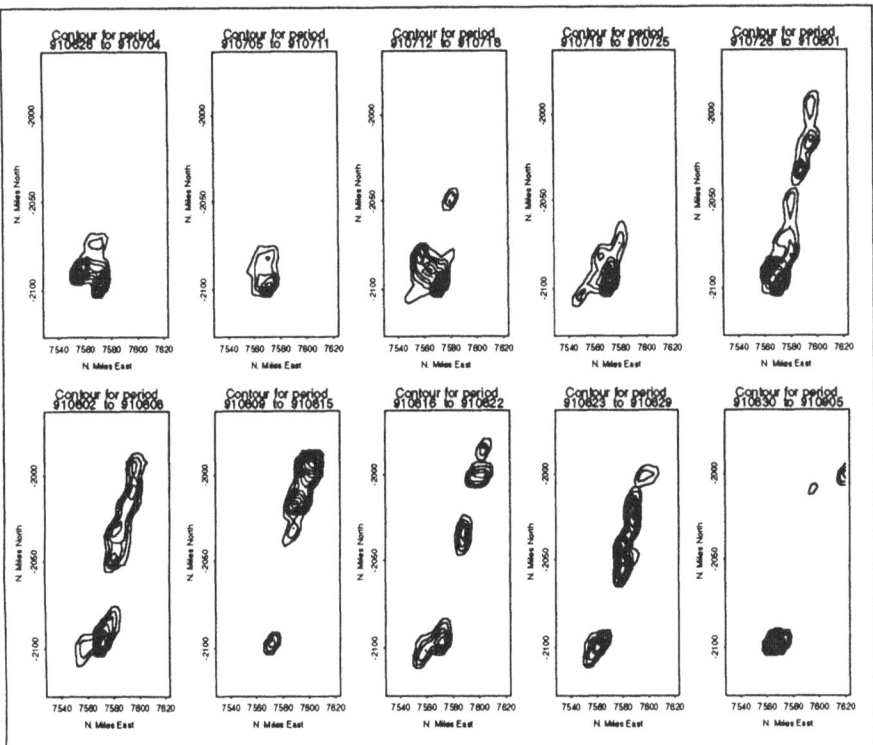

Figure 5.   Contour plots of sequential 7-day periods for the 1991 season using 8 × 8 n. mile blocks.

To illustrate the type of analysis for which the MaxEnt densities will be used data from the 1991 season was examined. The results can be seen in Figure 5. Traditionally, fishers start the season trawling in the deep waters of the southern part of the fishery. As the season progresses they move in a northerly direction to shallower water. In 1991, fishing started in the south and continued there until the last week of July when the shift to the north began. From the beginning of August, the majority of tows were in the northern fishing region.

Figure 5 suggests a high density conglomeration of fish in the south at the start of the season which grows, spreads north and then splits into two. The southern concentration of fish remains throughout the season, but seems to contract.

4. Conclusions

Despite the noisiness of the data and the problem of tows which have not been specially arranged to provide the best patterns, the method appears to provide images of fish density which correspond to real phenomena.

It is difficult for conventional methods to obtain high resolution spatial pictures of the fishery due to the extreme length of tows which may reach 50 n. miles. The MaxEnt method manages to tease out density information on a smaller scale than the measurement tool.

The CPUE method is currently used throughout the world to estimate the biomass of fisheries. It is based on an untested assumption about the relationship between catch rates and biomass. Detailed information of the type produced in this paper is required to understand the relationship between the fishers and the fish and beween fishery management and sustainable yield.

References

1. M. Vignaux, "Analysis of spacial structure in fish distributions using commercial catch and effort data from the New Zealand hoki fishery," *Canadian Journal of Fishery and Aquatic Science*, 1995. (submitted).
2. M. Vignaux, "Catch per unit effort (CPUE) analysis of West Coast South Island and Cook Strait spawning hoki fisheries, 1987-93," Fisheries Assessment Research Document 94/11, NZ MAF Fisheries, MAF Fisheries, Greta Point, Wellington, 1994.
3. S. F. Gull and G. J. Daniell, "Image reconstruction from incomplete and noisy data," *Nature*, **272**, pp. 686–690, 1978.
4. V. A. Macaulay, *Bayesian inversion with applications to physics*. PhD thesis, Theoretical Physics, University of Oxford, 1992.
5. J. Skilling and R. K. Bryan, "Maximum entropy imge reconstruction: general algorithm," *Mon. Not. R Astr. Soc*, **211**, pp. 111–124, 1984.

TOWARD OPTIMAL OBSERVER PERFORMANCE OF DETECTION AND DISCRIMINATION TASKS ON RECONSTRUCTIONS FROM SPARSE DATA

R.F. WAGNER, K.J. MYERS, D.G. BROWN AND M.P. ANDERSON
Center for Devices & Radiological Health, Rockville MD 20857

AND

K.M. HANSON
Los Alamos National Laboratory, Los Alamos NM 85745

Abstract. It is well known that image assessment is task dependent. This is demonstrated in the context of images reconstructed from sparse data using MEM-SYS3. We demonstrate that the problem of determining the regularization- or hyperparameter α has a task-dependent character independent of whether the images are viewed by human observers or by classical or neural-net classifiers. This issue is not addressed by Bayesian image analysts. We suggest, however, that knowledge of the task, or the use to which the images are to be put, is a form of prior knowledge that should be incorporated into a Bayesian analysis. We sketch a frequentist approach that may serve as a guide to a Bayesian solution.

Key words: observer performance, tomographic reconstruction, MEMSYS, hyperparameter selection

1. Introduction

Images are generally produced for the purpose of performing visual tasks. A visual task typically consists of an interpretation of an image and a decision about its content. The major premise in the field of image assessment is that the ranking of imaging systems is *task dependent*. This point was demonstrated in the literature about twenty-five years ago [1] and a consensus on the issue developed almost immediately. Nevertheless, the point is rarely addressed in the Bayesian and Maximum Entropy communities. It is the purpose of the present work to demonstrate the task dependence of image assessment in the context of image reconstruction from sparse data using a maximum entropy method. We shall see that the search for an optimal regularization parameter in this method does not have a unique solution: the solution depends on the task. We shall compare frequentist and Bayesian

K. M. Hanson and R. N. Silver (eds.), Maximum Entropy and Bayesian Methods, 211–220.
© 1996 *Kluwer Academic Publishers.*

approaches to the issue and will offer a frequentist approach to the problem that
may serve as a guide to a Bayesian solution.

2. Tasks and observers

We consider two very broad categories of imaging tasks: lesion or target detection,
and target discrimination or classification. A simple example of a detection task
in medical imaging is the task of determining whether a site in an image contains
a lesion or is only representative of normal background. A simple example of a
discrimination task is the task of viewing a blood vessel and determining whether
it has significant narrowing (stenosis) or not. For the special case where an imaging
system is linear and shift-invariant, the detection task can be considered as a task
that is concentrated in the low spatial frequencies. The discrimination task, however, requires no low spatial-frequency information: it is a mid- to high-frequency
task.

In the field of image assessment a number of classes of *image observers* are
considered. An observer is defined in terms of how the task is implemented or
performed. Here we limit ourselves to binary tasks and the case where the signal
is known exactly (SKE). In this paradigm the observer is focused on the region of
interest and has to decide which of two hypothesized states gave rise to the data,
e.g., lesion present, or only background present. (Performance of some complex
tasks is related to performance of SKE tasks in Refs. 2 and 3.) The most common
observer is the human, but there are other ways of realizing an observer using a
machine or computer algorithm. These include various matched filters, classical
decision rules based on the likelihood function (the foundation for the matched
filters [4]), the "proper" Bayesian rule based on the ratio of posterior proabilities
for the two hypotheses given the data [5], and a growing number of classical and
neural-net extensions of these decision rules.

Once a task and an observer are defined, standard methods from statistical
decision analysis can be applied to assess the performance of the task by the
observer. In the present work we investigated the performance of human and machine observers on lesion detection and classification tasks as a function of the
regularization parameter in the reconstruction algorithm.

3. Image "acquisition" and reconstruction

The present work is an extension and interpretation of previous work in this series
that considered limited angle tomography (eight views over 180 degrees). The data
are derived from a simulation of parallel-projection image acquisitions, 128 samples
per projection, and additive Gaussian noise. Images reconstructed from such data
will be corrupted by artifacts due to the small number of projections or views, as
well as by the additive measurement or detection noise. For additional details, see
[5-7].

We concentrate here on MEMSYS3, a Bayesian method of regularized reconstruction due to the Cambridge school of Gull and Skilling [8,9]. The algorithm

includes a numerically efficient method for minimizing a functional F, where

$$F = \tfrac{1}{2}\chi^2 - \alpha S \ . \tag{1}$$

The term involving χ^2 is the usual measure of misfit between the data and the reconstruction $\hat{f}$ (referred to the data domain). The quantity S in the second term is the Cambridge adaptation of the Shannon entropy $(-\sum_i \hat{f}_i \ln \hat{f}_i)$. The second term may thus be thought of as the exponent in an entropic prior probability distribution or, rather, a family of entropic prior probability distributions in the (hyper-) parameter α. The parameter α may be thought of as a regularization parameter that determines the degree of smoothness in the final reconstruction. We shall now give several views on the determination of the value of α.

4. Frequentist and Bayesian perspectives

In classical statistics, or the so-called frequentist perspective, one is interested in the long-term average behavior of a method of estimation or inference. In the context of image assessment, one assumes that the imaging system or reconstruction algorithm is expected to be used repeatedly under similar conditions, and that there is an opportunity to simulate or experiment with the system to determine its long-term average performance. One then chooses the regularization parameter, α in the present case, that gives the best long-term performance. In our approach, performance is specified in terms of how well an observer of the images performs the task of interest. There is no controversy over such an approach to experimental design when one is in the repeated-use mode and has the opportunity to study average performance [10].

In the Bayesian perspective, one usually does not have the luxury of long-term experience. One has only the present data set and one's principles, i.e., that the only rational approach to such limited-data-set problems is by way of probability theory. In the Gull and Skilling approach, the regularization parameter is determined from the current data set by maximizing the posterior probability of α. In the "classic" implementation of MEMSYS3 this gives rise to the relationship

$$\chi^2 + G = N \ . \tag{2}$$

The interpretation of this expression is as follows. The aimed-for value of the misfit, χ^2, plus a parameter G, matches the number of independent measurements, N. The quantity G is referred to as the number of "good" measurements and is determined with α in a self-consistent way from the number of significant eigenvalues of a weighted version of the matrix $\mathbf{H}^t\mathbf{\Sigma}^{-1}\mathbf{H}$. The weighting will be given in Section 10 below. Here, $\mathbf{H}$ is the system response matrix that characterizes the forward problem, $\mathbf{\Sigma}$ is the covariance matrix of the measurements, and the value of α is used in the determination of the significance of the eigenvalues.

Setting $G = 0$ in Eq. (2) yields the early "historic" implementation of MaxEnt where the target value of χ^2 is the number of independent measurements. Skilling and Gull [9] point out that the historic version is based on the frequentist consideration that the expected value of χ^2 over the ensemble of experiments is N. They and other investigators realized that the historic approach often led to underfitting of the data. In the classic version, the higher the quality of the measurements in terms of the parameter G, the closer one is permitted to fit the data. The classic solution seems intrinsically reasonable, even setting aside its assumptions and subtlety.

5. Decision-theoretic measures of task performance

When studying binary tasks, for example decisions of "normal" vs "abnormal" or patent vs occluded artery, one is able to use the accepted, in fact now required, standard of image assessment in the medical imaging community, namely, the curve of true-positive fraction vs false-positive fraction of responses in the binary classification. This curve is referred to as the receiver operating characteristic (ROC) curve. We follow the usual approach of taking the area under the ROC curve, A_z, as a summary figure of merit for our studies. This measure is equivalent to the true-positive fraction averaged over all false-positive fractions. It may also be obtained as the percent correct in a two-alternative forced-choice experiment. It is convenient to use an inverse error function to convert from A_z to the detectability index, d_a, which may be thought of as the integrated signal strength in units of the standard deviation of its underlying noise distribution (a decision-theoretic signal-to-noise ratio). The uncertainty in the determination of d_a due to the use of a finite number of image samples is usually estimated from the number of image samples and the sampling statistics of the binomial distribution. See [5-7] for further details.

6. Results: Detection Task

The detection task we studied is an idealization of a low-contrast lesion-detection task in a background of measurement noise and artifacts generated by the presence of high-contrast structures. In Fig. 1a we show one realization of a phantom for generating images of this class. Ten randomly placed low-contrast "lesions" can be seen as faint disks. These are the lesions or disks to be detected. Ten randomly placed high-contrast disks or lesions, which serve as the major source of reconstruction artifacts, are easily seen. In Figs. 1b through 1e we show the results of reconstructing images over a range of values of α. At the highest value of α the reconstructions are smooth; at the lowest value the reconstruction is pointillist in texture. In the limit of very small α the reconstruction may be thought of as the maximum likelihood reconstruction with a positivity constraint.

In Fig. 2 we give results for d_a for three classes of observers. Among the classical and Bayesian observers, those derived from the likelihood function and those derived from the ratio of posterior probabilities performed similarly [6,7]; the best average results are shown and are labelled "machine observer". Results for machine, neural-net, and human observer show a similar falloff from the maximum level of performance at low values of α to the weakest level of performance at high values of α. The "classic" implementation of MEMSYS3 yields a value of α that corresponds to a point on the shoulder of the performance curves near the plateau.

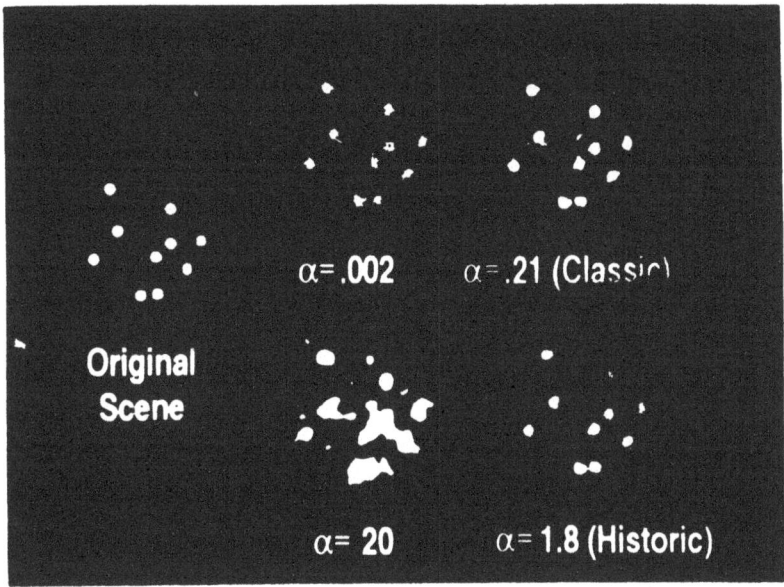

Figure 1. (a) One realization of a scene used for generating images for the detection task. (b) through (e) - MEMSYS3 reconstructions with α = 0.002, 0.21, 1.8, and 20.

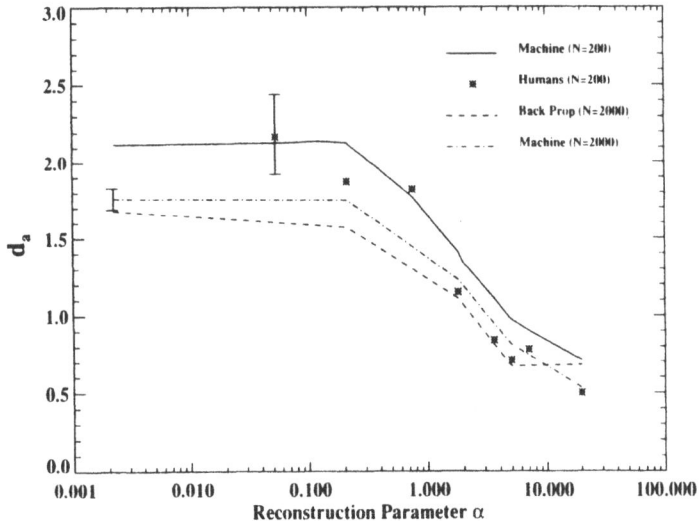

Figure 2. Decision-theoretic signal-to-noise ratio (d_a) for performance of the detection task by machine, back-propagation neural network, and human observers. The total number of independent image trials for a given observer is the value N shown. The neural network was trained on $\frac{N}{2}$ and tested on the other $\frac{N}{2}$. The error bars represent $\pm 1\sigma$.

7. Results: Discrimination Task

We selected a discrimination task that is an elaboration of the Rayleigh task of discriminating between a single star or object and a doublet. This task also serves as an idealization of the task of determining whether a blood vessel is unobstructed or is narrowed. In Fig. 3a we show one realization of a phantom for generating images of this class. There are eight cigar-shaped singlets and eight doublets. Each class of objects serves as a source of reconstruction artifact generation for the detection of the other. In Figs. 3b through 3e we show the results of reconstructing images for various values of α.

In Fig. 4 we give results for d_a for the three classes of observers. These results all show a similar maximum of performance in the neighborhood of $\alpha = 1.0$, with a fall-off from this maximum for smaller and larger values of α. It is of interest that the "classic" value of α occurs close to the position of the peak of these curves. The small difference between the classical or Bayesian machine and the neural-net observers is within the error bars.

8. Summary from frequentist perspective

These results reinforce and extend the remarkable similarity in performance noted in earlier work among all three classes of observers for a given task. This is satisfying because an original goal of these investigations was to find visual-like machine observers, i.e., algorithms that performed similarly to human observers.

More important for our present purposes, however, is the comparison between performance on different tasks. Comparing Figs. 2 and 4 we see that the α-dependence of the task performance for the detection task is *qualitatively different* from the α-dependence of the task performance for the Rayleigh-like discrimination task for all three classes of observers. (Similar differences on a related problem have been observed by Abbey and Barrett [11].) From the frequentist perspective, then, we can say that the problem of selecting the optimal regularization parameter has a task-dependent character. This task-dependence is not addressed in the Bayesian formulations of the regularization problem. The knowledge of the task required of an imaging system or image reconstruction algorithm, however, is a form of prior information that is suitable for inclusion in a proper Bayesian formulation. We are not yet able to present such a Bayesian formulation. We are, however, able to analyze the structure of the present frequentist treatment with a view toward an ultimate Bayesian solution. The remainder of this paper summarizes our present understanding of the relevant issues.

9. The importance of eigenvectors

The *eigenvalues* of the matrix $\mathbf{H}^t \mathbf{\Sigma}^{-1} \mathbf{H}$ were seen in Sec. 4 to play an important role in the Bayesian solution to the regularization problem. We contend, however, that the *eigenvectors* are the relevant quantities when task performance is brought into consideration. We briefly review our experience with a broad class of tasks.

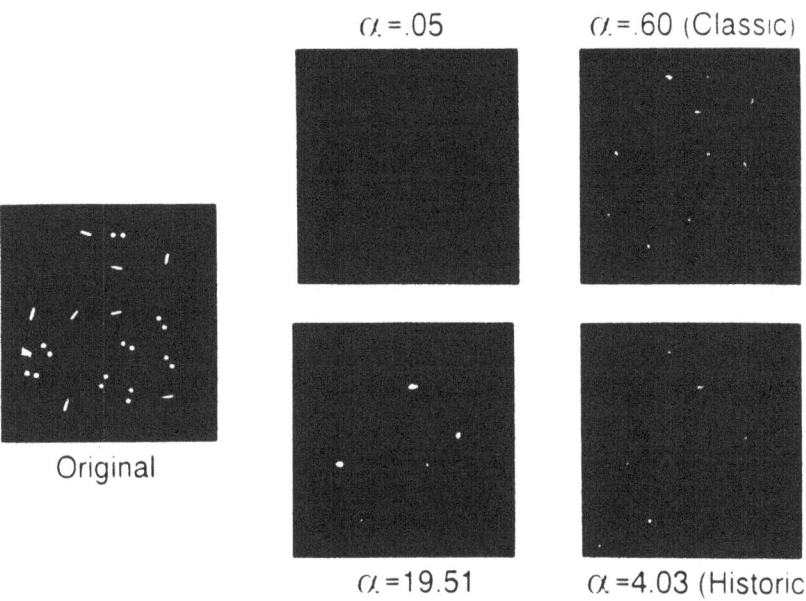

Figure 3. (a) One realization of a scene used for generating images for the Rayleigh discrimination task. (b) through (e) - MEMSYS3 reconstructions with $\alpha = 0.05, 0.6, 4.03$, and 19.51.

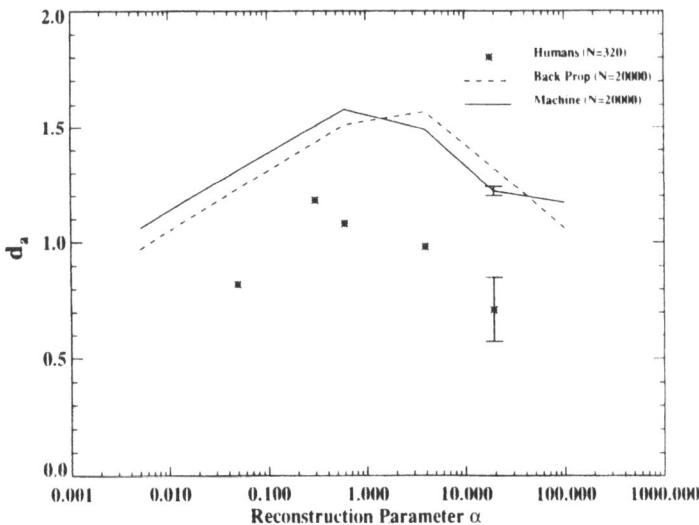

Figure 4. Decision-theoretic signal-to-noise ratio (d_a) for performance of the discrimination task by machine, back-propagation neural network, and human observers. Otherwise, as in Fig. 2.

Hanson [12] considered a wide class of problems where an ideal observer would suppress the low spatial-frequency components of an image. Images in which the background is inhomogeneous (variable, lumpy, etc.) are members of this class because inclusion of the low-frequency components would only lead to the accumulation of irrelevant noise. It is easy to show that the task of determining the separation of two lesions, the task of determining whether a stenosis (a vessel narrowing) is present, and many other Rayleigh-like tasks are also members of this class. Their performance not only requires no low spatial-frequency information, it is even impeded by the presence of such information. A task representative of some linear shift-invariant members of this class was analyzed by Myers et al. [13]. They showed that the optimal linear filter for the task is a bandpass filter that maximizes the detection of Fourier-domain eigenvectors within a task-dependent band of frequencies; it suppresses the detection of eigenvectors outside this band so as to maximize noise rejection in irrelevant bands. Optimal signal detection here does not depend on any measure of eigenvalue number. It depends on optimal detection of the eigenvectors of interest. This feature of the problem is ignored in all Bayesian approaches that we are aware of.

In particular, the allowable stopping points of the MEMSYS3 algorithm in its search for an optimal α proceed from very large values where, in effect, only low spatial frequencies are reconstructed, to very small values where, in effect, low, intermediate, and high spatial frequencies are reconstructed. However, what is needed for Rayleigh-like tasks is a means to concentrate on a mid-band of frequencies. Allowing very low and very high spatial frequencies into the reconstruction will swamp all of the visual-like image observers (in ours and related work) with irrelevant noise.

The eigenvector issue is not unique to the imaging problem. The issue should be treated in neural-net design and in countless other signal and image processing problems where principles for optimal order determination or optimal stopping parameters are sought.

10. Toward a solution

We end with a sketch of a solution to this problem. The solution will require some fundamentals from statistical decision theory.

The frequentist figure of merit for ideal detection of a difference signal Δf (a *vector*) by an imaging system whose measurements are characterized by the matrix $\Sigma_m^{-1} = H^t \Sigma^{-1} H$, in the case where there is no appreciable artifactual noise and the measurement noise is additive and Gaussian, is a detection theoretic signal-to-noise ratio (SNR) given by

$$d_a^2 = \Delta f^t H^t \Sigma^{-1} H \Delta f = \Delta f^t \Sigma_m^{-1} \Delta f . \qquad (3)$$

(See, e.g., [14, 15].) This figure of merit may be recognized as the Hotelling trace; it is proportional to the Mahalanobis distance between the two classes whose mean difference image is Δf. It may be generalized for a Gaussian ensemble of signals with known covariance matrix Σ_p by replacing Σ_m with $[\Sigma_m^{-1} + \Sigma_p^{-1}]^{-1}$.

In the present paper the mean difference signal for the detection task is a template with a profile given by that of the expected lesion; in the discrimination task it is a template with a profile given by the expected difference of the mean signals of the two classes. The observer mask that will achieve the optimal figure of merit in the limit of many tomographic views is the prewhitening matched filter given by $\Delta \mathbf{f}^t \mathbf{H}^t \mathbf{\Sigma}^{-1}$. This filter selects out from the data only the vectors (eigenvectors or singular vectors) required for its task. It discards all others. The present paper is a demonstration of a frequentist search for optimal linear and nonlinear observer decision functions (as well as regularization parameters) for the sparse-view version of this problem.

Some of the machinery for evaluating and optimizing such figures of merit already exists in the powerful MEMSYS packages. There, Bayesian figures of merit play a central role. For example, the matrix designated $\mathbf{A}$ that generates the eigenvalues contributing to G has structure similar to Eq. (3):

$$\mathbf{A} = [\hat{\mathbf{f}}^{\frac{1}{2}}]^t \mathbf{H}^t \mathbf{\Sigma}^{-1} \mathbf{H} [\hat{\mathbf{f}}^{\frac{1}{2}}] \ . \tag{4}$$

Here, $[\hat{\mathbf{f}}]$ is a diagonal *matrix* containing the current estimate of the object; $[\hat{\mathbf{f}}^{\frac{1}{2}}]$ is the corresponding square-root matrix. The number of significant eigenvalues of $\mathbf{A}$, on a scale determined by α, is the parameter G referred to earlier. In fact, $\alpha \mathbf{A}^{-1}$ is the MEMSYS analog of the quotient $\mathbf{\Sigma}_p^{-1} \mathbf{\Sigma}_m$ that determines the degree of regularization in Gaussian *MAP* estimation [4].

A fundamental difference between Eqs. (3) and (4) can be noted in the first and last factors of these expressions. In the case of Eq. (3) these factors are frequentist or long-term average quantities that specify the task. In the case of the Bayesian result in Eq. (4) they are estimates from the present data set alone. From the point of view of the present paper, optimal reconstruction must address the problem of preferentially visualizing the eigenvectors required for the task, in the spirit of the optimal filter behind Eq. (3); it is not sufficient to count the eigenvalues of a regularization matrix such as $\mathbf{A}$ in Eq. (4). The details remain to be worked out.

Acknowledgements

Over many years the authors have enjoyed discussions and collaborations on the present and related issues with Prof. H.H. Barrett and colleagues at the University of Arizona.

References

1. K. Rossmann, "Image quality and patient exposure," *Current Problems in Radiology*, **2**(2), pp. 1-34, Year Book Medical Publishers, Chicago, 1972.
2. D.G. Brown, M.F. Insana, and M. Tapiovaara, "Detection performance of the ideal decision function and its McLaurin expansion: Signal position unknown," J. Acoust. Soc. Am. **97**, pp. 379-398, 1995.
3. R.F. Wagner, K.J. Myers, A.E. Burgess, D.G. Brown, and M.J. Tapiovaara, "Maximum a posteriori detection: Figures of merit for detection under uncertainty," Proc. SPIE **1232**, pp. 195-204, (Bellingham WA) 1990.
4. A.D. Whalen, *Detection of Signals in Noise*, (Electrical Science Series) Academic Press, New York, 1971.

5. K.M. Hanson, "Making binary decisions based on the posterior probability distribution associated with tomographic reconstructions," in *Maximum Entropy and Bayesian Methods*, C.R. Smith, G.J. Erickson, and P.O. Neudorfer, eds., pp 313-326, Kluwer Academic, Dordrecht, 1992.

6. K.J. Myers, R.F. Wagner, and K.M. Hanson, "Binary task performance on images reconstructed using MEMSYS3: Comparison of machine and human observers," in *Maximum Entropy and Bayesian Methods*, A. Mohammad-Djafari and G. Demoment, eds., pp. 415-421, Kluwer Academic, Dordrecht, 1993.

7. K.M. Hanson and K.J. Myers, "Rayleigh task performance as a method to evaluate image reconstruction algorithms," in *Maximum Entropy and Bayesian Methods*, W.T. Grandy Jr., and L.H. Schick, eds., pp. 303-312, Kluwer Academic, The Netherlands, 1991.

8. S.F. Gull and J. Skilling, *Quantified Maximum Entropy "MEMSYS 3" Users' Manual*, Maximum Entropy Data Consultants Ltd., Royston England, 1989.

9. J. Skilling and S.F. Gull, "Bayesian maximum entropy image reconstruction," in *Spatial Statistics and Imaging*, A. Possolo, ed., pp. 341-367, Institute of Mathematical Statistics Lecture Notes-Monograph Series **20**, IMS, Hayward CA, 1991.

10. J.O. Berger, *Statistical Decision Theory and Bayesian Analysis*, Springer-Verlag, New York, 1985.

11. C.K. Abbey and H.H. Barrett, "Linear iterative reconstruction algorithms: Study of observer performance," in *Information Processing in Medical Imaging*, Y. Bizais et al. eds., pp. 65-76, Kluwer Academic, The Netherlands, 1995.

12. K.M. Hanson, "Variations in task and the ideal observer," Proc. SPIE **419**, pp. 60-67, 1983.

13. K.J. Myers, J.P. Rolland, H.H. Barrett, and R.F. Wagner, "Aperture optimization for emission imaging: Effect of a spatially varying background," J. Opt. Soc. Am. **A7**, pp. 1279-1293, 1990.

14. H.H. Barrett, "Objective assessment of image quality: Effects of quantum noise and object variability," J. Opt. Soc. Am. **A7**, pp. 1266-1278, 1990.

15. H.H. Barrett, J.L. Denny, R.F. Wagner, and K.J. Myers, "Objective assessment of image quality. II. Fisher information, Fourier crosstalk and figures of merit for task performance," J. Opt. Soc. Am. **A12**, pp. 834-852, 1995.

ENTROPIES FOR DISSIPATIVE FLUIDS AND MAGNETOFLUIDS WITHOUT DISCRETIZATION

DAVID MONTGOMERY

Dept. of Physics and Astronomy
Dartmouth College Hanover, NH 03755-3528, U.S.A.

Abstract.
    Decaying high-Reynolds-number turbulence in some fluid and magnetofluid systems can be long-lived and can robustly produce recognizable geometric features ("coherent structures"), largely independently of the initial conditions. On time scales short compared to the viscous and/or resistive decay time of the turbulence, the geometric structures can become quite few and symmetric. The impression of thermodynamic or statistical-mechanical behavior that this suggests can be sharpened by defining suitable entropies for the dissipative continua, using Jaynesian information-theoretic ideas as a guide. This amounts to a partial reformulation of statistical mechanics to include dissipation and includes a replacement of ensemble averages by "most probable states" as the quantity of most physical interest. The principal tests of the reformulations are numerical solutions of the relevant partial differential equations.

1. Introduction

The injection of statistical mechanics into theories of relaxing turbulence in fluid and magnetofluid systems has developed piecemeal, and has involved departures from the classical methods. Numerical solution of the applicable partial differential equations has usually served as the principal reality check.[1-4]

Those partial differential equations are, so far, the Navier-Stokes (NS) equation for a neutral fluid and the magnetohydrodynamic (MHD) equations for electrically-conducting fluids. Our efforts have revolved around trying to define entropies that can be used to predict the properties of the solutions to these partial differential equations in somewhat the same way that Boltzmann's H-function predicts features of the solutions to Boltzmann's equation in gas kinetic theory. This is an application of maximum entropy methods that is somewhat outside the mainstream of the data-analyzing uses to which "MaxEnt" methods are currently being put, but fits comfortably inside Jaynes's original program as put forward in his 1957 papers.[5]

K. M. Hanson and R. N. Silver (eds.), Maximum Entropy and Bayesian Methods, 221–227.
© 1996 *Kluwer Academic Publishers.*

Our efforts have been driven by the robustness with which certain computed turbulent phenomena recur for very wide classes of initial conditions. Three obvious examples are long-lived merging vortices in two-dimensional (2D) NS flow, long-lived magnetic islands in 2D MHD, and long-lived helical configurations in magnetized 3D MHD. In what follows, we will summarize the current status of the first of these, making reference to our recent publications for supporting numerical data and graphical displays. Then we will briefly remark upon work in progress connected with the second. For a (somewhat dated) review, see Ref. 6.

2. Navier-Stokes (2D) turbulence: present perspective

The core task has been to characterize the fluid system with an entropy whose maximization, subject to whatever constants of the motion one can identify, will predict the evolution of the system. The test is, Does the maximum entropy state turn out to be the one toward which the computed solutions evolve? This is an answerable question.

"Constants of the motion," here, must be interpreted to include approximate as well as exact constants, since it is decaying turbulence we are talking about, and if we wait long enough, the fluid or magnetofluid will return to its uniform laminar state. Any maximum entropy states which occur must appear on time scales which are still short compared to global energy decay times. Only for systems for which "slowly-decaying constants" (i.e., ideal invariants, which in the presence of dissipation, persist over long intervals on some dynamically-important time scale) exist does it make sense to attempt to apply maximum-entropy methods. To repeat: to be considered interesting, maximum entropy states must occur on time scales short compared to global energy decay times, but long compared to the time scales of some other important dynamical process, such as vortex merger.

Ideal (non-dissipative) continuum mechanics, despite recent brave assertions[7,8] to the contrary, is not very susceptible to statistical or maximum-entropy methods. The reason is that ideal continua typically have far too many conservation laws for the necessary "mixing" or quasi-ergodic behavior to be plausible. For example, in a 2D NS flow, the vorticity is a convected, pointwise constant of the motion. An implication is that closed contours of constant vorticity can never cross or intersect, no matter how much they may move or deform. An infinite number of "topological" constants of the motion freezes the configuration of the system to an extent that all the local maxima and minima must be preserved, for the entire spatial vorticity distribution. The repeated (and unsupported in detail) invocations of "coarse graining" that have been used to justify ignoring these topological constants in Euler equation calculations appear unpersuasive, to us.

Illustrating the maximum entropy methods for 2D NS flows, we must work in high Reynolds number regimes. For reasons which are well known in 2D NS turbulence theory, the global "energy decay time" scales linearly with the Reynolds number. Much shorter than the energy decay time is the large-scale eddy-turnover time, which is independent of Reynolds number. There is a third time scale, that required for two nearby like-sign vortices to merge or combine into a single vortex. No sharply-defined Reynolds number dependence for the like-sign vortex-merger time scale has been identified, but experience shows that it is much closer to the eddy-turnover time than it is to the energy decay time. Thus a great deal of dynamical activity of the vortex merger type may occur while the energy is decaying only a little. In particular, the computations of Matthaeus and collaborators have shown that for Reynolds numbers of the order of a few thousand, all of the possible like-signed vortex mergers can occur while most of the initial energy remains. On the other hand, many other ideal invariants (such as enstrophy, or mean-square vorticity) may decay considerably while the vortex-merger process goes to completion, and therefore cannot be considered as "approximate constants of the motion" in the same sense as the energy is.

We illustrate the way an entropy may be defined for an evolving 2D NS fluid. Consider the 2D NS dynamics in the vorticity representation,

$$\frac{\partial \omega}{\partial t} + \boldsymbol{v} \cdot \nabla \omega = \nu \nabla^2 \omega . \tag{1}$$

Here, ω is the vorticity and $\boldsymbol{v}$ is the fluid velocity, with $\boldsymbol{v} = (v_x, v_y, 0)$ in Cartesian coordinates, and $\nabla \times \boldsymbol{v} = \omega \hat{e}_z$. For all variables, $\partial/\partial z = 0$. Since $\nabla \cdot \boldsymbol{v} = 0$, we may write $\boldsymbol{v} = \nabla \times \psi \hat{e}_z$, where ψ is the stream function and obeys the 2D Poisson equation, $\nabla^2 \psi = -\omega$. Equation (1) is written in dimensionless variables in which ν^{-1} appears as a large-scale Reynolds number. The case of interest is $0 < \nu \ll 1$. The present discussion is restricted to spatially periodic boundary conditions, for which the boundary is a square of edge 2π in the x,y plane.

The vorticity ω has most of the properties one might like in order to treat it as a probability distribution except positive-definiteness. (In spatially doubly-periodic boundary conditions, the kinematic requirement $\int \omega \, d^2 x = 0$ over the square must hold for all time.) This difficulty may be remedied by embedding Eq. (1) in a two-fluid dynamics for ω^+ and ω^-, two positive semi-definite vorticities that obey

$$\left(\frac{\partial}{\partial t} + \boldsymbol{v} \cdot \nabla \right) \omega^\pm = \nu \nabla^2 \omega^\pm \tag{2a, b}$$

separately. The physical vorticity ω is understood as $\omega = \omega^+ - \omega^-$, with $\nabla^2 \psi = -\omega$, still, and the same field $\boldsymbol{v}$ entering into both Eq. (2a) and Eq. (2b): $\boldsymbol{v} = \nabla \psi \times \hat{e}_z$. Here, $\omega^\pm \geq 0$.

It is apparent by simple subtraction that any time one solves Eqs. (2a,b), a solution to Eq. (1) is simultaneously being generated. It is also easy to prove that if $\omega^\pm \geq 0$ initially, they remain so for all time. In this formulation, viscous dissipation appears as the gradual interpenetration of the ω^+ and ω^- vorticities; they both approach a spatially uniform state. Once $\omega^+ = \omega^-$ in a region, $\omega = 0$ there, and when they have become equal everywhere, $\boldsymbol{v} = 0$ and $\psi = 0$. Nevertheless, the <u>fluxes</u> of ω^+ and ω^- are rigorous constants of the motion: $\int \omega^\pm d^2 x \equiv \Omega = $ constant, all t. This can readily be proved by integrating Eqs. (2a,b).

It is now natural to introduce as an entropy for the system, the following Jaynesian expression

$$S = - \int (\omega^+ \ln \omega^+ + \omega^- \ln \omega^-) d^2 x + \text{const.} \tag{3}$$

Equation (3) appears as a natural generalization of Boltzmann's expression, the original model for all microscopic entropy functionals. One can almost prove an H-theorem for it. Using Eqs. (3), it is readily shown that $dS/dt \geq 0$ with zero being achieved only when $\omega^+ = \omega^- = \langle \omega \rangle = $ const. The only thing that stops us short of asserting an H-theorem is that other functionals, such as the enstrophy, $\frac{1}{2}\int(\omega^+ - \omega^-)^2 d^2x$, also vary monotonically and approach extremal values as $t \to \infty$. Subtle arguments to the effect that the limit of S is approached rigorously in some sense more rapidly than the limits of such functionals as the enstrophy (energy is no problem) are approached are required and have not yet been possible to sort out.

S as given by Eq. (3) has one remaining obvious drawback: it is dependent in an unsatisfactory way on the conserved spatial mean $\langle \omega^+ \rangle = \langle \omega^- \rangle = $ const. (the angle brackets $\langle \rangle$ mean averages over the basic square). However, Eqs. (2a,b) as well as ω, v, and ψ, are all invariant under the addition of the same additive constant to ω^+ and ω^-. It seems desirable to choose an S, if possible, which does not depend upon the spatial averages $\langle \omega^\pm \rangle$, since nothing in the physics depends upon them. A further, and final, subdivision of the vorticity seems desirable:

$$\omega^+ = \langle \omega^+ \rangle + \omega^{++} - \omega^{+-} \tag{4a}$$

$$\omega^- = \langle \omega^- \rangle + \omega^{--} - \omega^{-+} , \tag{4b}$$

where all four of the fields ω^{ij} $(i, j = +$ or $-)$ are ≥ 0. The rearrangment of the four ω^{ij} is what distinguishes one physical configuration from another and they seem like the natural variables in terms of which the entropic counting problem should be formulated. A specific four-fluid dynamics for the ω^{ij} can be written down if desired, but its content is identical to that of Eqs. (2). A natural generalization of Eq. (2) is now

$$S = -\sum_{ij} \int \omega^{ij} \ln \omega^{ij} d^2x + \text{const.} \tag{5}$$

Maximizing S must now be done subject to the constancy of the four $\int \omega^{ij} d^2x = $ const., and the (approximate) constancy of the energy

$$E \equiv \frac{1}{2} \int \psi(\omega^{++} - \omega^{+-} - \omega^{--} + \omega^{-+}) d^2x , \tag{6}$$

and is as nearly constant in time (at small ν) as it was in the original representation, Eq. (1).

The result of the maximization is a set of four "most probable" ω^{ij} given by

$$\omega^{ij} = \exp[-\alpha^{ij} \mp \beta\psi] . \tag{7}$$

In the ω^{ij} of Eq. (7), the α^{ij} are Lagrange multipliers that go with the conservation of the four fluxes, the $-\beta$ goes with ω^{++} and ω^{-+}, while the $+\beta$ goes with ω^{--} and ω^{+-}. All the Lagrange multipliers are independent of $\langle \omega^+ \rangle = \langle \omega^- \rangle$, and the most probable vorticity is now $\omega = \omega^+ - \omega^-$, where

$$\omega^+ = \langle\omega^+\rangle + \exp[-\alpha^{++} - \beta\psi] - \exp[-\alpha^{+-} + \beta\psi] \qquad (8a)$$

$$\omega^- = \langle\omega^-\rangle + \exp[-\alpha^{--} + \beta\psi] - \exp[-\alpha^{-+} - \beta\psi] \ . \qquad (8b)$$

The difference of Eqs. (8a), (8b), the physical vorticity, does not involve $\langle\omega^+\rangle = \langle\omega^-\rangle$ at all.

If complete $+/-$ symmetry is assumed, the pairs of α^{ij} Lagrange multipliers are both the same, and the most probable ω reduces to

$$\omega = [\exp(-\alpha^{++}) + \exp(-\alpha^{-+})]\exp(-\beta\psi)$$

$$- [\exp(-\alpha^{--}) + \exp(-\alpha^{+-})]\exp(\beta\psi)$$

$$= -c^{-1} \sinh \beta\psi \qquad (9)$$

where c^{-1} is a constant that has replaced the α^{ij} Lagrange multipliers.

The long-time 2D NS solutions of Matthaeus et al. have been well fit by Eq. (9), and Shan's long-time solutions to Eqs. (2) are better fit by Eqs. (8) without the assumption of $+/-$ symmetry. The result of Eq. (9), substituted into Poisson's equation,

$$\nabla^2\psi = -c^{-1} \sinh \beta\psi \qquad (10)$$

was derived long ago as a mean-field limit for the most probable distribution of ideal, discrete, line vortices. It was something of a surprise to find it reappearing in viscous continuum computations, a situation in which it had not been thought to be relevant.

Actually, the symmetry necessary to achieve the hyperbolic sinusoidal dependence of Eq. (10) is not accurately fulfilled. In the vortex merger process, there is no guarantee that energy will be equally shared between the positive and negative halves of the vorticity distribution, and typically one final-state vortex turns out to be slightly "hotter" than the other. This variation, typically in the range of 5 to 10 percent in associated energy, is regarded as a natural thermodynamic fluctuation. Its presence is why Shan found that he could fit the computed final-state vortices more accurately if he used a slightly different set of Lagrange multipliers in Eq. (8a) than in Eq. (8b).

We consider that for 2D NS decaying turbulence, the maximum entropy agenda has been substantially validated. Interesting issues are raised by trying to formulate a generalized agenda for 2D MHD turbulence, a topic to which we now turn our attention.[9,10]

3. 2 MHD decaying turbulence: a crucial test

In 2D MHD, the geometric convention is again a velocity field $\boldsymbol{v} = (v_x, v_y, 0) = \nabla\psi \times \hat{e}_z$ along with a magnetic field $\boldsymbol{B} = (B_x, B_y, 0) = \nabla A \times \hat{e}_z$, with $\partial/\partial z \equiv 0$.

$A = (0,\ 0,\ A)$ is a one-component vector potential. $B = \nabla \times j$, where $j = (0,\ 0,\ j) = -\nabla^2 A$, and ω is still $-\nabla^2 \psi$. Equation (1) generalizes to

$$\frac{\partial \omega}{\partial t} + \boldsymbol{v} \cdot \nabla \omega = \boldsymbol{B} \cdot \nabla j + \nu \nabla^2 \omega \ , \tag{11}$$

while A advances according to

$$\frac{\partial A}{\partial t} + \boldsymbol{v} \cdot \nabla A = \eta \nabla^2 A \ . \tag{12}$$

Equations (11), (12) are a natural generalization of the NS dynamics described by Eq. (1), and indeed contain it as a special case. Quite a few numerical studies of turbulence in periodic systems described by Eqs. (11) and (12) have been carried out. Though none of them are of as long duration or at as high resolution as those described by Eq. (1) to which we made allusion in the last section, enough have been done to know that once again, large-scale coherent structures of great regularity and simplicity will emerge on time scales short compared to the time necessary for the magnetofluid to become uniform, without flow, and without magnetic inhomogeneities.

Equations (11), (12) provide much more slippery terrain for defining entropies than does Eq. (1), both because the choice of fields is essentially wider and because there is no underlying ideal discrete-particle model underlying 2D MHD in the way that the discrete line vortex model underlies 2D NS flow in the limit of zero viscosity.

There are several imaginable informed guesses one may make for an entropy. One natural choice might be to try to define an S in terms of the "sources" of the fields, replacing the integrals $\int \omega \ln \omega \, d^2x$ with additive combinations of $\int \omega \ln \omega \, d^2x$ and $\int j \ln j \, d^2x$, perhaps after a decomposition into positive and negative parts in the manner of Eqs. (2). However, "natural" this may look, it does not lead to an invariant decomposition at the equation of motion level. That is, ω^+ may go negative and $-\omega^-$ may go positive, according to the decompositions that would appear in the 2D dynamics, and similarly for j.

On the other hand, A does provide, according to Eq. (12), an ideal, pointwise, convected invariant in the way that ω provides one in Eq. (1). We would be tempted to use $\int A \ln A \, d^2x$ integrals except for the absence of any mechanicallly analogous quantity.

We are at present conducting numerical computations involving decays of turbulent MHD fields according to Eqs. (11) and (12), and comparing the results with maximum entropy predictions using a variety of entropies for comparison. It was hoped to have definitive results to report at this meeting, but they are not yet available. What can be said is that in the absence of significant "cross helicity" in the initial conditions (i.e., $\int \omega A \, d^2x \cong 0$), the turbulence decays to a state with very little kinetic energy, and a sharp pointwise relation between A and j. As noted previously by Kinney and McWilliams[11], the configuration corresponds to two large, opposing, current filaments that are of more or less circular shape. One waits with great anticipation to see how accurately this configuration can be predicted as a maximum-entropy one.

4. Summary

There are two necessary steps, if one wishes to define an entropy that is useful for describing dissipative continuum systems with many degrees of freedom. First, one must write down a functional that plausibly can be said to measure the "probability" of a state of the system. Second, one must verify, most likely numerically, that the system in question approaches the "most probable state" obtained by maximizing that entropy functional, subject to whatever constraints the dynamics provide. The only point of view it would seem to be defensible to take in such an undertaking at present is a sort of "experimental" one, modelled on Boltzmann's investigations of dilute monatomic gases.

It is considered that a substantial measure of success has been achieved in this direction for 2D NS turbulence. The 2D MHD case is under active investigation, but substantial conceptual puzzles remain. An exciting beginning on the 3D NS case has been made by Chorin.[12] 3D MHD is also a likely area for future progress.[10]

We are persuaded that, in addition to their quite widespread application to data processing, maximum entropy formulations have a great deal yet to say about turbulent relaxation processes in dissipative continua. It is easy to forget that Boltzmann invented them for just such a problem.

5. Acknowledgments

This work has been supported in part by the U.S. Department of Energy under grant DE-FG02-85ER53194 and in party by the U.S. Naval Research Laboratory under grant N00014-95-1-G001.

Valuable discussions with Drs. W. H. Matthaeus, X. Shan, and M. Hossain are gratefully acknowledged.

References

1. Matthaeus, W. H., Stribling, T. W., Martinez, D., Oughton, S., and Montgomery, D., *Phys. Rev. Lett.* **66**, 2731 (1991a).
2. Matthaeus, W. H., Stribling, W. T., Martinez D., Oughton, S., and Mongomery, D., *Physica* **D51**, 531 (1991b).
3. Montgomery, D., Matthaeus, W. H., Stribling, W. T., Martinez, D., and Oughton, S., *Phys. Fluids* **A4**, 3 (1992).
4. Montgomery, D., Shan, X., and Matthaeus, W. H., *Phys. Fluids* **A5**, 2207 (1993).
5. Jaynes, E. T., *Phys. Rev.* **106**, 620 & **108**, 171 (1957a, b).
6. Montgomery, D., in *Maximum Entropy and Bayesian Methods in Inverse Problems*, ed. C. Ray Smith and W. T. Grandy, Jr., Dordrecht, D. Reidel Pub. Co., pp. 455-468 (1985).
7. Robert, R., and Sommeria, J., *J. Fluid Mech.* **229**, 291 (1991).
8. Robert, R., and Sommeria, J., *Phys. Rev. Lett.* **69**, 2776 (1992).
9. Montgomery, D., Turner, L., and Vahala, G., *J. Plasma Phys.* **21**, 239 (1979).
10. Ambrosiano, J., and Vahala, G., *Phys. Fluids* **24**, 2253 (1981).
11. Kinney, R., and McWilliams, J., private communication; to be published (1995).
12. Chorin, A. J., *Vorticity and Turbulence*, (New York, Springer-Verlag, Inc.) (1994).

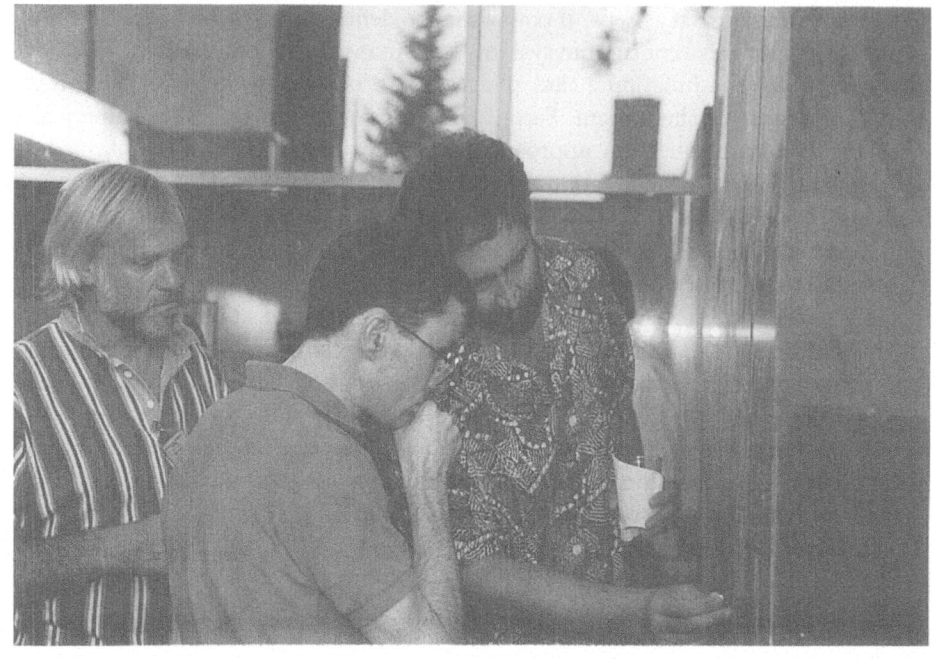

Julian Besag, David Madigan, and Michael Miller using the blackboard at the Santa Fe Institute.

ON THE IMPORTANCE OF α MARGINALIZATION IN MAXIMUM ENTROPY

R. FISCHER, W. VON DER LINDEN AND V. DOSE

Max-Planck-Institut für Plasmaphysik, EURATOM Association
POB 1533, D-85740 Garching, Germany[†]

Abstract. The correct entropic prior, computed by marginalization over the regularization parameter α, is used to invert photoemission data and to restore the famous "Susie" image. Comparison with the conventional maximum entropy procedure shows less overfitting of noise and demonstrates the residual ringing which is intrinsic to ill-posed inversion problems. An improvement to the steepest descent approximation reveals the reason for the overfitting. On top of that, the correct treatment of the regularization parameter is vital for the existence of the continuum-limit of MaxEnt.

Key words: Inverse Problem, Regularization, Entropic Prior, Image Processing

1. Introduction

The maximum entropy (MaxEnt) method is well established to recover "hidden information" from incomplete and noisy data [1]. In spite of the generality and power of MaxEnt there are still persisting problems concerning the numerical treatment. As Strauss, Wolpert and Wolf [2] pointed out, the reliability of the conventional MaxEnt procedure [3,4] depends on several conditions. In the classic MaxEnt formalism the regularization parameter α, entering the entropic prior for an N-dimensional positive additive distribution f

$$P(f|\alpha) = \frac{\exp[\alpha S(f)]}{Z_S(\alpha)}, \quad \text{where} \quad S(f) = \sum_{i=1}^{N} f_i - m_i - f_i ln(\frac{f_i}{m_i}) \quad , \quad (1)$$

has to be marginalized

$$P(f) = \int_0^\infty d\alpha P(f|\alpha) P(\alpha) \tag{2}$$

since α is a nuisance parameter [5]. m is the default model and $Z_S(\alpha)$ is the normalization. According to the marginal prior (2) the posterior $P(f|D)$ does not

---

[†]Email: rrf@ipp-garching.mpg.de

K. M. Hanson and R. N. Silver (eds.), Maximum Entropy and Bayesian Methods, 229–236.
© *1996 Kluwer Academic Publishers.*

depend on the regularization parameter

$$P(f|D) = \frac{P(D|f)P(f)}{P(D)} \tag{3}$$

In most MaxEnt-applications, however, the α-marginalization is replaced by the posterior distribution $P(f|\hat{\alpha}, D)$ where $\hat{\alpha}$ is determined by the maximum of the α-evidence $P(\alpha|D)$ [3,4]:

$$P(f|D) = \int_0^\infty d\alpha P(f|\alpha, D)P(\alpha|D) \approx P(f|\hat{\alpha}, D) = \frac{P(D|f, \hat{\alpha})P(f|\hat{\alpha})}{P(D|\hat{\alpha})} \tag{4}$$

For the evidence approximation to be valid, $P(\alpha|D)$ has to be sharply peaked in a region where $P(f|\alpha, D)$, viewed as a function of α, varies slowly. There is a further crucial approximation employed in standard MaxEnt not pointed out in [2]. Multidimensional integrals over image space f of the form

$$I = \int_0^\infty e^{\phi(f)} \frac{df^N}{\prod_i \sqrt{f_i}} \approx \frac{e^{\phi(f^*)}}{\prod_i \sqrt{f_i^*}} \int_{-\infty}^\infty e^{\frac{1}{2}(f-f^*)^T(\nabla\nabla^T\phi|_{f^*})(f-f^*)} df^N \tag{5}$$

are routinely approximated by 'steepest descent', i.e. by expanding ϕ to second order about its maximum f^*, extending the lower integration limits to $-\infty$, and replacing the integration measure by a constant. This approximation tends to overfit the noise for the following reason: The α-evidence $P(\alpha|D)$ is proportional to I with the exponent $\phi = \alpha S + L$ with L being the log-likelihood. Due to the divergent prefactor $1/\prod_i \sqrt{f_i^*}$ images with values close to zero are erroneously preferred, which is achieved by reducing α. The effect is amplified by the modified integration limits: the integral is overestimated by a factor 2 for each component $f_i^* \to 0$. The shortcomings of the approximate evaluation of I can be overcome by an integral-correction factor [6] which accounts for both, the integration limits and the measure.

The necessity of the explicit marginalization over α is emphasized for the first time by Bryan [5]. Strauss et al. [2] proposed a straightforward α-marginalization to find $P(f)$. In this approach the α-evidence is not required and the only remaining multidimensional integral, namely the determination of $Z_S(\alpha)$, factors into one-dimensional integrals. $P(f)$ can be tabulated efficiently since it depends only via the entropy $S(f)$ on f.

The conventional MaxEnt procedure has the further disastrous property [7] that the continuum limit does not exist. It appears that this failure occurs since α is kept fixed while changing the cell size. By marginalization over α and using Jeffrey's prior for α it can be shown that macroscopic expectation values, e.g. $\sum_{\text{domain D}} < f_i >$, exist in the continuum limit.

In the first section we apply the α-marginalization to mock data to distinguish between overfitting caused by the approximations and ringing which is intrinsic to ill-posed inversion problems. In the second section we discuss the ill-posed inversion of island-size distributions from photoemission data which depends strongly on the treatment of the regularization-parameter. Finally, we demonstrate the applicability of the α-marginalization on large data sets in the reconstruction of the "Susie" image [3].

2. Convolution data

In the upper panels of Fig. 1 a mock data set is shown consisting of a Lorentzian and a step blurred by a Gaussian. The data were distorted by Poisson noise with high (low) variance in Fig. 1a (Fig. 1c), respectively. The lower panels depict the reconstructions with the conventional MaxEnt procedure using the evidence approximation and the steepest descent procedure, the steepest descent procedure with integral corrections and the α-marginalization. On first sight the recon-

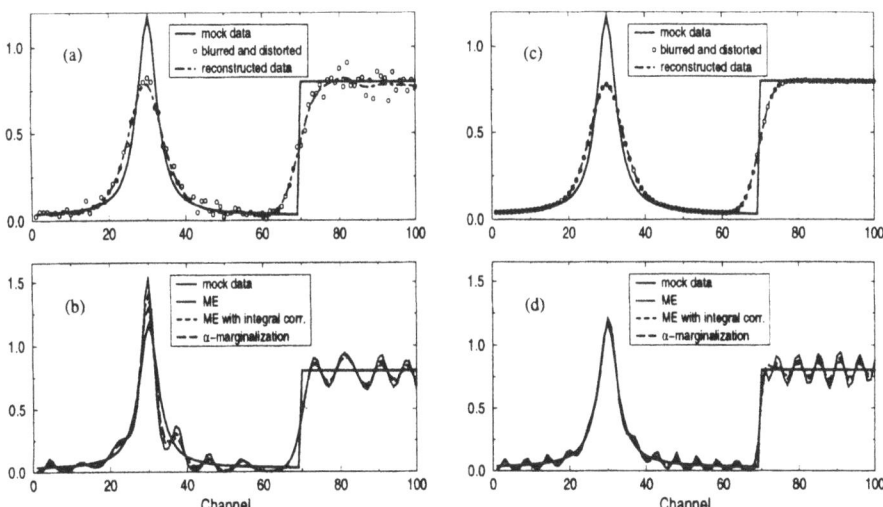

Figure 1. Upper panels: mock data (solid line) convolved with a Gaussian and distorted by Poisson noise (circles) with high (low) variance in Fig. 1a (Fig. 1c), respectively. Lower panels: MaxEnt deconvolutions with α-marginalization (long-dashed line), conventional MaxEnt procedure (dotted line) and steepest descent with integral corrections (short-dashed line, lies very close to the dotted line).

structions look frightening, the step is not reproduced properly and the image is superposed by considerable ringing.

The ringing on the plateau and in the slopes of the Lorentzian can be reduced by introducing a preblur on a "hidden image" as proposed by Skilling [8] (not shown here). But our calculations showed that the preblur has the property to set as many variables of the hidden image to very small values. It appears that this tendency is a consequence of the steepest descent approximation. Small values are erroneously preferred due to the divergent prefactor and the modified integration limits. Furthermore, sharp structures like the step cannot be reproduced with a constant preblur necessary for reducing the high ringing level. For the future, the introduction of a preblur with variable width over the image following the pixon approach [9] may be promising.

An integral-correction factor to the steepest descent approximation demonstrates the essential contribution to the overfitting. The reconstruction obtained by the integral-correction factor (short-dashed line) hardly deviates from that obtained by the α-marginalization (long-dashed line).

The reason for the disappointing ringing which even resists the α-marginalization is intrinsic to the problem. To show this analytically the problem is simplified. A Gaussian likelihood distribution $P(g|f)$ is chosen with a linear model Af and a quadratic prior distribution $P(f|\alpha,m)$

$$P(g|f) \ \propto \ \exp\left\{-\frac{(g-Af)^2}{2\sigma^2}\right\} \tag{6}$$

$$P(f|\alpha,m) \ \propto \ \exp\left\{-\frac{\alpha}{2}(f-m)^2\right\} \quad. \tag{7}$$

The posterior mode is then given by

$$-A^T g + A^T A f + \lambda(f-m) = 0 \quad \text{with} \quad \lambda = \alpha\sigma^2 \quad. \tag{8}$$

Splitting the data $g = g^x + \Delta g$ and the reconstruction $f = f^x + \Delta f$ into exact values and noise, and denoting $m = f^x + \Delta m$, the equation for the deviations reads:

$$(A^T A + \lambda)\Delta f = A^T \Delta g + \lambda\Delta m \tag{9}$$

The first term on the rhs describes the influence of the noise on the reconstruction f. For example, without regularization ($\lambda = 0$) small variations in the data result in huge variations in f if the eigenvalues of $A^T A$ fall off very fast. The ringing in Fig. 1b is mainly due to the large noise level in the data (Fig. 1a). The second term, denoting the difference between the default model m and the exact solution f^x, causes ringing even for $\Delta g \to 0$. The ringing caused by the Δm-term is shown in Fig. 1d where the noise level is very low (Fig. 1c). The ringing occurs for small λ in this limit and for $\Delta g = 0$ equation (9) yields

$$\Delta f_i \approx \sum_{\nu,\varepsilon_\nu<\lambda} \phi_i^{(\nu)}\left(\sum_j \phi_j^{(\nu)}\Delta m_j\right) \quad, \tag{10}$$

with $\phi^{(\nu)}$ being the ν-th eigenvector of $A^T A$ and ε_ν the corresponding eigenvalue. This expression predicts considerable ringing if $A^T A$ has eigenvalues smaller than λ and if the overlap between the corresponding eigenvectors and Δm, the discrepancy between default model and exact solution, is large. This is the case if the apparatus function (described by A) is much broader than the structures in Δm, i.e. structures in the image which are not accounted for by the default model. Ringing is therefore mainly due to the ill-posed nature of the inversion problem and could only be overcome by an improved default model m. Yet to a small extend, it is also influenced by the treatment of the regularization parameter (λ) as can be seen in Fig. 1c,d.

3. Island-size distributions from photoemission data

Our actual problem was the reconstruction of island-size distributions from photoemission data of ultra-thin Ag coverages on a Pd(111) surface. Without going

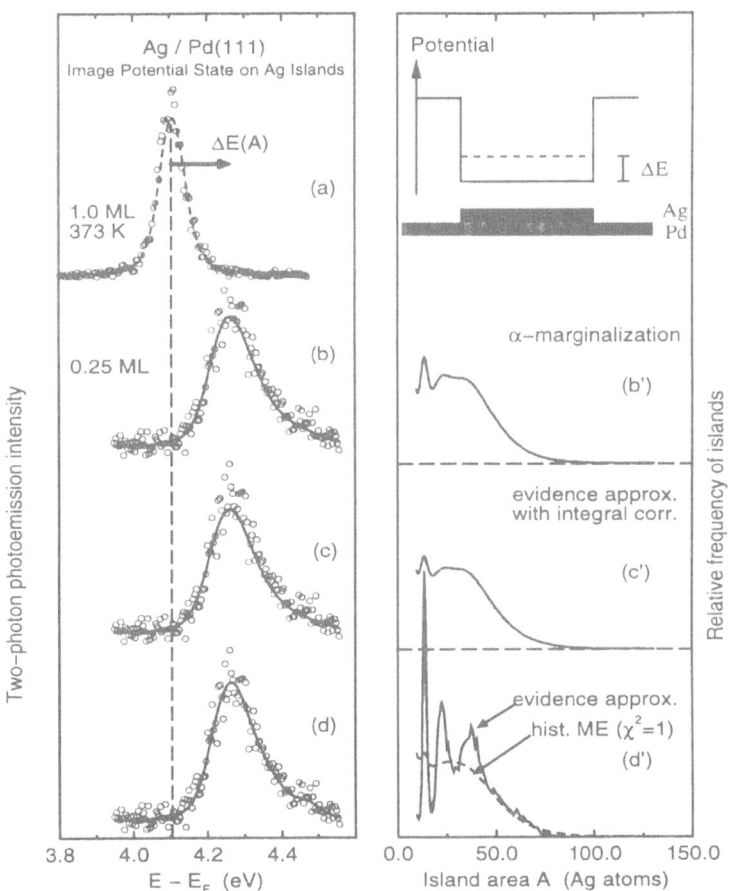

Figure 2. Photoelectron spectra of a surface state on the homogeneous Ag monolayer (a) and on one atomic layer high Ag islands at a Ag coverage of 0.25 monolayer (b)-(c) (left panel). The surface state on islands is shifted in energy relative to the state of the homogeneous Ag layer due to confinement on Ag-islands (right-panel inset). Island-size distributions derived from the experimental data are shown in the right panel for different approaches: the established evidence approximation (d), historic MaxEnt where α is determined via $\chi^2 = 1$ (dashed), the evidence approximation with integral corrections (c), and the α-marginalization (b).

into details a short summary should highlight the problem. Small amounts of Ag adsorbed on a Pd single crystal surface form islands with sizes of a few nm. On the homogeneous Ag surface an electronic surface state exists with a well defined energy and width (Fig. 2a). On small Ag islands the surface state is shifted to higher energy due to the localization of the state on these islands (Fig. 2b-d). Like a quantum mechanical particle in a potential box, here the electron of the surface state on the Ag island, the energy of the state increases with decreasing box size (Fig. 2 right-panel inset). Approximately, the energy shift is inversely proportional to the island area. This surface state is a very sensitive tool to study the surface morphology of the sample [10,11]. The energy and width of the surface state gives us information about nucleation and growth of the Ag adsorbate. In a photoemission experiment the surface state is measured with the energy distribution curves of the emitted photoelectrons.

The surface state on small Ag islands [(b)-(d)] is energetically shifted and asymmetrically broadened. The shift of the surface state is assigned to the lateral confinement and the broadening is due to the existence of a distribution of different island sizes. Hence, the energy distribution curve bears the information about the size distribution of the Ag islands.

To reconstruct the island-size distribution we model the photoemission spectrum by an integral over the energy shifted spectrum $K(E, A)$ of the homogeneous Ag film (Fig. 2a) times the island-size distribution function $f(A)$

$$G(E) = \int_0^\infty dA K(E, A) f(A) \tag{11}$$

The energy shift is calculated for a finite two-dimensional circular potential well. Further details can be found in [10-12]. The island-size distributions derived from the energy distribution curve of the experimental data are shown in the right panel of Fig. 2. Four different approaches were employed to invert eq. (11): conventional MaxEnt (d'), historic MaxEnt where α is determined via $\chi^2 = 1$, MaxEnt plus integral-correction (c'), and α-marginalization (b'). The reconstruction in data-space to the various approaches are shown as solid lines in the left panel. The conventional MaxEnt approach (d') clearly yields an overfitted solution. The exact approach (b') is much smoother. The integral corrected version of the conventional MaxEnt (c') is close to the exact result. Even historic MaxEnt yields a much better result than the approximate classic MaxEnt.

The exact approach allows to determine an effective α_P, although it marginalizes over α [2]. Relative to α_P the regularization parameter is for (d') a factor 17 smaller and for (c') a factor 1.8 larger. The surprising fact was that if we take the effective α_P of the α-marginalization as the optimum $\hat{\alpha}$ in the conventional MaxEnt approach we get the result of the α-marginalization. From this we conclude that the α-evidence $P(\alpha|D)$ is indeed sharply peaked and all the problems are caused by the approximate treatment of the evidence-integrals. The conditions for the evidence analysis are, at least in this case, fulfilled. The small remaining discrepancy between MaxEnt plus integral correction and the exact approach is supposedly due to the Taylor-expansion of the exponent Φ.

The dominant peak at an island size of 14 Ag atoms is due to the small structure in the data at an energy of 4.47 eV. The result obtained by the α-marginalization does not emphasize further island sizes. The existence of a sub-structure on a broad distribution was surprising and is interpreted as a favored magic island size. The overfitted solution of the conventional MaxEnt approach yields two more magic island sizes due to the approximations. This example shows the necessity for the correct regularization with the α-marginalization.

4. Image reconstruction with "Susie"

To assess the α-marginalization we restored the famous "Susie"-image which is digitized on a 128×128 pixel grid with grey-level values between 40 and 255 (Fig. 3a). This is a traditional example for MaxEnt image processing [3, and references therein]. For comparison with previous approaches we blurred the picture by a 6-pixel radius Gaussian point-spread function and added noise of unit variance (Fig. 3b). The reconstruction obtained by the conventional MaxEnt approach shows nasty overfitting (Fig. 3d). An additional numerical problem of the conventional MaxEnt procedure is the evaluation of the logarithm of the determinant of the 16384×16384 Hessian matrix which is numerically rather cumbersome. This task can be tackled either by Monte Carlo techniques or it can be avoided [3] by using a quadratic prior. A great advantage of the α-marginalization is that this difficulty does not arise in the first place.

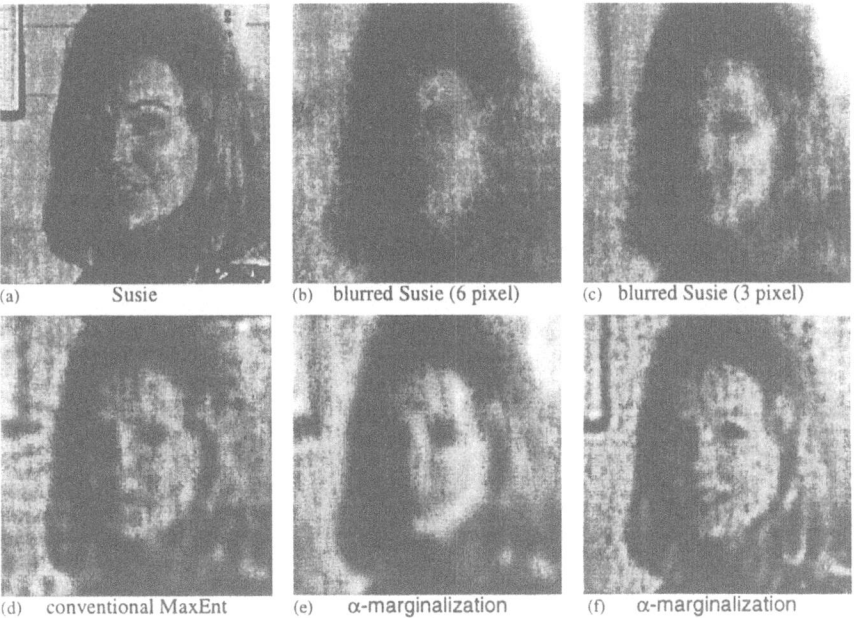

(a)   Susie   (b)   blurred Susie (6 pixel)   (c)   blurred Susie (3 pixel)

(d)   conventional MaxEnt   (e) α-marginalization   (f) α-marginalization

Figure 3.   Upper panel: image of "Susie" (a), blurred by a 6[3]-pixel Gaussian PSF (b) [(c)], respectively. Lower panel: MaxEnt reconstructions using the evidence approximation (d) and the α-marginalization [(e), (f)].

The result obtained by the α-marginalization (Fig. 3e) is much more pleasant due to the reduced ringing level. As Strauss et al. supposed the reconstruction of Susie based on the evidence approximation is overfitted. Nevertheless, the inversion-intrinsic ringing is still present. The intrinsic ringing increases if the structures get sharper. To show this we performed a 3-pixel blurred image of Susie

(Fig. 3c). The optimal reconstruction (Fig. 3f) obtained by the α-marginalization is still superposed by considerable ringing. This can only be avoided by a more elaborate default model.

5. Summary

We have demonstrated that a great deal of overfitting, generally observed in Max-Ent images, is due to the approximate evaluation of the α-evidence and the replacement of the α-marginalization by α-evidence analysis. It appears that the overfitting of noise can be eliminated by the α-marginalization while the ringing, caused by the ill-posed inversion problem, is retained. The correct treatment of the regularization parameter has also a strong impact on adjusting other hyper-parameters, e.g. the width of the preblur proposed by Skilling. It is also vital for the correct continuum-limit of the theory. For the reconstruction of island-size distributions from photoemission data the exact regularization turned out to be essential for the physical problem under consideration. The applicability of the α-marginalization to large data sets is demonstrated by the reconstruction of the Susie image.

References

1. S. F. Gull and J. Skilling, "Maximum entropy method in image processing," *IEE Proc.*, **131 F**, pp. 646–659, 1984.
2. C. E. M. Strauss, D. H. Wolpert and D. R. Wolf, "Alpha, evidence, and the entropic prior," in *Maximum Entropy and Bayesian Methods (1993)*, G. Heidbreder, ed., Kluwer Academic, Dordrecht, 1995.
3. S. F. Gull, "Developments in maximum entropy data analysis," in *Maximum Entropy and Bayesian Methods*, J. Skilling, ed., pp. 53–71, Kluwer Academic, Dordrecht, 1989.
4. J. Skilling, "Quantified maximum entropy," in *Maximum Entropy and Bayesian Methods*, P. F. Fougère, ed., pp. 341–350, Kluwer Academic, Dordrecht, 1990.
5. R. Bryan, "Maximum entropy analysis of oversampled data problems," *Eur. Biophys. J.*, **18**, pp. 165–174, 1990.
6. W. von der Linden, R. Fischer, and V. Dose, "Evidence integrals," *In these proceedings*, 1995.
7. J. Skilling and S. Sibisi, "Prior distributions on measure space," *In these proceedings*, 1995.
8. J. Skilling, "Fundamentals of MaxEnt in data analysis," in *Maximum Entropy in Action*, B. Buck and V. A. Macaulay, eds., pp. 19–40, Clarendon Press, Oxford, 1991.
9. R. Puetter, "Pixon-based image reconstruction in high energy astrophysics," *In these proceedings*, 1995.
10. R. Fischer, S. Schuppler, N. Fischer, Th. Fauster and W. Steinmann, "Image states and local work function for Ag/Pd(111)," *Phys. Rev. Lett.*, **70**, pp. 654–657, 1993.
11. R. Fischer, Th. Fauster and W. Steinmann, "Threedimensional localization of electrons on Ag islands," *Phys. Rev. B*, **48**, pp. 15496–15499, 1993.
12. R. Fischer, "submitted to Surf. Sci."

QUANTUM MECHANICS AS AN EXOTIC PROBABILITY THEORY

SAUL YOUSSEF

Supercomputer Computations Research Institute
Florida State University
Tallahassee, FL 32306-4052[†]

Abstract. Recent results suggest that quantum mechanical phenomena may be interpreted as a failure of standard probability theory and may be described by a Bayesian complex probability theory.

Key words: Quantum Mechanics, Bayesian, Complex Probability Theory

There is more to probability theory than proving theorems in a particular mathematical system. One is also in a position to make predictions about real physical systems by adding extra assumptions to the standard axioms. Such predictions are necessarily subject to experimental test, and, to the extent that one believes in the extra assumptions, such tests may be interpreted as testing the correctness of probability theory itself. Now this may already seem like an odd point of view, especially here, since this conference series itself provides a most impressive record of success for probability theory in a vast array of situations with no indication of a problem – so why is there any reason to doubt probability theory? Here I think that there is a historical effect: probability theory may actually be failing all the time, it's just that the situations where a failure occurs are called "quantum mechanical phenomena" and thus appear in physics conferences instead of in probability theory conferences. This suggests that perhaps there is something wrong with probability theory after all, and that this may be where quantum mechanical effects come from. Let's adopt this point of view and see where it leads [1–3].

An obvious place to test our new point of view is the two–slit experiment where, as everyone knows, the fact that an interference pattern is observed even if one particle is sent through at a time, forces us to conclude that it is not true that a particular particle either goes through *slit* #1 or through *slit* #2; in general, then, a particle cannot be said to follow a path through space. This is the "wave–particle duality," the basic effect in quantum mechanics. Notice, however, that from our new point of view, the standard argument has a hole in it due to it's essential reliance on probability theory. For a position x on the screen where there is a dip

---

[†] Email: youssef@scri.fsu.edu

K. M. Hanson and R. N. Silver (eds.), Maximum Entropy and Bayesian Methods, 237–244.
© *1996 Kluwer Academic Publishers.*

in the interference pattern, one reaches a contradiction by noticing that

$$P(x) = P(x \ via \ slit\#1) + P(x \ via \ slit\#2) \geq P(x \ via \ slit\#1) \qquad (1)$$

means that opening the second slit should not cause the probability to arrive at x to decrease. But if we are willing to modify probability theory, then the standard argument and it's surprising conclusions do not necessarily follow. In fact it is clear that in order to escape the standard conclusions, a modified probability theory must provide a way for probabilities to cancel each other and so an obvious first guess is to allow probabilities to be complex numbers. Here, of course, the argument grinds to a halt for a frequentist since frequencies are not the complex numbers. However, as Bayesians, we are not completely out of options because for us, probabilities start out only as (real and non–negative) measurements of "likelihood" where the frequency meaning for this likelihood is derived after the fact. Similarly, we might consider a complex "likelihood" and see if a frequency meaning can be found for this as well. In fact, the simplest thing to do is to take Cox's assumptions[4] and just drop the restriction that probabilities be real and non–negative. In this case, it turns out that Cox's entire argument follows as before and one ends up with "complex probability theory" having the same form as the probability theory that you're used to except that probabilities are complex. For any propositions a, b and c,

$$(a \rightarrow b \wedge c) = (a \rightarrow b)(a \wedge b \rightarrow c) \qquad (I.a)$$

$$(a \rightarrow b) + (a \rightarrow \neg b) = 1 \qquad (I.b)$$

$$(a \rightarrow false) = 0 \qquad (I.c)$$

where I have written the complex probability that proposition b is true given that proposition a is known as "$(a \rightarrow b)$" (to be read: "a goes to b") reserving the more familiar notation $P(b|a)$ for standard $[0, 1]$ probabilities.

   Given our complex probability theory we would like to continue in parallel with the Bayesian development and construct a frequency meaning for complex probabilities. Recall that for standard probability, this works by supposing that the probability of something is p and considering N copies of that situation with $f = n/N$ successes. Using the central limit theorem f is asymptotically gaussian with mean p. Then, since the probability for f to be in any interval not containing p can be made arbitrarily small by increasing N, this fixes the frequency meaning for p. Essentially, the frequency meaning of p rests on the extra assumption that an arbitrarily small probability for f to be in some interval means that f in fact will never be observed to be in that interval in a real experiment. The situation is not quite so simple in complex probability theory because a zero complex probability does not in general mean that the corresponding event will never happen. However, we can proceed by assuming that this extra condition is true for a special set of propositions U called the "state space." Let's also assume that this U satisfies the following for $x, y \in U$, propositions a, b, time $t \leq t' \leq t''$:

$$x_t \wedge y_t = false \ if \ x \neq y \qquad (II.a)$$

$$(a_t \rightarrow b_{t''}) = (a_t \rightarrow U_{t'} \wedge b_{t''}) \qquad (II.b)$$

$$(a_t \wedge x_{t'} \rightarrow b_{t''}) = (x_{t'} \rightarrow b_{t''}) \qquad (II.c)$$

where subscripts denote time, as in "a_t" meaning "a is true at time t," and where a set of propositions with a subscript denotes the *or* of each element with the same subscript, as in $U_t = \vee_{x \in U} x_t$. These are just Markovian style axioms intuitively corresponding to "the system has a state." Roughly, the system cannot be in two different states at the same time (II.a), the system is in some state at each intermediate time (II.b) and the knowledge that a system is at some point in the state space makes all previous knowledge irrelevant (II.c). We assume that U is a measure space with $(a_t \rightarrow U_{t'}) = \int_{x \in U}(a_t \rightarrow x_{t'})$. Note the clash of terminology where the Hilbert space of standard quantum mechanics is sometimes also called the "state space." Here U is only a measure space of propositions.

Given I and II, one can repeat the standard argument for the expression

$$Prob(a_t, b_{t'}) = \frac{\int_{x \in U} |a_t \rightarrow b_{t'} \wedge x_{t'}|^2}{\int_{x \in U} |a_t \rightarrow x_{t'}|^2}$$

which predicts the frequency that b is found to be true at time t' given that a is known at a previous time t. Although *Prob* as defined is able to predict the frequencies for outcomes of any experiment, it fails to extend to propositions involving mixed times (e.g. $b_{t'} \wedge c_{t''}$, $t'' > t'$). This is an interesting point because it is exactly this failure that allows complex probability theory to escape Bell's theorem[3]. Also, although I don't have a sharp result, it is seems likely that this effect disappears in a classical limit, thus explaining why standard probability theory works in the classical domain.

Immediate consequences of axioms I.a–I.c and II.a–II.c include facts familiar from probability theory such as $(a \rightarrow true) = 1$, $(a \rightarrow b \vee c) = (a \rightarrow b) + (a \rightarrow c) - (a \rightarrow b \wedge c)$, and if $(a \rightarrow b) \neq 0$, then $(a \wedge b \rightarrow c) = (a \rightarrow c)(a \wedge c \rightarrow b)/(a \rightarrow b)$ *(Bayes Theorem)*. Following standard probability theory, propositions a and b are said to be *independent* if $(q \wedge a \rightarrow b) = (q \rightarrow b)$ for all q and, just as in standard probability theory, "locality" enters via assumptions of independence. For instance, if experiments e_1 and e_2 have possible results r_1 and r_2 respectively, then the assumptions that $\{r_1, r_2\}$, $\{e_1, r_2\}$, and $\{e_2, r_1\}$ are independent imply

$$(e_1 \wedge e_2 \rightarrow r_1 \wedge r_2) = (e_1 \rightarrow r_1)(e_2 \rightarrow r_2)$$

as one would expect from, for example, two experiments which have nothing to do with each other. Other simple consequences of the axioms are described in references 1-4 including

- The Path Integral
- The Superposition Principle
- The Expansion Postulate
- The Schrödinger/Klein–Gordon Equations for $U = R^d$

where the standard wavefunction $\Psi(x, t)$ is proportional to the complex probability

$$(Everything\ that\ you\ know\ about\ the\ system \rightarrow x_t)$$

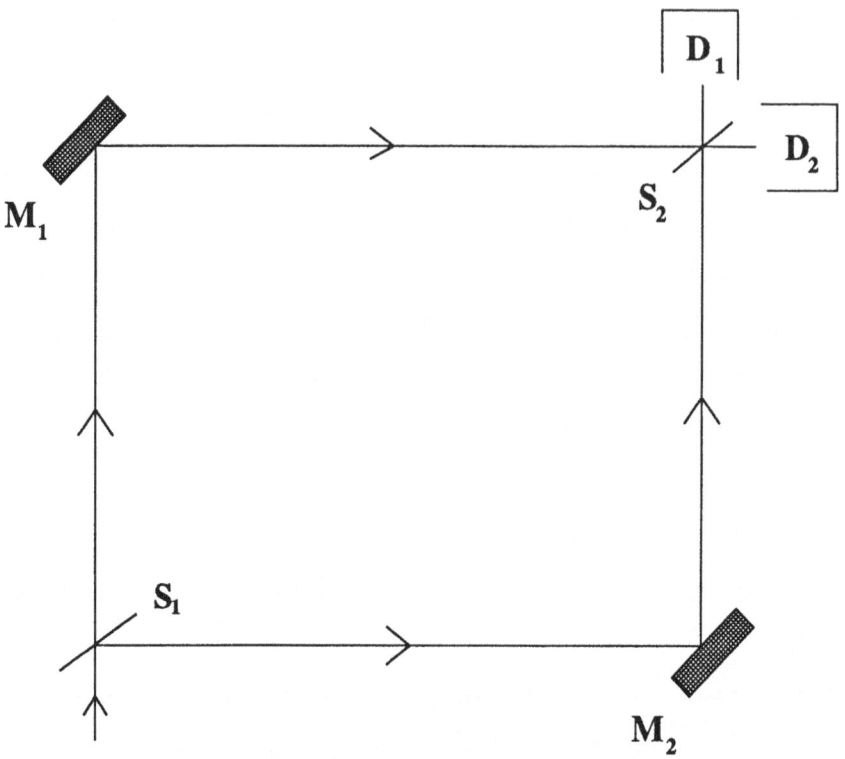

Figure 1.   A simple interferometer where a particle enters as indicated and encounters a beam splitter (S_1), a mirror (M_1 or M_2), a second beam splitter (S_2) ending up either in detector D_1 or in detector D_2.

making the Bayesian status of the wavefunction obvious. In particular, the same system may be described by different wavefunctions depending upon what is known and such wavefunctions can clearly not be "the state of the system" in any reasonable sense.

To get a feeling for how things work, let's consider a typical interferometer as shown in figure 1. Particles enter the device one at a time, pass through a beam splitter, hit one of two mirrors and pass through a second beam splitter ending up either in detector D_1 or in D_2. Although it looks perfectly possible for a particle to end up in D_2, experimentally we mysteriously find that this never happens. All the particles register in D_1. In standard quantum theory, one describes this situation by saying that there are two paths for a particle to go from the source to D_2. The amplitude for these paths have opposite signs and since the probability is the square of the total amplitude, this explains why particles never enter D_2. Now let's consider a modified situation (figure 2) where a device is attached to one of the mirrors which is able to detect if the mirror was struck by the particle. After each particle passes through, the device either indicates "hit" or "no hit." Experimentally, we find that the results are now different with about half of the

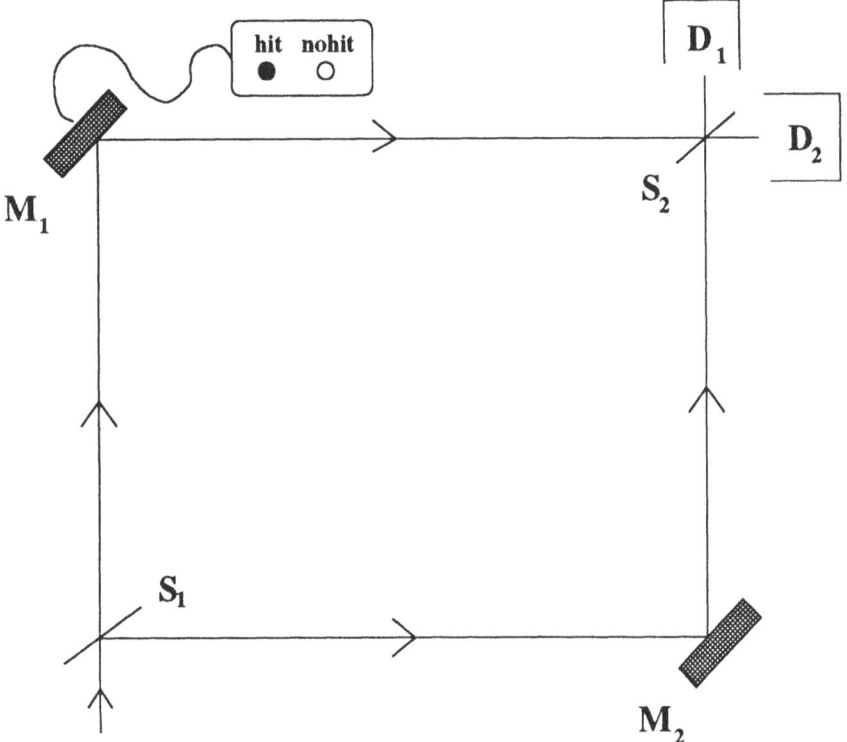

Figure 2.   A interferometer similar to the one shown in figure 1. The device is attached to the mirror M_1 records whether M_1 was struck by the particle or not.

particles ending up in detector D_2. How can we explain this? In standard quantum mechanics there are still two paths for a particle to reach D_2 as before. However, since we can now tell which path was taken by the particle by inspecting the hit/no–hit device, the amplitudes for the two paths no longer interfere. This is a special case of the general principle:

   – *Paths interfere only if there is no way of knowing which path was taken, even in principle.*

This is a rather mysterious statement since it suggests that whether or not you can deduce which path was taken somehow affects the behavior of the particle. Also, it is not clear what "even in principle" means here or what happens if, for example, the hit/no–hit device works with, say, 99% efficiency. Even so, the basic prediction of this principle is correct and this raises the question of whether there is an analogous principle in complex probability theory and whether the predictions are the same. You can indeed easily deduce from the axioms and the definition of *Prob* that

   – *Paths interfere if and only if they end at the same point in U.*

and this appears to give the same predictions as the more mysterious sounding standard quantum mechanical principle. For instance, within complex probability theory, the situation of figure 1 could be described by a state space R^3 (assuming that the particle is spinless) in which case the complex probability to arrive at D_2 is the sum of the complex probabilities for two paths, which also cancel, just as in standard quantum mechanics. In the situation shown in figure 2, one simply notes that the state space R^3 is evidently no longer sufficient to describe the system. If one extends the state space to, say, $U = R^3 \times \{hit, nohit\}$, then the interference is lost because the two paths for reaching D_2 now end at different points in U. There is also a continuum between this result, where the hit/no–hit device is assumed to work perfectly, and a situation where the device works so badly that the propositions "hit" and "no–hit" are independent of which path the particle is taking. In this case, you can easily show that the original effect is restored[2].

Of course, standard quantum mechanics is perfectly capable of handling a situation like that of figure 2. It's just that a rigorous treatment of the problem with a Hilbert space including the hit/no–hit device where the state vector evolves under the action of a Hamiltonian would be rather difficult, especially considering the simplicity of the answer. This helps to explain the popularity of "which path" style arguments in spite of their ambiguity. Here complex probability theory has the advantage that simple assumptions about a system can rigorously be encorporated without having to decide what the assumptions mean in terms, for example, of solutions to the Schrödinger equation. A rigorous treatment of these problems within quantum mechanics would also have to address the issue of whether the initial state is "mixed" or not since not all situations in quantum mechanics can be described by a vector in a Hilbert space, some require "statistical mixtures" of vectors in a Hilbert space. This provides an interesting test for complex probability theory. Since ordinary "statistical mixtures" are no longer available to us, situations requiring "mixed states" had better be handled within the existing axioms. These situations appear to indeed be handled quite smoothly and naturally within the complex probability theory described here[2].

To take an example with more detailed predictions, consider a single scalar particle with $U = R^d$ and consider a sequence of propositions $x_o, x_1, \ldots, x_n$ in $U = R^d$ where each x_j implicitly has a time subscript t_j with $t_{j+1} - t_j = \tau > 0$. As always, $(x_o \to x_n)$ is given by the "path integral"

$$(x_o \to x_n) = \int_{x_1} \ldots \int_{x_{n-1}} (x_o \to x_1)(x_1 \to x_2) \ldots (x_{n-1} \to x_n).$$

Of course, there is a path integral in standard quantum mechanics as well[5] where one proceeds to dynamics by assuming that the amplitude for a path is proportional to e^{iA} where the "action" A is the time integral of a classical Lagrangian for the system. Here we can avoid these extra assumptions by repeating the same argument within each τ sized interval making n sub–path integrals (figure 3) with time step $\epsilon = \tau/N$. By letting both τ and ϵ go to zero, one can extract a central–limit–theorem–like result where for small $|z|$ and small τ, $(x_t \to (x + z)_{t+\tau})$ is

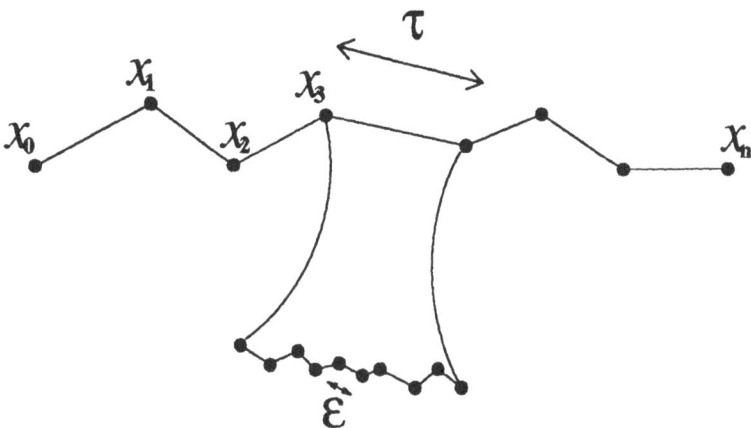

Figure 3. For $x_0, x_n \in U$, the complex probability $(x_0 \rightarrow x_n)$ is equal to a "path integral" over $x_1, x_2, \ldots, x_{n-1}$ with time step τ. The argument can be repeated making n sub-path integrals with time step ϵ.

given by

$$\frac{1}{(2\pi\tau)^{\frac{d}{2}}\sqrt{\det[\nu]}}\exp(-\tau[\frac{1}{2}(\frac{z_j}{\tau} - \nu_j)W_{jk}(\frac{z_k}{\tau} - \nu_k) + \nu_0])$$

where $\nu_o(x)$, $\nu_j(x)$, $\nu_{jk}(x)$ and $W_{jk}(x)$ are moments of $\mu(x, z, \tau) \equiv (x_t \rightarrow (x + z)_{t+\tau})$ defined by

$\nu_0(x) = \int_{z \in U} \mu_\tau(x, z, 0)$
$\nu_j(x) = \int_{z \in U} \mu_\tau(x, z, 0)z_j$
$\nu_{jk}(x) = \int_{z \in U} \mu_\tau(x, z, 0)z_j z_k$

with $W_{jk} = M_{jl}M_{lk}^T\omega_l$ where M is the matrix which diagonalizes ν_{jk} such that $M_{lj}^T\nu_{jk}M_{km} = \delta_{lm}/\omega_l$. With velocity v_j given by the limit of z_j/τ, the above propagator is equivalent to the Lagrangian

$$\mathcal{L}(x, v) = \frac{i}{2}(v_j - \nu_j)W_{jk}(v_k - \nu_k) - i\nu_0$$

where we recognize $\nu_j(x)$ and $\nu_o(x)$ as the electromagnetic fields and where $W_{jk}(x)$ contains the particle mass and space–time metric. Notice that we have not assumed Lorentz or gauge invariance to get this result. The claim is that this is the only Lagrangian consistent with the state space R^d.

Since our complex probability theory is both "realistic" in the sense of assuming that a particle does come through one slit or the other in the two slit experiment (II.b) and local in the sense of accepting locality assumptions as assumptions of "complex statistical independence," you might think that we would run afoul of Bell's theorem or other more recent limitations on local realistic theories. As I've already mentioned, Bell's result does not follow in complex probability theory. This means that although Bell's result has almost universally been interpreted as ruling out local realistic theories, from our more general point of view, it forces a choice between local realism and standard probability theory. In fact, Bell's result can be interpreted as another little hint that there is something wrong with probability theory. Besides Bell, there are a large number of more recent limitations on local realistic theories, each of which provides a test of the complex probability formulation. Details are available only for three representative results of this type (including Bell) where complex probability theory appears not to be excluded[3].

You might expect that if quantum mechanical phenomena can be described by complex probability theory, the Bayesian view might help in understanding some of the long standing semi–paradoxical measurement and observer questions in quantum mechanics. Here, it's helpful to first think about a purely classical experiment where a single coin is flipped and then uncovered, revealing that it landed "heads." From the Bayesian point of view, of course, the situation before the observation could be described by the distribution $(1/2, 1/2)$ and after observing heads our description would be adjusted to $(1, 0)$. The problem is, what would you say to a student who then asks:

— *Yes, but what causes* $(1/2, 1/2)$ *to evolve into* $(1, 0)$*? How does it happen?*

Here we recognize a victim of a severe form of "Mind Projection Fallacy"[6] where the person asking this question has confused what they know about the system with the system itself. With the Bayesian view of complex probabilities, it is clear that this same mistake is possible in quantum mechanics as well, where one would now be mistaking a wavefunction for the state of the system. This very view, however, is the standard picture of quantum mechanics and so it is hardly surprising that similar mysteries arise. This view is also implicit in questions such as

— Can the wavefunction be measured?
— What is the source of the non–local effects in EPR?
— Can macroscopic superpositions be created?
— Is the Universe in a pure state?

Although these are active research questions, it seems inescapable to me that if quantum phenomena are correctly described by a Bayesian probability theory, then all of these questions have trivial answers and they all ultimately fall into the same category as the student's question about coin flipping experiments.

For this audience, I hardly need to point out that the ideas that we have discussed here may not only clarify the meaning of quantum mechanics, but may also lead to ways of improving quantum mechanical calculations using prior knowledge in the same sense that prior knowledge is used to improve probability calculations in Bayesian Inference.

References

1. S.Youssef, *Mod.Phys.Lett.* **A6**, 225 (1991).
2. S.Youssef, *Mod.Phys.Lett.* **A9**, 2571 (1994).
3. S.Youssef, *Phys.Lett.* **A204**, 181(1995).
4. R.T.Cox, *Am.J.Phys.* **14**, 1(1946).
5. R.P.Feynman and H.R.Hibbs, *Quantum Mechanics and Path Integrals* (McGraw–Hill, 1965).
6. E.T.Jaynes, in *Maximum Entropy and Bayesian Methods*, ed. J.Skilling(Kluwer, 1989).

BAYESIAN PARAMETER ESTIMATION OF NUCLEAR FUSION CONFINEMENT TIME SCALING LAWS

V. DOSE AND W. VON DER LINDEN

Max-Planck-Institut für Plasmaphysik, EURATOM Association
D-85740 Garching b. München, Germany [¶]

AND

A. GARRETT

63 High Street, Grantchester, Cambridge CB3 9NF, England

Abstract. A data set on the confinement time of the toroidal magnetic confinement stellarator type fusion experiment $W7AS$ has been analyzed in terms of three scaling laws derived from the more advanced tokamak line of fusion research. The analysis suggests stellarator performance to follow closely the well known Lackner-Gottardi scaling.

Key words: parameter estimation, model comparison

1. Introduction

Nuclear fusion is the process whereby a heavier nucleus is formed upon fusing two lighter ones. The reaction is accompanied by a considerable release of energy. An example is the presently most important fusion reaction

$$^{2}H_{1} + {}^{3}H_{1} \longrightarrow {}^{4}He_{2} + {}^{1}n_{0} + 17.6 \text{MeV} \qquad . \qquad (1)$$

Its net effect is to burn hydrogen to helium ash and neutrons. In view of the ample abundance of $^{2}H_{1}$ (and easy production of $^{3}H_{1}$) there rests some hope on controlled nuclear fusion to satisfy energy needs of coming generations.

The fusion process does not occur spontaneously. The hydrogen nuclei must be forced against their Coulomb repulsion to approach each other to a very short critical distance before the reaction occurs. The experimental way to achieve this is the production of violent collisions. Such collisions occur when a mixture of deuterium and tritium is raised to extremely high temperatures of the order of 10^{8} K. At such temperatures the particles do no longer exist as neutral atoms but rather form a plasma of nuclei and electrons. Keeping in mind that not every

---

[¶]e-mail: wvl@ibmop5.ipp-garching.mpg.de

K. M. Hanson and R. N. Silver (eds.), Maximum Entropy and Bayesian Methods, 245–250.

collision but only the most violent ones lead to nuclear fusion, the necessity arises to confine the plasma spatially in order to give the fusion process multiply repeated trials. Since the particles constituting the plasma are charged, magnetic fields may be used to form cages which keep the particles together for a 'sufficiently' long time at the necessary high temperature. Confinement is never perfect and therefore plasma cooling occurs due to energy diffusion and convection. This energy loss must be compensated by external supply of heating. The energy content per unit volume

$$W = \frac{3}{2} k_B T n \qquad (2)$$

is proportional to n, the joint density of electrons and nuclei of the plasma, and the plasma temperature T via the Boltzmann constant k_B. For the externally supplied power per unit volume p we then have at equilibrium

$$p = \frac{W}{\tau} = \frac{3}{2} k_B T n / \tau \qquad . \qquad (3)$$

It follows from Eq.3 that the energy confinement time τ is a quantity of considerable interest since it determines at a given supply of external power p the plasma temperature and hence the rate of violent collisions. The energy confinement time τ has in recent years been assessed empirically on the basis of a power law dependence on the plasma and confinement relevant variables [1–3]

$$\tau = e^{\mu_1} n^{\mu_2} B^{\mu_3} p^{\mu_4} a^{\mu_5} \iota^{\mu_6} \qquad , \qquad (4)$$

where B denotes the confining magnetic field, a the minor radius in a toroidal plasma device and ι a dimensionless quantity characterizing the structure of the magnetic flux surfaces in a toroidal configuration. The general form of Eq.4 can also be justified from a more fundamental point of view exploiting dimensional analysis [4]. Magnetic confinement fusion research has so far developed along two conceptually different lines. Several explicit coefficient vectors have been derived from tokamak type of magnetic confinement experiments. The stellarator line of research is not yet as far developed and the question arises to what extent confinement data from stellarator experiments can be reconciled with existing tokamak experience. This is an implicit formulation of the Bayesian problem of parameter estimation and model comparison[5] which we shall now outline explicitly.

The probability of a model M as specified by a particular set of coefficients $\vec{m}$ given a set of N_D measurements D is, following Bayes' theorem

$$P(M|D) = \frac{P(M)}{P(D)} P(D|M) \qquad . \qquad (5)$$

$P(M)$ is the prior probability for a particular model M, $P(D|M)$ the likelihood function and $P(D)$ the normalizing factor. $P(D)$ needs not to be calculated explicitly if several models are to be compared since we can resort to the odds ratios

$$\frac{P(M_i|D)}{P(M_k|D)} = \frac{P(M_i)}{P(M_k)} \frac{P(D|M_i)}{P(D|M_k)} =: O_{ik} \qquad (6)$$

which factorize into the prior odds $P(M_i)/P(M_k)$ and the likelihood ratio $P(D|M_i)/P(D|M_k)$. Using the normalization, Eq.6 can be solved for the individual probabilities in terms of the odds ratios as

$$P(M_i|D) = O_{ik} / \sum_{j=1}^{N_{\text{mod}}} O_{jk} \quad . \tag{7}$$

We therefore need to calculate

$$P(D|M) = \int_{-\infty}^{\infty} P(\vec{\mu}|M) P(D|\vec{\mu}, M) \, d^N \mu \quad , \tag{8}$$

where N is the number of parameters. The seemingly simple piece of the integrand is the likelihood function

$$P(D|\vec{\mu}, \omega, M) = (\frac{\omega}{2\pi})^{N_D/2} e^{-\frac{\omega}{2} L^2} \tag{9}$$

with

$$L^2 := \sum_{i=1}^{N_D} (D_i - f(\vec{\mu}, \vec{x}_i))^2 \quad . \tag{10}$$

L^2 measures the misfit between experimental data and the values predicted by the scaling relation Eq.4. The latter depends on the suggested parameters $\vec{\mu}$ and the 'input variables' $\vec{x} = (n, B, p, a, \iota)^T$. Yet we have chosen to introduce a new hyper-parameter ω which characterizes the precision of the data. Contrary to expectation, this quantity is in the overwhelming number of cases not accurately known though an experienced experimenter may have a rather good estimate of the error level, which we will assume to be the case subsequently. Employing the marginalization rule $P(D|\vec{\mu}, M)$ may be expressed as

$$P(D|\vec{\mu}, M) = \int_0^{\infty} P(\omega) P(D|\vec{\mu}, \omega, M) \, d\omega \quad . \tag{11}$$

For $P(\omega)$ Jeffreys' prior could be used. However, we need not pretend to be that ignorant. In fact, a sensible estimate ω_0 of the errors is frequently available. The maximum entropy estimate of the ω-prior is then the exponential

$$P(\omega) = \frac{1}{\omega_0} e^{-\frac{\omega}{\omega_0}} \quad . \tag{12}$$

The integration in Eq.11 can now be carried out easily and yields

$$P(D|\vec{\mu}, M) = (\frac{\omega_0}{2\pi})^{\frac{N_D}{2}} \Gamma(\frac{N_D + 2}{2}) \left(1 + \frac{\omega_0}{2} L^2\right)^{-\frac{N_D + 2}{2}} \quad , \tag{13}$$

a Student's t-distribution with $N_D + 1$ degrees of freedom. In order to carry out the integration in Eq.8 we need to specify the prior for $\vec{\mu}$. We choose to express it quite generally as

$$P(\vec{\mu}|\alpha, M) = (\frac{\alpha}{2\pi})^{\frac{N}{2}} e^{-\frac{\alpha}{2} Q^2} \tag{14}$$

with

$$Q^2 = (A\vec{\mu} - \vec{b})^T (A\vec{\mu} - \vec{b}) \tag{15}$$

The matrix A will in our particular application simplify to the identity matrix and, consequently, the vector $\vec{b}$ is identical to the 'default' vector $\vec{m}$ specified by the model under consideration. In the case of complete ignorance Jeffreys' prior would be the appropriate prior for the hyper-parameter α. If, however, a sensible guess α_0 is given for α, one is prompted to use an exponential prior, like in Eq.12, resulting in a Student's t-distribution with $N + 1$ degrees of freedom

$$P(\vec{\mu}|M) = (\frac{\alpha_0}{2\pi})^{\frac{N}{2}} \Gamma(\frac{N+2}{2}) \left(1 + \frac{\alpha_0}{2} Q^2\right)^{-\frac{N+2}{2}} , \tag{16}$$

with $N = 6$ being the number of exponents in the scaling law. The integrand in Eq.8 is thereby fully specified by Eq.13,16. In order to carry out the integration we determine first the maximum of the integrand, calculate the curvature and employ the Gaussian approximation.

As a check we have considered one dimensional integrals of the form

$$\int_{-\infty}^{\infty} (\frac{1}{1+y^2})^r dy \tag{17}$$

which can be carried out exactly. Comparison with the Gaussian approximation shows that the error decreases rapidly as a function of r and is only 4% for $r = 10$ and 0.8% for $r = 50$. For the case at hand, $r > 100$, the Gaussian approximation is very well justified.

It is quite interesting to write down the condition for the maximum of the integrand which becomes

$$\nabla_\mu L^2 + \frac{(L^2 + 2/\omega_0)/(N_D + 2)}{(Q^2 + 2/\alpha_0)/(k + 2)} \nabla_\mu Q^2 = 0 . \tag{18}$$

This equation has the form of a Tikhonov regularization problem [6,7]. Retaining only the $\nabla_\mu L^2$ term results in the least squares solution for the parameters $\vec{\mu}$, while the second term in Eq.18 yields the default model $\vec{\mu} = \vec{m}$. The prefactor of $\nabla_\mu Q^2$ constitutes the regularization parameter which has emerged explicitly from the above analysis. Its interpretation is quite simple. It is essentially the ratio of the mean L^2 per degree of freedom in the data and the mean Q^2 per degree of freedom in parameter space. This view of Eq.18 yields a sensible guess of the scale of the hyper-parameter α_0. Viewed as a regularization problem α_0 should certainly lie between the largest and the smallest eigenvalue of the Hessian $\nabla\nabla^T L^2$. A plot of the size of the eigenvalues in increasing order, a so called scree diagram, will define an appropriate value for α_0 at that level where the sequence shows a sharp somewhat irregular falloff.

The analysis outlined above was applied to a data set comprising 199 measurements of τ on specified values of n, B, p, a, ι. Contrary to the usual situation in Bayesian analysis the problem is therefore heavily over-determined. The structure of this data set has been analyzed and described previously [8] and it was

TABLE 1. Result of Bayesian model comparison.

	μ_1	μ_2	μ_3	μ_4	μ_5	μ_6	Odds
M_1	-1.75	0.60	0.80	-0.60	2.0	0.40	0
M_2	-2.41	0.60	0.80	-0.60	2.0	0.40	-4.8
M_3	-1.67	0.00	1.05	-0.50	2.0	0.85	-8.5
$W7 - AS$	-1.76	0.59	0.73	-0.64	1.98	0.53	

concluded that the set of 'input variables' n, B, p, a, ι is sufficiently robust in order to perform a meaningful analysis. Three different models were compared in the Bayesian analysis. Model M_1 associated with the names Lackner-Gottardi [3] was derived semi-empirically from a theoretically guided analysis of tokamak data. Model M_2 is a theoretical one based on ion plateau neo-classical transport theory[9] and model M_3, the so called ITER 89-P scaling was derived from a wealth of experimental data from several tokamak experiments in the ITER conceptual design activity 1989[10]. M_1 and M_2 differ only in the numerical constant $\exp(-\mu_1)$ entering the scaling law, while M_3 exhibits strongly different dependencies on the other variables.

Whether one of the three models applies better to the stellarator situation than the others is an open question. In particular the present authors have no preference for a particular model and therefore assign equal prior probabilities $P(M_\nu) = 1/3$ to them. Table 1 presents in the first three lines the coefficient vectors of the three models; the last line shows the coefficient vector $\vec{\mu}$ resulting from the actual data set using M_1 which is strongly superior to M_2 and M_3. The last column in table I shows the odds ratios in Neper relative to model M_1. The Bayesian community has sometimes chosen to express odds ratios in decibels $= 10 \log_{10}(x1/x2)$. This measure is derived from the field of communication techniques. Another measure for amplification or attenuation often employed in communication techniques is Neper $= \ln(x1/x2)$ which recommends itself even more strongly as a measure of odds ratios in Bayesian theory.

In conclusion the analysis of a 'real world' data set for the energy confinement time in the W7-AS stellarator on the background of three different models has singled out the Lackner-Gottardi scaling significantly as describing stellarator confinement best. The result is of significance for the next generation stellarator WVII-X, roughly twice as large in linear dimension and presently entering the construction phase, since confinement times for this device predicted on the basis of Lackner-Gottardi scaling are longer than those predicted from the other two models.

References

1. J. Hugill and J. Sheffield, "Empirical tokamak scaling," *Nucl. Fusion*, **18**, p. 15, 1978.
2. S. Kaye and R. Goldston, "Global energy confinement scaling for neutral-beam-heated tokamaks," *Nucl. Fusion*, **25**, p. 65, 1985.
3. K. Lackner and N. Gottardi, "Tokamak confinement in relation to plateau scaling," *Nucl. Fusion*, 30, p. 767, 1990.
4. J. Connor and J. Taylor, "Scaling laws for plasma confinement," *Nucl. Fusion*, **17**, p. 1047, 1977.
5. M. Tribus, *Rational Descriptions, Decision, and Design*, Pergamon of Canada, Toronto, 1969.
6. A. Tikhonov and V. Arsenin, *Solution of Ill Posed Problems*, Wiley, New York, 1977.
7. D. Keren, "A bayesian framework for regularization," *these proceedings*, 1995.
8. V. Dose, W. von der Linden and A. Garrett, "A bayesian approach to the global confinement time scaling in W7-AS," *Nucl. Fusion*, **submitted**, 1995.
9. Wendelstein Project Group. "Wendelstein WVII-X application for preferential support", IPP EURATOM Association, 1990.
10. P.N. Yushmanov, et al., "Scaling for tokamak energy confinement," *Nucl. Fusion*, **30**, p. 1999, 1990.

HIERARCHICAL SEGMENTATION OF RANGE AND COLOR IMAGE BASED ON BAYESIAN DECISION THEORY

P. BOULANGER
National Research Council of Canada
Institute for Information Technology
Ottawa, Canada, K1A-0R6[†]

Abstract. This paper describes recent work on hierarchical segmentation of registered color and range images. The algorithm starts with an initial partition of small first order regions using a robust fitting method constrained by the detection of depth and orientation discontinuities in the range signal and color edges in the color signal. The algorithm then optimally group these regions into larger and larger regions until an approximation limit is reached. The algorithm uses Bayesian decision theory to determine the local optimal grouping and the complexity of the models used to represent the range and color signals. Experimental results are presented.

Key words: Color,Range, Sensor Fusion, Segmentation

1. Introduction

The ability to integrate and represent in a coherent maner muliple source of sensor data is at the base of understanding images in terms of intrinsic physical properties. This paper describes a method to segment range and color produced by a newly developed range sensor. This sensor can measure in perfect synchronisation a range signal corresponding to the distance between the sensor and the surface of an object and a corresponding color signal corresponding to the reflectivilty of the surface to three distinct laser frequency.

Many references to the problem of sensor fusion can be found in the litterature. A survey and analysis of multi-sensor fusion and integration methods can be found in Luo and Kay [9], Bajcsy and Allen [1], and Mitiche and Agarwal [10].

Koezuka et al. [8] used range to determine shape and the intensity to recognize marking and object features. Gil et al. [6] also use registered intensity and range information produce by a range sensor to obtain a reliable and complete edge map of the scene. Edge map in both signals are used to fill the gaps in each edge map using edges from the other edge map. Duda et al. [5] segment a registered

---

[†]Email: Boulanger@iit.nrc.ca

K. M. Hanson and R. N. Silver (eds.), Maximum Entropy and Bayesian Methods, 251–259.
© 1996 *Kluwer Academic Publishers.*

range and intensity map to extract regions that correspond to planar surfaces in the scene. Hackett and Shanh [7] segment range and intensity images by using a split and merge method. The algorithm consist of two steps. First, the initial seed regions are determined by using the most dominant sensor at a given time. Second, the initial segmentation is refined by using region merging based on if the strength of range and intensity boundaries are low.

In the case of color and range sensor fusion very few paper can be found. Baribeau et al. [2] discuss the problem of estimating the bidirectional reflectance-distribution function (BRDF) for each pixel from the fusion of the range and color information and a sensor model. Shirai [12] integrate sparce range data obtained by stereo vision with color data, using color stereo pair. The color is used to assist in model-based classification and object recognition to build a rich description of the scene. Regions are first classified using color and further split based on edge information and range information obtained from stereo. The scene is eventually represented in a series of 3-D planner patches and its relationship between them.

In this paper, one will analyze a new segmentation algorithm based on a hierarchical grouping of an initial partition based on a Bayesian criteria. The algorithm starts with an initial partition of the range and color images constrained by the detection of depth and orientation discontinuities for the range images and of color edges for the color information.

From this initial partition the algorithm start grouping these regions into larger and larger one until the approximation error in one of the regions is greather than a predetermined threshold. The algorithm then try to transform these primitives into more complex ones by using higher order models. The key idea behind the algorithm is that one should start with the simplest hypothesis about the model of the data, and gradually increase the complexity of the hypothesized form as statistical evidence grows. This paper present a consistent view of the grouping criterion and the generalization process based on Bayesian decision theory. The end result of this segmentation process is a compact representation of a scene composed of continuous surface patches with a constant color model.

2. Problem Definition

In this approach to segmentation, the relevant structure of a range and color image is viewed as a piecewise smooth parametric polynomial contaminated by noise. A piecewise smooth parametric surface $\vec{\eta}_r(u, v)$ and its corresponding color signal $\vec{\eta}_c(u, v)$ can be partitioned into N smooth surface model $\vec{f}_{rl}(u, v; \mathbf{A}_{rl})$ and color model $\vec{f}_{cl}(u, v; \mathbf{A}_{cl})$ over a connected support region Ω_l:

$$\vec{\eta}_r(u, v) = \sum_{l=1}^{N} \vec{f}_{rl}(u, v; \mathbf{A}_{rl})\xi(u, v, \Omega_l) \tag{1}$$

$$\vec{\eta}_c(u, v) = \sum_{l=1}^{N} \vec{f}_{cl}(u, v; \mathbf{A}_{cl})\xi(u, v, \Omega_l) \tag{2}$$

where $\xi(u, v, \Omega_l)$ is the characteristic function of the region Ω_l, which is equal to unity if $(u, v) \in \Omega_l$ and zero otherwise. The arrays $\mathbf{A}_{rl}$ and $\mathbf{A}_{cl}$ are the model parameters for each signals. The function $\vec{\eta}_r(u, v) = (x, y, z)^T$ is a three dimentional signal corresponding to the x, y, z component of the range signal and the function $\vec{\eta}_c(u, v) = (r, g, b)^T$ to the red r, green g, and blue b component of the color signal.

The segmentation problem can be stated as following: given a discrete range $\vec{r}(u_i, v_i) = (x_i, y_i, z_i)^T$ and color image $\vec{C}(u_i, v_i) = (r_i, g_i, b_i)^T$ and an approximation thresholds ε_t find the N image regions Ω_l approximated by N statistically reliable functions $\vec{f}_{rl}(u_i, v_i; \mathbf{A}_{rl})$ and $\vec{f}_{cl}(u_i, v_i; \mathbf{A}_{cl})$ subject to:

$$\chi^2 = \frac{1}{n_l} \sum_{(u_i, v_i) \in \Omega_l} (\vec{f}_{rl}(u_i, v_i; \mathbf{A}_{rl}) - \vec{r}(u_i, v_i))^T \Sigma_r^{-1} (\vec{f}_{rl}(u_i, v_i; \mathbf{A}_{rl}) - \vec{r}(u_i, v_i)) \qquad (3)$$

$$+ (\vec{f}_{cl}(u_i, v_i; \mathbf{A}_{cl}) - \vec{C}(u_i, v_i))^T \Sigma_c^{-1} (\vec{f}_{cl}(u_i, v_i; \mathbf{A}_{cl}) - \vec{C}(u_i, v_i)) < \varepsilon_t$$

$$\forall\ l = 1, \cdots, N$$

The parameter n_l is equal to the number of pixels in the region Ω_l. The matrices Σ_r and Σ_c are the covariance matrices of the noise associated to the range and color signal.

The basic steps of the algorithm are the following:

1. Normalize the color by computing the bidirecional reflectance- distribution function (BRDF).

2. Do an initial partitioning of the data set based on a first order parametric model using a robust fitting technique constrained by depth and orientation discontinuities for the range signal and color edges for the corrected color signal.

3. Group adjacent first order regions with other first order regions or points to produce a larger first order region. Validate the grouping corresponding to the one which is the most similar based on a Bayesian criterion.

4. Loop until the similarity criterion is smaller than a certain threshold.

5. Generalize the first order regions to second order one if the decision is supported by statistical significance test.

6. Group adjacent first or second order regions to other points, first, or second order regions to produce a larger region corresponding to the highest order of the two. Validate the grouping corresponding to the one which is the most similar.

7. Generalize second order regions into higher order parametric polynomials using geometrical heuristics if it is supported by Bayesian decision.

8. Proceed with more grouping until no more regions are generalized.

3. Color Reflectance Modelling

Ideally the color signal should be independent of the senor parameters and should depend only on basic physical properties such as the material pigmentation. Baribeau et al. [2] discuss such possibilities by modeling the physical image formation of the color range sensor using a Lambertian model. Using a similar physical model, the total power recieved at the detector at time t is equal to:

$$P_r(u, v; \lambda) = \left(\frac{1}{4\pi} S_r T_r \exp(-\alpha r) T_x P_x(u, v; \lambda) \right) \left(\frac{\rho_\lambda(u, v)\vec{\mu} \cdot \vec{n}(u, v)}{r^2} \right) \quad (4)$$

where $P_x(u, v; \lambda)$ is equal to the intentanious total power of the transmitted laser in the direction $\vec{\mu}$ and a wavelength λ; T_x is the transmission coefficent of the incident beam; T_r is the transmission coefficient of the reflected signal produced by a Lambertian surface of area S_r with a normal equal to $\vec{n}(u, v)$ and a albedo equal to $\rho_\lambda(u, v)$ at a distance r form the measuring sensor; α is the coefficient of attenuation in the atmosphere and is assumed constant for all wavelength used. The relative power between the incicent beam and the measured one is equal to:

$$\frac{P_r(u, v, t; \lambda)}{P_x(u, v; \lambda)} = K \left(\frac{\rho_\lambda(u, v)\vec{\mu} \cdot \vec{n}(u, v)}{r^2} \right) \quad (5)$$

where $K = \frac{1}{4\pi} S_r T_r \exp(-\alpha r) T_x$ and is assumed to be a constant for all wavelength. An estimate of the incident laser power $P_w(u, v, \lambda)$ can be performed by scanning a white Lambertian surface for each similar u, v. The power is equal to:

$$P_x(u, v; \lambda) = \frac{r_w^2 P_w(u, v, \lambda)}{\vec{n}_w(u, v) \cdot \vec{\mu}}. \quad (6)$$

Using this estimated laser power one can compute the bidirectional reflectance distribution function (BRDF) with the following equation:

$$\frac{P_r(u, v, t; \lambda)}{P_x(u, v; \lambda)} \left(\frac{r^2}{\vec{\mu} \cdot \vec{n}(u, v)} \right) = \rho_\lambda(u, v). \quad (7)$$

The white standard used as a reference is a flat bar coated with barium sulphite paint. This material is known to have very good Lambertian behavior. The computed BRDF for the three wavelength is an intrinsic property corresponding to the material pigmentation.

4. Multidimentional Signal Representation

The type of model used to represent the shape of the range data and the spatial evolution of the intrinsic color is highly constrained by the feasibility of the corresponding segmentation algorithm. A parametric Bézier polynomial is used to represent both signals and is defined as:

$$\vec{f}_{sl}(u, v; \mathbf{A}_{sl}) = \sum_{i=0}^{k} \sum_{j=0}^{k} \vec{a}_{ij} B_i(u) B_j(v) \quad (8)$$

$B_m(t)$ is a Bernstein polynomial defined as:

$$B_m(t) = \frac{k!}{(k - m)!m!} t^m (1 - t)^{k-m}. \quad (9)$$

The subscript s can be equal to r or c depending on if one wants to represent the range or color information. This equation can be represented in matrix form by:

$$\vec{f}_{sl}(u, v; \mathbf{A}_{sl}) = \mathbf{A}_{sl}\mathbf{M}_{l|uv} \tag{10}$$

where the array $\mathbf{A}_{sl} = [\vec{a}_{00}, \vec{a}_{10}, \vec{a}_{01}, \cdots, \vec{a}_{kk}]$ is the coefficients array of size $3 \times (k+1)^2$ and

$$\mathbf{M}_{l|uv} = [B_0(u)B_0(v), B_1(u)B_0(v), B_0(u)B_1(v), \cdots, B_k(u)B_k(v)]^T$$

is the basis function matrix of size $(k+1)^2 \times 1$.

If one assume that the range $\vec{r}(u_i, v_i)$ and color $\vec{C}(u_i, v_i)$ image data is corrupted by Gaussian noise of means $\vec{\mu}_r = \vec{\mu}_c = \vec{0}$ and for which the covariance matrices are equal to $\mathbf{\Sigma}_r$ and $\mathbf{\Sigma}_c$, then the optimal model coefficients are the ones which minimize the log-likelihood function of the observations corresponding to the minimum of the standard least squared metric L^2 given by equation (3).

The minimum occurs when $\nabla_{\mathbf{A}_{rl}} \chi^2 = 0$ and $\nabla_{\mathbf{A}_{cl}} \chi^2 = 0$ which correspond in matrix form to:

$$\mathbf{T}_{sl} = \mathbf{A}_{sl}\mathbf{L}_l \tag{11}$$

where $\mathbf{L}_l = [\mathbf{M}_{l|u_1v_1}, \cdots, \mathbf{M}_{l|u_{n_l}v_{n_l}}]$ is a matrix of size $(k+1)^2 \times n_l$ and $\mathbf{T}_{sl} = [\vec{s}(u_1, v_1), \cdots, \vec{s}(u_{n_l}, v_{n_l})]$ a matrix of size $3 \times n_l$ corresponding to the sensor measurements. The solution correspond to the normal equation equal to:

$$\mathbf{A}_{sl} = \mathbf{T}_{sl}\mathbf{L}_l^T(\mathbf{L}_l\mathbf{L}_l^T)^{-1} = \mathbf{V}_{sl}\mathbf{R}_l^{-1} \tag{12}$$

where $\mathbf{R}_l$ is an Hermitian matrix of size $(k+1)^2 \times (k+1)^2$ corresponding to the covariance matrix of the basis functions and $\mathbf{V}_{sl}$ is a matrix corresponding to the correlation between the basis functions and the measurements. Using this notation the covariance matrix of the approximation error is equal to:

$$\hat{\mathbf{\Sigma}}_{sl} = \mathbf{T}_{sl}\mathbf{T}_{sl}^T - 2\mathbf{A}_{sl}\mathbf{V}_{sl}^T + \mathbf{A}_{sl}\mathbf{R}_l\mathbf{A}_{sl}^T \tag{13}$$

The average error on the model parameters $\delta\mathbf{A}_l$ is proportional to the diagonal element of the inverse of the matrix $\mathbf{R}_l$, i.e.,

$$\delta\mathbf{A}_{sl} = \frac{\mathbf{D}_{sl}^T\mathbf{U}_l}{n_l - (k-1)^2} = [\delta\mathbf{A}_{sl|i}, \delta\mathbf{A}_{sl|j}, \delta\mathbf{A}_{sl|k}]^T \tag{14}$$

where $\mathbf{D}_l = [\text{Diag } \hat{\mathbf{\Sigma}}_{sl}] = [\hat{\sigma}_{sl|i}^2, \hat{\sigma}_{sl|j}^2, \hat{\sigma}_{sl|k}^2]$ is a 1×3 matrix corresponding to the variance of the fitting error in each orthogonal directions and is equal to the diagonal elements of the matrix $\hat{\mathbf{\Sigma}}_{sl}$. The matrix $\mathbf{U}_l = [\text{Diag } \mathbf{R}_l^{-1}]$ is a $1 \times (k+1)^2$ matrix where each element is the diagonal element of the covariance matrix $\mathbf{R}_l^{-1}$

5. Initial Partition Method

Like many region growing techniques, one needs to make an initial guess of the primitives and then iteratively refine the solution. Besl [3] used the topographic map to determine seed points where his algorithm grows regions of increasing size and complexity. There is a relationship between the quality of the initial guess and the number of iterations required to converge to the final region size. Because of the importance of the initial partition, the algorithm use a robust fitting technique constrained by previously detected depth and orientation discontinuities in the range image and edges in the color image. The algorithm uses a Least Median Square (LMS) fitting method developed by Rousseeuw and Leroy [11] which allows up to 50% outliers. The algorithm to find the initial partition is the following:

– Set the window size $L = L_{\max}$ to the maximum window size.

– Find a square neighbourhood of size $L \times L$ where there is no depth nor orientation discontinuities nor color edges present.

– Do least median square fitting and detect the outliers (not sensitive to 50% of outliers).

– Eliminate the outlier from the window by releasing their availability to other regions.

– Compute the least square model without the outliers for the range and color information.

– Proceed for the whole image with the same window size.

– Do the same operation with a reduced window size $L = L - 2$ until the minimum window size $L_{\min}$ has been reached.

This new initial partition technique is not sensitive to impulse noise (up to 50% of outliers) and is capable of producing excellent seed regions even for a large neighbourhood.

6. Compatibility Function

A similarity function is a predicate that determine if two regions can be merged into one. Let Ω_i be a region composed of n_i points defined by the maximum likelihood model parameters $\mathbf{A}_{ri} = (\vec{b}_{00}, \vec{b}_{10}, \cdots, \vec{b}_{kk})^T$ for the range signal and $\mathbf{A}_{ci} = (\vec{c}_{00}, \vec{c}_{10}, \cdots, \vec{c}_{kk})^T$ for the color signal. Each region is also caracterized by the covariance matrices $\hat{\boldsymbol{\Sigma}}_{ri}$ and $\hat{\boldsymbol{\Sigma}}_{ci}$. Let $\delta\mathbf{A}_{ri} = (\delta\vec{b}_{00}, \delta\vec{b}_{10}, \cdots, \delta\vec{b}_{kk})^T$ and $\delta\mathbf{A}_{ci} = (\delta\vec{c}_{00}, \delta\vec{c}_{10}, \cdots, \delta\vec{c}_{kk})^T$ be the margin of error on the model parameters estimated by equation (14). Let $\{\Omega_m\}$ be the set of N_t regions adjacent to the region Ω_i and defined by the models $\mathbf{A}_{rm}$ and $\mathbf{A}_{cm}$ with a margin of error equal to $\delta\mathbf{A}_{rm}$ and $\delta\mathbf{A}_{cm}$. The best grouping of region Ω_i with one of its neighbours correspond to the one for which:

$$P(\Omega_b \wedge \Omega_i | \Omega_i) = \max_b \frac{\displaystyle\prod_{u,v \in \Omega_i} p_t(u,v|\mathbf{A}_{rb}; \mathbf{A}_{cb}) p(\delta\mathbf{A}_{rb}) p(\delta\mathbf{A}_{cb})}{\displaystyle\sum_{j=1}^{N_t} \prod_{u,v \in \Omega_i} p_t(u,v|\mathbf{A}_{rj}; \mathbf{A}_{cj}) p(\delta\mathbf{A}_{rj}) p(\delta\mathbf{A}_{cj})} \tag{15}$$

where $p_t(u,v|\mathbf{A}_{rb}; \mathbf{A}_{cb})$ is equal to the probability that a point in region Ω_i with coordinate u and v would be predicted by one of the models $\mathbf{A}_{rb}$ and $\mathbf{A}_{rb}$ adjacent to Ω_i. The likelyhood of grouping the region Ω_i with Ω_b is given by:

$$P(\Omega_i|\mathbf{A}_{rb};\mathbf{A}_{cb}) = \alpha \, \exp\left(-\frac{\hat{\sigma}_{ib}^2}{2}\right) = \prod_{u,v\in\Omega_i} p_t(u,v|\mathbf{A}_{rb};\mathbf{A}_{cb}) \qquad (16)$$

where

$$\hat{\sigma}_{ib}^2 = \mathrm{Tr}[(\mathbf{T}_{ri}\mathbf{T}_{ri}^T - 2\mathbf{A}_{rb}\mathbf{V}_{ri}^T + \mathbf{A}_{rb}\mathbf{R}_{ri}\mathbf{A}_{rb}^T)\Sigma_r^{-1} + \qquad (17)$$
$$(\mathbf{T}_{ci}\mathbf{T}_{ci}^T - 2\mathbf{A}_{cb}\mathbf{V}_{ci}^T + \mathbf{A}_{cb}\mathbf{R}_{ci}\mathbf{A}_{cb}^T)\Sigma_c^{-1}]$$

is the sum of the square difference between the functions representing region Ω_b extrapolated to predict region Ω_i. The functions $p(\delta\mathbf{A}_{rb})$ and $p(\delta\mathbf{A}_{cb})$ is the *a priori* probability of the region Ω_b and can be evaluated by the following equation:

$$p(\delta\mathbf{A}_{rb}) = \omega_1 \exp -\frac{1}{2}\left[\delta\mathbf{A}_{rb|x}\,\Sigma_{Ar}^{-1}\,\delta\mathbf{A}_{rb|x}^T + \delta\mathbf{A}_{rb|y}\,\Sigma_{Ar}^{-1}\,\delta\mathbf{A}_{rb|y}^T + \delta\mathbf{A}_{rb|z}\,\Sigma_{Ar}^{-1}\,\delta\mathbf{A}_{rb|z}^T\right]$$
$$(18)$$

$$p(\delta\mathbf{A}_{cb}) = \omega_2 \exp -\frac{1}{2}\left[\delta\mathbf{A}_{cb|r}\,\Sigma_{Ac}^{-1}\,\delta\mathbf{A}_{cb|r}^T + \delta\mathbf{A}_{cb|g}\,\Sigma_{Ac}^{-1}\,\delta\mathbf{A}_{cb|g}^T + \delta\mathbf{A}_{cb|b}\,\Sigma_{Ac}^{-1}\,\delta\mathbf{A}_{cb|b}^T\right]$$
$$(19)$$

where Σ_{Ar} and Σ_{Ac} are equal to the true covariance matrix of the parameters $\mathbf{A}_{rb}$ and $\mathbf{A}_{rb}$ and represents the strength of the belief that the coefficients of the matrices $\mathbf{A}_{rb}$ and $\mathbf{A}_{cb}$ are the true value of the coefficients. In practice, the covariance matrix cannot be evaluated, but in our implementation, we artificially set the diagonal elements of the matrix Σ_{Ar} and Σ_{Ar} are equal to β^2 and the off-diagonal to zero.

From equations (17), (18), and (19) the *a posteriori* probability of a grouping correspond to the one which maximize the numerator. If one compute the log of the numerator of equation (15), one obtain a grouping coefficient equal to:

$$c_{ib} = \hat{\sigma}_{ib}^2 + \frac{(\hat{\sigma}_{rb|x}^2 + \hat{\sigma}_{rb|y}^2 + \hat{\sigma}_{rb|z}^2)\mathrm{Tr}\,\mathbf{R}_{rb}^{-1}}{\beta^2(n_b - (k-1)^2)} + \frac{(\hat{\sigma}_{cb|r}^2 + \hat{\sigma}_{cb|g}^2 + \hat{\sigma}_{cb|b}^2)\mathrm{Tr}\,\mathbf{R}_{cb}^{-1}}{\beta^2(n_b - (k-1)^2)} \qquad (20)$$

Using this compatibility coefficient, one can select the best groupings, by selecting from all the possible grouping the one corresponding to the minimum value.

7. Geometrical Generalization

The problem of segmentation is to find the most reliable minimal description of an image. This statement implies that the complexity of the model used by the segmentation algorithm must only be increased if there is a strong statistical evidence. Let $\hat{\sigma}_t^2$ be the approximation error of the model with the larger number of parameters $p_{\max} = (k+1)^2$ as computed by equation (13). Its value is kept as a comparison basis. In order to validate a parameter in one of the coefficient matrices $\mathbf{A}_{rl}$ and $\mathbf{A}_{cl}$ the algorithm first eliminate this element from the coefficient matrix by setting it equal to zero and then compute the new approximation error $\hat{\sigma}_n^2$. The variation of the relative error is given by:

$$\frac{\hat{\sigma}_n^2 - \hat{\sigma}_t^2}{\hat{\sigma}_t^2} = \frac{\hat{\sigma}_n^2}{\hat{\sigma}_t^2} - 1 = r - 1. \qquad (21)$$

The variables $\hat{\sigma}_n$ and $\hat{\sigma}_t$ are equal to the sum of the squared error for the reduced model and the full model respectively.

If the statistics r is close to unity, one may conclude with confidence that the ith component of one of coefficient matrices is not statistically significant. The statistical distribution of the variable r is distributed as a Snedecor's F distribution with $\nu_1 = 1$ and $\nu_2 = n_l - p_{\text{max}}$ degrees of freedom. The decision to reject the parameter i from the coefficient matrix with a degree of confidence α is given by:

$$P_F(r \geq r_o) = \int_{r_o}^{\infty} p_F(r) dr \geq \alpha. \tag{22}$$

In the algorithm the parameter α is set equal to 0.1.

8. Experimental Results

In order to illustrate the segmentation process, a billard ball composed of different color region was scanned using the National Reserach Council color range sensor. One can see in Figure 1a a 3-D display of the color range image. In the present implementation, the position of depth and orientation discontinuities and color edges was computed by using a morphological method developed by Boulanger [4]. One can see in Figure 1b the combined range and color edges. After the initial partition of the scene and the generation of the graph structure to represent this partition, the algorithm start the grouping process as described above. The grouping process is terminated when the average extrapolation error $\hat{\sigma}_{ib}$ of the best grouping is over a threshold of 0.5 mm in the range image and 10.0 in the color image. The model generalization method is then applied and the grouping process resumed for the same threshold. One can see in Figure 1c the partition of the scene produced by the algorithm and in Figure 1d the reconstructed color range image from the segmentation model.

9. Conclusion

The present algorithm produces a quasi-optimal partition of colored range images even when the images are corrupted by a high level of noise. At the end of the hierarchical grouping, one has a model where one can analyze color range images at different levels of representation. The data structure produced by the algorithm is directly accessible to high level tasks such as model building, identification, and pose determination. The algorithm is also clean in the sense that there is no ad-hoc threshold (beside resolution ε_t and the *a priori* distribution parameter β) that would make the algorithm hard to tune. Bayesian decision criterion always makes sure that one take the best grouping or generalisation decision.

References

1. Bajcsy, R. and Allen, P. (1986) Multisensor integration, *Encyclopedia of Artificial Intelligence*, N.Y. Wiley, pp 632–638

2. Baribeau, R., Rioux, M., and Godin, G. (1992) Color reflectance modeling using a polychromatic laser range sensor, *IEEE Trans. on Pattern Analysis and Machine Intelligence*, Vol. **PAMI–14, no. 2**, pp. 263–269

3. Besl, P.J. and Jain, R.C. Segmentation through variable–order surface fitting, *IEEE Trans. on Pattern Analysis and Machine Intelligence*,Vol. **PAMI-10, no. 2**, pp. 167– 192

4. Boulanger, P., Blais, F., and Cohen P. (1990) Detection of depth and orientation discontinuities in range images using mathematical morphology, *10th International Conference on Pattern Recognition*, Atlantic City.

5. Duda, R.O., Nitzan, D., and Barret, P. (1979) Use of range and reflectance data to find planer surface regions, *IEEE Trans. Pattern Analysis and Machine Intelligence* Vol. **PAMI-1 no. 3**, pp. 259–271

6. Gil, B., Mitiche, A., and Aggarwal, J.K. (1983) Experiments in combining intensity and range edge maps, *Computer Vision, Graphics, and Image Processing*,Vol. **21 no. 3**, pp. 395–411

7. Hackett, J.K. and Shah, M. (1989) Segmentation using intensity and range data, *Optical Engineering*, Vol. **28, no. 6**, pp. 667–674

8. Koezuka, T., Kakinoki, Y., Suto Y., Nakashima, M., and Inagaki, T. (1989) High-speed 3–D vision system using range and intensity images covering a wide area, *Application of Artificial Intelligence VII*,Vol. **SPIE–1095**, pp. 1134–1141

9. Luo, R.C. and Kay, M.G. (1989) Multisensor integration and fusion in intelligent systems, *IEEE Trans. on System, Man, and Cybernetics*, Vol. **SMC-19. no. 6**, pp. 901–931

10. Mitiche, A. and Aggarwal, J.K. (1986) Multiple sensor integration/fusion through image processing: A review, *Optical Engineering*,Vol. **25 no. 3**, pp. 380–386

11. Rousseeuw, P.J. and Leroy, A.M. Robust (1987) regression and outlier detection. J. Wiley & Sons

12. Shirai, Y. (1992) 3–D computer vision and applications, *IEEE International Conf . on Pattern Recognition*, pp. 236–245

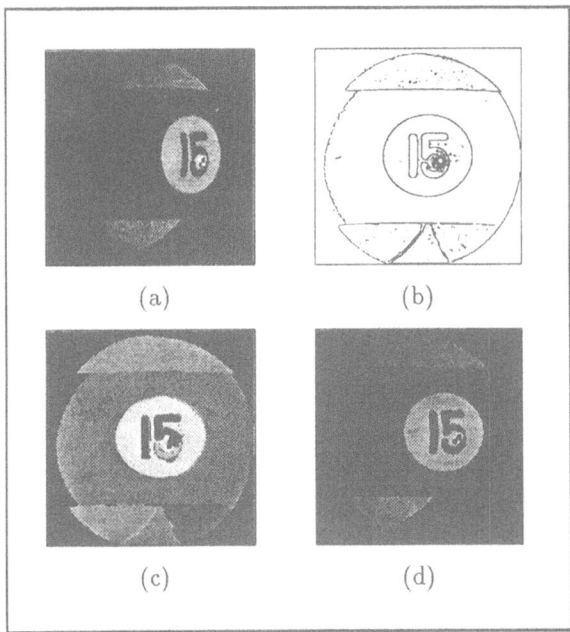

(a)   (b)

(c)   (d)

Figure 1. Segmentation of color range images: (a) 3–D view of the color range image (b) detected range and color edges, (c) segmented image regions, (d) reconstructed color range image.

Imre Csiszár at the podium.

PRIORS ON MEASURES

JOHN SKILLING AND SIBUSISO SIBISI
University of Cambridge, Cavendish Laboratory
Madingley Road, England CB3 0HE

Abstract. A "measure" is the mathematical concept underlying distributed quantities such as images, spectra, and probability distributions. Inference about a measure requires a suitable Bayesian prior. If the prior is to remain valid on indefinitely small cells, it must be "infinitely divisible", and in consequence samples from it will be "atomic" (being essentially a limited sum of delta functions). Infinitely divisible priors are described in terms of the distribution of delta function strengths via the Lévy-Khinchin representation. Such priors include the Poisson process, but not the Quantified Maximum Entropy prior that has been used as its continuous reformulation.

Key words: process, infinitely divisible, Lévy measure, Quantified Maximum Entropy

1. Introduction

Many practical problems of inference involve estimating the distribution of some non-negative and finite but continuously variable quantity Φ across a domain. Images (of intensity), spectra (of amounts), and probability distributions are among the applications. In each case, the relevant quantity is the amount of Φ in a sub-domain or cell. This is additive when cells are combined. Formally, if A and B are disjoint sub-domains ($A \cap B = \emptyset$), then

$$\Phi_{A \cup B} = \Phi_A + \Phi_B \quad \text{and} \quad 0 \le \Phi_C < \infty \quad \forall C \tag{1}$$

Mathematicians call such a quantity a "measure", and (whenever the domain $\{x\}$ is continuous) physicists often describe it in terms of its density $\phi(x) = d\Phi/dx$. Note that Φ remains a measure under general transformation of x, but not of Φ. For example, a map of light amplitude ($= \sqrt{\text{intensity}}$) is not a measure. Measures have a natural coordinate representing "additive quantity" that we ought not destroy.

For Bayesian inference, we need to begin by assigning a prior distribution over the range of conceivable measures. In the absence of any structured model for Φ, we wish to treat the problem directly in nonparametric form, as opposed to imposing any particular parametric form defined by a finite number of parameters.

261

K. M. Hanson and R. N. Silver (eds.), Maximum Entropy and Bayesian Methods, 261–270.
© 1996 *Kluwer Academic Publishers.*

In this introductory paper, we shall take our prior to be uncorrelated, so that knowledge of Φ in one domain does not alter the prior elsewhere (save perhaps for normalisation which can be included as a later conditioning). When the overall domain is partitioned into M cells, this implies that the density w.r.t. $d\Phi$ factorises as

$$\Pr(\Phi) = P_1(\Phi_1)P_2(\Phi_2)\dots P_M(\Phi_M) \tag{2}$$

In the common case that the cells are a priori equivalent, all the P_i are the same function. It transpires, though, that the P_i are not fully arbitrary.

2. Development of Quantified Maximum Entropy

The simplest uncorrelated prior is the Poisson process, modelled by the canonical team of monkeys [1] throwing quanta of pre-ordained quantity q at the cells. Taking equivalent cells for simplicity, the Poisson formula is

$$\Pr(\Phi|\mu, q, M) = \prod_{i=1}^{M} \sum_{r_i=0}^{\infty} \delta(\Phi_i - r_i q)e^{-\mu}\mu^{r_i}/r_i! \tag{3}$$

where μ is the expectation number of quanta per cell. Though the simplicity of this model has its attractions, the discrete character of the intensity scale is a critical disadvantage. The prior is inconsistent with any incommensurable constraint such as $\Phi_2/\Phi_1 = \sqrt{2}$ that some future measurement might suggest.

The natural fixup for this, inspired by statistical mechanics [2], is to allow an indefinitely large number of tiny quanta, $q \to 0$ at fixed expectation $\overline{\Phi} = \mu q$. Stirling's formula approximates the Poisson factorials to give

$$\Pr(\Phi|\overline{\Phi}, q, M) \approx \exp(S(\Phi; \overline{\Phi})/q)\Big/\prod \sqrt{(2\pi q \Phi_i)} \tag{4}$$

$$S(\Phi; \overline{\Phi}) = \sum_{i=1}^{M}(\Phi_i - \overline{\Phi} - \Phi_i \log(\Phi_i/\overline{\Phi})) \tag{5}$$

in terms of the entropy S. Unfortunately, the approximation only holds properly in the $q \to 0$ limit, in which case the prediction is arbitrarily sharp:

$$\Pr(\Phi|\overline{\Phi}, 0, M) = \prod \delta(\Phi_i - \overline{\Phi}) \tag{6}$$

Just as happens in statistical mechanics with large Avogadro's number, the structure is set up at the outset to be almost surely almost uniform. Future data would need to be absurdly definitive in order to override this prior assignment, so the $q \to 0$ limit is inadmissible.

A fixup for this is to adopt the suggestion

$$\Pr(\Phi|\overline{\Phi}, q, M) \propto \exp(S(\Phi; \overline{\Phi})/q)\Big/\prod \sqrt{\Phi_i} \tag{7}$$

from Stirling but to allow q to remain finite, accepting that there is no closed form for the normalisation integral. This is known as the "Quantified Maximum Entropy", or QME, prior [3].

Practical programs have not used the QME prior directly, but have turned to a second order Gaussian approximation about the maximum a posteriori solution. Skilling [4] attempted to derive this approach from practical desiderata. Essentially, the central most probable result was required to be maximum entropy because of the attractive symmetry properties that then hold [5], and quantitative inferences from the surrounding probability cloud were required to hold in the continuum limit $M \to \infty$ of very many, very small cells. Indeed, this combination of properties has demonstrable value.

Yet the Gaussian form is definitively wrong insofar as it necessarily allows negative intensities to occur in samples of Φ. In fact, negative excursions in Φ approach certainty as the otherwise desirable continuum limit is approached. Correspondingly, third and higher order corrections to the Gaussian diverge upwards as $M \to \infty$, so that the Gaussian approximation fails to represent the bulk of the distribution $\Pr(\Phi)$. Closer analysis shows that the QME prior too becomes almost surely almost uniform over macroscopic domains computed in the continuum limit, so it too is inadmissible.

3. Infinitely Divisible Processes

Failure in the continuum limit indicates that more care is needed when defining the cells. Specifically, we need a prior defined consistently on all partitions of the full domain into sets $A, B, \ldots$. Statisticians call such a prior, defined to be one for which the densities obey the Kolmogorov condition

$$\Pr(\Phi_A, \Phi_B, \ldots) = \int \Pr(\Phi_A, \Phi_B, \ldots, \Phi_X) d\Phi_X \tag{8}$$

a "process". If Φ is to be a measure, we also require additivity which involves a Laplace convolution for the densities on disjoint subdomains:

$$\Pr(\Phi_{A \cup B}) = \int \int \delta(\Phi_A + \Phi_B - \Phi_{A \cup B}) \Pr(\Phi_A) \Pr(\Phi_B) d\Phi_A d\Phi_B \tag{9}$$

Indeed, given any partitioning π, we need to be able to start with an arbitrary subdivision π' of it, and recover the prior on π by any path of additive cell combination. We call such a prior a "path-independent process", or π-process for short. Any π-process obeys the Kolmogorov condition [6] so the terminology is consistent.

Conversely, we should be able to divide the cells of an existing partition and still retain a valid formulation. Sometimes, we may simply want a direct geometrical subdivision. At other times, we may have a physical reason, as when an apparently 6-sided die is revealed to be a 60-facetted polyhedron. Even in apparently discrete cases, it might be unduly blinkered to deny the possibility of such underlying hidden variables.

Suppose cell $A \cup B$ is divided into equal halves A and B. Equation (9) can be used to construct the prior density P for $\Phi_{A \cup B}$ from the common sub-prior density p for Φ_A and Φ_B. Alternatively, (9) can be solved for the sub-prior, either directly by computer or algebraically by Laplace transforms where the relationship is

$$\mathcal{L}P = (\mathcal{L}p)^2 \qquad \text{or} \qquad \mathcal{L}p = (\mathcal{L}P)^{1/2} \tag{10}$$

This works straightforwardly for the Poisson process. If $\Phi_{A \cup B}$ has a Poisson distribution with mean μ quanta, then Φ_A and Φ_B have Poisson distributions with mean $\mu/2$, and conversely. Figure 1 illustrates this for $\mu = 6 \cdot 5$.

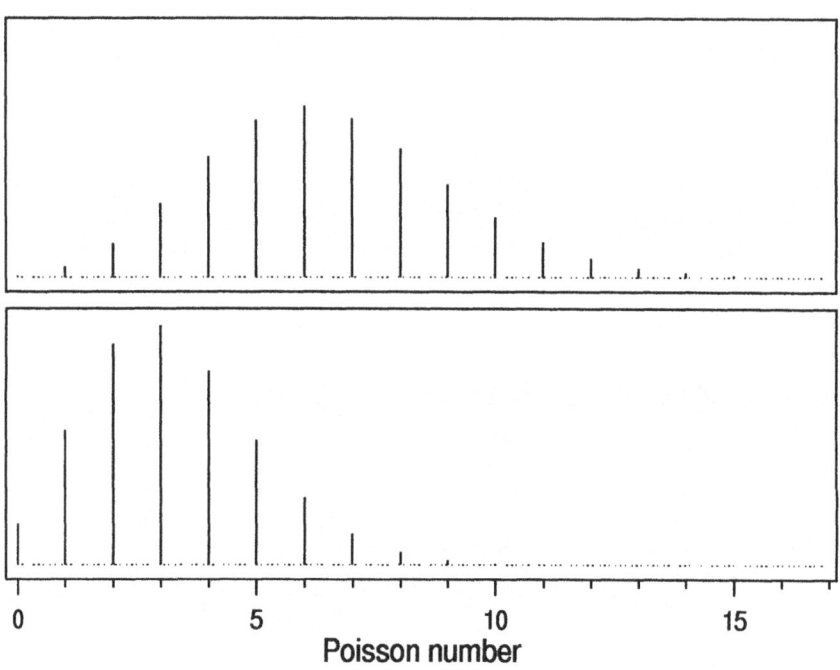

Figure 1.   Poisson process of mean 6·5 (above) halved into another Poisson process of mean 3·25 (below).

However, if a QME prior is used for Φ_A and Φ_B, then the consequential prior for $\Phi_{A \cup B}$ is not of this form. That may already be disconcerting, but worse follows. If QME is used for $\Phi_{A \cup B}$ instead, the solution for the sub-prior is partly negative, which is meaningless. Figure 2 illustrates this for a typical $q = 1/6 \cdot 5$. So the QME prior cannot even be halved, let alone divided indefinitely.

To avoid such contradictions, we require our prior P on Φ to be "infinitely divisible" (*id*). In Laplace transform terms, we require

$$\mathcal{L}^{-1}((\mathcal{L}P)^{1/r}) \geq 0 \qquad \forall r \geq 1 \tag{11}$$

There is a formal test for this [7]. Let

$$\nu(s) = -\log \int_0^\infty P(\Phi) e^{-s\Phi} d\Phi \tag{12}$$

Then P is *id* if and only if

$$\nu(0) = 0 \qquad \text{and} \qquad (-)^{r+1} d^r \nu / ds^r \geq 0 \quad \forall r \geq 1, \forall s \geq 0 \tag{13}$$

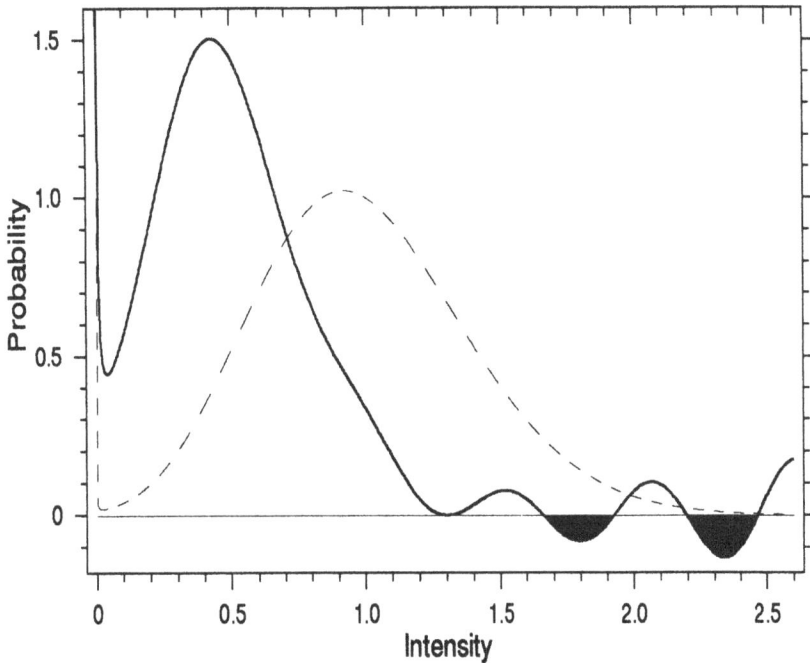

Figure 2. QME prior for $q = 1/6.5$ (dashed) halved into a non-QME distribution (solid) with impossible negative excursions (shaded).

The Poisson process passes this test, as do some others such as an exponential prior. But QME fails, producing (for example) a negative 12th derivative for $s < 0.13$ in the case $q = 1/6.5$. QME is not alone: the flat prior

$$P(\Phi) = \{1 \quad \text{for} \quad \Phi < 1, \quad 0 \quad \text{otherwise}\} \tag{14}$$

also fails (Figure 3).

Infinite divisibility of a macroscopic prior is a non-trivial property. However, more insight can be gleaned from the microscopic limit.

4. Atomicity

It happens that typical samples Φ from an *id* prior are "atomic", meaning that most of the total quantity involved is likely to be found in a limited number of cells, no matter how finely divided the domain.

More explicitly, "Given any fraction $0 < f < 1$ and any certainty $0 < c < 1$, there exists $N(f, c) < \infty$, independent of the partitioning into cells, for which the strongest N cells carry (with certainty c) at least the fraction f of the total quantity". (Technically, a definitive background can also form part of an *id* sample,

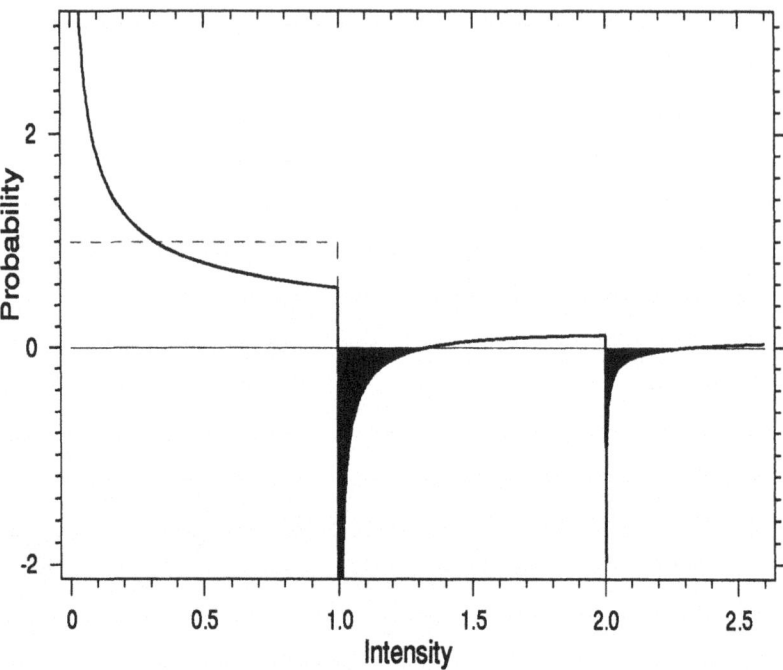

Figure 3.   Flat prior (dashed) halved into a distribution (solid) with impossible negative excursions (shaded).

but for our purposes we choose to ignore this extension.) The atomic property was proved by Ferguson [6] and Blackwell [8] for a special case, and more generally by Kingman [9]. These proofs are terse, but it is not too difficult to construct longer proofs that are less formal.

Far from being in any sense smooth, typical samples from an *id* prior are like a scattering of delta functions or "atoms" when viewed at sufficiently high resolution (Figure 4). Going to yet higher resolution merely localises the atoms more precisely - they do not continue to break apart. A small cell has a correspondingly small probability of holding a finite quantity (though is more likely to be almost empty). This is opposite to the finite probability of small quantity that might have occurred if the overall quantity had been more uniformly spread. Of course, atomicity refers to samples, and means and higher moments can easily average to smooth functions.

The atomic property is perfectly reasonable. If, contrariwise, all the small cells were expected to hold comparable (though correspondingly small) quantities, then the Law of Large Numbers would make macroscopic sums almost determinate, so that the prior at finite scale would be almost surely almost uniform. This observation is not quite strong enough to prove the atomic "Law of Small Numbers" property, but it demonstrates plausibility.

5. Microscopic Formulation

Because an indefinitely small cell will almost never contain more than one significant atom (and will usually contain none), it is simpler to set up the prior at the
microscopic level of description.

Let a small cell Δx have prior density w.r.t. $d\Phi$

$$\Pr(\Phi|\Delta x) = N(\Delta x, \Phi), \quad \Phi > 0 \tag{15}$$

Here N is the basic "atomic number density" that defines and bounds our freedom
in setting an id prior. Formal expositions of this [7] use the Lévy measure L, related
to N by

$$L(\Delta x, d\Phi) = \Phi N(\Delta x, \Phi)d\Phi \tag{16}$$

Nevertheless, the atomic number density N may be the clearer concept for many
scientists. Clearly, N must be non-negative, and in physical applications we also
require a finite expectation of Φ, so that

$$N \geq 0 \quad \text{and} \quad \langle \Phi \rangle = \int_0^\infty \Phi N(\Delta x, \Phi)d\Phi < \infty \tag{17}$$

Taking an infinitesimal cell, the bulk of the probability is assigned to $\Phi = 0$, so
that

$$\Pr(\Phi|dx) = \left(1 - \int_0^\infty N(dx, \Psi)d\Psi\right)\delta(\Phi - 0) + N(dx, \Phi) \tag{18}$$

which correctly sums to unity. Its Laplace transform is

$$\begin{aligned}
\widetilde{P}(s|dx) &= \int_0^\infty d\Phi e^{-s\Phi} \Pr(\Phi|dx) \\
&= 1 - \int_0^\infty d\Phi(1 - e^{-s\Phi})N(dx, \Phi) \\
&= \exp\left(-dx \int_0^\infty d\Phi(1 - e^{-s\Phi})n(x, \Phi)\right) \tag{19}
\end{aligned}$$

In order to generate the macroscopic prior $P(\Phi|\Delta x)$ for a finite cell Δx, we convolve
by multiplying the Laplace transforms

$$\begin{aligned}
\widetilde{P}(s|\Delta x) &= \exp\left(-\int_0^\infty d\Phi(1 - e^{-s\Phi}) \int_{\Delta x} dx\, n(x, \Phi)\right) \\
&= \exp\left(-\int_0^\infty d\Phi(1 - e^{-s\Phi})N(\Delta x, \Phi)\right) \tag{20}
\end{aligned}$$

which is the transform of the required finite distribution. In terms of the Lévy
measure, the formula is [7]

$$\widetilde{P}(s|\Delta x) = \exp -\int_0^\infty \frac{1 - e^{-s\Phi}}{\Phi}L(\Delta x, d\Phi) \tag{21}$$

which is known as the Lévy-Khinchin representation. By construction, any finite-scale prior derived in this way must be a valid probability distribution. However, we have seen earlier that not all finite-scale distributions can take this form. In particular, the construction shows that all *id* distributions extend to indefinitely large Φ when evaluated at finite scale, which is why the bounded flat prior could not be *id*.

6. Examples

The simplest case has N factorised in terms of a base measure α on x. Here the Poisson process with quantum size q has

$$N(\Delta x, \Phi) = \alpha(\Delta x)\delta(\Phi - q) \qquad \text{or} \qquad L(\Delta x, d\Phi) = \alpha(\Delta x)q\delta(\Phi - q)d\Phi \qquad (22)$$

$$\Pr(\Phi) = \sum_{r=0}^{\infty} \delta(\Phi - rq)e^{-\alpha}\alpha^r/r! \qquad (23)$$

All the atoms hold the same quantity q, and the finite-scale prior for the cell just reflects the Poisson expectation αq.

Generalising slightly, a double Poisson process has two quantum sizes q_1 and q_2, possibly over different base measures α_1 and α_2:

$$N(\Delta x, \Phi) = \alpha_1(\Delta x)\delta(\Phi - q_1) + \alpha_2(\Delta x)\delta(\Phi - q_2)$$
$$L(\Delta x, d\Phi) = (\alpha_1(\Delta x)q_1\delta(\Phi - q_1) + \alpha_2(\Delta x)q_2\delta(\Phi - q_2))d\Phi$$

The corresponding finite-scale prior density is the convolution

$$\Pr(\Phi) = \sum_{j=0}^{\infty}\sum_{k=0}^{\infty} \delta(\Phi - jq_1 - kq_2)\left(e^{-\alpha_1}\alpha_1^j/j!\right)\left(e^{-\alpha_2}\alpha_2^k/k!\right) \qquad (24)$$

Any atomic number density can be built up from suitably many quantum sizes, so that the general *id* prior can be represented as a compound Poisson process. However, not every atomic number density induces a finite-scale prior with tractable algebraic form.

Perhaps the simplest tractable example with continuous Φ is the Gamma process:

$$N(\Delta x, \Phi) = \alpha(\Delta x)e^{-\lambda\Phi}/\Phi \qquad \text{or} \qquad L(\Delta x, d\Phi) = \alpha(\Delta x)e^{-\lambda\Phi}d\Phi \qquad (25)$$

This is about as close to the Jeffreys scale-invariant atomic number density $1/\Phi$ as is consistent with normalisation of Φ. Its finite-scale prior density is

$$\Pr(\Phi|\alpha, \lambda) = \frac{\lambda^\alpha}{\Gamma(\alpha)}e^{-\lambda\Phi}\Phi^{-1+\alpha} \qquad (26)$$

7. Conclusions

We have seen that some of the priors that have historically been proposed for inferring measures, including the Quantified Maximum Entropy prior, lack the crucial property of "infinite divisibility". Requiring this property leads to a family of admissible priors that are "atomic".

Atomicity lets us make an interesting connection with parametric inference. Although our approach is nonparametric, we have found that typical samples of Φ can be accurately modelled by a bounded number of significant delta functions. It would be possible to assign a parametric prior directly upon these delta functions, letting the locations be uniform over x-measure and letting the amplitudes be distributed according to the atomic number density. Having done this, though, the number of significant delta functions is distributed in a known way according to the compound Poisson process implicit in the Lévy-Khinchin representation. Requiring the prior to be an uncorrelated process has given extra information about it, which is not obvious from the direct parametric approach.

ACKNOWLEDGMENTS This work was supported by MaxEnt Solutions Limited.

References

1. S. F. Gull and G. J. Daniell, "Image reconstruction from incomplete and noisy data," *Nature*, **272**, pp. 209–230, 1978.
2. E. T. Jaynes, "Concentration of distributions at entropy maxima," in *E. T. Jaynes Papers on Probability, Statistics and Statistical Physics*, R. D. Rosenkrantz, ed., pp. 315–336, Reidel, Dordrecht, 1983.
3. J. Skilling, "Classic maximum entropy," in *Maximum Entropy and Bayesian Methods*, J. Skilling, ed., pp. 45–52, Kluwer Academic, Dordrecht, 1989.
4. J. Skilling, "Fundamentals of MaxEnt in data analysis," in *Maximum Entropy in Action*, B. Buck and V. A. Macaulay, eds., pp. 19–40, Clarendon, Oxford, 1991.
5. J. Skilling, "The axioms of maximum entropy," in *Maximum Entropy and Bayesian Methods in Science and Engineering*, C. R. Smith and G. J. Erickson, eds., pp. 173–188, Kluwer Academic, Dordrecht, 1988.
6. T. S. Ferguson, "A Bayesian analysis of some nonparametric problems," *Ann. Statist.*, **1**, pp. 209–230, 1973.
7. W. Feller, *An Introduction to Probability Theory and its Applications, Vol. II*, Wiley, New York, 1971.
8. D. Blackwell, "Discreteness of Ferguson selections," *Ann. Statist.*, **1**, pp. 356–358, 1973.
9. J. F. C. Kingman, *Poisson Processes*, Oxford University. Oxford, 1993.

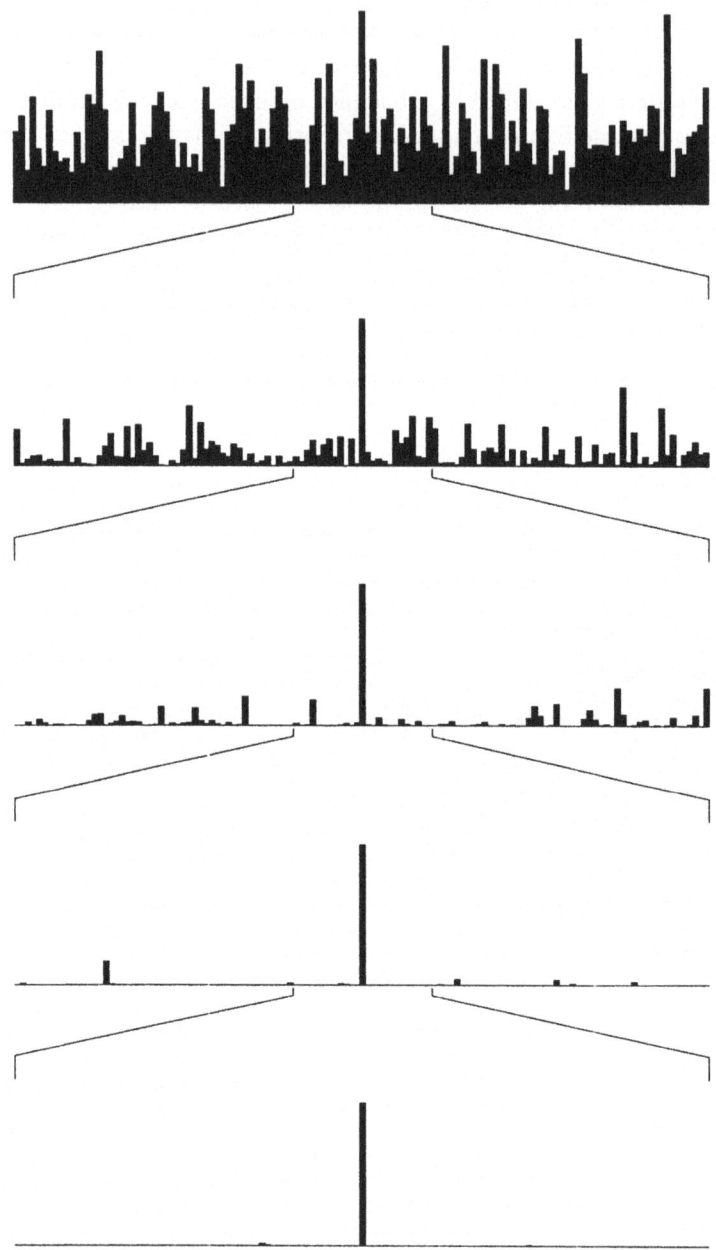

Figure 4.   Subdivision of an *id*-process into atoms. A sample from a Gamma process with measure $\alpha = 5$ per cell is shown (top) with successive five-fold expansions of the central fraction.

DETERMINING WHETHER TWO DATA SETS ARE FROM THE SAME DISTRIBUTION

DAVID H. WOLPERT
The Santa Fe Institute
1399 Hyde Park Road
Santa Fe, NM 87501[†]

Abstract. This paper presents two Bayesian alternatives to the chi-squared test for determining whether a pair of categorical data sets were generated from the same underlying distribution. It then discusses such alternatives for the Kolmogorov-Smirnov test, which is often used when the data sets consist of real numbers.

Key words: Chi-squared test, Kolmogorov-Smirnoff test

1. Introduction

Let d_1 and d_2 be two sets of elements from a space X, with cardinalities N_1 and N_2, respectively. View d_1 and d_2 as "samples" of two distributions over X, p_1 and p_2 respectively. Based on d_1 and d_2, do we believe that p_1 and p_2 are equal (or at least close approximations of each other), and how confident are we in this belief?

When X is finite, the traditional approach to this common problem is the chi-squared test. When X is the set of real numbers, the traditional approach is instead the Kolmogorov-Smirnoff test. (See [2] and references therein for discussions of both tests.) Both of these tests can be viewed as types of null-hypothesis tests. Accordingly, they suffer from a number of defects: they are average-data rather than this-data; they can (sort of) rule out the null, but not "rule it in"; they are very dependent on issues like the "power" and "size" of the statistic, etc.

Clearly a Bayesian alternative to these tests, properly constructed, would be preferable. Some have confused such an alternative with, for example, proofs (like in [1]) that in the proper limit the chi-squared test approximates a Bayesian procedure of some sort. What is instead needed is a first-principles Bayesian approach to the problem, which in general may have no relation to tests like chi-squared.

In this paper two such approaches are worked out in detail for the finite X case, and some possible approaches are mapped out for the uncountable X case. See also [3] for related work.

[†] Email: dhw@santafe.edu

K. M. Hanson and R. N. Silver (eds.), Maximum Entropy and Bayesian Methods, 271–276.

2. Finite X—the posterior expected difference approach

Consider the case where X has a finite number of possible values, m. For this scenario, both p_1 and p_2 consist of m real numbers, all of which are non-negative, and which sum to 1. I will indicate the i'th component of p_j by $p_j(i)$. Also I will indicate the histograms over the X values induced by d_1 and d_2 as $d_1(i)$ and $d_2(i)$ respectively. So $d_j(i)$ is the number of elements in d_j that have the i'th X value.

Let $S(p_1, p_2)$ be a measure of the similarity between p_1 and p_2. For simplicity, in this paper I will concentrate on the quadratic distance measure, $S(p_1, p_2) \equiv \sum_{i=1,m} (p_1(i) - p_2(i))^2$. However the calculations presented here can be applied to any analytic S by expanding that S in a Taylor series. Moreover, as is illustrated below, for many non-quadratic S simple tricks allow one to perform the calculations without resorting to such an expansion.

In this section I will show how to calculate the posterior expected value of S, S_1, and the posterior expected value of S^2, S_2. The formula for S_1 provides a measure of how much the data indicates that p_1 and p_2 differ, and in $\sqrt{S_2 - (S_1)^2}$ we have an error bar for that measure.

First note that

$$S_1 \equiv \int dp_1 dp_2 \ S(p_1, p_2) \ P(p_1, p_2 \,|\, d_1, d_2), \quad \text{and}$$

$$S_2 \equiv \int dp_1 dp_2 \ S^2(p_1, p_2) \ P(p_1, p_2 \,|\, d_1, d_2).$$

By Bayes' theorem $P(p_1, p_2 \,|\, d_1, d_2) \propto P(d_1 \,|\, p_1) \, P(d_2 \,|\, p_2) \, P(p_1, p_2)$. Now assume the d_i were created by IID sampling of p_i: $P(d_i \,|\, p_i) \propto \prod_{j=1}^{m} p_i(j)^{d_i(j)}$. All that remains to fully fix the integrals for S_1 and S_2 is the prior, $P(p_1, p_2)$.

2.1. CALCULATING THE FIRST MOMENT

Given the preceding, up to an overall proportionality constant, S_1 is given by

$$J \equiv \int dp_1 dp_2 \ [\prod_{j=1}^{m} p_1(j)^{d_1(j)} \ p_2(j)^{d_2(j)}] \ P(p_1, p_2) \ S(p_1, p_2). \tag{1}$$

The proportionality constant is set by normalization and equals

$$K \equiv \int dp_1 dp_2 \ \prod_{j=1}^{m} p_1(j)^{d_1(j)} \ p_2(j)^{d_2(j)} \ P(p_1, p_2). \tag{2}$$

To proceed further we must specify the prior, $P(p_1, p_2)$. Write

$$P(p_1, p_2) = F(p_1, p_2) \times \delta\{(\sum_{j=1}^{m} p_1(j)) - 1\} \times \prod_{j=1}^{m} \theta(p_1(j))$$

$$\times \delta(\{\sum_{j=1}^{m} p_2(j)) - 1\} \times \prod_{j=1}^{m} \theta(p_2(j)), \tag{3}$$

where the (Dirac) delta functions force each p_i to have its components sum to 1, and the (Heaviside) theta functions force all such components to be non-negative.

Consider the case where $F(p_1, p_2)$ is analytic, i.e., where it can be written as a sum of products of powers of the components of p_1 and p_2. Note that the likelihood term in our integrals is simply a product of powers of the components of p_1 and p_2. Accordingly, if we know how to calculate S_1 and S_2 for the case where $F(p_1, p_2)$ is a constant, by using appropriate modifications of the d_i we can calculate S_1 and S_2 for any analytic F. (I.e., the Dirichlet prior is the conjugate prior for this problem.) Accordingly, without loss of generality, from now on I will take $F(p_1, p_2) = 1$.

The integrals J and K are relatively straight-forward to evaluate [4]:

$$K(d_1, d_2) = \frac{\prod_{i=1}^{m} \Gamma(d_1(i) + 1) \ \Gamma(d_2(i) + 1)}{\Gamma(N_1 + m) \ \Gamma(N_2 + m)}. \tag{4}$$

where "$\Gamma(.)$" is the gamma function. (For current purposes, where the $d_j(i)$ are integers, the gamma function is just a factorial.) Next, use the expansion $S(p_1, p_2) = \sum_{i=1}^{m}[p_1(i) - p_2(i)]^2 = \sum_{i=1}^{m}\{[p_1(i)]^2 + [p_2(i)]^2 - 2p_1(i)p_2(i)\}$. This gives

$$J = \sum_{i=1}^{m} \{K[d_1 + 2(i), d_2] + K[d_1, d_2 + 2(i)] - 2K[d_1 + 1(i), d_2 + 1(i)]\}. \tag{5}$$

where by "$d_i + t(j)$" is meant the histogram d_i with t extra counts added to the j'th bin. Note that the total number of counts in $d_i + t(j)$ is $N_i + t$; this must be taken into account when plugging the formula for K into the formula for J.

With Eq.'s (4) and (5) we can calculate the posterior expected value of S, J/K.

2.2. CALCULATING THE SECOND MOMENT

S_2 is calculated in a similar way to the calculation of S_1. It can be written as $S_2 = \frac{Z_1 + Z_2}{K}$, where K is the same as in Eq. (4), and Z_1 and Z_2 are linear combinations of K's. To evaluate this ratio first define $g(x, y) \equiv (x + y)!/x!$ and $n_i \equiv N_i + m - 1$. Then

$$\frac{Z_1}{K} = \sum_{i=1}^{m} \{ \frac{g(d_1(i), 4)}{g(n_1, 4)} - 4\frac{g(d_1(i), 3) \ g(d_2(i), 1)}{g(n_1, 3) \ g(n_2, 1)} +$$

$$6 \frac{g(d_1(i), 2) \ g(d_2(i), 2)}{g(n_1, 2) \ g(n_2, 2)} - 4\frac{g(d_1(i), 1) \ g(d_2(i), 3)}{g(n_1, 1) \ g(n_2, 3)} + \frac{g(d_2(i), 4)}{g(n_2, 4)} \}. \tag{6}$$

Note that this can be broken up into five separate sums - doing that, you only need to perform five separate divisions (no denominator term involves the summation variable i). Moreover, the $g(., .)$ terms can be pre-calculated.

A similar calculation gives the following:

$$\frac{Z_2}{K} = 2\sum_{i<j} \{ \frac{g(d_1(i), 2) \ g(d_1(j), 2)}{g(n_1, 4)} + 2\frac{g(d_1(i), 2) \ g(d_2(j), 2)}{g(n_1, 2) \ g(n_2, 2)} +$$

$$\frac{g(d_2(i),2)\, g(d_2(j),2)}{g(n_2,4)} - 4\frac{g(d_1(i),2)\, g(d_1(j),1)\, g(d_2(j),1)}{g(n_1,3)\, g(n_2,1)} -$$
$$4\,\frac{g(d_2(i),2)\, g(d_1(j),1)\, g(d_2(j),1)}{g(n_2,3)\, g(n_1,1)} +$$
$$4\,\frac{g(d_1(i),1)\, g(d_2(i),1)\, g(d_1(j),1)\, g(d_2(j),1)}{g(n_1,2)\, g(n_2,2)} \}. \qquad (7)$$

As in evaluating Z_1/K, it makes sense to break up this expression into a set of sums (to reduce the number of divisions to six) and precompute quantities like g's. We would still seem to have an m^2 calculation though.

To get around this, use the following identity:

$$\sum_{i\neq j} u(j)\, v(j)\, U(i)\, V(i) = [\sum_j u(j)\, v(j)]\, [\sum_i U(i)\, V(i)] - \sum_i u(i)\, v(i)\, U(i)\, V(i).$$

Next note that each of the g's in the numerators in Eq. (7) is a product of no more than two terms. This allows us to evaluate all products of those g's using Eq. (8), and thereby make the entire calculation linear in m. More precisely, for

i) the first term in Eq. (7), set $u = U, v = V$, $u(j) = d_1(j)+2$, and $v(j) = d_1(j)+1$;
ii) the second term, set
$u(j) = d_2(j) + 2, v(j) = d_2(j) + 1, U(j) = d_1(j) + 2$, and
$V(j) = d_1(j) + 1$;
iii) the third term, set $u = U, v = V$, and $u(j) = d_2(j) + 2, v(j) = d_2(j) + 1$;
iv) the fourth term, set $u(j) = d_1(j) + 1, v(j) = d_2(j) + 1, U(j) = d_1(j) + 2$, and
$V(j) = d_1(j) + 1$;
v) the fifth term, set $u(j) = d_1(j) + 1, v(j) = d_2(j) + 1, U(j) = d_2(j) + 2$, and
$V(j) = d_2(j) + 1$;
vi) the sixth term, set $u(j) = d_1(j) + 1, v(j) = d_2(j) + 1, U(j) = d_1(j) + 1$, and
$V(j) = d_2(j) + 1$.

The details of plugging all of this into Eq. (7) are not too illuminating, and in the interests of space are left as an exercise for the reader.

2.3. COMMENTS

Adding data doesn't change the number of operations needed to calculate S_1 and S_2. On the other hand, despite the large degree of cancellation in our equations (e.g., when one divides Eq. (5) by Eq. (4)), things do get more expensive as one increases m, the number of bins. This is because we have many products over bins. Here conventional tricks like operating in logarithm space (so products become sums) are needed to keep the computational time (not to mention underflow and overflow problems) tractable.

Finally, simple tricks allow evaluation of S_1 and S_2 for some non-quadratic choices of $S(p_1, p_2)$, without going to the trouble of Taylor expanding such an S. For example, to evaluate the posterior moment of the Kullback-Leibler distance between p_1 and p_2, one has to be able to evaluate integrals of the form

$\int dp_1 dp_2 \ln(p_1(i)) \prod_{j=1}^{m} [p_1(j)^{\hat{d}_1(j)} p_2(j)^{\hat{d}_2(j)}] P(p_1, p_2)$ (where $\hat{d}_1$ and $\hat{d}_2$ are in general slight variants of d_1 and d_2). To do this we can use Eq. (4) and the simple identity $x^a \ln(x) = (\partial_n x^n)|_{n=a}$ to get logarithms into the integrands [4].

3. Finite X—the ratio of posteriors approach

The technique outlined above assigns measure 0 to the set of events $p_1 = p_2$. I.e., it says that it is impossible for p_1 to equal p_2, regardless of the data. It is possible to use a Bayesian technique that instead assigns comparable probabilities to the two models $M_1 \equiv \{p_1 = p_2\}$ and $M_2 \equiv \{p_1 \neq p_2\}$. To see how first write

$$P(d_1, d_2 | M_1) = \int dp_1 dp_2 \, P(d_1, d_2 | M_1, p_1, p_2) \, P(p_1, p_2 | M_1), \quad \text{and then}$$

$$P(p_1, p_2 | M_1) = P(p_2 | p_1, M_1) \, P(p_1 | M_1) = \delta(p_1 - p_2) \, G(p_1) \, stuff(p_1),$$

where "$stuff(p_1)$" is the usual expression forcing p_1 to be a probability distribution.
Write $P(d_1, d_2 | M_1, p_1, p_2) = P(d_1 | p_1) \, P(d_2 | p_2)$, so

$$P(d_1, d_2 | M_1) = \int dp_1 \, G(p_1) \, stuff(p_1) \prod_{j=1}^{m} p_1(j)^{d_1(j) + d_2(j)} \frac{N_1! \, N_2!}{\prod_{j=1}^{m} (d_1(j))! \, (d_2(j))!}.$$

As usual, to analyze the analytic G case it suffices to consider the case where G is a constant. Being careful to maintain normalization, for this case

$$P(d_1, d_2 | M_1) = \frac{(m-1)! \, (N_1)! \, (N_2)!}{(N_1 + N_2 + m - 1)!} \times \frac{\prod_{j=1}^{m} (d_1(j) + d_2(j))!}{\prod_{j=1}^{m} (d_1(j))! \, (d_2(j))!}. \quad (9)$$

Next write $P(d_1, d_2 | M_2) = \int dp_1 dp_2 \, P(d_1 | p_1) \, P(d_2 | p_2) \, F(p_1, p_2) \, stuff(p_1, p_2)$. Again, take F constant. (As an aside, if one wants non-constant F and G, it probably makes sense to have them "correspond" in some way.) This gives

$$P(d_1, d_2 | M_2) = \frac{[(m-1)!]^2 \, (N_1)! \, (N_2)!}{(N_1 + m - 1)! \, (N_2 + m - 1)!}. \quad (10)$$

This depends only on N_1, N_2 and m; no other aspects of the d_i are relevant.
Finally, use Eq.'s (9) and (10) to get the ratio of the posteriors of the models:

$$\frac{P(M_1 | d_1, d_2)}{P(M_2 | d_1, d_2)} = \frac{P(M_1)}{P(M_2)} \times \frac{P(d_1, d_2 | M_1)}{P(d_1, d_2 | M_2)}. \quad (11)$$

This posterior ratio for the uniform F and G case is extremely quick to evaluate and in many respects is at least as "reasonable" in its behavior as the traditional chi-squared test. Nonetheless, one may want to consider non-uniform F and G. In particular, non-uniform F raises/lowers the probabilities of $p_1 - p_2$ pairs for which $p_1 \neq p_2$ but which lie close to $\{p_1 = p_2\}$. So for example, if $d_1 = d_2$, then having F favor $p_1 - p_2$ pairs that lie close to $\{p_1 = p_2\}$ will "leach" some of the posterior probability of M_1 into the posterior probability of M_2. This is because F will be favoring $p_1 - p_2$ pairs that can reasonably explain the data.

4. The Uncountable X Case

For real-valued X, binning X (so that the techniques of the previous section can be applied) is sometimes problematic. That is because the final result can depend on the binning scheme used. One obvious potential solution to this problem is to take inspiration from the Kolmogorov-Smirnov test: have the statistic concern differences in the cumulative distribution functions (CDF's) rather than the density functions directly. For example, one might define $S(p_1, p_2) \equiv \sum_{i=1}^{m} [CDF_1(i) - CDF_2(i)]^2$. Since the CDF's tend to be relatively insensitive to the precise binning, with this scheme how you bin should not be a big problem.

Another possibility is to use a prior that favors smooth p_i, so that $p_i(j)$ is close to $p_i(k)$ if bin j is close (in X) to bin k. Such a prior can be used with either of the posterior ratio or statistic moments approaches. For the latter a CDF-based statistic is not needed; a conventional (e.g., quadratic) S could be used.

A third possibility is not to bin, but rather consider a parameterized set of p_i. Under this scheme one could use either of the posterior ratio or moments of S approaches. However now the integrals would be over the parameters of the p_i rather than over the p_i directly.

Finally, there are some schemes that involve neither binning nor parameters. For example, one could define a new space $Y \equiv d_1 \cup d_2$ and do the analysis in that space. So the p_i are now distributions over Y, and the values in the histograms of the d_i are all 0's and 1's (assuming there are no delta functions in $P(p_1(X), p_2(X))$, so there are no duplicates in $d_1 \cup d_2$). The idea would be to have the prior favor smooth p_i, where the degree to which $p_i(j)$ is pushed towards $p_i(k)$ depends on the distance between the X values corresponding to elements j and k of Y.

Future work involves comparing these schemes to other Bayesian procedures (F. Ruggeri—private communication) related to the Kolmogorov-Smirnov test.

Acknowledgements: I would like to thank Tony Begg and especially Hank Vaccaro for helpful discussions. I would also like to thank Ken Hanson for performance well beyond the call of duty in helping me with Latex problems. This work was supported in part by TXN.

References

1. D. Lindley, *Introduction to probability and statistics 2*, Cambridge University Press. (1965).
2. W.H. Press et al, *Numerical Recipes in C*, Cambridge University Press. (1992).
3. D.R. Wolf, *Mutual Information as a Bayesian Measure of Independence*, send email to "comp-gas@xyz.lanl.gov" with subject "get 9511002".
4. D.H. Wolpert, D.R. Wolf, *Estimating functions of probability distributions from a finite set of samples*, Physical Review E, in press. (1995).

OCCAM'S RAZOR FOR PARAMETRIC FAMILIES
AND PRIORS ON THE SPACE OF DISTRIBUTIONS

VIJAY BALASUBRAMANIAN
Princeton University
Physics Department - Jadwin Hall
Princeton, NJ 08544, U.S.A.[†]

Abstract. I define the *razor*, a natural measure of the complexity of a parametric family of distributions relative to a given true distribution. I show that empirical approximations of this quantity may be used to implement parsimonious inference schemes that favour simple models. In particular, the *razor* is seen to give finer classifications of model families than the Minimum Description Length principle as advocated by Rissanen. In a certain strong sense it is shown that the logarithm of the Bayesian posterior probability of a model family given a collection of data converges in the large sample limit to the logarithm of the *razor* of the family. This provides the most accurate asymptotics to date for Bayesian parametric inference. These results are derived by treating parametric families as manifolds embedded in the space of probability distributions. In the course of deriving a suitable integration measure on such manifolds, it is shown that, in a certain sense, a uniform prior on the space of probability distributions would induce a Jeffreys' Prior on the parameters of a parametric family of distributions.

Key words: Occam's Razor, Jeffreys' Prior, Fisher Information, Bayes Rule, MDL.

1. Introduction

In many statistical situations we are given a collection of data $E = \{e_1 \cdots e_N\}$ drawn independently from some true distribution t, and we wish to build a model of t from the data. Often the model is taken to be a member of a parametric model family. For example, we may choose to model data from a distribution on the real line by choosing a member of the family of Gaussians. Once a parametric model family is chosen, the task of statistical inference is reduced to one of parameter estimation for which there are many standard techniques whose individual pitfalls are well understood. In this paper we will be concerned with the question "How do we choose the model family to describe a particular true distribution?".

[†]Email: vijayb@phoenix.princeton.edu

K. M. Hanson and R. N. Silver (eds.), Maximum Entropy and Bayesian Methods, 277–284.
© *1996 Kluwer Academic Publishers.*

We will begin by constructing a quantity $R_N(A)$, called the *razor* of the model family A given N data points. The *razor* will be shown to be an index of the simplicity and accuracy of A as a description of t, given N data points drawn independently from t with which to pick a member of A to model t. It will be shown that lower-dimensional, more robust models of t are deemed "simpler" by the razor. (A more "robust" model family is less sensitive to the precise choice of parameters because a larger fraction of its volume is concentrated close to the true distribution.) The razor of a parametric family is derived by treating the family as a manifold embedded in the space of distributions. While deriving an integration measure on such parameter manifolds we find that the uniform prior distribution, in a certain sense, on the space of distributions induces a Jeffreys' prior on the parameters of a model family. The argument proceeds by explicitly counting the number of distinguishable distributions indexed by a model family and provides a novel justification for a choice of Jeffreys' prior for Bayesian inference. Finally, we show that the logarithm of the Bayesian posterior probability of a model family given data drawn from t converges in a certain strong sense to the logarithm of the razor. This result provides the most accurate asymptotics known to date for Bayesian parametric model selection. An incidental result is that the Minimum Description Length principle in its stochastic complexity incarnation amounts to a truncation of the full Bayesian procedure which is justified by the axioms of probability theory ([1],[2],[3],[4]).

Throughout this paper we will treat parametric model families as manifolds embedded in the space of distributions so that each choice of parameters indexes a distribution. For convenience, we will suppose that only one parameter patch is necessary to cover the manifold and that the parametrization embeds the d dimensional family as a compact subspace of R^d. We will also use the conventions that $\Theta = \{\theta_1 \cdots \theta_d\}$ labels the parameters of a model family and that $d^d\Theta$ represents the Lebesgue measure on a d dimensional parameter space. Finally, we will use the summation convention that repeated indices in any products are summed over unless otherwise specified $(J_{ij}I^i \equiv \sum_i J_{ij}I^i)$. Due to the brevity of this paper, proofs of many assertions will merely be sketched here. The detailed arguments are available in [1].

2. Definition of The Razor

In this section we will motivate the definition of the razor of a model by appealing to the Bayesian approach to model family inference. Suppose we are given a collection of outcomes $E = \{e_1 \dots e_N\}$, $e_i \in X$ drawn independently from a true density t. In the Bayesian approach, we choose between families A and B by computing the posterior conditionals $\Pr(A|E)$ and $\Pr(B|E)$ and picking the family with the higher probability. Since these conditional probabilities depend on the specific outcomes, we must analyze their statistics to understand the typical behaviour of Bayesian model selection. Let A be parametrized by a set of parameters $\Theta = \{\theta_1, \dots \theta_d\}$. Then Bayes Rule tells us that $\Pr(A|E) = (\Pr(A)/\Pr(E)) \int d^d\Theta \; w(\Theta) \Pr(E|\Theta)$ where $\Pr(A)$, $w(\Theta)$ and $\Pr(E)$ are priors on the model family, the parameters and the N outcome sample space respectively. Since we are interested in comparing

$Pr(A|E)$ with $Pr(B|E)$, the prior $Pr(E)$ is a common factor that we may omit and for lack of any better choice we take the prior probabilities of A and B to be equal and omit them.

We now write $Pr(E|\Theta) = \prod_{i=1}^{N} Pr(e_i|\Theta) = \exp(\sum_{i=1}^{N} \ln Pr(e_i|\Theta)) \equiv e^{F(\Theta)}$ where the exponent, $F(\Theta)$, is a sum of N independent, identically distributed random variables of the form $\ln Pr(e_i|\Theta)$. So, as N grows large, the Central Limit Theorem applies and $F(\Theta)$ is distributed as a Gaussian with mean $N\mu$ where $\mu = \int dx\, t(x) \ln Pr(x|\Theta) = -D(t||\Theta) - h(t)$. In this expression the differential entropy of the true distribution $h(t) = -\int dx\, t(x) \ln t(x)$ is assumed finite, as is the relative entropy between t and the distribution indexed by Θ, $D(t||\Theta) = \int dx\, t(x) \ln[t(x)/Pr(x|\Theta)]$. By putting the expected value of $F(\Theta)$ into $Pr(A|E)$ we obtain our candidate for the *razor*: $R_N(A) = \int d^d\Theta\, w(\Theta) \exp{-N(D(t||\Theta) - h(t))}$. We now drop the factor $\exp{-Nh(t)}$ which is common to the probabilities of all model families and fix the prior $w(\Theta)$.

It is commonly supposed that the most conservative prior on a parameter space is the uniform prior since that reflects complete ignorance. In fact that choice of prior suffers from a serious deficiency. The uniform priors relative to different parametrizations assign different probabilities to the same set of parameters ([5],[6]). Consequently, if $w(\Theta)$ was uniform in the parameters, the probability of a model family would depend on the arbitrary parametrization. The problem can be cured by requiring that all *distributions* rather than all *parameters* are equally likely. In the next section we will argue that the uniform prior (in a suitable sense) in the space of probability distributions induces a Jeffreys' Prior on a parameter manifold. The definition of the Jeffreys' Prior in terms of the determinant of a two-form on the parameter manifold also guarantees the reparametrization invariance of the razor ([5],[6]). Letting $w(\Theta)$ be Jeffreys' prior we define the razor of a model A, given N outcomes drawn from t, to be:

$$R_N(A) = \frac{\int d^d\Theta \sqrt{\det J(\Theta)}\, e^{-ND(t||\Theta)}}{\int d^d\Theta \sqrt{\det J(\Theta)}} \tag{1}$$

where J, the Fisher Information matrix at Θ is defined as $J_{ij} = \partial_{\phi_i}\partial_{\phi_j} D(\Theta||\Theta + \Phi)|_{\Phi=0}$. This is our candidate for an intrinsic measure of the simplicity and accuracy of the family A as a model of t, given N data points from t with which to pick a member of A as the model distribution. In Section 5 the precise connection betweeen the razor and the typical behaviour of Bayes Rule will be analyzed.

3. Priors on Parameter Manifolds

A choice of Jeffreys' prior for model estimation has been previously suggested from varying points of view ([5], [7],[8],[9], [10], [11]). In this section we argue that in a certain sense the uniform prior on the space of probability distributions induces a Jeffreys' prior on a parameter manifold. The argument begins by defining a statistical measure of distinguishability of probability distributions. We argue that a measure on a parameter manifold should only count distinguishable distributions to avoid overcounting. Giving equal weight to each distinguishable distribution in a parametric family yields Jeffreys' prior on a parameter manifold. This argument provides a novel justification for a choice of Jeffreys' prior.

Let Θ_p and Θ_q index two distributions in a parametric family and let $E = \{e_1 \cdots e_N\}$ be drawn independently from one of Θ_p or Θ_q. In the context of model estimation, a suitable measure of distinguishability can be derived by asking how well we can guess which of Θ_p or Θ_q produced E. Let α_N be the probability that Θ_q is mistaken for Θ_p and let β_N be the probability that Θ_p is mistaken for Θ_q. Let β_N^ϵ be the smallest possible β_N given that $\alpha_N < \epsilon$. Then Stein's Lemma tells us that $\lim_{N \to \infty}(-1/N)\ln\beta_N^\epsilon = D(\Theta_p \| \Theta_q)$ ([12]).

From the proof of Stein's Lemma, it can be shown that the minimum error β_N^ϵ exceeds some fixed β^* in the region where $\kappa/N \geq D(\Theta_p \| \Theta_q)$ with $\kappa \equiv -\ln\beta^* + \ln(1-\epsilon)$.([1],[12]). As N grows large for fixed κ, any Θ_q in this region is necessarily close to Θ_p. Therefore, setting $\Delta\Theta = \Theta_q - \Theta_p$, Taylor expansion gives $D(\Theta_p \| \Theta_q) \approx (1/2)J_{ij}(\Theta_p)\Delta\Theta^i\Delta\Theta^j + O(\Delta\Theta^3)$ where J is the Fisher Information defined below Equation 1. [1] We define the *volume of indistinguishability* at levels ϵ, β^*, and N to be the volume of the region of high probability of error and find, to leading order:

$$V_{\epsilon,\beta^*,N} = \left(\frac{2\pi\kappa}{N}\right)^{d/2} \frac{1}{\Gamma(d/2+1)} \frac{1}{\sqrt{\det J_{ij}(\Theta_p)}} \tag{2}$$

If β^* is very close to one, the distributions inside $V_{\epsilon,\beta^*,N}$ are not very distinguishable and should not be counted separately in the razor. (Equivalently, the Bayesian prior should not treat them as separate distributions.) We wish to construct a measure on the parameter manifold that reflects this indistinguishability. We also assume a principle of "translation invariance" by supposing that volumes of indistinguishability at given values of N, β^* and ϵ should have the same measure regardless of where in the space of distributions they are centered. An integration measure reflecting these principles of indistinguishability and translation invariance can be defined at each level β^*, ϵ, and N by covering the parameter manifold economically with volumes of indistinguishability and placing a delta function in the center of each element of the cover. This definition reflects indistinguishability by ignoring variations on a scale smaller than the covering volumes and reflects translation invariance by giving each covering volume equal weight in integrals over the parameter manifold.

The measure described above essentially discretizes the parameter manifold to reflect indistinguishability at levels β^*, ϵ and N. The continuum limit ($\beta^* \to 1$, $N \to \infty$) can be obtained by considering integrals over the parameter manifold defined with respect to a sequence of measures at each β^*, ϵ and N. By paying careful attention to technical difficulties involving sets of measure zero and certain sphere packing problems, it can be shown that the normalized continuum measure on a parameter manifold that reflects indistinguishability and translation invariance is Jeffreys' prior ([1],[13]). In essence, the derivation in [1] shows how to "divide out" the volume of indistinguishable distributions on a parameter manifold and

---

[1] We have used the fact that $D(\Theta_p \| \Theta_q)$ attains a minimum of zero when $\Theta_p = \Theta_q$. We have also assumed that the derivatives with respect to Θ commute with expectations taken in the distribution Θ_p.

hence gives equal weight to equally distinguishable volumes of distributions. In this sense, Jeffreys' prior is seen to be a uniform prior on the *distributions* indexed by a parametric family.

4. Asymptotics of The Razor

In this section we will show that the razor constructed in previous sections is an index of the simplicity and accuracy of a model family as a description of a given true distribution. The analysis is made simpler by assuming that $D(t\|\Theta)$ is a smooth function of Θ that attains its unique global minimum at Θ^*, an interior point of the compact parameter manifold.[2] We will take the metric on the parameter manifold to be the Fisher Information since the Jeffreys' prior has the form of a measure derived from such a metric.[3] We will use ∇_μ to indicate the covariant derivative with respect to Θ_μ with a flat connection for the Fisher Information metric.[4] Readers who are unfamiliar with covariant derivatives may read ∇_μ as the partial derivative with respect to Θ_μ since we will not be emphasizing the geometric content of the covariant derivative.

An asymptotic series in powers of $1/N$ can be derived for the razor by observing that for large N the integrand of the razor is narrowly peaked around $\Theta^* = \arg\min_\Theta D(t\|\Theta)$. Writing $(\det J)^{1/2}$ as $\exp(1/2)Tr\ln J$, we Taylor expand the exponent in the integrand of the razor around Θ^* and shift and rescale the integration variable to $\Phi = N^{1/2}(\Theta - \Theta^*)$ to arrive at:

$$R_N(A) = \frac{e^{-(ND(t\|\Theta^*)-\frac{1}{2}Tr\ln J(\Theta^*))}N^{-d/2}\int d^d\Phi\, e^{-((1/2)\tilde{J}_{\mu\nu}\phi^\mu\phi^\nu+G(\Phi))}}{\int d^d\Theta\sqrt{\det J_{ij}}} \quad (3)$$

where $\tilde{J}_{\mu\nu} = \nabla_\mu\nabla_\nu D(t\|\Theta)|_{\Theta^*}$ and $G(\Phi)$ contains higher powers of Φ multiplied by higher derivatives of $D(t\|\Theta)$ and $Tr\ln J$. All terms in G can be shown to be suppressed by powers of N. Next we introduce a source $h = \{h_1\cdots h_d\}$ so that $\int d^d\Phi \exp((-1/2)\tilde{J}_{\mu\nu}\Phi^\mu\Phi^\nu+G(\Phi)) = \exp(G(\nabla_h))\int d^d\Phi \exp((-1/2)\tilde{J}_{\mu\nu}\Phi^\mu\Phi^\nu + h_\mu\Phi^\mu)$ where the argument of G, $\Phi = \{\phi_1\cdots\phi_d\}$, has been replaced by $\nabla_h = \{\partial_{h_1}\cdots\partial_{h_2}\}$ to turn G into a differential operator. Subject to one final assumption that the bounds of the integration can be extended to infinity with negligible error, we can do the Gaussian integral with the source and then apply $\exp G(\nabla_h)$ to find an asymptotic series in $1/N$. Defining $V = \int d^d\Theta\sqrt{\det J_{ij}}$ and $\chi_N(A) = -\ln R_N(A)$ we find that:

$$\chi_N(A) = ND(t\|\Theta^*) + \frac{d}{2}\ln N - \frac{1}{2}\ln\left[\frac{\det J_{ij}(\Theta^*)}{\det \tilde{J}_{\mu\nu}}\right] - \ln\left[\frac{(2\pi)^{d/2}}{V}\right] + O(1/N) \quad (4)$$

The terms of $O(1/N)$ and smaller can be explicitly evaluated with some labour and are omitted here for the sake of brevity. (See [1].) This method of asymptotic

---

[2] For brevity we will omit mention of some exponentially small terms arising from the local minima of $D(t\|\Theta)$.

[3] This choice of metric also follows the work described in [7] and [8].

[4] See [7], [8] and [1] for further details on metrics and connections in a statistical setting.

expansion can find fruitful application in other statistical problems where we desire to find the asymptotics of quantities whose integrands are dominated by narrow maxima. A limited version of this method used commonly in statistical mechanics has appeared before in Information Theory ([14],[9]).

We can see why the razor measures simplicity and accuracy of a parametric family by examining Equation 4 and noting that models with larger $R_N(A)$ and hence smaller $\chi_N(A)$ are better. The $O(N)$ term which dominates asymptotically, $ND(t\|\Theta^*)$, measures the relative entropy distance between the true distribution and the best model distribution on the manifold and is, therefore, a measure of accuracy. The $O(\ln N)$ term penalizes models with many degrees of freedom and is a measure of simplicity. This term arises geometrically because the volume of a peak in the integrand of the razor falls more rapidly as a function of N in higher dimensions. The $O(1)$ term, is even more interesting. The determinant of $\tilde{J}^{-1}$ is proportional to the volume of the ellipsoid in parameter space around Θ^* where the value of the integrand of the razor is significant.[5] The scale for determining whether $\det \tilde{J}^{-1}$ is large or small is set by the Fisher Information on the surface whose determinant defines the volume element. Consequently the term $\ln(\det J/\det \tilde{J})^{1/2}$ can be understood as measuring the robustness of the model in the sense that it measures the relative volume of the parameter space which provides good models of the true. More robust models in this sense will be less sensitive to the precise choice of parameters. We also observe from the discussion regarding Jeffreys' prior that the volume of indistinguishability around Θ^* is proportional to $(\det J)^{-1/2}$. So the quantity $(\det J/\det \tilde{J})^{(1/2)}$ is essentially proportional to the ratio V_{large}/V_{indist}, the ratio of the volume where the integrand of the razor is large to the volume of indistinguishability introduced earlier. Essentially, a model family is better (more natural or robust) if it contains many distinguishable distributions that are close to the true. The term $\ln(2\pi)^d/V$ can be understood as a preference for models that have a smaller invariant volume in the space of distributions and hence are more constrained. While accuracy is favoured asymptotically by the term of $O(N)$, for a small amount of data, the preference for "simplicity" embodied in the $O(\ln N)$ and $O(1)$ terms can be more important in comparing the razors of model families. Hence, parsimonious model selection can be implemented by estimating the razors of different families and picking the one with the largest estimate.

5. Asymptotics of Bayes Rule

In this section we derive the precise relationship between the razor and the Bayesian posterior probability of a model family given a set of outcomes $E = \{e_1 \cdots e_N\}$ drawn from t. As discussed in Section 2, this probability is given by $R_E(A) = \int d^d\Theta (\det J(\Theta))^{1/2} \exp \ln(\Pr(E|\Theta))/\int d^d\Theta(\det J(\Theta))^{1/2}$ where we have dropped priors on the model family and the data and adopted a Jeffreys' prior on the parameters. The asymptotics of $R_E(A)$ can be computed with the aid of a Taylor expansion around the maximum likelihood parameter $\hat{\Theta} = \arg\max_\Theta \ln \Pr(E|\Theta)$ in

---

[5] If we fix a fraction $f < 1$ where f is close to 1, the integrand of the razor will be greater that f times the peak value in an elliptical region around the maximum.

exact analogy with the analysis of $R_N(A)$. Defining $\chi_E(A) = -\ln R_E(A)$ we find that to $O(1/N)$:

$$\chi_E(A) = -\ln \Pr(E|\hat{\Theta}) + \frac{d}{2}\ln N - \frac{1}{2}\ln\left[\frac{\det J_{ij}(\hat{\Theta})}{\det \tilde{I}_{\mu\nu}(\hat{\Theta})}\right] - \ln\left[\frac{(2\pi)^{d/2}}{V}\right] + O(\frac{1}{N}) \quad (5)$$

where we have defined $\tilde{I}_{\mu\nu} = -\nabla_\mu\nabla_\nu \ln \Pr(E|\Theta)/N$. It can be shown that every term in the asymptotic expansion of $\chi_E(A)$ can be arrived at by substituting derivatives of $\ln\Pr(E|\Theta)/N|_{\hat{\Theta}}$ for corresponding derivatives of $D(t|\Theta)|_{\Theta^*}$ in the expansion of $\chi_N(A)$. (See [1].)

We will assume that the maximum likelihood estimator $\hat{\Theta}$ is *consistent* in that it converges to $\Theta^* = \arg\min_\Theta D(t\|\Theta)$ in probability. Assuming consistency and certain other technical conditions, it is shown in [1] that any finite collection of terms in the asymptotic expansion of $\chi_E(A) - Nh(t)$ converges in probability to the corresponding sequence of terms in the expansion of $\chi_N(A)$, where $h(t)$ is the entropy of the true distribution.[6] More specifically, let $T_E(A, k, k')$ consist of the terms of $O(1/N^k)$ to $O(1/N^{k'})$ in $\chi_E(A) - Nh(t)$ and let $T_N(A, k, k')$ be the corresponding sequence in $\chi_N(A)$.[7] Then for any $\epsilon > 0$ and $0 < \delta < 1$, for all sufficiently large N, $\Pr(N^k|T_E(A, k, k') - T_N(A, k, k')| < \epsilon) > \delta$. This provides the most accurate asymptotics to date for Bayesian model selection.

6. Comparison With The Minimum Description Length Principle

The Minimum Description Length (MDL) priciple for model family selection states that the best model family given a collection data is the one that minimizes the code length of the data given the model. It has been shown by Rissanen that the model that attains the minimum $\min_{\Theta, d}\{-\log\Pr(E|\Theta) + (d/2)\log N\}$ gives the most efficient coding rate possible for the observed data amongst all universal codes ([3],[4]). The quantity minimized by Rissanen has been called the *stochastic complexity* of a collection of data relative to a given model family and comparison with Equation 5 readily shows that the stochastic complexity amounts to a truncation of the logarithm of the Bayesian posterior probability of a model family at the second term in its asymptotic expansion. Therefore the stochastic complexity converges in the large sample limit to the leading terms in the razor, showing that MDL in the guise of stochastic complexity is an empirical approximation of the razor of a model family. The results discussed in this paper concerning the meaning of the razor and the fact that it reflects the typical asymptotics of Bayes Rule strongly suggest that the definition of stochastic complexity ought to be extended to include all the subleading terms in the expansion listed in Equation 5. Indeed, Rissanen has recently used the work of Clarke and Barron on the asymptotics of Bayes Rule to add the terms of $O(1)$ in Equation 5 to the definition of stochastic complexity

---

[6] The main additional condition required is that the for any $\epsilon > 0$ there is a neighbourhood M of Θ^* such that, for every outcome e_i and $\Theta \in M$, $|\ln\Pr(e_i|\Theta) - \ln\Pr(e_i|\Theta^*)| < \epsilon$. This property is known as *equicontinuity*.

[7] Consider the asymptotic expansion $\chi_E(A) - Nh(t) = Nc_{-1} + (d/2)\ln N + c_1/N + c_2/N^2 + \cdots$. Then, for example, $T_E(A, 4, 6) = c_4/N^4 + c_5/N^5 + c_6/N^6$.

([2],[9]). He finds that this extension removes a redundancy in the earlier definition. Our results suggest a further extension to include the terms of $O(1/N)$ and smaller in Equation 5. Some recent work by Yamanishi in a general decision-theoretic setting yields the same suggestion for an extended definition of stochastic complexity ([15]).

7. Conclusion

In this paper we have introduced the *razor* of a parametric model family A as an index of the simplicity and accuracy of the family as a model of the true distribution t, given a certain amount of data from t. We have shown that the razor favours simpler, more robust models when the amount of data is small but asymptotically picks the most accurate model in relative entropy sense. The logarithm of the Bayesian posterior probability of a model family given the data converges in a certain strong sense to the logarithm of the razor. This result provides the most accurate asymptotics to date for Bayesian model selection. Our results were arrived at by treating parametic families as manifolds embedded in the space of distributions and it was shown that the uniform prior on the space of distributions, in a certain sense, yields a Jeffreys' prior on the parameters. More detailed discussion and analysis may be found in [1].

References

1. V. Balasubramanian, "A geometric formulation of occam's razor for inference of parametric distributions." Available as preprint number adap-org/9601001 from http://xyz.lanl.gov/ and as Princeton University Physics Preprint PUPT-1588, January 1996.
2. J. Rissanen, "Fisher information and stochastic complexity." Submitted to the IEEE Transactions of Information Theory, 1994.
3. J. Rissanen, "Universal coding, information, prediction and estimation," *IEEE Transactions on Information Theory*, **30**, pp. 629–636, July 1984.
4. J. Rissanen, "Stochastic complexity and modelling," *The Annals of Statistics*, **14**, (3), pp. 1080–1100, 1986.
5. H. Jeffreys, *Theory of Probability*, Oxford University Press, 3rd ed., 1961.
6. P. Lee, *Bayesian Statistics: An Introduction*, Oxford University Press, 1989.
7. S. Amari, *Differential Geometrical Methods in Statistics*, Springer-Verlag, 1985.
8. S. Amari, O. Barndorff-Nielsen, R. Kass, S. Lauritzen, and C. Rao, *Differential Geometry in Statistical Inference*, vol. 10, Institute of Mathematical Statistics Lecture Note-Monograph Series, 1987.
9. B. Clarke and A.R.Barron, "Information-theoretic asymptotics of bayes methods," *IEEE Transactions on Information Theory*, **36**, pp. 453–471, May 1990.
10. A.R.Barron and T. Cover, "Minimum complexity density estimation," *IEEE Transactions on Information Theory*, **37**, pp. 1034–1054, July 1991.
11. C. Wallace and P. Freeman, "Estimation and inference by compact coding," *Journal of The Royal Statistical Society*, **49**, pp. 240–265, July 1987.
12. T. Cover and J. Thomas, *Elements of Information Theory*, Wiley, New York, 1991.
13. J. Conway and N. Sloane, *Sphere Packings, Lattices and Groups*, Springer-Verlag, New York, 2nd ed., 1988.
14. A. Barron, *Logically Smooth Density Estimation*. PhD thesis, Stanford University, August 1985.
15. K. Yamanishi, "A decision-theoretic extension of stochastic complexity and its applications to learning." Submitted to the IEEE Transactions of Information Theory, June 1995.

SKIN AND MAXIMUM ENTROPY:
A HIDDEN COMPLICITY ?

B. DUBERTRET
Department of Civil and Mechanical Engineering,
M.I.T., 02139 Cambridge, U.S.A.[‡]

N. RIVIER
Laboratoire de Physique Théorique,
Université Louis Pasteur, Strasbourg, France.[§]

AND

G. SCHLIECKER
Theoretische Physik,
Freie Universität, Berlin, Germany.

Abstract. Random cellular structures (froths, foams, undifferentiated biological tissues) are in statistical equilibrium thanks to elementary local transformations. They form a statistical ensemble, with universal properties (structural equation of state, and distribution of cell shapes, up to priors). Notably, *all* natural random cellular structures in two dimensions follow a unique relation -that can be obtained by maximum entropy inference- between the variance of the cell shape distributions, and the probability that a cell has six neighbours (Lemaitre's Law).By obtaining the distributions of cell shapes thanks to coupled rate equations, one is able to propose a mechanism for the renewal of biological tissues like the mammal epidermis. No prior probabilities are necessary, but some assumptions on the division kernel are needed. The solutions are very restricted, and agree with those obtained by MaxEnt and with experimental data.Random cellular structures offer therefore an excellent testing ground for Maxent inference.

Key words: Maximum Entropy, cellular structures, biological tissues, priors, statistics

[‡]permanent address: Université Louis Pasteur, Strasbourg, France.
[§]E-mail: nick@fresnel.u-strasbg.fr

K. M. Hanson and R. N. Silver (eds.), Maximum Entropy and Bayesian Methods, 285–294.

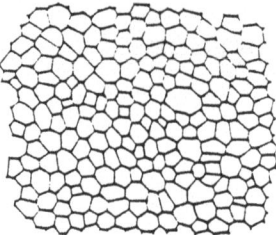

Figure 1.   cut of the epithelium of the cucumber

1. Introduction

Random space-filling cellular networks are often encountered in engineering as
well as natural materials such as bones, cork, plants, food (bread, meringue,...),
cosmetics [1]. Because of their particular mechanical properties and their theoretically challenging structure, they have been studied in scientific fields as various
as biology (epidermal tissues [2,3]), metallurgy (two dimensional (2D) section of
polycrystals [4], fluid mechanics (soap froths [5,6]) and geography (administrative
division [7])).

Because of the complexity of three dimensional structures, it is often preferred
to tackle the problem of their evolution by the bias of 2D cuts. The similarities
of those cuts (despite the different driving forces that dictate their evolution) has
led many workers to set up descriptive methods which are independent of the
forces involved [8]. For the last 10 years this kind of approach has been the basis
of many experimental and theoretical studies germane to the description of the
topological properties of 2D cellular structures [9-11]. Our purpose in this paper
is to combine such topological methods with the principle of maximum entropy to
model the statistical equilibrium and the renewal of biological tissues such as the
human epidermis or the epithelium of the cucumber.

2. Topological representation of biological tissues

The fact that tissue geometry and evolution can be described by simple physical
or mathematical (filling space at random) models was recognized early by Hales
[12], Errera [13], Matzke [14], Lewis [2], and others [15,16].

As seen by a physicist, a cut of a cellular undifferentiated biological tissue appears structurally as a random tessellation of space by cells, which are n-sided
topological polygons (cf Fig. 1). The cells explore the configuration space through
topological "collisions", which are *mitosis* (fragmentation or division) and *cell disappearance* (detachment from the basal layer). These two topological transformations govern the evolution of the tissue and are responsible for its random structure and its invariant statistical state. The statistical equilibrium, characterized
by maximum entropy, is the macroscopic cellular space-filling that can be realized
by the largest number of microscopic configurations of cells and their neighbours.
This macroscopic geometry is characterized by a distribution of cells $\{p_n\}_n$ and by
equations of state, familiar in statistical mechanics (Boltzmann distribution and
ideal gas law).

2.1. INFORMATION AVAILABLE

A cell in 2D is described by the number n of its sides. Introduce $A_n \geq 0$, the average area of n-sided cells, and the correlator $M_k(n) = p_k N_{kn}$, number of k-sided cells neighbouring a n-sided cell (a cell with n sides has n neighbours hence, $\sum_k p_k N_{kn} = n$). The inevitable restrictions on p_n are :

normalization:

$$\sum_n p_n = 1$$

topology:

$$\sum_n n p_n = 6 \tag{1}$$

space-filling:

$$\sum_n n A_n p_n = A_{tot}/N$$

topological correlations:

$$\sum_n N_{kn} p_n = k$$

Beside those topological constraints, the tissue is subject to a physical metric constraint resulting from the energetical correlations between the cells. An n-sided cell can be assimilated to a disclination of charge $(n - 6)$ [9]. As a consequence, two neighbouring cells of the same charge are increasing the energy whereas two neighbouring cells of opposite sign bind together to form a dislocation defect costing less energy. It thus appears that one can associate with topological correlations an energy per edge:

$$
\begin{aligned}
U_{corr}/E &= \varepsilon \sum_{n,k} p_n p_k (n-6)^2 \sigma (k-6)^2 \\
&= \varepsilon \sigma \mu_2 \sum_n p_n (n-6)^2 = \varepsilon \sigma \mu_2^2
\end{aligned}
\tag{2}
$$

Where σ is a parameter defined by $M_k(n)$ (see below), ε is a positive constant, and $\mu_2 = \sum_n p_n (n-6)^2$ is the second moment of the distribution. One should add to this energy a disclination self energy $U_{self} = \eta \sum_n p_n (n-6)^2 = \eta \mu_2$, so that the net energy due to cells with 6 sides is always positive.

Each pair of neighbouring disclinations contributes by the product of their charges. Thus $U_{corr} = 0$ for hexagonal structures ($\mu_2 = 0$) and is maximum for uncorrelated froths ($\sigma = 1/6$). A disclinated cell has elastic energy $\varepsilon \sigma \mu_2 (n-6)^2$ proportional to the square of the gaussian curvature (6-n) it imparts. (By the theorem of Gauss-Bonnet, the average gaussian curvature is a topological constant: $\langle n \rangle = 6$ on a plane). There is no topological correlation beyond nearest neighbours and elastic stress is screened by the disorder.

2.2. STRUCTURAL SIGNATURE OF DISORDER: LEWIS' LAW AND ABOAV'S LAW

The maximum entropy theory of Rivier and his co-workers [17,18] predicts a linear variation of N_{kn} and A_n with n (third and fourth constraints (1) are linear combination of the first two and duplicate them, thus effectively maximizing the entropy) for structure in statistical equilibrium:

$$A_n = (A_{tot}/N)\lambda(n - (6 - 1/\lambda)) \tag{3}$$
$$M_k(n) = p_k(k - 6)\sigma(n - 6) + (n + k - 6) \tag{4}$$

Equation 3, a relation between average sizes and shapes discovered empirically by Lewis in 1930 [19], has been observed throughout the biological world. The lagrange multiplier λ is a structural parameter imposing the linear relationship. Equation 4 yields Aboav's law:

$$nm(n) = \sum_k kp_k N_{kn} = (6 - a)n + \mu_2 + 6a \tag{5}$$

which implies that the total number of edges of the cells adjacent to a n-sided cell varies linearly with n. Because of Weaire's sum rule $\langle nm(n) \rangle = \langle n^2 \rangle$ (another consequence of Maxent and equation 4), this equation depends on a unique parameter $a = -\sigma\mu_2$, which is in general positive and of the order of 1 in natural structures [18].

2.3. INVARIANT DISTRIBUTION OF CELLS SHAPE

Maximizing the entropy (by the mean of the two Lagrange multipliers γ and β) of a system subject to the energetical constraint (2) and to the independent constraints (1-i and 1-ii) and some priors $\{q_n\}_n$ yields the most probable distribution of cells shape:

$$p_n = \underbrace{q_n \exp(-\beta(n - 6))}_{r_n} \exp(-\gamma(n - 6)^2) \tag{6}$$

A priori probabilities can be derived from maximum entropy considerations ([9]) by exploiting the symmetries of the constraints. The topological energy and the topological constraint are even in $i = n - 6$. This symmetry enables us to bin together the r_n when they are factor of the same exponential $\exp(-\gamma(n - 6)^2)$. The application of Laplace's principle of indifference (equal probability a priori) yields: $r_5 + r_7 = r_4 + r_8 = r_3 + r_9 = 2$ and $r_n = 1$ for $(n \geq 10)$. The distribution predicted by maximum entropy is thus the binned gaussian distribution ([9]):

$$\forall i \in \mathcal{N}, i \in]0, 3] \Longrightarrow p_{6+i} + p_{6-i} = 2\exp(-\gamma i^2)/S$$

$$p_6 = 1/S \tag{7}$$

$$\forall i \in \mathcal{N}, i \geq 4 \Longrightarrow p_{6+i} = \exp(-\gamma i^2)/S$$

It is in good agreement with experimental and numerical distributions. S, a normalization constant related to the Lagrange paramater γ is the even moment

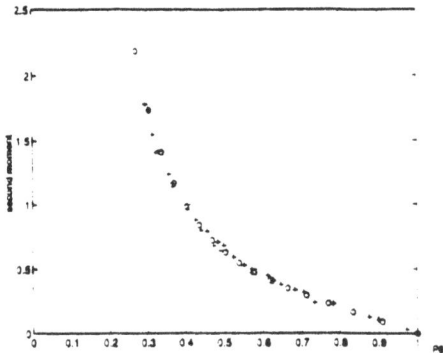

Figure 2. Lemaitre's law:μ_2 as a function of p_6. $\bigcirc$: Maxent theory. +: experimental (disks on air table, botanical tissues) and numerical data (Lemaitre *et al.* 1991, 1993)

generating function. Accordingly, there is a relation between the second moment or variance μ_2 of the distribution and p_6, Lemaitre's law (Figure 2), which was discovered empirically by Lemaitre *et al.* [20,11]. A guess of its functional form was made by LeCaër and Delannay [21]. The functional form,

$$p_6 \in [0.3, 0.7] \Longrightarrow \mu^2 p_6^2 \simeq 1/(2\pi) = 1.59$$

$$(p_6 > 0.7) \Longrightarrow \mu_2 \simeq 1 - p_6 \tag{8}$$

and proof of its universality through the Maxent argument sketched above, were derived in ([9]).

Lemaitre's law is the equivalent of the virial equation of state in liquids and gases. The second expression $\mu_2 = 1 - p_6$ corresponds to the ideal gas law and $p_6 > 0.7$ is the condition for the froth to be "ideal". The astonishing fact of natural froths (and some numerically simulated) is that they <u>all</u> obey (at first approximation) the same , universal Lemaitre law. It is as if all liquids satisfied the same equation of state.

3. Coupled rate equations. The human epidermis.

An alternative method (using some of the results found above thanks to maxent) searches for a stationary distribution, solution of rate equations. Applied success-fully to the cucumber's epithelium ([3]) it also allows a realistic description of the human epidermis.

The human epidermis is in general and at the lowest level of sophistication a fluid of cells, filling space at random and transiting through the malphigii layer from the one-cell deep basal layer where they are born (through mitosis) to the corneum layer where they die ([22]) (Figure 3). Each layer is in statistical equilib-rium: this is the realm of statistical mechanics. Since the basal layer is nearly the only place where cells divide, it can be assumed that the renewal of the epidermis depends solely on its evolution.

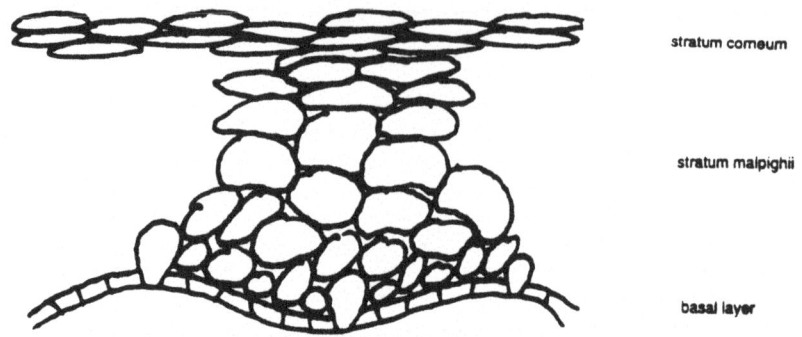

Figure 3.   vertical schematic cut of the human's epidermis

Consider the basal layer, we have a tissue with N_s s-sided cells, distributed as $p_s = N_s/N, N = \sum_s N_s$. It is in statistical equilibrium, thus,

$$0 = \frac{dp_s}{dt} = \frac{1}{N}[\frac{dN_s}{dt} - p_s\frac{dN}{dt}] \qquad (9)$$

The population of s-sided cells is affected by both cell division and cell disappearance.

3.1. RATE EQUATION DUE TO MITOSIS

When a cell divides, the population of s-sided cells is affected if:
 (i) an s-sided cell divides.
 (ii) a dividing k-cell yields an s-sided daughter.
 (iii) an s-sided cell is an affected neighbour of the mother
 (iv) the affected neighbour had (s-1) sides before division
   The distribution of the number of sides of the cells of the tissue in statistical equilibrium is then a function of $P_m(k)$, the conditional probability that an existing k-sided cell divides, the break-up kernel $\Gamma(k \rightarrow s)$, and the rate of cell division $D_m(k)$. Putting together, equation (9) reads:

$$0 = \sum_k p_k P_m(k) D_m(k) [\overbrace{-\delta_{ks}}^{i} + \overbrace{\Gamma(k \rightarrow s)}^{ii} + (2/k)\overbrace{(M_{s-1}(k) - M_s(k))}^{iii \quad iv} - p_s] \quad (10)$$

where the first four terms in the brackets correspond respectively to the topological mechanisms (i) through (iv) described above. The last term expresses the production of one extra cell during mitosis.

3.2. RATE EQUATION OF THE SKIN

Apart from cell division, there is another topological process which allows the tissue to reach statistical equilibrium or respond to local demand in cells: the departure from the basal layer. When a s-sided cell leaves the basal layer, the trace of its neighbours may be topologically affected by gaining or losing sides: the departing cell takes exactly six sides with her, insuring the global flatness and the conservation of the energy of the tissue. The departure can be modeled by a succession of T1 processes ended by a T2 [23]. It yields a topological scar on the basal membrane. In the mean field approximation, the pressure surrounding the leaving cell is isotropic. It is therefore possible to calculate analytically the probability $a_i(k)$ that a k-sided cell gives i sides $(-1 \leq i \leq k-3)$ to one of its neighbours ([23]):

$$\forall k > 4, a_{-1}(k) = \frac{(k-3)a_{-1}(k-1)+1}{k} \tag{11}$$

$$\forall k > i+5, a_i(k) = \frac{(k-3)a_i(k-1)+2a_{i-1}(k-1)}{k} \tag{12}$$

The steady state of the basal layer is then described by the rate equation:

$$
\begin{aligned}
0 = {} & \sum_k p_k P_m(k) D_m(k)[-\delta_{ks} + \Gamma(k \to s) + (2/k)(M_{s-1}(k) - M_s(k))] + \\
& \sum_k p_k P_d(k) D_d(k)[\underbrace{-\delta_{ks}}_{i} - \underbrace{M_s(k)(1 - a_0(k))}_{ii} + \underbrace{\sum_{i=1}^{s-3} M_{s-i}(k) a_i(k)}_{iii} + \\
& \underbrace{M_{s+1}(k) a_{-1}(k)}_{iv}]
\end{aligned}
\tag{13}
$$

where the first part of the equation is the one due to mitosis (the last term has disappeared because the number of cells leaving the tissue is equal to the number of extra cells produced by mitosis) and the second describes the influence of the departure of a cell. The term (i) is produced by the departure of an s-sided cell; the residual terms are due to the redistribution of sides while k-sided cells leave the basal layer. The equilibrium distribution is now function of the parameters defined for the cucumber plus $P_d(k)$, the probability that an existing k-sided cell leaves the basal layer, and $D_d(k)$ the departure rate for k-sided cells.

3.3. RESOLUTION IN THE CASE OF A FLAT TISSUE

It is possible to solve the nonlinear system 13 numerically. It turns out that only very few parameters give a physical solution. Among the possible set of parameters, if one assumes $D_m(k)P_m(k) = (5.1 - k)^{10}$, $D_d(k)P_d(k) = (k - 7)^{10}$, a complete symmetric kernel $\Gamma(k \to s)$, and $\sigma = 0.1$, the theoretical distribution fits very well the experimental one as shown on Figure 3. If the assumed mitotic kernel $\Gamma(k \to s)$ is not complete symmetric, the distributions obtained are too wide.

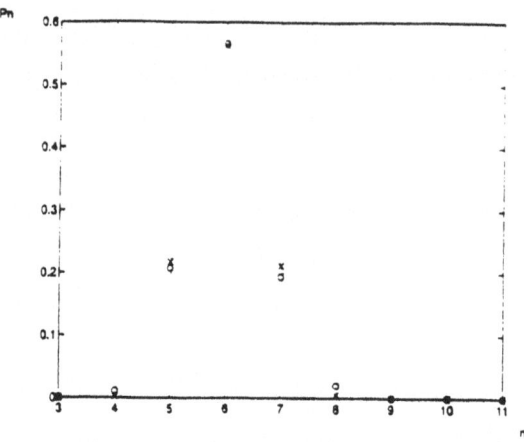

Figure 4. distribution of the basal cells for a flat tissue. ×: Experimental data, ○ :theory.

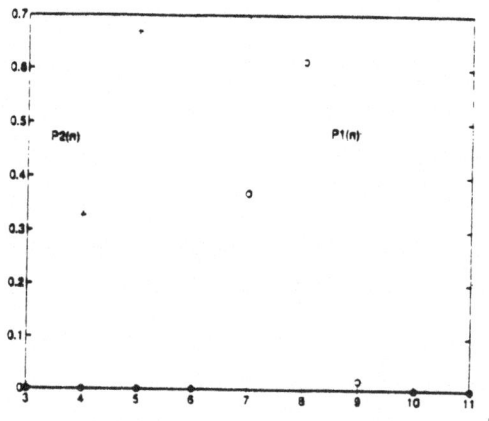

Figure 5. predicted distribution of the cells about to divide $P_1(n)$ and the cells about to leave $P_2(n)$.

3.4. CONSEQUENCE

Putting $\mu = (\#\text{dividing cells})/(\#\text{cells})$, we can calculate the probability $P_1(k) = P_k P_m(k)/\mu$ (resp. $P_2(k) = P_k P_d(k)/\mu$) that a cell about to divide (resp. to leave) has a k-sided trace on the basal membrane.

As shown on Figure 5, according to the model, the bigger the trace of a basal cell on the basal membrane, the more likely it is to divide. Cells with a smaller trace than average are likely to detach and ascend in the epidermis. This behaviour (similar to Von Neumann's law) can be explained thanks to the surface tension of the cells. As pointed out by one of us in [18], if the horizontal pressure on the basal layer is isotropic (mean field approximation), it is energetically favorable for a cell with more than six sides to increase its surface attachment on the basal layer whereas it is the opposite if it has less than six sides.

4. Conclusions

We have shown that the epidermis is ruled, at the lowest level and notwithstanding some interesting deviations, by maximum entropy, and is therefore in statistical equilibrium. This can be verified by structural equations of state such as Lewis', Aboav's, and especially Lemaitre's laws, which indicate that entropy is maximized both by making inescapable constraints redundant, and by selecting least informative priors through binning the topological random variables in groups imposed by symmetry [24].

Acknowledgments:B.D. would like to thank Lorna Gibson (M.I.T.) for greeting him in her lab, and the research team I.N.S.E.R.M. U312 for biological data and useful discussions. B.D. is funded by La Société Christian Dior. This work is supported by the EU Mobility Program, FOAMPHYS Network, Contract ER-BCHRXCT940542.

References

1. L. J. Gibson and M. Ashby, *Cellular Solids*, Pergamon, New-York, 1988.
2. F. Lewis, "The correlation between cell division and the shapes and sizes of prismatic cells in the epidermis of the cucumis," *Anat. Records*, **38**, (341), 1928.
3. N. Rivier, X. Arcenegui Siemens, and G. Schliecker, "Cell division and evolution of biological tissues," in *Fragmentation physics*, D. Beysens, X. Campi, and E. Pefferkorn, eds., World Scientific, 1995.
4. D. A. Aboav, "The arrangement of grains in a polycrystal," *Metallogr.*, **3**, (383), 1970.
5. D. Weaire and R. Phelan, "The structure of monodisperse foam," *Phil. Mag. Letters*, **70**, (345), 1994.
6. J. Glazier and D. Weaire, "The kinetics of cellular patterns," *J. Phys. Condens. Matter*, **4**, (1867), 1992.
7. V. Pignol, R. Delannay, and F. Le Caër *Acta Stereol.*, **12**, (149), 1993.
8. D. Weaire and N. Rivier, "Soap cells and statistics -random patterns in two dimensions," *Contemp. Phys.*, **25**, (59), 1984.
9. N. Rivier, "Maximum entropy for random cellular structures," in *From statistical physics to statistical inference and back*, Kluwer, ed., P. Grassberger and J.-P. Nadal, 1994.
10. H. Telley, *Modélisation et simulation de la croissance des mosaiques polycristallines.* PhD thesis, EPFL, Lausanne, Suisse, 1989.
11. J. Lemaitre *et al.*, "Arrangements of cells in voronoi tesselations of monosize packings of discs," *Phil. Mag. B*, **67**, (347), 1993.
12. S. Hales, *Vegetable staticks*, Innis and Woodwark, 1727.
13. L. Errera, "Sur une condition fondamentale d'équilibre des cellules vivantes," **103**, (822), 1886.
14. E. B. Matzke, "In the twinkling of an eye," **27**, (222), 1950.
15. K. J. Dormer, *Fundamental tissue geometry for biologists*, Cambridge university press, 1980.
16. V. V. Smoljaninov, "Mathematical models of tissues," tech. rep., Nauka, 1980. in Russian.
17. N. Rivier and A. Lissowski, "On the correlation between sizes and shapes of cells in epithelial mosaics," *J. Phys. A*, **15**, (L 143), 1982.
18. N. Rivier, "Order and disorder in packings and froths," in *Disorder and granular media*, D. Bideau and A. Hansen, eds., North Holland, 1993.
19. F. T. Lewis, "A volumetric study of the growth and cell division in two types of radially prismatic epidermal cells of cucumis," *Anat. Records*, **47**, (59), 1930.
20. J. Lemaitre, J.-P. Troadec, *et al.*, "Experimental study of densification of disk assemblies," *Europhys. Lett.*, **14**, (77), 1991.
21. G. Le Caër and R. Delannay, "Correlations in topological models of 2D random cellular structures," *J. Phys. A*, **26**, (3931), 1993.

22. W. Montagna and P. Parakkal, *The structure and function of skin*, ch. 2. Academic Press, 3rd ed., 1974.

23. B. Dubertret, "Etude de l'influence de la mort sur l'équilibre statistique de l'épiderme des mammifères," Master's thesis, Université Louis Pasteur, Strasbourg, France, 1994.

24. E. Jaynes, "The well-posed problem," *Found. Physics*, **3**, (477), 1973.

PREDICTING THE ACCURACY OF BAYES CLASSIFIERS

ROBERT R. SNAPP

Computer Science and Electrical Engineering Department
University of Vermont
Burlington, VT 05405[†]

Abstract. A statistical procedure is presented that estimates the probability of error of a Bayes pattern classifier from experimental data. The algorithm is demonstrated on two pattern classification problems, including one that involves multispectral satellite images.

Key words: Pattern recognition, Bayes classifier, Bayes error

1. Introduction

The study of decisions that accurately assign observed phenomena (or patterns) to states of nature is a central activity of statistical pattern recognition [4]. Under the conventional pattern classification paradigm, these decisions are implemented in two stages (see Fig. 1). First, each observable pattern is represented by a *feature vector* $\mathbf{x} \in \mathbb{R}^n$ that is the Cartesian product of n features or measurements. Then each feature vector is assigned to a *pattern class*, e.g., an element of the set $\mathbb{L} = \{1, 2, \ldots, C\}$, that represents one of C distinct states of nature. A function $\theta : \mathbb{R}^n \to \mathbb{L}$ that implements the second stage is called a *pattern classifier*. As a practical example, consider the classification of cells taken from a cancer biopsy. Here, each individual cell can be regarded as a pattern. For each cell, n descriptive features, such as the mean diameters of the cell body and nucleus, are extracted.

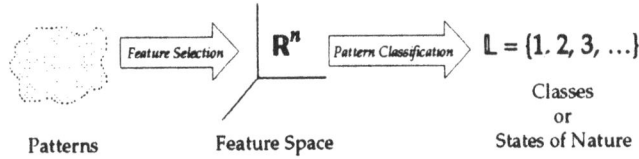

Figure 1. The conventional paradigm for pattern classification.

[†] Email: snapp@emba.uvm.edu

K. M. Hanson and R. N. Silver (eds.), Maximum Entropy and Bayesian Methods, 295–302.

The resulting feature vector is then assigned to one of the three pattern classes *benign, malignant,* or *healthy* by an appropriate mapping θ.

In the following, we let the j-th *decision region* $\mathcal{R}_j = \{\mathbf{x} \in \mathbb{R}^n : \theta(\mathbf{x}) = j\}$ denote the set of feature vectors in $\mathbb{R}^n$ that are assigned to class j. Note that the set of decision regions $\{\mathcal{R}_j\}$ forms a complete partition of the feature space, viz. $\mathbb{R}^n \subseteq \cup_j \mathcal{R}_j$, and $i \neq j \Rightarrow \mathcal{R}_i \cap \mathcal{R}_j = \emptyset$.

If the patterns are generated by a time-invariant random process, then the accuracy of the pattern classifier can be represented by its risk, e.g., the probability that a random input feature vector is misclassified.[1] Thus, let P_ℓ denote the prior probability that a pattern from class ℓ occurs; and $f_\ell(\mathbf{x})$, the probability density of the feature vectors that originate from patterns from class ℓ. We assume that $\sum_{\ell=1}^C P_\ell = 1$. We also let $f(\mathbf{x}) = \sum_{j=1}^C P_j f_j(\mathbf{x})$ denote the total probability density of the feature vectors, and $\mathcal{S} = \{\mathbf{x} \in \mathbb{R}^n : f(\mathbf{x}) > 0\}$, the support of f. If $\mathbf{x} \in \mathcal{S}$, then the posterior probability that an observed feature vector $\mathbf{x}$ belongs to class ℓ can be obtained from Bayes' theorem as

$$\hat{P}_\ell(\mathbf{x}) = \Pr(\ell \mid \mathbf{x}) = \frac{f_\ell(\mathbf{x}) P_\ell}{f(\mathbf{x})}.$$

The probability that a random pattern is assigned to an incorrect class is thus

$$R = \sum_{i=1}^C \int_{\mathcal{R}_i \cap \mathcal{S}} [1 - \hat{P}_i(\mathbf{x})] f(\mathbf{x}) \, d\mathbf{x}.$$

A *Bayes classifier* minimizes this probability by assigning patterns to the class with maximum posterior probability, i.e. $\theta(\mathbf{x}) = \arg\max_{\ell \in \mathbb{L}} \hat{P}_\ell(\mathbf{x})$. In this case the misclassification probability, or *Bayes error* is given by

$$R_B = \int_{\mathcal{S}} \left[1 - \max_{1 \leq \ell \leq C} \hat{P}_\ell(\mathbf{x})\right] f(\mathbf{x}) \, d\mathbf{x}. \tag{1}$$

In practice the prior probabilities $\{P_\ell\}$ and class-conditional densities $\{f_\ell(\mathbf{x})\}$ are rarely known. Thus, most pattern classifiers are designed from incomplete information that may primarily consist of a random set of m labeled feature vectors $\mathcal{X}_m = \{(\mathbf{x}^j, \ell^j) \in \mathbb{R}^n \times \mathbb{L} : j = 1, \ldots, m\}$. (Here the label ℓ^j represents the correct pattern class from which the feature vector $\mathbf{x}^j$ originated.) Popular classifiers include nearest-neighbor classifiers [4], neural network classifiers, and classification trees [1]. Let R_m denote the probability that a practical pattern classifier, trained on a random labeled sample $\mathcal{X}_m$ of size m, misclassifies a random input feature vector. Then from the discussion above, $R_m \geq R_B$. Because R_B measures the intrinsic accuracy of the assumed feature representation, an estimate of its value can be useful even if a Bayes classifier cannot actually be constructed.

---

[1]To simplify our discussion, we assume that every misclassification incurs an equal cost against the classifier's utility. More generally, if c_{ij} denotes the cost of assigning a pattern that originates from class i to class j, then the risk is $R = \sum_{i,j} c_{ij} \int_{\mathcal{R}_j \cap \mathcal{S}} \hat{P}_i(\mathbf{x}) f(\mathbf{x}) \, d\mathbf{x}$.

In the following, we describe and demonstrate a new method for estimating R_B from a finite random sample. Like other studies of this problem [7–9], the new method uses empirical estimates of the finite-sample risk R_m of many k-nearest-neighbor classifiers. (The k-nearest-neighbor algorithm is useful because its infinite-sample risk, $R_\infty = \lim_{m \to \infty} R_m$, can be forced to be arbitrarily close to R_B by selecting k appropriately.) The novel feature of the new method for estimating R_B is that it assumes a parametric model of the finite-sample risk that is valid for a smooth family of classification problems. Fitting this model to empirical measurements of R_m for several values of m, can yield efficient estimates of R_∞ and R_B.

Section 2 describes the k-nearest-neighbor classifier, and its infinite and finite sample properties. The new method for estimating R_B is described in Section 3. Section 4 illustrates the method using two pattern classification problems: a synthetic problem described by two normally distributed pattern classes; and a pixel classification problem with feature vectors taken from a multispectral satellite image of the earth's surface.

2. k-Nearest-Neighbor Classifiers

The k-nearest-neighbor classifier [5,6] is popular for its simplicity and analytic tractability. This algorithm requires four ingredients: (i) a positive integer k, (ii) a metric, or pattern similarity function $d(\mathbf{x}, \mathbf{y})$, (iii) a reference sample of m labeled feature vectors $\mathcal{X}_m = \{(\mathbf{x}^j, \ell^j) \in \mathbb{R}^n \times \mathbb{L} : j = 1, \ldots, m\}$, and (iv) a tie-breaking procedure. Given a new feature vector $\mathbf{x}$, the algorithm identifies the k feature vectors from $\mathcal{X}_m$ that lie closest to $\mathbf{x}$ with respect to the given metric. (These k labeled feature vectors are called the k nearest neighbors of $\mathbf{x}$.) The algorithm then assigns $\mathbf{x}$ to the class that appears most frequently amongst these k nearest neighbors. (If two or more classes occur with greatest frequency, then the tie-breaking procedure is invoked.)

Below, we discuss the infinite and finite sample properties of the risk of the k-nearest-neighbor classifier. Thus, we let R_m denote the probability that a random feature vector $\mathbf{x}$ is misclassified if the reference sample $\mathcal{X}_m$ is constructed by selecting m i.i.d. labeled feature vectors. (The evaluation of R_m is computed by averaging over the ensemble of input feature vectors $\mathbf{x}$, and the ensemble of reference samples $\mathcal{X}_m$ of size m.) We also let $\mathbf{l} = (\ell^1, \ldots, \ell^k) \in \mathbb{L}^k$ denote the ordered k-tuple of class labels of the k-nearest neighbors of $\mathbf{x}$. Here the indices are ordered according to the distances between the k nearest neighbors and $\mathbf{x}$:

$$d(\mathbf{x}, \mathbf{x}^1) \le d(\mathbf{x}, \mathbf{x}^2) \le \cdots \le d(\mathbf{x}, \mathbf{x}^k).$$

The action of the classifier, including the tie-breaking strategy, can be represented by an *assignment partition*, $\mathcal{L}_1, \ldots, \mathcal{L}_C$, of the space $\mathbb{L}^k$, where the element $\mathcal{L}_i$ contains every ordered k-tuple of class labels, $\mathbf{l} = (\ell^1, \ldots, \ell^k)$, that yields an assignment to class i.

2.1. THE INFINITE-SAMPLE LIMIT

Cover and Hart [2] showed that under weak conditions R_m tends to the limit

$$R_\infty(k) = \sum_{\ell=1}^{C} \sum_{1 \notin \mathcal{L}_\ell} \int_S dx\, f(\mathbf{x})\, \hat{P}_\ell(\mathbf{x})\, \hat{P}_{\ell^1}(\mathbf{x}) \cdots \hat{P}_{\ell^k}(\mathbf{x}), \qquad (2)$$

as $m \to \infty$. Moreover they obtained the following bounds on $R_\infty(k)$:

$$R_B \leq R_\infty(k) \leq R_\infty(1) \leq R_B\left(2 - \frac{C}{C-1}R_B\right). \qquad (3)$$

This result was later extended by Stone [13] who showed that if $k \to \infty$, such that $k/m \to 0$ as $m \to \infty$, then $R_m \to R_B$, almost surely. Thus, the k-nearest-neighbor classifier is asymptotically consistent with a Bayes classifier.

As a practical application of asymptotic consistency, Devroye [3] derived the following inequalities: if $k \geq 5$, and $C = 2$, then there exist universal constants $\alpha = 0.3399\cdots$, and $\beta = 0.9749\cdots$ such that $R_\infty(k)$ is bounded by

$$R_B \leq R_\infty(k) \leq (1 + a_k)R_B, \quad \text{where} \quad a_k = \frac{\alpha\sqrt{k}}{k-3.25}\left(1 + \frac{\beta}{\sqrt{k-3}}\right). \qquad (4)$$

More generally with $C = 2$,

$$R_B \leq R_\infty(k) \leq \left(1 + \sqrt{\frac{2}{k}}\right)R_B. \qquad (5)$$

2.2. THE FINITE-SAMPLE PROBABILITY OF ERROR

The probability that a k-nearest-neighbor classifier, using a random reference sample of m labeled feature vectors, misclassifies a random feature vector can be expressed exactly as

$$R_m = \int_S dx \int_S dx^k \int_{B^k} dx^{k-1} \cdots \int_{B^3} dx^2 \int_{B^2} dx^1\, g(\mathbf{x}, \mathbf{x}^1, \ldots, \mathbf{x}^k)\, e^{-(m-k)\, h(d(\mathbf{x},\mathbf{x}^k),\mathbf{x})}, \qquad (6)$$

where $B^j = \{\mathbf{y} \in S : d(\mathbf{x}, \mathbf{y}) \leq d(\mathbf{x}, \mathbf{x}^j)\}$ represents the intersection of the ball of radius $d(\mathbf{x}, \mathbf{x}^j)$ centered at the point $\mathbf{x}$ with S, and the functions g and h are represented as follows. Let

$$\psi(\rho, \mathbf{x}) = \int_{d(\mathbf{y}, \mathbf{x}) < \rho} f(\mathbf{y})\, d\mathbf{y}$$

denote the probability that a random feature vector falls within a distance ρ from $\mathbf{x} \in \mathbb{R}^n$, and let $(m)_k = m(m-1)\cdots(m-k+1)$. Then g and h are given by,

$$g(\mathbf{x}, \mathbf{x}^1, \ldots, \mathbf{x}^k) = (m)_k \sum_{\ell=1}^{C} P_\ell f_\ell(\mathbf{x}) \sum_{1 \notin \mathcal{L}_\ell} \prod_{j=1}^{k} P_{\ell^j} f_{\ell^j}(\mathbf{x}^j),$$

and $h(\rho, \mathbf{x}) = -\log(1 - \psi(\rho, \mathbf{x}))$.

A general evaluation of Eqn. (6) does not appear to be possible. However, if (i) f is finite and uniformly bounded away from zero over S, (ii) the class-conditional densities $f_\ell(\mathbf{x})$ have uniformly bounded partial derivatives up through order $N+1$, with $N \geq 2$, at every point in S, and (iii) the contributions to the integral from the boundary of S are negligible, then the integral in Eqn. (6) can be evaluated using a multidimensional generalization of Laplace's method [10,11]. In this case, R_m can be represented by the truncated asymptotic expansion

$$R_m = R_\infty(k) + \sum_{j=2}^{N} c_j m^{-j/n} + O\left(m^{-(N+1)/n}\right). \tag{7}$$

For problems that satisfy hypotheses (i) – (iii), the coefficients c_2 through c_N can be expressed in terms of k, the chosen metric d, the prior probabilities P_ℓ and the class-conditional densities $f_\ell(\mathbf{x})$. (In particular, $c_3 = 0$ if $n \geq 2$, and $c_5 = 0$ if $n = 2$ or $n \geq 4$.) If the distributions are unknown, the coefficients in Eqn. (7) may be inferred empirically.

3. Empirical Estimations of the Bayes Risk

Eqn. (7) describes how the finite-sample risk R_m converges to its infinite-sample limit $R_\infty(k)$ for a smooth family of classification problems. As an application, this formula can be used to predict the infinite-sample risk of a k-nearest-neighbor classifier from empirical estimates of R_m for different sample sizes. For example, let $\{\hat{R}_{m_i}\}$ denote a sequence of estimates of the risk for the sample-size sequence $\{m_i : i = 1, \ldots, M\}$. Each estimate, $\hat{R}_{m_i}$, might represent the relative frequency that labeled random feature vectors are misclassified by an ensemble of k-nearest-neighbor classifiers that use independent random reference samples of size m_i. The source of labeled feature vectors could be a real-time random process, or a fixed pool of random data that is sampled with replacement.

Once the set of estimates of $\hat{R}_{m_i}$ are determined, the values of $R_\infty(k)$, and c_2 through c_N can be estimated by finding the values of these parameters that minimize

$$\sum_{i=1}^{M} \left(\hat{R}_{m_i} - R_\infty(k) - \sum_{j=2}^{N} c_j m_i^{-j/n} \right)^2.$$

The resulting estimate of $R_\infty(k)$ is especially interesting in that it can be used to predict the Bayes error. For example, if $k = 1$, then (3) implies that $R_\infty(k)/2 \leq R_B \leq R_\infty(k)$. Tighter bounds follow from Devroye's inequalities (4) and (5). To obtain an estimate of R_B with precision ϵ, choose $k > 2/\epsilon^2$, and estimate $R_\infty(k)$ by the above method. Then by (5), $R_\infty(k) - \epsilon \leq R_B \leq R_\infty(k)$. The next two examples demonstrate these methods.

TABLE 1. Estimates of the model coefficients and Bayes error for a classification problem with two normal classes.

k	$R_\infty(k)$	$n = 1$ $(N = 2)$	$n = 5$ $(N = 6)$
1	0.2248	$R_m = 0.2287 + \dfrac{0.6536}{m^2}$ $R_B = 0.172 \pm 0.057$	$R_m = 0.2287 + \dfrac{0.1121}{m^{2/5}} + \dfrac{0.2001}{m^{4/5}} - \dfrac{0.0222}{m^{6/5}}$ $R_B = 0.172 \pm 0.057$
7	0.1746	$R_m = 0.1744 + \dfrac{4.842}{m^2}$ $R_B = 0.152 \pm 0.023$	$R_m = 0.1700 + \dfrac{0.2218}{m^{2/5}} - \dfrac{1.005}{m^{4/5}} + \dfrac{3.782}{m^{6/5}}$ $R_B = 0.148 \pm 0.022$
63	0.1606	$R_m = 0.1606 + \dfrac{20.23}{m^2}$ $R_B = 0.157 \pm 0.004$	$R_m = 0.1595 + \dfrac{0.1002}{m^{2/5}} - \dfrac{1.426}{m^{4/5}} + \dfrac{10.96}{m^{6/5}}$ $R_B = 0.156 \pm 0.004$

4. Experimental Results

4.1. TWO NORMAL PATTERN CLASSES

Consider a two-class problem with $P_1 = P_2 = 1/2$, and

$$f_\ell(\mathbf{x}) = \frac{1}{(2\pi)^{n/2}} e^{-\frac{1}{2}\left((x_1 + (-1)^\ell)^2 + \sum_{i=2}^n x_i^2\right)},$$

for $\ell = 1$ and 2. Pseudorandom labeled feature vectors $(\mathbf{x}, \ell)$ were numerically generated in accordance with the above for dimensions $n = 1$ and $n = 5$. Twelve sample sizes between 10 and 3000 were examined. For each dimension and sample size the risks R_m of many independent k-nearest-neighbor classifiers with $k = 1, 7$, and 63 were empirically estimated. (Because Eqn. (7) does not accurately describe the very small sample behavior of the k-nearest-neighbor classifier, sample sizes smaller than $2k$ were not included in the fit.)

Estimates of the coefficients in Eqn. (7) for six different fits appear in the first equation of each cell in Table 1. For reference, the second column contains the value of $R_\infty(k)$ obtained by numerically integrating Eqn. (2). Estimates of the Bayes risk appear in the second equation of each cell. Cover and Hart's inequality (3) was used for $k = 1$, and Devroye's inequality (5) was used if $k \geq 7$. For this problem, Eqn. (1) evaluates to $R_B = (1/2)\,\mathrm{erfc}(1/\sqrt{2}) = 0.15865$.

4.2. A PIXEL CLASSIFICATION PROBLEM

In the second experiment, described in Snapp and Xu [12], a large random sample (or pool) of 2^{22} labeled feature vectors were extracted from the pixels in a

TABLE 2. Coefficients that minimize the squared error fit for different values of N.

N	$R_\infty(1)$	c_2	c_4	c_6
2	0.0757133	0.126214		
4	0.0757846	0.124007	0.0132804	
6	0.0766477	0.0785847	0.689242	-2.68818

multispectral satellite image of the earth's surface. Each pixel was represented by five spectral components, $\mathbf{x} = (x_1, \ldots, x_5)$, each in the range $0 \leq x_\nu \leq 255$. (Thus, $n = 5$.) The class label of each pixel was determined by one of the remaining spectral components, $0 \leq y \leq 255$. Two pattern classes were defined: $\{y < \gamma\}$, and $\{y \geq \gamma\}$, where γ was a predetermined threshold. With $k = 1$, 200 – 1000 Bernoulli trials were performed for each value of m_i. For each trial a reference sample of m_i labeled feature vectors was constructed by selecting vectors from the pool with replacement. The risk induced by each reference sample was then estimated by classifying between 2000 and 20000 labeled "test" vectors that were also randomly selected with replacement from the pool. The empirical risk $\hat{R}_{m_i}$ was computed as the relative frequency that test vectors were misclassified, averaged over all reference samples of size m_i. (The number of experiments performed for each value of m_i, and the number of test vectors, were chosen to ensure that the sample variance of $\hat{R}_{m_i}$ was less than 10^{-4}.) This process was repeated for 33 different values of m_i in the range $100 \leq m_i \leq 15000$.

The coefficients obtained are displayed in Table 2. Note that the robustness of the fit begins to dissolve, for this data, at $N = 6$. However, the estimate for $R_\infty(1)$ appears to be stable. Using the $N = 4$ estimate for $R_\infty(1)$, Eqn. (3) predicts that $R_B = 0.0568 \pm 0.0190$. A plot of the best fit with $N = 4$ against the experimental data appears in Fig. 2. The close agreement between the data points and the solid curve suggests that the parametric finite-sample risk model (7) is valid for this classification problem.

5. Conclusion

Efficient predictions of the Bayes error from sample data may guide the development of more accurate classifiers and feature representations for practical pattern recognition problems. By fitting a parametric model of the finite-sample risk of the k-nearest-neighbor classifier to a set of empirical risk estimates, the infinite-sample risk of this nonparametric classifier can be estimated. Predictions of the Bayes risk can then be obtained from inequalities that are based on the Bayes consistency of this algorithm. Greater efficiency may result from the theory of optimal experimental design and a careful application of Bayesian methods.

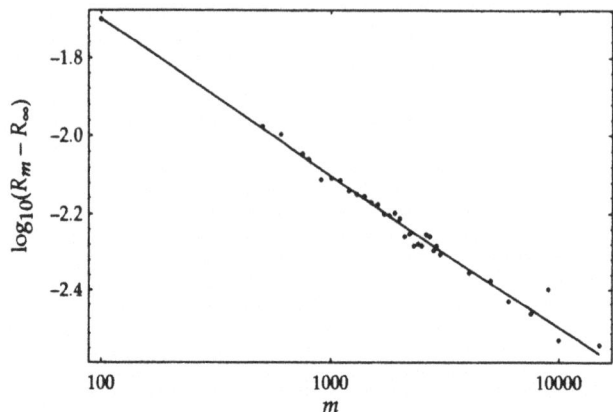

Figure 2. The best fourth-order ($N = 4$) fit of Eqn. (7) to 33 empirical estimates of $\hat{R}_{m_i}$ for a pixel classification problem obtained from a multispectral satellite image. Using $R_\infty = 0.0758$, the fourth-order fit, $R_m = 0.0758 + 0.124m^{-2/5} + 0.0133m^{-4/5}$, is plotted as a solid curve on a log-log scale to reveal the significance of the $j = 2$ term.

Acknowledgments. This work was supported in part by Rome Laboratory, Air Force Material Command, USAF, under grant number F30602-94-1-0010, and by the National Science Foundation, under grant number No. NSF OSR-9350540.

References

1. L. Breiman, J. H. Friedman, R. Olshen, and C. J. Stone, *Classification and Regression Trees*, Wadsworth & Brooks, Pacific Grove, CA, 1984.
2. T. M. Cover and P. E. Hart, "Nearest neighbor pattern classification," *IEEE Trans. Inform. Theory*, IT–13, 1967, pp. 21–27.
3. L. Devroye, "On the asymptotic probability of error in nonparametric discrimination," *Ann. Statist.*, vol. 9, 1981, pp. 1320–1327.
4. R. O. Duda and P. E. Hart, *Pattern Classification and Scene Analysis*, John Wiley & Sons, New York, 1973.
5. E. Fix and J. L. Hodges, "Discriminatory Analysis: Nonparametric Discrimination: Consistency Properties," from *Project 21-49-004, Report Number 4*, UASF School of Aviation Medicine, Randolf Field, Texas, 1951, pp. 261–279.
6. E. Fix and J. L. Hodges, "Discriminatory Analysis: Nonparametric Discrimination: Small Sample Performance," from *Project 21-49-004, Report Number 11*, UASF School of Aviation Medicine, Randolf Field, Texas, 1951, pp. 280–322.
7. K. Fukunaga and D. M Hummels, "Bayes error estimation using Parzen and k-NN procedures," *IEEE Trans. Pattern Anal. Mach. Intell.*, PAMI-9, 1987, pp. 634–643.
8. J. M. Garnett III and S. S. Yau, "Nonparametric estimation of the Bayes error of feature extractors using ordered nearest neighbor set," *IEEE Trans. Comp.*, C-26, 1977, pp. 46–54.
9. G. Loizou and S. J. Maybank, "The nearest neighbor and the Bayes error rates," *IEEE Trans. Pattern Anal. Mach. Intell.*, PAMI-9, 1987, pp. 254–262.
10. D. Psaltis, R. R. Snapp, and S. S. Venkatesh, "On the finite sample performance of the nearest neighbor classifier," *IEEE Trans. Inform. Theory*, IT-40, 1994, pp. 820–837.
11. R. R. Snapp and S. S. Venkatesh, "k Nearest Neighbors in Search of a Metric," 1995, submitted.
12. R. R. Snapp and T. Xu, "Estimating the Bayes Risk from Sample Data," 1995, submitted.
13. C. J. Stone, "Consistent nonparametric regression," *Ann. Statist.*, vol. 5, 1977, pp. 595–645.

MAXIMUM ENTROPY ANALYSIS OF GENETIC ALGORITHMS

JONATHAN L. SHAPIRO AND MAGNUS RATTRAY
Department of Computer Science
University of Manchester
Manchester, M13 9PL, UK

AND

ADAM PRÜGEL-BENNETT
NORDITA
Blegdamsvej 17,
K-2100 Copenhagen Ø, Denmark

Abstract.
Genetic algorithms are widely-used search techniques which have been applied to many problems in optimization, machine learning, design, and many other domains. However, genetic algorithms are not well-understood theoretically, because the dynamics of genetic algorithms are mathematically intractable in even the most simple situations. We show how maximum entropy assumptions can be used to derive a set of equations describing the evolution of the algorithm on simple model problems.

Key words: Genetic Algorithms, Optimization, Statistical Mechanics, Maximum Entropy Distributions

1. Introduction

Genetic Algorithms (GAs) are a class of stochastic search techniques which have been widely used to seek solutions to hard problems. They have been applied in a wide range of domains, including: optimization problems, machine learning, training neural networks or evolving neural network architectures, design of electronic and gas networks, and many other problems. Perhaps the most important reason for the popularity of GAs is the simple fact that they have been found to work in so many applications. For introductions to genetic algorithms, see [1-3].

Although they have been widely studied empirically, genetic algorithms are not well understood theoretically. One reason is that their dynamics is extremely difficult to solve in even the simplest situations. This is in contrast with other search techniques. Consider, for example, gradient descent search. Although one obviously cannot predict how gradient algorithms will perform on arbitrary com-

K. M. Hanson and R. N. Silver (eds.), Maximum Entropy and Bayesian Methods, 303–310.

plex search problems, one *can* predict how it will perform on simple problems, such as a multidimensional quadratic minimum. Solving such problems helped researchers understand how the method worked and led to such improvements as conjugate gradient methods or Newton's method [4]. With regard to genetic algorithms, however, there are no solvable systems. Theoretical techniques which have been proposed to describe GAs are either intractable [5], fail to make quantitative predictions, or make predictions which are not realized experimentally [6,7].

The goal of our research is to develop a formalism which allows one to solve GA dynamics in simple systems. We desire methods which can work on problems of realistic problem size and finite population size. The actual optimization problems we have studied are clearly toys, however they contain some features in common with more realistic problems. Our hope is that the study of such systems will lead to a better understanding of how genetic algorithms work and how they can be better applied.

However, we will not solve this system from first principles; a maximum entropy assumption is essential to our approach. We will pick a few degrees of freedom which we believe characterize the system. All other degrees of freedom will be assumed to follow a maximum entropy distribution. From this, we can derive a simple set of iterative equations which when iterated numerically predict the properties of the best solution found in terms of the search parameters and the starting state, and the time it takes to find such a solution. Comparison with simulations show that the method is accurate on the problems we have studied.

2. The Basic Genetic Algorithm and Why it is Hard to Analyze

The basic idea of a genetic algorithm is very simple. First, one encodes candidate solutions to the problem one is trying to solve as strings. We consider binary strings here, although it is often useful to employ strings of characters drawn from a larger alphabet. Genetic algorithms act on a population of strings.

This population is acted upon by three genetic operators. Mutation makes small local changes to the potential solutions; for instance by replacing each bit with its compliment with a small probability. Recombination (crossover) combines parts of pairs of solutions to produce new solutions. For each recombination two parents are chosen from the population at random. There are numerous ways of recombining. For example, in single-point crossover, a crossing point i^* is chosen at random. The new string is produced by taking all bits to the right of i^* from one parent and all bits to the left from the other. Multiple crossing points may also be used (multi-point crossover) or the bits can be selected from each parent at random (uniform crossover). The final operator, selection, introduces no new solutions; rather it changes the proportions of strings in the population: the proportion of higher quality solutions is increased and the proportion of lower quality solutions is decreased. These three operators are applied to the population repeatedly until the population stops improving or a solution of a desired quality is found.

The system described above is very simple to simulate; why is it so difficult to analyze mathematically? The fact that the GA acts on a population and the binary operation of crossover make this system very difficult. For example, a genetic algorithm is a Markov process, and Markov chain analysis has been applied to it, both in a biological context [8] and in engineering applications [5,9]. In this formulation, the state of the system can be described by the probability of every possible string being in the population — 2^N numbers for binary strings of length N. The transition probabilities between states is a quadratic function of the current state, due to the fact that recombination requires a pair of parents. Thus, although one can describe the transition of the current state under the genetic search operators formally, it can only be solved in special circumstances such as very small strings (one or two bits), or in the infinite population limit (which is not a good approximation because in an infinite population the optimal solution is guaranteed to be present in the population; one only requires selection to amplify it), or in selection free "neutral theory". Attempts to solve this stochastic dynamics directly have also only been accomplished in very small systems [10].

3. The Maximum Entropy Ansatz

If one tries to solve GA dynamics rigorously, it usually means attempting to solve for all the details of the system. For example, the Markov chain formulation, if solved, would give the probability of every possible string. This is an incredible amount of information; this superfluity of information contributes to the complexity of the approach. This is much more information than one is actually interested in. What one really wants to know, typically, are answers to questions like the following: what is the fitness of the best member of the population as a function of time, what is the average fitness is as a function of time, and when do these stop improving. Thus, what is required are dynamical equations describing the evolution of a few statistical measures of the population.

We have developed an approach which does this [11-13]. It is analogous to what one does in statistical mechanics: rather than solving the detailed dynamics of all of the molecules in a system, one treats the behavior of a few macroscopic quantities and assumes that the distribution of all degrees of freedom maximize entropy subject to the known macroscopic quantities. In this system, however, maximum entropy will have to be applied at each timestep.

It may not be obvious why maximum entropy assumptions are required. Why not chose macroscopic order parameters and integrate out the remaining degrees of freedom? The reason is that it is hard to find order parameters which characterize all search operators. Selection acts directly on fitness, so the if the system is described by characteristics of the fitness distribution, the effect of selection can be found directly. However, mutation and crossover effects depend upon the configurations of the strings and the inter-string correlations in the population. These cannot be inferred from fitness statistics. Conversely, if we use statistical measures of the string configurations and correlations, we will be unable to infer the fitnesses properties of the strings (except for very special fitness functions), and will be unable to calculate the effect of selection. Maximum entropy assumptions allow us to close the loop by inferring the unknown quantities from the known ones. This is illustrated with a simple example in the next section.

4. Derivation of Dynamical Equations in a Simple Problem

The method will be illustrated with a very simple problem, the optimization of a linear, spatially inhomogeneous function. In this problem, each bit of the string contributes individually an arbitrary amount. The fitness is,

$$F[\vec{\tau}] = \sum_{i=1}^{N} J_i \tau_i.$$

Here $\vec{\tau}$ represents the string of ± 1 of length N, and J_i is a random number (usually drawn from zero-mean, unit-variance Gaussian distribution). This is obviously a trivial problem to solve and a GA is not required to solve it (simply set all τ_i's to the sign their corresponding J_i). It is a useful problem, because it has a non-trivial relationship between string configuration and fitness.

We will take as order parameters the first four cumulants of the fitness distribution in the population. Cumulants are statistical properties of a distribution: the first two are the mean and variance respectively, the third is related to the skewness and measures the asymmetry of the distribution, and the fourth is related to the kurtosis; it measures whether the distribution falls off faster or more slowly than a Gaussian [14]. The ith cumulant is denoted κ_i. We will show how to derive equations describing the cumulants after the application of each search operator in terms of those before.

4.1. SELECTION

We consider Boltzmann selection in which the probability of selecting a string with fitness F is proportional to $\exp(\beta F)$, where the normalization factor is chosen to insure that some string is chosen at each selection. Here β controls the strength of selection.

One might think that the new distribution of fitness in terms of the old $\rho(F)$ would simply be $\exp(\beta F)\rho(F)$ up to a normalization factor. In a finite population, however, there will be fluctuations around this. The way to treat this is to draw P levels from ρ. The "partition function" is the generating function for moments,

$$Z(\gamma) = \sum_{\alpha=1}^{P} \exp(\beta F^{\alpha}) \exp(-\gamma F^{\alpha}).$$

The generating function for cumulants is the log of this, which we average over selections of the levels from the distribution ρ

$$G(\gamma) = \left[\prod_{\alpha=1}^{P} \int_{-\infty}^{\infty} \rho(F^{\alpha}) dF^{\alpha} \right] \log(Z(\gamma)).$$

Derivatives of G in terms of γ will yield the cumulants. This problem is analogous to the Random Energy Model proposed and studied by Derrida [15], and can be computed by using the same trick — represent the log as an integral,

$$\log(Z) = \int\limits_{0}^{\infty} \frac{\exp(-t) - \exp(-t\,Z)}{t} dt. \tag{1}$$

and average over the energy levels.

For Boltzmann selection, the cumulant expansion for the distribution after selection in terms of selection before can be found as an expansion in the selection parameter β,

$$
\begin{aligned}
\kappa_1^s &= \kappa_1 + \beta\left(1 - \frac{1}{P}\right)\kappa_2 + \frac{\beta^2}{2}\left(1 - \frac{3}{P}\right)\kappa_3 + \frac{\beta^3}{3!}\left[\left(1 - \frac{7}{P}\right)\kappa_4 - \frac{6}{P}\kappa_2^2\right] + \cdots, \\
\kappa_2^s &= \left(1 - \frac{1}{P}\right)\kappa_2 + \beta\left(1 - \frac{3}{P}\right)\kappa_3 + \frac{\beta^2}{2}\left[\left(1 - \frac{7}{P}\right)\kappa_4 - \frac{6}{P}\kappa_2^2\right] + \cdots, \\
\kappa_3^s &= \left(1 - \frac{3}{P}\right)\kappa_3 + \beta\left[\left(1 - \frac{7}{P}\right)\kappa_4 - \frac{6}{P}\kappa_2^2\right] + \cdots.
\end{aligned}
\tag{2}
$$

Or the cumulants can be found for arbitrary β numerically.

Thus, the cumulants after selection can be written in terms of those before with no assumptions required. Some implications of these equations to chosen appropriate selection parameters was discussed in earlier work [12,13].

4.2. MUTATION

Mutation and crossover differ from selection in that selection acts on the fitness of the string, while mutation and crossover act on the configuration of the string. Thus, the effects of these later two operators will depend on statistics of the configuration as well as statistics of the fitness distribution. For example, mutation (as we will see shortly) depends on whether the fitness occurs by aligning many $\tau's$ with small J_i's or a few τ's with large J_i's; a characteristic which cannot be inferred from the fitness distribution alone. This is where a maximum entropy ansatz is required.

To study the effect of mutation, we introduce a set of mutation variables m_i^α, one for each site of each string. This is 1 if the site is mutated on that string; 0 otherwise, where m is the mutation probability. In terms of these variables, the fitness after mutation is

$$F = \sum_i J_i \left(1 - 2m_i^\alpha\right) \tau_i^\alpha. \tag{3}$$

Averaging over all variables gives the cumulants after mutation. This yields,

$$
\begin{aligned}
\kappa_1^m &= (1 - 2m)\kappa_1 \\
\kappa_2^m &= \kappa_2 + \left(1 - (1 - 2m)^2\right)\left(\sum_i J_i^2 - \kappa_2\right)
\end{aligned}
\tag{4}
$$

$$\kappa_3^m = \kappa_3(1 - 2m)^3 - 2(1 - 2m)\left(1 - (1 - 2m)^2\right) \sum_i J_i^3 \langle \tau_i \rangle$$

where $< \tau >$ denotes population average. (I have ignored $1/P$ corrections which are easily worked out [13]). The fourth cumulant can also be worked out; it depends upon an interstring correlation $\sum_i J_i^4 < \tau_i^\alpha \tau_i^\beta >_{\alpha \neq \beta}$ as well as the fourth cumulant before mutation.

The first two equations express obvious effects — mutation brings the mean and variance back towards values of a random population. This means that the mean decreases towards zero and the variance increases toward $\sum_i J_i^2$. The higher cumulants depend not only on fitness cumulants before mutation, however, they depend upon the configurations of strings in the population. For example, the third cumulant depends upon the correlation between the bits and the $J's$ at each site, the terms which involve configurational average of the strings $< \cdots >$. We cannot from the fitness distribution alone.

4.3. CROSSOVER

Crossover can be treated in a similar manner to mutation. Introduce a set of crossover variables

$$\chi_i^{\alpha\beta} = \begin{cases} 1 & \alpha \text{ crossed with } \beta \text{ at site } i \\ 0 & \text{otherwise} \end{cases} \tag{5}$$

the statistics of which are determined by the precise form of crossover.

If crossover produces two complementary children, fitness is conserved in such a way that both the mean and the variance remains unchanged. The third and higher cumulants are reduced and brought to natural values. For simplicity, the equations are not expressed here, but can be find in [13]. However, these involve similar configurational averages.

4.4. THE MAXIMUM ENTROPY CALCULATION OF THE CONFIGURATIONAL AVERAGES

We need to know the average bit configuration per site

$$t_i \equiv < \tau_i > = \frac{1}{P} \sum_{\alpha=1}^{P} \tau_i^\alpha \tag{6}$$

where α indexes the population member. This is allowed to fluctuate subject to the constraints that the mean is fixed and the correlation is fixed. The maximum entropy distribution [16] is proportional to

$$\exp\left[\lambda_1 P \sum_i J_i t_i + \lambda_2 P^2 \sum_i t_i^2 + S(t_i)\right], \tag{7}$$

where the entropy is (after Stirling's approximation)

$$S(t_i) = -\frac{P}{2} \log\left(1 - t_i^2\right) + \frac{P}{2} t_i \log\left(\frac{1 - t_i}{1 + t_i}\right). \tag{8}$$

The macroscopics are found by uncompleting the square and optimizing. The cumulants are found to be

$$\kappa_1 = \sum_i J_i \int_{-\infty}^{\infty} \tanh(\lambda_1 J_i + \lambda_2 x) \exp\left(-\frac{x^2}{2}\right) \frac{dx}{\sqrt{2\pi}}. \tag{9}$$

$$\kappa_2 = \sum_i J_i^2 \left[1 - \int_{-\infty}^{\infty} \tanh^2(\lambda_1 J_i + \lambda_2 x) \exp\left(-\frac{x^2}{2}\right) \frac{dx}{\sqrt{2\pi}} \right]. \tag{10}$$

These set the values of the Lagrange multipliers λ_1 and λ_2. The configurational averages can be found from these. For example, to compute the third cumulant after mutation, we required

$$t_i = \int_{-\infty}^{\infty} \tanh(\lambda_1 J_i + \lambda_2 x) \exp\left(-\frac{x^2}{2}\right) \frac{dx}{\sqrt{2\pi}}. \tag{11}$$

5. Results and Application to More Complex Problems

The maximum entropy assumptions can be tested directly by measuring the average bit as a function of J_i or by determining how well the method predicts the behavior of the genetic algorithm. Both tests suggest that the approach is reasonably accurate. For example, figure 1 shows a comparison of a simulation with the theory described here iterated numerically. These results are typical — the method gets the asymptotics and the time to stasis accurately, but is off for intermediate times.

This approach has been applied to more complex problems: a spin-glass chain, a problem which contains a large number of local minima, but is trivial to solve [13]; and the subset-sum problem [17], a task which is NP-complete, but can be solved by pseudopolynomial methods.

We find that the dynamics of genetic algorithms can be solved by choosing a few macroscopic quantities and inferring the constituents by a maximum entropy assumption. This allows us to derive a set of equations for the macroscopic quantities which predict the behavior of the algorithm on simple problems. Since the maximum entropy assumption appears to be a good approximation in the systems we have studied, we have learned something about genetic algorithms, namely they are sufficiently mixing to make all populations with given fitnesses statistics equally likely.

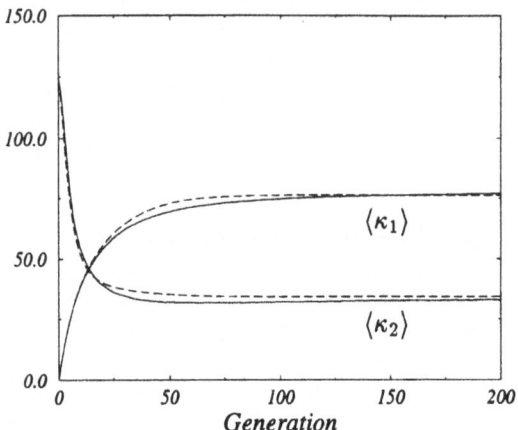

Figure 1. The solid curves show the evolution of the first two cumulants for $N = 127$, $P = 50$, and $m = 1/(2N)$. The dashed curve shows the theoretical prediction.

References

1. D. E. Goldberg, *Genetic Algorithms in Search, Optimization & Machine Learning*, Addison-Wesley (Reading, Mass), 1989.
2. L. Davis, *Handbook of Genetic Algorithms*, Van Nostrand Reinhold (New York), 1991.
3. M. Mitchell, *An Introduction to Genetic Algorithms*, MIT Press, Cambridge, MA, 1995.
4. W. H. Press, S. A. Teukoldky, W. T. Vetterling, and B. P. Flannery, *Numerical Recipes in C*, Cambridge University Press, second edition ed., 1992.
5. A. Nix and M. D. Vose, "Modeling genetic algorithms with markov chains," *Annals of Mathematics and Artificial Intelligence*, 5, pp. 79–88, 1991.
6. J. J. Grefenstette, "Deception considered harmful," in *Foundations of Genetic Algorithms 2*, L. D. Whitley, ed., Morgan Kaufmann (San Mateo), 1993.
7. S. Forrest and M. Mitchell, "What makes a problem hard for a genetic algorithm? some anomalous results and their explanation," *Machine Learning*, 13, (2/3), 1993.
8. W. Ewens, *Mathematical Population Genetics*, Springer, New York, 1972.
9. H. Mühlenbein, "Evolution in space and time – the parallel genetic algorithm," in *Foundations of Genetic Algorithms*, G. Rawlins., ed., Morgan-Kaufman, 1991.
10. D. Whitley, "An executable model of a simple genetic algorithm," in *Foundations of Genetic Algorithms 2*, L. D. Whitley, ed., Morgan Kaufmann (San Mateo), 1993.
11. A. Prügel-Bennett and J. L. Shapiro, "An analysis of genetic algorithms using statistical mechanics," *Phys. Rev. Letts.*, 72, (9), pp. 1305–1309, 1994.
12. J. L. Shapiro, A. Prügel-Bennett, and M. Rattray, "A statistical mechanical formulation of the dynamics of genetic algorithms," *Lecture Notes in Computer Science*, 864, pp. 17–27, 1994.
13. A. Prügel-Bennett and J. L. Shapiro, "Dynamics of genetic algorithms for simple random ising systems." Submitted for publication, 1995.
14. M. Abramowitz and I. Stegun, *Handbook of Mathematical Functions*, Dover Publications, Inc., 1964.
15. B. Derrida, "Random-energy model: An exactly solvable model of disordered systems," *Phys. Rev.*, **B24**, pp. 2613–2626, 1984.
16. T. Cover and J. Thomas, *Elements of Information Theory*, Wiley, 1991.
17. M. Rattray, "An analysis of a genetic algorithm solving the subset-sum problem." Submitted for publication, 1995.

DATA FUSION IN THE FIELD OF NON DESTRUCTIVE TESTING

S. GAUTIER[1,2], G. LE BESNERAIS[2,3]

AND

A. MOHAMMAD-DJAFARI[2], B. LAVAYSSIÈRE[1]

[1] *EDF, Direction des Études et Recherches,*
6 quai Watier, 78401 Chatou, France
[2] *Laboratoire des Signaux et Systèmes (CNRS-ESE-UPS),*
Supélec, Plateau de Moulon, 91192 Gif-sur-Yvette, France [†]
[3] *Presently : ONERA, 29 avenue de la Division Leclerc,*
92320 Chatillon, France

Abstract. This paper deals with the reconstruction of metallic blocks from gammagraphic and ultrasonic data. The inverse problems that have to be faced are solved within a Bayesian framework. Firstly, the problem of 3D reconstruction from gammagraphies is discussed. Then, we propose a new deconvolution method for ultrasonic traces. Finally, we present a reconstruction method that accounts for both sets of data: the ultrasonic data is used to detect breaks in the object and those bounds are then incorporated in the prior model for the reconstruction from gammagraphies.

1. Introduction

As a part of its researches in non destructive testing, Électricité de France is interested in the evaluation of metallic blocks that might include air flaws. Following this goal, we deal with the reconstruction of metallic blocks from X-ray and ultrasonic imaging. The actual conditions of control for gammagraphy prevent the gamma source from being turned around the inspected object: to be more precise, the source must be placed above the block and can only be orientated within a cone whose top-angle is reduced to 20 degrees (see Fig.1a). Such conditions greatly reduce the amount of information from the data along the vertical direction, and thus disable any good quality reconstruction from the gammagraphies alone. To overcome this lack of information, ultrasonic traces that provide information along the vertical direction are collected on the upper face of the object (see Fig.1b). We

---

[†] Email: gautier@lss.supelec.fr

K. M. Hanson and R. N. Silver (eds.), Maximum Entropy and Bayesian Methods, 311–316.
© 1996 *Kluwer Academic Publishers.*

aim at fusing these complementary types of data in order to obtain a 3D image of the object.

Firstly, we state the notations and describe the mathematical models that are chosen for the observation processes. Then, we present Bayesian reconstruction methods that fit separately to each inspection system. Finally, taking advantage of this work, we deduce a sequential reconstruction method that takes both X-ray and ultrasonic data into account.

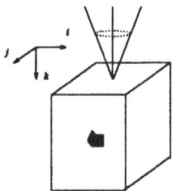

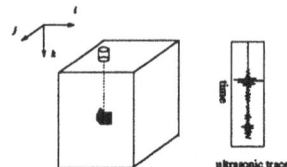

Fig.1a: conditions of control for gammagraphy.   Fig.1b: conditions of control for ultrasonic imaging.

2. Models for the observation processes

The observed radiation, that is formed into the vector y after being sampled, is supposed to be described by the equation $y = Ax + b$, where x stands for the sampled attenuation of the object, A is the observation matrix (that results from the Beer-Lamber law) and b is a noise process that is assumed to be centered, white, Gaussian and independant of x.

For the ultrasonic data, we presume that for each sensor position, the sampled trace z results from the discrete convolution between the wavelet h, that is emitted along the vertical direction by the sensor, and a sequence of reflectivity r that represents the inhomogeneities of the medium under the considered sensor position. The convolution process is supposed to be corrupted by an additive, centered white and Gaussian noise n that is assumed to be independant of r. Practically, h has to be identified: a minimal-phase wavelet is estimated by linear prediction of the trace z. This minimal phase hypothesis may be innacurate, but it seems that no robust identification method is known for our problem, and the phase estimation would need a special study. Finally, in matrix notation, the sampled trace z is given by the equation $z = Hr + n$, where H is a convolution matrix whose elements are those of the estimated wavelet.

Recovering the attenuation from the radiographs or estimating the reflectivity from an ultrasonic trace are both ill-posed problems. In order to obtain realistic solutions, we introduce prior knowledge in a Bayesian framework, that is, we regularize the problem. The choice of the prior model is then essential, as will be seen in the two next sections.

3. 3D reconstruction from gammagraphies

The blocks that are to be inspected consist *a priori* of homogeneous zones (flaw or metal). Such a prior knowledge about local smoothness can be modeled through a Markov random field. The main and well-known difficulty is then to define a Gibbs potential that allows the recovery of sharp edges between homogeneous zones. In such a view, some authors introduce a line process, either explicitly [1] or implicitly [2]. We believe that such prior models do not suit our problem: firstly, the data are too poor for recovering edges at the right positions; secondly, the introduction of line processes leads to complex minimization algorithms that would be too costly to cope with our 3D reconstruction problem. Therefore, we choose a convex Gibbs potential as those proposed by Bouman and Sauer [3]. This still makes it possible to have discontinuities appeared, and leads to simple minimization algorithms.

Following the observation model and the assumptions about the noise, the maximum *a posteriori* estimate is defined by:

$$\widehat{x}_{MAP} = \arg \min_{x} \left\{ \| y - Ax \|_2^2 + \lambda \sum_{s,t \in \Gamma} |x_s - x_t|^p \right\}, \quad (p > 1),$$

where λ is a regularization parameter and Γ refers to the set of second order cliques. Taking p equal to 2 results in Gaussian regularization that is well-known to provide over-smoothed reconstructions. We are most interested in the cases where the order p is close to 1, which favour the emergence of boundaries in the solution because big differences between neighbour sites are not penalized too much. Since $p > 1$, the criterion to be minimized is convex and differentiable and so, the search for the solution can be carried out through a gradient method.

We studied this reconstruction method for an austeno-ferritic block . This block (see Fig. 2) has been electro-eroded to obtain a cylindrical flaw along j, that is 2mm in diameter. We present a transverse cut of a 3D reconstruction (see Fig. 3) that has been obtained from seven real radiographs, for p equal to 1.3. The result is pretty good, since the flaw appears clearly in the center part of the cut. Still, as expected, one can notice that the flaw is extended along k, whereas it should be prefectly circular: the reconstruction undoubtly suffers from the lack of information of the data along the vertical direction k. To check the influence of p, we also give the reconstruction associated with a Gaussian regularization (see Fig.4): the result is indeed over-smoothed.

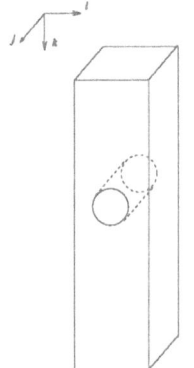

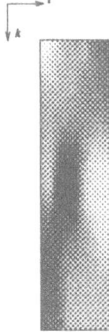

Fig.2: block made of austeno-ferritic steel, with a cylindrical electro-eroded flaw.

Fig.3: transverse cut of a reconstruction for $\lambda = 10$ and $p = 1.3$.

Fig.4: transverse cut of a reconstruction with a Gaussian prior ($\lambda = 30$ and $p = 2$).

4. Deconvolution of ultrasonic traces

To choose a prior model for the reflectivity r, we have to face about the same problems as for the attenuation. On the one hand, we are tempted into modeling the reflectivity as a Gauss-Bernoulli process that enables the recovery of spiky sequences. But Gauss-Bernoulli deconvolution [4] turns out to be poorly robust with respect to badly known wavelet, which is just the case here. We then decided on a weaker model: a sequence of reflectivity r is modeled as a white process whose marginal density is a generalized Gaussian of order γ ($\gamma > 1$).

For a given standard deviation, the smaller γ is, the sharper around zero and the more long-tailed the generalized Gaussian is, that is, the sharper any realization of a sequence should be. Therefore, here again, we are most interested in the cases where γ is close to 1. According to the assumptions and the notations given in section 2, the maximum *a posteriori* estimate is given by :

$$\widehat{r}_{MAP} = \arg \min_{r} \left\{ ||z - Hr||_2^2 + \mu||r||_\gamma^\gamma \right\}, \quad (\gamma > 1),$$

where μ is a regularization parameter. The computation of the solution demands the minimization of a convex and differentiable criterion; it can be performed by a gradient method. Practically, we have to make a compromise for the choice of p: the closer to 1 p is, the sharper the solution is likely to be, but as well, the harder it is to compute the solution because the "less convex" the criterion is. Deconvolution results are shown for the processing of a trace (see Fig. 5) that has been collected on the upper face of the block in Fig. 2, above the hole. The observation is pretty noisy, which is typical of the traces that are obtained when inspecting austeno-ferritic blocks. Classical predictive deconvolution fails to estimate the reflectivity (see Fig. 6). We present the result given by the proposed method, which is much better: one can successively notice the flaw, an echo due to a rebound in the sensor and the bottom of the block (see Fig. 7). Those good results are still corrupted by double spikes that may be explained by a bad identification of the wavelet.

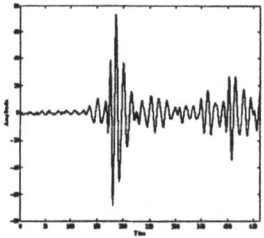 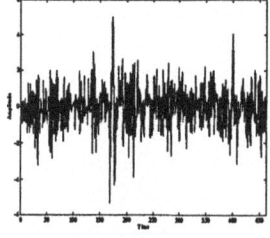 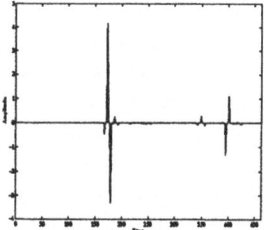

Fig.5: trace collected on the upper face of the block schematized in Fig. 2.

Fig.6: predictive deconvolution of the trace in Fig. 5.

Fig.7: deconvolution obtained by the proposed method ($\mu = 1000, \gamma = 1.1$) for the trace in Fig. 5.

5. Data fusion

For the data fusion, the ideal case would be to estimate jointly the attenuation and the reflectivity from both sets of data, as already pointed out by Marroquin [5]. Still, because of the difficulty in defining a joint prior density, we have been working on a sequential method.

In a first step, the ultrasonic data is used to detect the presence of vertical breaks; we form a binary vector q that represents those breaks: $q_{s,t}$ is set to 1 if there is a horizontal boundary between sites s and t. This step has not been achieved yet for real data, but we believe the processing proposed in section 4 should make it possible to estimate correctly q, though a bad time-depth conversion could harden that task.

Secondly, so as to take advantage of the information held by q, we modify the prior Markov field for the attenuation: the interaction between voxels that are separated by a horizontal boundary is suppressed. Calling H and V the sets of vertical and horizontal second-order cliques, the Gibbs potential for x is given by:

$$U(x) = \sum_{s,t \in H} |x_s - x_t|^p + \sum_{s,t \in V} |x_s - x_t|^p (1 - q_{s,t}), \quad (p > 1).$$

That kind of approach has already been proposed in medical imaging [6] for Gaussian prior models. Here, we reduce the order p close to 1 so that homogeneous zones are favoured. As previously, the solution can be achieved *via* a gradient algorithm.

We give 2D simulation results, supposing that the first step of the fusion method has been completed. The studied object is a square image with a square flaw at its center (see Fig. 8). So as to approach the experimental conditions given in introduction, we simulate 3 projections by -45, 0 and 45 degrees that were corrupted by an additive, white and Gaussian noise, with a 10dB SNR. Firstly, a reconstruction that uses no information about the boundaries is shown (see Fig.9): as in section 3, it suffers from the lack of information of the data along the vertical direction. Secondly, we present the result of the fusion reconstruction method (see Fig.10), when the right information about the horizontal breaks is introduced (see Fig.11): the use of such a prior knwoledge undoubtly serves the solution.

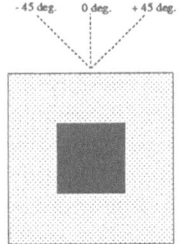

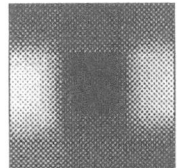

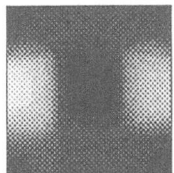

Fig.8: original object used for simulations.

Fig.9: reconstruction without any information about bounds ($p = 1.2$).

Fig.10: fusion reconstruction, taking advantage of the right information about upper bounds ($p = 1.2$).

Fig.11: accurate upper bounds used for fusion.

6. Conclusion

We have studied the problem of 3D reconstruction from gammagraphies and proposed a new deconvolution method for ultrasonic traces. Those processings, whose links have been stressed, both give good results for industrial applications, with small computational costs. This might suggest that, in the future, generalized Gaussian priors should replace the classical Gaussian regularization for applications. It still has to be noticed, that automated hyperparameter estimation has not been achieved yet.

Furthermore, we described a sequential fusion reconstruction method, that seems attractive and is now being tested on real data. The next step is to develop more complex fusion schemes. Indeed, we think that the choice of a joint prior density should allow a better use of the complementary information given by the data. To conclude, we believe that the field of application of this work may be wider than non destructive testing, and could for instance include medical imaging.

References

1. F. C. Jeng, and J. W. Woods, "Compound Gauss-Markov random fields for image estimation," *IEEE Transactions on Signal Processing*, vol. 39, pp. 683–697, 1991.
2. M. Nikolova, A. Mohammad-Djafari, and J. Idier, "Inversion of large-support ill-conditioned linear operators using a Markov model with a line process," *Proceedings IEEE ICASSP*, vol. V, pp. 357–360, Adelaide, South Australia, 1994.
3. C. Bouman, and K. Sauer, "A generalized image model for edge-preserving MAP estimation," *IEEE Transactions on Image Processing*, vol. 2, pp. 296–310, 1993.
4. F. Champagnat, J. Idier, and G. Demoment, "Deconvolution of sparse spike trains accounting for wavelet phase shifts and colored noise," *Proceeedings IEEE ICASSP*, vol. III, pp. 452–455, Minneapolis, Minnesota, 1993.
5. J. Marroquin, S. Mitter, and T. Poggio, "Probabilistic solution of ill-posed problems in computational vision." *Journal of the American Statistical Association*, vol. 82, pp. 76–89, 1987.
6. C. T. Chen, X. Ouyang, W. H. Wong, X. Hu, and V. E. Johnson, "Sensor fusion in image reconstruction," *IEEE Transactions on Nuclear Science*, vol. 38, pp. 687–692, 1991.

DUAL STATISTICAL MECHANICAL THEORY FOR UNSUPERVISED AND SUPERVISED LEARNING

G. DECO AND B. SCHÜRMANN
Siemens AG, Central Research and Development
Otto-Hahn-Ring 6, 81739 Munich, Germany

1. Abstract

We derive a statistical-mechanics-based ensemble model of unsupervised learning defined by redundancy reduction between the output components of neural nets and entropy conservation from inputs to outputs. The theory derived for unsupervised learning results in one for supervised learning by using ensemble theory based on the maximum-likelihood principle.

2. Introduction

The problem of learning and generalization from examples by using neural networks has been treated both in the frame of statistics [1] and of statistical physics [2]. In the statistical physics approach an ensemble of neural networks is used to address the problem of generalization of learning from a finite number of noisy training examples. The ensemble treatment of neural networks [2] assumes the final model to be probabilistic, built by an integration of single models weighted with the corresponding probability distribution. Here we focus our attention on the problem of both unsupervised and supervised learning by neural networks from given examples. Unsupervised learning was formulated [3] by a neural implementation of the biological principle of redundancy reduction. The brain performs a statistical decorrelation of the input environment in order to extract statistically independent relevant information. The goal of redundancy reduction is to factorize the output probability distribution without loosing information. Deco and Schürmann [3] devised an architecture and a learning paradigm for the unsupervised extraction of statistical correlations, performing Barlow's unsupervised learning in the most general fashion and implementing a *nonlinear independent component analysis*. In this paper we formulate a

317

K. M. Hanson and R. N. Silver (eds.), Maximum Entropy and Bayesian Methods, 317–321.
© *1996 Kluwer Academic Publishers.*

statistical mechanics-based theory for unsupervised learning as modeled in
[3]. Furthermore, we may at the same time pose the problem of supervised
learning as an unsupervised learning one, such that we obtain an ensemble
theory for supervised learning based on the maximum-likelihood principle.
An upper bound for the prediction probability of a new point not included
in the training data is found. This upper bound is determined essentially
by the rate between the Fisher Information for the training set and the one
for a set including the training data and the new point. It is possible to
use this upper bound as a mechanism to decide actively on the novelty of
a new data and therefore to use it as a mechanism of query learning.

3. Statistical Theory of Unsupervised Factorial Learning

In this section, we formulate an ensemble theory for unsupervised learning. We employ a single-layer architecture that attempts to extract correlations. The architecture is always reversible, conserves the volume and
therefore the transmitted information. In general the environment is non-
Gaussian distributed and non-linearly correlated. The learning rule decor-
relates statistically the elements of the output by minimizing the mutual
information between the output components. The aim of this section is
to derive a statistical-mechanics-based model of the unsupervised learning
mechanism devised by Deco and Schürmann [3]. We concentrate on the
one-layer volume-conserving triangular architecture of Figure 1-a. Let us
denote the n-dimensional input and output vectors for the unsupervised
architecture of Figure 1 by $\vec{\xi}$ and $\vec{\Upsilon}$, respectively. The output vector is
defined by $\Upsilon_i = \xi_i + f_i(\xi_0, \cdots, \xi_j, \vec{\omega}_i)$, with $j < i$, and $\vec{\omega}_i$ being the parameter vector which determines the parametrical function (for example
a neural network) f_i. Let us denote the entropy of a random variable X
by $h(X)$. Unsupervised learning minimizes the redundancy (i.e. the mutual
information) at the output components given by $R = \sum_{j=1}^{n} h(\Upsilon_j) - h(\vec{\Upsilon})$.

By using the fact that entropy is conserved (due to the fact that the
transformation $\vec{\Upsilon} = \vec{U}(\vec{\xi})$ has a Jacobi determinant equal to unity), i.e.
$h(\vec{\Upsilon}) = h(\vec{\xi})$ =constant, minimization of the redundancy is reduced to
minimizing the term $\sum_{j=1}^{n} h(\Upsilon_j)$. We now are able to formulate an ensemble theory for unsupervised learning. The training set consists of example
patterns, $T^{(P)} = \{\vec{\xi}^{(q)}, 1 \leq q \leq P\}$. The ensemble of networks given by
different parameters $\vec{\omega} = \{\vec{\omega}_1, \cdots, \vec{\omega}_n\}$ is weighted by using again the maximum entropy principle. In this case the macroscopic constraint is the
minimization of redundancy, so that the Gibbs function now is defined by

$$p(\vec{\omega}/T^{(P)}) = \frac{e^{\beta \sum_{i=1}^{P} \sum_{j=1}^{n} \log(p(\Upsilon_j^{(i)}/\vec{\omega}))}}{Z(P)} \tag{1}$$

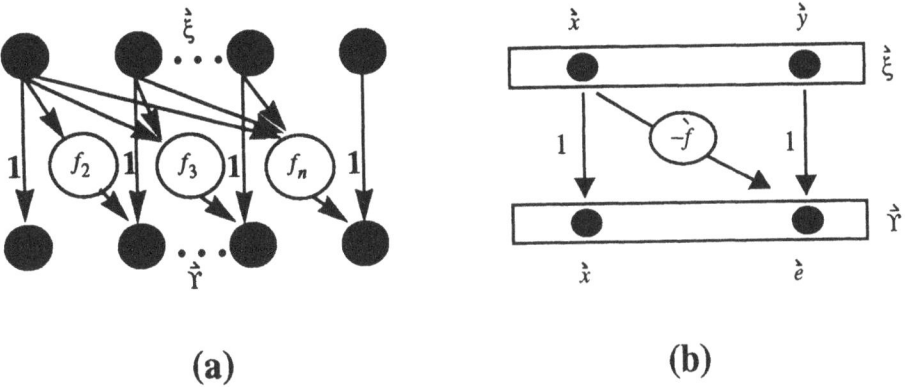

Figure 1. (a) Volume-Conserving Triangular Architecture; (b) Unsupervised Architecture for Supervised Learning.

with the partition function $Z(P) = \int e^{\beta \sum_{i=1}^{P} \sum_{j=1}^{n} \log(p(\Upsilon_j^{(i)}/\vec{\omega}))} d\vec{\omega}$. We perform a Taylor expansion of the exponent around $\vec{\omega}_P$ defined as that point where the empirical entropy multiplied by the number of patterns P, i.e. $H^{(P)} = \sum_{i=1}^{P} \sum_{j=1}^{n} \log(p(\Upsilon_j^{(i)}/\vec{\omega}))$, is minimal. Hence, the integrand in the definition of $Z(P)$ adopts a Gaussian form and therefore the integral can be easily calculated yielding

$$Z^{(P)} \cong e^{\beta H^{(P)}|_{\vec{\omega}_P}} (2\pi)^{\frac{D}{2}} (\det(\beta F^{(P)}))^{\frac{-1}{2}} \qquad (2)$$

where $D = dim(\vec{\omega})$ and the non-negative definite matrix $F^{(P)}$ is defined by $F^{(P)} = -\nabla\nabla H^{(P)}|_{\vec{\omega}_P} = -\sum_{i=1}^{P} \sum_{j=1}^{n} \nabla\nabla \log(p(\Upsilon_j^{(i)}/\vec{\omega}_P))$. The term $F^{(P)}$ has the form of an empirical Fisher Information. We need to calculate also $Z(P+1)$ because we want to study the effect of adding a new pattern. To do so we make a Taylor expansion of $H^{(P+1)}$ around the point defined above. The integrand of $Z(P+1)$ is again a Gaussian, so:

$$Z^{(P+1)} \cong e^{\beta H^{(P+1)}|_{\vec{\omega}_P}} (2\pi)^{\frac{D}{2}} (\det(\beta F^{(P+1)}))^{\frac{-1}{2}} (e^{\beta H|_{\vec{\omega}_P} (\vec{g}^T (H^{(P+1)})^{-T} \vec{g})} \qquad (3)$$

where $\vec{g} = \nabla H|_{\vec{\omega}_P}$ and $H = \sum_{j=1}^{n} \log(p(\vec{\Upsilon}_j/\vec{\omega}_P))$ are defined at the new point. Assuming for the probability model of each network the escort distribution $p(\vec{\Upsilon}/\vec{\omega}, \beta) = exp(\beta \log(p(\vec{\Upsilon}/\vec{\omega})))/z$ we obtain for the probability distribution $p(\vec{\Upsilon}, T^{(P)})$ of the new point:

$$p(\vec{\Upsilon}, T^{(P)}) = \int p(\vec{\omega}/T^{(P)}) p(\vec{\Upsilon}/\vec{\omega}, \beta) d\vec{\omega} = \frac{Z(P+1)}{zZ(P)} \qquad (4)$$

$$\cong (\frac{e^{\beta H(1-\frac{1}{2}\vec{g}^T(F^{(P+1)})^{-T}\vec{g})}}{z})\det^{\frac{1}{2}}(F^{(P)}(F^{(P+1)})^{-1}). \tag{5}$$

Thus we have derived an approximate expression for the probability distribution of the output components which is essentially based on the probability distribution given by the best network (the one with parameter vector $\vec{\omega}_P$) and by the square root of the rate between the determinants of the Fisher Information without and with including the new point. In the next section we will use these results for obtaining an ensemble theory of supervised learning based on the maximum-likelihood principle.

4. Duality Between Unsupervised and Supervised Learning

We pose the problem of supervised learning as an unsupervised one. As we will demonstrate, for unsupervised learning the ensemble theory derived above then results in a theory for supervised learning by making use of ensembles based on the maximum-likelihood principle. The input vector $\vec{\xi}$ for the unsupervised architecture of Figure 1-b is defined to be composed of two components $\vec{x}$ and $\vec{y}$ of dimensions N and M, respectively, i.e. $\vec{\xi} = \{\vec{x}, \vec{y}\}$. These two vectors are related through a probability distribution $\rho(\vec{x}, \vec{y})$, so that they can be regarded as input and output of a relation to be learned by supervised learning. The input vector $\vec{\xi}$ is given empirically by the set of the training data, as defined in section 3. Let us denote the output of the triangular architecture by $\vec{\Upsilon}$ which is also composed of two vectors such that $\vec{\Upsilon} = \{\vec{x}, \vec{e}\}$. The network output component $\vec{e}$ is defined by $\vec{e} = \vec{y} - \vec{f}(\vec{x}, \vec{\omega})$ where in supervised learning $\vec{e}$ is the error and $\vec{\omega}$ is the parameter vector which describes the general function $\vec{f}$. The maximum-likelihood principle for supervised learning requires to be chosen so that the empirical likelihood $L = \frac{1}{P}\sum_{i=1}^{P}\log(p(\vec{y}^{(i)}/\vec{x}^{(i)}, \vec{\omega}))$ is maximal. The conditional probability $p(\vec{y}^{(i)}/\vec{x}^{(i)}, \vec{\omega})$ should be regarded as a measure of the compatibility of the pairs $(\vec{y}^{(i)}, \vec{x}^{(i)})$. On the other hand, as discussed above the goal of unsupervised learning is redundancy minimization. The architecture of Figure 1-b can only minimize the redundancy inherent in the relation between the vectors $\vec{x}$ and $\vec{y}$, i.e. it aims to extract the correlations between these vectors which is the goal of supervised learning. In fact, unsupervised learning minimizes the redundancy at the output components given by $R = \sum_{j=1}^{N} h(x_j) + \sum_{j=1}^{M} h(e_j) - h(\vec{\Upsilon})$. By using the fact that entropy is conserved, i.e. $h(\vec{\Upsilon}) = h(\vec{\xi})$ =constant, and assuming that the distribution of $\vec{x}$ is stationary, i.e. $\sum_{j=1}^{N} h(x_j)$ =constant, minimization of redundancy is reduced to minimization of the term $\sum_{j=1}^{M} h(e_j)$. Due to the fact that $h(\vec{e}) \leq \sum_{j=1}^{M} h(e_j)$ and taking into account that $h(\vec{e}) = -L$ minimization of redundancy of the output components $\vec{\Upsilon}$ is equivalent to maximization of the likelihood L. Put differently, maximum-likelihood supervised learning is equivalent to minimizing the entropy of the error which is the goal of unsupervised reduction of redundancy. This is the dual formulation of supervised and unsupervised learning. We now make use of the results previously obtained for the ensemble theory of unsupervised learning. We obtain for the distribution of $\vec{y}$

$$p(\vec{y}/\vec{x}, D^{(P)}) \cong (\frac{e^{\beta H'(1-\frac{1}{2}\vec{g}'^T(F'^{(P+1)})-T\vec{g}')}}{z})\det^{\frac{1}{2}}(F'^{(P)}(F'^{(P+1)})^{-1}) \qquad (6)$$

and therefore the probabilities

$$p(\vec{y}/\vec{x}, D^{(P)}) \leq p(\vec{y}/\vec{x}, \vec{\omega}_P, \beta)(\frac{\det(F'^{(P)})}{\det(F'^{(P+1)})})^{\frac{1}{2}}. \qquad (7)$$

In the last equations, $F'^{(P)} = -\sum_{i=1}^{P} \nabla\nabla \log(p(\vec{e}_j^{(i)}/\vec{x}^{(i)}, \vec{\omega}_P))$, $H' = \log(p(\vec{e}/\vec{x}, \vec{\omega}_P))$, $\vec{g}' = \nabla H'|_{\vec{\omega}_P}$. Hence we have obtained an upper bound for the prediction probability of new data given P training data. The upper bound is essentially determined by the square root of the rate between the determinants of the Fisher Information without and with including the new point. The factor $p(\vec{y}/\vec{x}, \vec{\omega}_P, \beta)$ is the probability of observing the new pair given by the best network (the one with parameter vector $\vec{\omega}_P$). The square root of the rate between the determinants of the Fisher Information provides us with a measure of how much the reliability of the best network should be reduced. We call the negative logarithm of this quantity the novelty measure $NW(P) = -0.5 \log(\det F'^{(P)}/\det F'^{(P+1)})$. This information is a consequence of the use of the ensemble approach. In summary, the statistical-mechanics-based theory of unsupervised learning by redundancy reduction with a volume-conserving network has been formulated and used for the improvement of the standard ensemble theory of supervised learning by exploiting the duality between the two learning paradigms.

References

1. MacKay, D. (1991) Bayesian Interpolation, *Neural Computation*,Vol. **4**, pp. 415-447.
2. Levin, E., Tishby, N. and Solla, S. (1990) A Statistical Approach to Learning and Generalization in Layered Neural Networks, *Proceedings of the IEEE*,Vol. **78**, pp. 1568-1574.
3. Deco, G. and Schürmann, B. (1995) Learning Time Series Evolution by Unsupervised Extraction of Correlations, *Physical Review E*,Vol. **51**, pp. 1780-1790.

Imre Csiszár, Roger Bilisoly, Michael Miller, and Pierre Boulanger at table at Richard Silver's home.

COMPLEX SINUSOID ANALYSIS BY BAYESIAN DECONVOLUTION OF THE DISCRETE FOURIER TRANSFORM

F. DUBLANCHET[1,2], P. DUVAUT[1] AND J. IDIER[2]
[1]*ETIS/ENSEA 6, av. du Ponceau 95014 Cergy-Pontoise, France.*
[2]*Laboratoire des Signaux et Systèmes (CNRS–ESE–UPS)*
Plateau de Moulon, 91192 Gif-sur-Yvette Cedex, France. [†]

Abstract. This paper addresses the mixed detection-estimation of complex sinusoids embedded in noise. Processing the Discrete Fourier Transform (DFT) of the noisy data is regarded as an inverse problem in the frequency domain. Its ill-posed nature may be coped in a Bayesian framework using prior information. To this end, the impulsive structure of the expected complex spectrum is described by a compound Bernoulli-Gaussian (BG) complex process. It is also shown how the Bernoulli process can be driven by Fermi-Dirac statistics.

1. Introduction

A poor robustness with respect to the unavailable knowledge of the number of sources is the main drawback shared by the classical high resolution spectral analysis methods, based upon either a singular value decomposition of the data matrix (such as ESPRIT, MUSIC, MIN-NORM, MATRIX-PENCIL) or a maximum likelihood estimation approach (BRESLER). Many criteria used to select the model order, such as the Akaike Information Criterion [1], suffer from a well-known lack of robustness. In such spectral analysis methods [2], the estimation of the number of sources is only lately accounted for, and the criteria that are used to separate noise and signal subspaces address only the estimation part of the problem [3]. To overcome this drawback, it is required to state the whole problem as a *detection-estimation* issue [4]. We show how the problem of analyzing complex sinusoids embedded in an additive Gaussian noise can be formulated as an *inverse problem* in the frequency domain whose ill-posed nature requires some regularization in a Bayesian framework. Processing the Discrete Fourier Transform of the signal is regarded as an inverse problem, which is linearized around a set of equally spaced discrete frequencies, provided a very fine sampling of the frequency domain can be obtained. The spectrum to be restored is modeled as a compound non-uniform Bernoulli-Gaussian complex process [5] [6] [7] : the BG *prior model* is a convenient way to include our prior knowledge about its impulsive structure.

[†]Email: dublanchet@lss.supelec.fr

K. M. Hanson and R. N. Silver (eds.), Maximum Entropy and Bayesian Methods, 323–328.
© 1996 *Kluwer Academic Publishers.*

A finer regularization stage consists in relating the Bernoulli parameter to the power spectral density of the signal. Interestingly, the so-called *stochastic regularization* is based on an analogy with a classical statistical physics problem, leading the Bernoulli process to be driven by Fermi-Dirac statistics.

Restoration of a BG process involves a mixed *detection-estimation* operation.

2. Problem formulation in the frequency domain

Consider N samples $Z[n]$ of a deterministic process $Y[n]$ that consists of the sum of p complex sinusoids embedded in a zero-mean white Gaussian complex valued circular noise $W[n]$, with known variance σ_w^2 :

$$Z[n] = Y[n] + W[n] = \sum_{i=1}^{p} A_i e^{(2j\pi\nu_i n + j\phi_i)} + W[n]. \qquad (1)$$

Assuming $W[n]$ to be statistically independent of $Y[n]$ and the frequencies ν_i to lie in $[0, 1[$, the problem consists in determining both the discrete variable p (detection) and the p sets $\{\nu_i, A_i, \phi_i\}$ of parameters (estimation).

Unlike the classical techniques which make use of either a time representation or a parametric modeling of the observed data, the proposed method works on a different representation of the same data, that is their (normalized) Fourier transform (FT) : $z(\nu) = \frac{1}{\sqrt{N}} \sum_{n=0}^{N-1} Z[n] e^{-2j\pi\nu n}$, $\nu \in [0, 1[$. Defining the complex amplitudes $A_i^c = A_i e^{j\phi_i}$, the FT of the noise-free sequence $Y[n]$ may be expressed as

$$y(\nu) = \sum_{i=1}^{p} A_i^c h(\nu - \nu_i) = \sum_{i=1}^{p} A_i^c \int_0^1 h(\nu - \nu') \delta(\nu' - \nu_i) d\nu' \qquad (2)$$

in which the continuous complex function $h(.)$ coincides with the well-known Dirichlet kernel $h(\nu) = e^{-j\pi\nu(N-1)} \sqrt{N} \operatorname{sinc}(\pi N\nu)/\operatorname{sinc}(\pi\nu)$. This kernel has a relatively narrow central lobe and very high side lobes. If the number of data samples were infinite, the side lobes would disappear and the central lobe would become a Dirac delta function, in which case convolution with such a function would simply reproduce the correct spectrum. For a finite number N of samples, spectral resolution is thereby limited to the width of the main lobe $1/N$. Note that the Dirac delta function $\delta(.)$ that appears in the circular convolution product (2) reflects explicity the impulsive structure of the true spectrum.

In practice, the DFT samples naturally (2) at N equally spaced discrete frequencies n/N, $n = 0, \ldots, N-1$. The general problem of estimating the set $\{\nu_1, \ldots, \nu_p\}$ $\in [0, 1[^p$ of continuous frequencies from the only available observed data or their DFT is highly nonlinear. We propose to linearize the problem around a grid of $N_F = \alpha N$ equally spaced discrete frequencies $k/\alpha N$, $k = 0, \ldots, \alpha N - 1$, provided α may take arbitrary integer values greater than one. Since each unknown frequency ν_i can be written as $\nu_i = k_i/N_F - \delta\nu_i$ with the constraints $|\delta\nu_i| < 1/2N_F$ and $0 \leq k_i \leq N_F - 1$, the estimation of any signal frequencies that occur between two discrete frequency samples may be further performed using a first order series expansion of the convolution kernel around the aforementioned grid. Thus, Eq. (2)

may be approximated as closely as desired (depending on the discretization order α) by the following double convolution product :

$$y[n] \simeq \sum_{k=0}^{N_F-1} h[n-k]s[k] + g[n-k]r[k], \text{ with } g(.) = \frac{dh}{d\nu}(.) \tag{3}$$

in which the p non-zero components $s[k_i]$ and $r[k_i]$ are related to the frequency shifts $\delta\nu_i$ by $r[k_i] = \delta\nu_i\, s[k_i]$. The modulus of the resulting approximation error is bounded by $\epsilon = \pi^2/6\alpha^2$ and the noisy version of Eq. (3) may be written in matrix form

$$z = Ax + w, \quad A = [H \,|\, G] \text{ and } x = [s^t | r^t]^t. \tag{4}$$

Both $(N \times N_F)$-submatrices H and G exhibit a circulant structure. As the DFT of $W[n]$, the noise w is still white of complex circular Gaussian distributed values, and independent of x. Recovering the sparse sequence x from the DFT z is stated as an inverse problem in the sampled spectral domain.

For the sake of convenience, every quantities referenced by k and n in the sequel will lie respectively in $[0, \dots, N_F-1]$ and $[0, \dots, N-1]$.

3.   Two regularization levels

3.1.   ILL-POSED PROBLEM

The convolution matrix A is singular. Even if we exclude the $(2\alpha - 1)N$ zero singular values which are associated to noninformative directions, there will still be singular values close to zero. More precisely, making use of both particular properties concerning circulant matrices and the relationship between $g(.)$ and $h(.)$, nonzero eigenvalues of the normal matrix $A^t A$ may be analytically evaluated. This yields the condition number of AA^t to be equal to $1 + 4\pi^2(N-1)^2$, not depending on α. The above considerations obviously prohibit any numerically unstable generalized inverse solution, so that the clearly *ill-posed* nature of the inverse problem (4) invokes some regularization.

3.2.   BAYESIAN BG REGULARIZATION

Consistently with our prior knowledge about the desired spectrum, a complex BG process is chosen as prior model to select an impulsive solution. The discrete frequencies at which the Bernoulli sequence takes 1 values locate the sinusoids to within about a frequency shift ; in agreement with the hypotheses that lead to (4) : (i) r and s are driven by the same white Bernoulli sequence q with parameter $\lambda \in [0,1]$: $\forall k = 0, \dots, N_F - 1$: $\Pr(q[k] = 1) = \lambda$ and $\Pr(q[k] = 0) = 1 - \lambda$; (ii) $(s[k]\,|\,q[k]=q)$ is assumed to be a zero-mean complex circular Gaussian variable, with variance $q\sigma^2$; (iii) given q, the p frequency shifts $\delta\nu_i$ are modeled as mutually independent random variables, also independent of the $s[k]$'s and uniformly distributed over the centered interval $[-\frac{1}{2N_F}, \frac{1}{2N_F}[$. Consequently, the conditional probability density of $(r[k]\,|\,q[k]=q)$ is shown to be expressed by the following exponential integral :

$$p_{r[k]\,|\,q[k]=1}(r) = \frac{N_F}{\pi\sigma^2} \int_1^{+\infty} \frac{1}{t} e^{-4N_F^2|r|^2 t/\sigma^2} dt \; ; \; p_{r[k]\,|\,q[k]=0}(r) = \delta(r).$$

The closest complex Gaussian distribution $\mathcal{N}(0, q\sigma^2/12N_F^2)$ is preferably substitued for $p_{r[k]\,|\,q[k]=q}(.)$ so as to remain in a Gaussian setting. The BG model is then fully characterized by the white process (q, s, r) and the probability laws of each quantity involved. The two hyperparameters (λ, σ^2) entirely define the problem in a Bayesian framework. The Bernoulli parameter λ is seen as the average probability of occurence of one particular frequency ν_i between two discrete frequencies k/N_F and $(k+1)/N_F$. Up to now, λ is assumed to be constant over the frequency range $[0, 1[$. Exploiting further information, we show how it is possible to relate the latter to the power spectral density of the observed signal, yielding a *nonuniform* Bernoulli process.

3.3. STOCHASTIC REGULARIZATION

The stochastic regularization is a finer stage that consists in assigning a suitable prior probability density $\lambda[k]$ to each candidate discrete frequency k/N_F. Using only the N observed data does not enable to select a solution of better resolution than the minimum norm one, *i.e* the periodogram. In that sense, the $\alpha-1$ discrete frequencies introduced between two consecutive $1/N$-spaced DFT samples are "degenerated". To lift this degeneracy requires to add information on the last, which should be consistent with the available one about the N DFT samples. By that way, we emphasize an analogy with the classical statistical physics problem [8] of finding the best repartition of particles over given energy levels. Thereby, all of the possible realizations of the Bernoulli process between two consecutive DFT samples are considered as *indistinguishable* particles (no particle has a distinctive tag by which it could be differenciated from others) to be distributed over $\alpha-1$ degenerated frequencies themselves similar to $\alpha-1$ energy levels. Several realizations are likely to occur between two consecutive DFT samples. However, no more than one realization can be in any one discrete frequency, which express the Pauli exclusion principle applied to our problem. The solution of this combinatorial problem involves the maximization of the statistical entropy in opened field, subject to the only energetic constraint directly stemmed from both the periodogram π (the square magnitude of the DFT) and Parseval's equality. Then invoking Lagrange multipliers and Stirling's approximation leads to the well-known Fermi-Dirac distribution $\lambda[k] = 1/1 + \exp\{(\pi[k] - \theta)/\rho\}$. Furthermore, given (ρ, θ) the stochastic regularization provides a subset $\mathcal{P}$ consisting of γ $(\gamma \leq N_F)$ more probable sites when λ is compared to a threshold. This yields computational savings by exploring $\mathcal{P}$ rather than the whole set of discrete frequencies.

4. Bayesian detection-estimation scheme

Restoration of a BG process is a simultaneous detection-estimation operation which may be performed sequentially. First, detection of the Bernoulli sequence

q through optimization of a suitable likelihood criterion, and second, estimation of x. Different strategies are proposed in BG deconvolution and the most reliable results are expected with a detection based upon a *marginal posterior* likelihood combined with a maximum *a posteriori* estimation of x [6] [7] :

$$\begin{cases} \widehat{q} = \arg\max p(q|z) = \arg\max p(z|q)\Pr(q) \\ \widehat{x} = \arg\max p(x|\widehat{q}, z) = \arg\max p(z|\widehat{q}, x)p(x,\widehat{q}) \end{cases} .$$

Detection of q requires to maximize the following criteria over $\mathcal{P}$:

$$\mathcal{C}_D(q) = -\ln|B(q)| - z^\dagger B(q)^{-1}z + \sum_{k\in Q_1}\ln\lambda[k] + \sum_{k\in Q_0}\ln(1-\lambda[k])$$

where
$$\begin{cases} R(q) = \begin{bmatrix} \sigma^2\mathrm{diag}\{q\} & 0 \\ 0 & \frac{\sigma^2}{12N_F^2}\mathrm{diag}\{q\} \end{bmatrix} \\ B(q) = \mathrm{E}\{zz^\dagger\} = AR(q)A^\dagger + \sigma_w^2 I ; Q_q = \{k/q[k] = q, q = 0\ \mathrm{ou}\ 1\}. \end{cases}$$

Exact maximization of $\mathcal{C}_D(q)$ may be too computationally demanding (2^γ evaluations). It is then replaced by a suboptimal iterative maximization based on the SMLR technique [5] : starting from an initial sequence q_0, the criterion is evaluated on a neighborhood of q_0 whose all elements q_k differ from q_0 only at site k. The low computational burden due to a simple relationship between $\mathcal{C}_D(q_k)$ and $\mathcal{C}_D(q_0)$ justify such an approach :

$$\mathcal{C}_D(q_k) = \mathcal{C}_D(q_0) + u_0^\dagger V_k\Theta_k^{-1}V_k^t u_0 - \ln\left(\frac{\sigma^4}{12N_F^2}|\Theta_k|\right) - \epsilon_k\ln\left(\frac{1}{\lambda[k]} - 1\right)$$

with :
$$\begin{cases} \epsilon_k = q_k[k] - q_0[k] = \pm1 ; \Theta_k = \epsilon_k\Pi^{-1} + V_k^t D_0 V_k \\ \Pi = \begin{bmatrix} \sigma^2 & 0 \\ 0 & \frac{\sigma^2}{12N_F^2} \end{bmatrix} ; V_k = \begin{bmatrix} v_k & 0 \\ 0 & v_k \end{bmatrix} ; v_k[l] = \delta[k-l] \\ D = A^\dagger(B(q))^{-1}A ; u = A^\dagger(B(q))^{-1}z \end{cases}$$

The Bernoulli sequence which maximizes the detection criterion over the whole neighborhood is then selected as the next initial sequence using the updated quantities : $D_k = D_0 - D_0 V_k\Theta_k^{-1}V_k^t D_0$ and $u_k = u_0 - D_0 V_k\Theta_k^{-1}V_k^t u_0$. MAP estimation of x relies on both the linearity of (4) and the normality of $(x|q)$. Using the classical MAP formulas in a linear and Gaussian setting, we obtain : $\widehat{x} = R(\widehat{q})A^\dagger B(\widehat{q})^{-1}z$. The $\widehat{p} = \sum_k\widehat{q}[k]$ nonzero components of $\widehat{s}$ et $\widehat{r}$ are selected by $\widehat{Q}_1 = \{\widehat{k}_i, i = 1, \ldots, \widehat{p}\}$ and provide a least squares estimate of the $\widehat{p}$ corresponding frequency shifts as : $\widehat{\delta\nu}_i = \{\mathrm{Re}(\widehat{r}_i)\mathrm{Re}(\widehat{s}_i) + \mathrm{Im}(\widehat{r}_i)\mathrm{Im}(\widehat{s}_i)\}/|\widehat{s}_i|^2$. Denoting $\widehat{h}_i[n] = h(n/N - \widehat{\nu}_i)$, it then follows $\forall i = 1, \ldots, \widehat{p} : \widehat{\nu}_i = \widehat{k}_i/N_F - \widehat{\delta\nu}_i$ and $\widehat{A}_i^c = \widehat{h}_i^\dagger(\widehat{h}_i\widehat{h}_i^\dagger + \sigma_w^2/\sigma^2 I)^{-1}z$. The SMLR procedure is initialized with a zero sequence. A synthetic form in which the Bernoulli parameter is assumed to be constant may be found in [6] [7]. When λ is constant, the probability threshold above which any realization occuring at site k increases the criterion value takes the following sigmoidal form, interestingly close to the one found in the stochastic regularization : $1/\left\{1 + \exp(\epsilon_k w_{0,k})\right\}$ with $w_{0,k} = \boldsymbol{u}_0^\dagger \boldsymbol{V}_k \boldsymbol{\Theta}_k^{-1} \boldsymbol{V}_k^t \boldsymbol{u}_0 - \ln\left(|\boldsymbol{\Theta}_k|\sigma^4/12N_F^2\right)$. However, in such a case the detection threshold is *updated* at each SMLR iteration.

5. Conclusion

Such a regularization, combined with the perfect knowledge of the convolution operator $\boldsymbol{A}$, yields a high resolution analysis with a satisfactory robustness (Fig.1), with the beneficial advantage of taking explicitly into account the *a priori* unknown number of sinusoids as shown in Fig.2. Solving the problem in the continuous frequency domain without any linearization may be performed by extending the BG model to Poisson-Gaussian. Further developments in that way are currently studied.

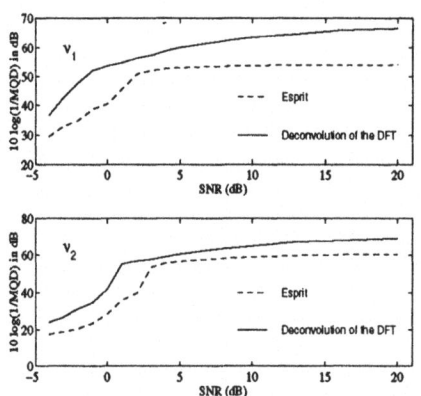

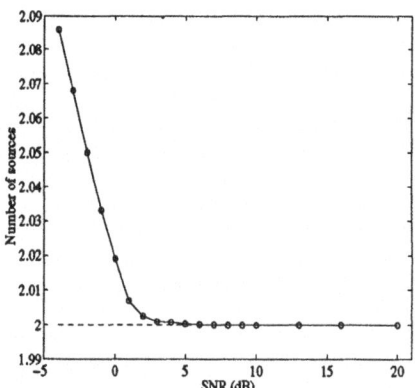

Fig.1 : conditionally to a correct detection, comparison of the Mean Quadratic Distances (MQD) concerning estimation of two signal frequencies (ν_1, ν_2) with corresponding powers $(1\,;\,0,5)$. $N = 64$; $\alpha = 2$; ν_1 and ν_2 are 1/128-spaced and both occur between discrete frequencies. The number of sources is not estimated by ESPRIT and then set to 2.

Fig.2 : estimation of the number of source averaged over 500 independent trials for each SNR value ; these results correspond to the signal mentionned in Fig.1.

References

1. M. Akaike, "A new look at statistical model identification," *IEEE Trans. Aut. Control*, **AC-19**, pp. 716–723, December 1974.
2. S. M. Kay and S. L. Marple, "Spectrum analysis - a modern perspective," *Proc. IEEE*, **69**, pp. 1380–1418, 1981.
3. H. L. VanTrees, *Detection, Estimation and Modulation Theory*, Wiley, Part1, New York, 1968.
4. P. Duvaut and F. Dublanchet, "EXPULSE, une nouvelle méthode d'analyse de raies fondée sur une approche inverse," *Traitement du Signal*, to appear.
5. J. Kormylo and J. M. Mendel, "Maximum-likelihood detection and estimation of bernoulli-gaussian processes," *IEEE Trans IT*, **IT-28**, pp. 482–488, 1982.
6. Y. Goussard, G. Demoment and J. Idier, "A new algorithm for iterative deconvolution of sparse spike trains," *Proc. of ICASSP*, pp. 1547–1550, 1990.
7. F. Champagnat, Y. Goussard and J. Idier, "Unsupervised deconvolution of sparse spike trains using stochastic approximation," *IEEE Trans. on Signal Processing*, submitted.
8. P. Duvaut, *Traitement du Signal*, Hermès, Chap.10, 1994.

STATISTICAL MECHANICS OF CHOICE

PETER S. FAYNZILBERG
Carnegie Mellon University
Graduate School of Industrial Administration
Schenley Park
Pittsburgh, Pennsylvania 15213–3890[†]

Abstract. We derive an equilibrium distribution for a population of decision makers, such as consumers or voters, in which each member makes choices not known with certainty. It is shown that, when the shares held by available choice alternatives are known, the MaxEnt distribution of the ensemble belongs to the product-form family.

Key words: Choice problem, statistical mechanics, product-form, partition function

1. Introduction

Traditionally, the subject of statistical mechanics has been concerned with a description of the state of a large system of physical particles (Boltzmann [2], Gibbs [8]; Balescu [1], Feynman [6], or Landau and Lifshitz [14] may be consulted for a modern exposition). In contrast, when addressing the statistical mechanics of choice, we attempt to describe a population of decision-makers. This paper contains a brief account of results related to the equilibrium state of such a population; see Faynzilberg [5] for more details.

The problem of choice arises when a decision-maker — human or an automaton — attempts to select an alternative from a specified set. In doing so, the decision-maker may be guided by various criteria. In particular, when he or she selects that alternative which maximizes preferences, the decision-maker is deemed to be rational (Fishburn [7], Kreps [13], Malinvaud [15], Varian [17]). In this paper we abstract away from the degree of the decision-maker's rationality and concentrate on the behavioral consequences of choice. In doing so, we assume that the choices that the individuals will make in the future are not necessarily known with certainty. Consequently, the choice profile of each decision-maker is given by a tuple $\mathbf{q} = (q_1, q_2, \ldots, q_n)$, in which n is the number of available alternatives, and for each

---

[†] Internet: petersf@andrew.cmu.edu

K. M. Hanson and R. N. Silver (eds.), Maximum Entropy and Bayesian Methods, 329–334.
© *1996 Kluwer Academic Publishers.*

$i \leq n$ the component q_i is the probability with which the alternative i is going to be chosen.

The contexts in which the choice behavior of large populations is of interest arise often in business and social sciences. Consider, for instance, a political party that develops a platform for an upcoming election. Each voter attaches a certain level importance to each bundle of issues [1] and may be expected to vote for the most important one. His preferences, however, may not be fully known to the candidates and even to himself. The distribution of choice profiles in the population is not known either — various opinion polls provide only *average* predisposition of the voters. Given the averages, the voter population may be relatively homogenous or, alternatively, rather polarized with equally large numbers of voters being for and against the issue in question. Clearly, the degree of success of the platform chosen by the party will differ significantly in these two cases.

Alternatively, consider a typical product market such as that for toothpaste, for example. Acting as a decision-maker, each consumer (or a household) chooses a brand of toothpaste from n alternatives offered by the producers. Some consumers exhibit a high degree of brand loyalty and choose only the favored brand with certainty: a consumer loyal to Brand 2, say, is described by the choice profile $\mathbf{q} = (0, 1, 0, \ldots, 0)$. In contrast, some consumers are influenced mostly by the price and choose that brand which is on sale (Kotler [12]). An ultimate "brand-switcher" will have the profile $\mathbf{q} = (1/n, \ldots, 1/n)$. The firm that markets a good in such a market is interested, of course, in the composition of the population with respect to the predisposition to buy its products, *i.e.*, the extent to which the consumer population is comprised of "loyals" and "switchers" of various kinds. Depending on the heterogeneity of the population, the firm will choose different pricing, advertising, and product offering strategies. In similarity to the electorate, averages are usually known in the form of market shares — the proportion that sales of a given company constitutes in the total sales of the product category. This information is often available from the distribution and market research companies.

Although the analysis presented below is motivated by the problem of choice in populations such as those described earlier, it has a rather high degree of generality. Note that the state of each member of the population is described by a point in the standard simplex S^n. In equilibrium, it is not relevant whether this point describes choices or some other behavior. Consequently, the analysis of the problem is applicable, without modifications, to all systems whose elements have S^n as their state spaces. A mixture of polarized light may serve as an example.

By applying the maximum entropy formalism of E.T. Jaynes [10,11], we show that, when the shares held by alternatives are known, the heterogeneity of the population (the macrostate) is described by a product-form distribution. As a statistical-mechanical ensemble, the population of decision-makers is rather unusual in that its partition function is known in closed form.

---

[1] We talk about bundles of issues and not the issues themselves because more than one issue may appeal to the voter and appear in the platform. That is, individual issues are not mutually exclusive and do not, therefore, represent alternative choices.

2. Problem Formulation

We are concerned with the (macro)state of a large population of decision-makers each of whom makes uncertain choices. That is, given $n > 1$ available alternatives, the decision-maker selects one of them according to his profile $\mathbf{q} \in S^n$, where S^n is the unit simplex in $\mathbf{R}^n$:

$$S^n = \{x \in \mathbf{R}^n : \sum_{i=1}^{n} x_i = 1\} \tag{1}$$

It may be that preferences of individuals are deterministic, and the behavior inherits randomness from external perturbations. Alternatively, individual preferences may be intrinsically stochastic. Regardless of its origin, the uncertainty in the state of the decision-maker is captured by a point in the simplex (1).

Typically, choice behavior of an individual is influenced by factors — notably advertising — that are external to the population, and by interactions among the members of the population. The latter is often referred to as the "word-of-mouth advertising." We assume that the external influences are captured by market shares, and that the interactions among the decision-makers leading to changes in their choice profiles are short-lived, *i.e.*, their duration is much smaller than inter-purchase or, respectively, inter-election times. This assumption is well supported behaviorally and enables us to view the population as, essentially, an "ideal gas" of individuals.

The macrostate of the population may then be found by maximizing entropy of the distribution with density (a Radon-Nikodym derivative with respect to the Lebesgue measure on $\mathbf{R}^n$) f subject to the constraints given by known market shares and normalization. In this formulation, however, the problem is not well posed: the simplex (1), which supports f, is not finite, and any choice of f will maximize entropy giving it the value $+\infty$. For this and practical purposes, we discretize the simplex by superimposing on it a lattice S_L^n of points, the Cartesian components of which differ by $1/L$, where $L \geq 2$ is an integer. Effectively, this amounts to "rounding off" the probabilities in a profile $\mathbf{q} = (q_1, \ldots, q_n)$ to the nearest $1/L$th. The lattice S_L^n contains a finite number of points only (given, in fact, by the number $\binom{n+L-1}{n-1}$ of weak compositions of a natural number L into n parts; a counting and other proofs are available — see Stanley [16]), so the entropy of f on S_L^n, which in the discrete version we denote by f_L, is finite. Note that even if the technical reasons for substituting the lattice S_L^n for the simplex this S^n were absent, the discrete version of the problem would be necessitated by practical considerations: a producing firm dealing with consumers, for example, cannot subdivide the market into an infinite number of segments that the points of S^n represent.

Thus, we determine f by maximizing entropy,

$$H[f] = - \sum_{\mathbf{q} \in S_L^n} f_L(\mathbf{q}) \log f_L(\mathbf{q}), \tag{2}$$

subject to constraint:

$$\sum_{q \in S_L^n} f_L(q) = 1 \tag{3}$$

$$\sum_{q \in S_L^n} q f_L(q) = Q, \tag{4}$$

where Q is the n-tuple of market shares. In the next section, we state the results of this maximization.

3. Main Results

A solution of the optimization problem (2)–(4) may be found by the classical method of Lagrange multipliers. To this effect we form an objective function by adding to the entropy functional (2) constraints (3) and (4) multiplied by undetermined factors $\log(e/Z)$ and β. This leads to

Proposition 1 *If an individual (particle) state is represented by a point in S_L^n and the mean is known, the ensemble distribution belongs to the product-form family:*

$$f_L(q) = \frac{1}{Z_L} e^{-\beta \cdot q} \chi_L^n(q), \tag{5}$$

where χ_L^n is the indicator function of the lattice S_L^n. The Lagrange multiplier β (inverse "temperatures") and the partition function

$$Z_L = \sum_{q \in S_L^n} e^{-\beta \cdot q} \tag{6}$$

are defined implicitly by the constraints (3)–(4).

We gave above a general formulation of Proposition 1 that is not contextually limited to the choice problem: as we mentioned earlier, the essential aspect of the system under investigation is that the states of its individual elements are described by points in the simplex (1).

Interestingly, the product form distributions (5) arise often as steady states of Markov processes. In particular, they are generically present in the description of the analysis of closed (so-called Jackson) queueing networks (Conway and Georganas [3]). Note, however, that we made no assumptions with regard to the equilibration process; in particular, no assumption of the Markov property has been made.

As is typical in statistical mechanics, the computation of the partition function (6) is not trivial. The system under consideration is rather unusual in that Z_L is available in closed form:

Proposition 2 *The partition function Z_L has the following representation (Harrison [9] , Faynzilberg [4]):*

$$Z_L = \sum_{i=1}^{n} e^{-\beta_i L} \prod_{j \neq i} \frac{1}{e^{\beta_j - \beta_i} - 1}. \tag{7}$$

The meaning of Lagrange multipliers β is revealed by the following comparative statics result, which is consistent with the usual interpretation of multipliers as the respective shadow costs:

Proposition 3 *Multipliers β are equal to the rates of change of the maximal (thermodynamic) value H^* of entropy with respective market shares:*

$$\beta = \frac{\partial H^*(\mathbf{Q})}{\partial \mathbf{Q}}. \tag{8}$$

Finally, as a solution of the continuos version of the problem, we may take the limit as the inter-cite distance $1/L$ diminishes to zero:

$$f(\mathbf{q}) = \lim_{L \to \infty} \frac{1}{V_n} \binom{n+L-1}{n-1} f \circ \kappa_L(\mathbf{q}), \tag{9}$$

where V_n is the volume of the simplex S^n and

$$\kappa_L(\mathbf{q}) = \frac{1}{L} \operatorname*{argmin}_{\mathbf{k} \in \mathbf{Z}^n} \|\mathbf{q} - L\mathbf{k}\|$$

is the approximation of $\mathbf{q}$ in S_L^n in terms of the Euclidean norm $\|\cdot\|$ on $\mathbf{R}^n$.

References

1. R. Balescu. *Equilibrium and Nonequilibrium Statistical Mechanics.* John Wiley & Sons, New York, 1975.
2. L. Boltzmann. *Lectures on Gas Theory.* 1896, 1898. A recent reprint: Berkeley, 1964.
3. A.E. Conway and N.D. Georganas. *Queueing Networks — Exact Computational Algorithms: A Unified Theory Based on Decomposition and Aggregation. Computer Systems Series,* The MIT Press, Cambridge: Mass, 1989.
4. P.S. Faynzilberg. *Normalization and Generation of Product-form Distributions.* Working Paper 1994–10, Graduate School of Industrial Administration, Carnegie Mellon University, Pittsburgh, Pennsylvania, March 1994.
5. P.S. Faynzilberg. *Statistical Mechanics of Choice: MaxEnt Estimation of Population Heterogeneity.* Working Paper 1995–06, Graduate School of Industrial Administration, Carnegie Mellon University, Pittsburgh, Pennsylvania, (forthcoming).
6. R.P. Feynman. *Statistical Mechanics: A Set of Lectures.* W.A. Benjamin, Inc., Reading: Mass., 1972.
7. P.C. Fishburn. *Utility Theory for Decision Making.* John Wiley & Sons, New York, 1970.
8. J.W. Gibbs. *Elementary Principles in Statistical Mechanics.* Yale University Press, New Haven, 1902.
9. P.G. Harrison. On normalizing constants in queueing networks. *Operations Research,* 33:464–8, 1985.
10. E.T. Jaynes. Information theory and statistical mechanics. *Physical Review,* 106(4):620–30; 108(2):171–90, 1957.
11. E.T. Jaynes. *Papers on Probability, Statistics, and Statistical Physics. Synthese Library,* D. Reidel Publishing Company, Dordrecht:Holland, 1983.
12. P. Kotler. *Marketing Management: Analysis, Planning, and Control.* Prentice-Hall, Englewood Cliffs, NJ, fifth edition, 1994.
13. D.M. Kreps. *A Course in Microeconomic Theory.* Princeton University Press, Princeton, NJ., 1990.
14. L. D. Landau and E. M. Lifshitz. *Statistical Physics.* Volume 5 of *Course of Theoretical Physics,* Pergamon Press, Oxford, 3rd edition, 1980.

15. E. Malinvaud. *Lectures on Microeconomic Theory*. Volume 2 of *Advanced Textbooks in Economics*, North-Holland, Amsterdam, 1985.

16. R.P. Stanley. *Enumerative Combinatorics*. Wadsworth & Brooks, Monterey, California, 1986.

17. H.R. Varian. *Microeconomic Analysis*. W. W. Norton & Company, Inc., New York, second edition, 1984.

RATIONAL NEURAL MODELS BASED ON
INFORMATION THEORY

ROBERT L. FRY
The Johns Hopkins University/
Applied Physics Laboratory
Laurel, MD 20723

Abstract: Rationality necessitates that the neuron must base its decisions on probability measures established on the interrogative proposition q which essentially defines the neuron as an observer and which is established during learning. This requirement is induced by the fact that probability is the only measure of degree of plausible belief which meets all logical consistency requirements. Similarly, the establishment of a probability measure by a rational neuron must entail the usage of a maximized entropy criterion since this is the only logically consistent means of doing so as based on observable information. A maximum entropy (ME) formulation can be shown to provide the basic functional form of the model neuron including synaptic weights and a sigmoidal transfer characteristic. However, this formulation requires the specification of linear constraints which are unavailable. Alternatively, a maximum mutual information (MMI) formulation is shown to be fully constrained in this regard and can make exclusive use of locally available information. Solutions take the form of the Hopfield neuron model with a requirement for Hebbian learning.

Key words: entropy, mutual information, directed divergence

1. Introduction and Overview

Biological organisms which possess a neurological system exhibit varying degrees of what can be termed rational behavior. One can hypothesize that rational behavior and thought processes in general arise as a consequence of the intrinsic rational nature of the neurological system and its constituent neurons. A similar statement may be made of the immunological system [1]. The concept of rational behavior can be made quantitative. In particular, one possible characterization of rational behavior is as follows:

(1) A physical entity (observer) must exist which has the capacity for both measurement and the generation of outputs (participation). Outputs represent decisions on the part of the observer which will be seen to be rational.

K. M. Hanson and R. N. Silver (eds.), Maximum Entropy and Bayesian Methods, 335–340.
© *1996 Kluwer Academic Publishers.*

(2) The establishment of the quantities measurable by the observer is achieved through learning. Learning characterizes the change in knowledge state of an observer in response to new information and is driven by the directed divergence information measure of Kullback [2].

(3) Output decisions must be made optimally on the basis of noisy and/or missing input data. Optimally here implies that the decision-making process must abide by the standard logical consistency axioms which give rise to probability as the only logically consistent measure of degree of plausible belief. An observer using decision rules based on such is said to be rational.

Information theory can be used to quantify the above leading to computational paradigms with architectures that closely resemble both the single cortical neuron and interconnected planar field of multiple cortical neurons all of which are functionally identical to one another. A working definition of information in a neural context must be agreed upon prior to this development, however. Such a definition can be obtained through the Laws of Form - a mathematics of observation originating with the British mathematician George Spencer-Brown [3].

It is commonly accepted that neurons process information although any precise definition of information within a neural context remains outstanding. Such a definition is fundamental to the analysis and understanding of biological neurons and their ability to process information. The following definition is proposed: *Information is that which is measurable by an observer.* The assumption of this definition means that what is information to one measurement device (observer) is not necessarily information to another physically independent system (observer) as related to a pervading state of nature. This definition implies that a quantity distinguishable by an observer is information hence information can be seen to be a unique property of the observer. The concept of distinguishability on the part of an observer can be seen to lie at the basis of the Maximum Entropy formulation and can be seen to lie at the foundation of information theory itself.

2. Distinguishability as Information

In George Spencer-Brown derived his *Calculus of Indications* as a means of quantifying the concept of "distinguishability" in his *Laws of Form (LoF)* [3]. Indicational expressions within the *Calculus of Indications* provide a quantitative means of describing an observer in terms of the quantities measurable by the observer. In fact, any system can be described using indicational expressions which explicitly delineate system observables and outputs. The most primitive act within the *LoF* is that of distinction which requires both a measurement device and corresponding measurable quantities and is enacted in physics through the Hermitian operator. The lack of any generated indication or observable on the part of an observer means that the observer is a closed physical system unobservable to any other physically independent system/observer and hence is not information as based on the previous definition and therefore is unimportant in our analysis.

The act of distinction on the part of an observer arises from symmetry-breaking on the part of the observer and is the dual-state system as created through learning. The defining distinction and the observer are synonymous. For instance, biological neurons observe spatio-temporal patterns of activity and must decide on which patterns it will fire - the outward manifestation of which is the action potential. Certain pattern classes will more likely give rise to an action potential which represents the *decision* of the neuron. As described here, the observing neuron can probabilistically decide in favor of one distinction over the its contradiction and hence either generate an action potential (indication) or not as based on measured spatio-temporal information. The phenomenology leading to the formation of the defining distinction is deemed to correspond to the process of learning in a neural context. Learning can then be interpreted as the search for the appropriate *question* as opposed to the appropriate *answer*. The use of probability by the neuron to perform its decision-making is what intrinsically makes the neuron rational.

Simple *Calculus of Indications* expressions can be used to describe information flow within a functional neuron q or neural field Q which extract information from the environment through a set of measurement functions F which depend logically on the input training ensemble $x=(x_1\ x_2\ x_3...x_N)$ and the outputs y where $x \epsilon B^N$, $y \epsilon B^M$, $B=\{0,1\}$ and in general $M \leq N$. The adoption of $B=\{0,1\}$ is in part arbitrary since the information measures used here are scale-invariant. Therefore, 0 and any numerical value distinguishable from 0 are acceptable although the choice of B lends itself to more concise derivations upon the application of the discussed information-theoretic measures. Boolean expressions arise as a consequence of one limited interpretation of the *LoF*. Below are the basic *LoF* forms for the noted Boolean expressions. The most important is the logical "implies" which mathematically corresponds to physical observation. For

instance, the notation $\overline{a}|b$ denotes the distinction of a measured quantity a by the

observer b. The situation $\overline{b}|b = \overline{}|$ means that the observer has observed a response

matched to its defining distinction b leading to the indication $\overline{}|$ which is an action potential in a neural context.

Literal	*LoF* Interpreted as Logic Boolean Expression	*LoF*
not a	~a	$\overline{a}|$
a or b	a∨b	ab
a and b	a∧b	$\overline{\overline{a}|\ \overline{b}|\ |}$
a implies b (b observes a)	a⊃b	$\overline{a}|\ b$

Neural indicational expressions include feedback of the indicational output y to the set of measurement functions internal to Q thereby providing a means of adapting the measurement functions in response to supplied elements $x \epsilon B^N$. Equation 1 shows the correspondence between q and the *calculus of indications* expressions for the sampling functions dictated by the set $F_1=\{x_1y, x_2y,...,x_Ny,y\}$. Literally, this particular set can be described as the case where the observer q observes "the conjunctions x_1y *with*[1] x_2y *with* ... *with* x_Ny *with* the indication y." Note that these are not just conjunctions of signals in space, but are also conjunctions of the respective signals in time. The *LoF* expression for the observer making measurements through the set F_1 is given in Equation 1.

$$\overline{x_1|\overline{y}}\ |\ \overline{x_2|\overline{y}}\ |\cdots \overline{x_N|\overline{y}}|\ \overline{y}|\quad |q = y \qquad (1)$$

This form can be mathematically simplified at the expense of loosing insight into the flow of information into and within the model neuron.

3. Probability as the Only Rational Basis for Decision-making

Probability is the only logically consistent measure of degree of plausible belief [4] on the part of an observer. An observer which uses such must be deemed rational. As such, let the joint probability $P(x,y)$ describe the internal statistical model of the environment for a rational neuron q which is developed through learning. Suppose that q uses the measurement set F_1 to extract information from environmental input in order to compute $P(x,y)$. The observer q is assumed to not be capable of maintaining a complete record of input, but rather can only estimate moments estimated through arithmetic averaging of the outputs of the respective measurement functions possessed by the observer.

The statistical model $P(x,y)$ is really a state of knowledge possessed by q. $P(x,y)$ provides q with the capacity of performing rational decision-making regarding the validity of q given the observable information x. It should be noted that $P(x,y)$ reflects knowledge of possible environmental states as possessed (stored) by the subject neuron. The evaluation of $P(x,y)$ can only practically take place in response to a supplied in x and the resulting self-generated output y at which time an action potential either is or is not generated.

4. Maximum Entropy (ME) Formulation

It is assumed that the neural field Q is capable of extracting information from external inputs $x \in B^N$ and its own outputs y through the use of a set the union of the individual neural measurement sets F_n and is given by $F = \bigcup\limits_{n=1}^{N} F_n$ The field Q uses the sampled data to estimate moments on the defined measurement functions F which it in turn uses to compute the joint distribution $P(x,y)$. The computed moments serve to realize $P(x,y)$ as a unique network ME distribution or equivalently a Gibbs distribution parameterized by synaptic connection weights which are in fact the Lagrange multipliers for the ME distribution as are all the decision thresholds μ_n. For instance, suppose that the

---

[1] Here, *with* means Boolean disjunction "or" and is more accurately used here to convey literal meaning.

measurement functions for a single neuron q consist of F_1 as defined earlier. In this case $P(\mathbf{x},y)=\exp[-\Sigma\lambda_n x_n y+\mu y]/Z$ where Z is the standard partition function and $\Lambda_1=\{\lambda_1,\lambda_2,...,\lambda_N, \mu\}$ is the Lagrange set for the corresponding moments on the elements of F_1. In this case, the classical Hopfield neuron arises [5],[6] and the maximum likelihood (ML) decision rule can be used to generate an action potential when $\Sigma\lambda_n x_n-\mu>0$, otherwise none is generated. If a stochastic decision rule is adopted, then an action potential can be generated according to the sigmoidal transition rule given by $P(y=1|\mathbf{x})=1/1+\exp[-\zeta(\mathbf{x})]$ where $\zeta(\mathbf{x})=\Sigma\lambda_n x_n-\mu$. A simple calculation will show that ζ is in fact the statistical *evidence* function which is defined to be the log of the odds function given by the likelihood ratio $P(y=1|\mathbf{x})/P(y=0|\mathbf{x})$. Computed statistical evidence is therefore used by the neural observer to perform rational decision-making.

There is a problem with this formulation, however. It is unrealistic to expect that a neuron or collection of neurons has to computationally solve the ME optimization problem to obtain the prerequisite Lagrange set. For one thing, this would imply that the moments of the measurement set either be specified by a supervisory source or that the moments be extracted and stored as computed from observed data. Secondly, ME optimization does not have a known simple biological plausible implementation which uses locally available information. Consider the following alternative.

It is a well known fact that the Lagrange multipliers and the distributional moments are completely conjugate to one another for an ME formulation with linear constraints. Knowing one set fully constrains the other. Therefore, the specification of the values for a Lagrange set Λ will simultaneously define the complete ME distribution and a corresponding set of moments. The question here then becomes that of how to specify values for Λ or what methodology to use for computing Λ directly from observable data. One possible strategy is to select values for Λ which serve to maximize the mutual information between the input $\mathbf{x}$ to the Q-field and output $\mathbf{y}$ (or in the case of an individual neuron - just y).

5. Maximized Mutual Information (MMI)

Suppose that the neural field Q rationalizes according to $P(\mathbf{x},\mathbf{y})=\exp[-\lambda\cdot\mathbf{f}]/Z$ where $\mathbf{f}$ is a vectorized representation of the set of measurement functions F defined architecturally within Q. In this case, one can dynamically compute the Lagrange values λ which maximize the mutual information between $\mathbf{x}$ and $\mathbf{y}$ through the use of the Gibbs Mutual Information Theorem (GMIT) as derived in [6] which states that the optimal λ is given by the extrema of $J=I(\mathbf{x};\mathbf{y})+\Sigma\alpha_m g_m(\lambda)$ where $I(\mathbf{x},\mathbf{y})$ is the mutual information between $\mathbf{x}$ and $\mathbf{y}$ and each $\alpha_m g_m(\lambda)$ represents architectural constraints placed on one or more members of the   Lagrange set - realistic biological constraints might include limited neurotransmitter resources, non-negativity of specific synaptic connections, etc. In this case, GMIT states that for all $n=1,2,...,N$ and each $f_n\in$ F.

$$\frac{\partial J}{\partial\lambda_n} = E\left\{\left[<f_n>-f_n(\mathbf{x},\mathbf{y})\right]\ln\left(\frac{P(\mathbf{x}|\mathbf{y})}{P(\mathbf{x})}\right)\right\}+\frac{\partial}{\partial\lambda_n}\Sigma\alpha_m g_m(\lambda) = 0 \qquad (2)$$

As an example, the set F_1 with an l_2-norm constraint on λ such that $|\lambda|^2=c$, leads to a Hopfield neuron model [5] which maximizes its output entropy over the ensemble of training patterns. A related results holds for the measurement set F as defined above. In this case the network can, with some further realizable architectural restrictions, be logically factored into functionally equivalent neural components q_n which correspond to the individual sets F_n, respectively for $n=1,2,...,N$ and $|Q|=N$.

6. Summary

Together ME and MMI lead to a neural field which is factorable in both form and function into component computational entities which correspond to the Hopfield neuron model   including decision threshold, action potential realization, Hebbian learning, sigmoidal transfer characteristic, and conditionalized principal component analysis using a simple modification of an equation originally described by Oja [7]. The resulting physical neuron paradigm can in many regards be considered rational and furthermore has biological plausibility [8],[9].

References

[1] Francisco J. Varela, *Principles of Biological Autonomy*, North Holland, 1979.

[2] Solomon Kullback, *Information theory and statistics*, Wiley, 1959 and Dover, 1968.

[3] George Spencer-Brown, *Laws of Form*, E. P. Dutton, New York 1979

[4] R.T. Cox, *The algebra of probable inference*, The Johns Hopkins Press, Baltimore, 1961.

[5] Hopfield, J. J., "Neural networks and physical systems with emergent collective computational abilities," *Proc. Nat. Acad. Sci. USA*, 79, 2554-2558, 1982.

[6] R. L. Fry, "Observer-participant models of neural processing," *IEEE Trans. Neural Networks*, July, 1995.

[7] E. Oja, "A simplified neuron model as a principal component analyzer," *J. of Math. Biol.* 15, 267-273., 1982

[8] William B. Levy, Costa M. Colbert, and Nancy L. Desmond, "Elemental adaptive processes of neurons and synapses: A statistical/computational perspective," in *Neuroscience and connectionists theory*, pp 187-235, 1990.

[9] William B. Levy, "Maximum entropy prediction in neural networks," in proc. *International Joint Conference on Neural Networks*, 1990.

A NEW ENTROPY MEASURE WITH THE EXPLICIT NOTION OF COMPLEXITY

WLODEK HOLENDER
Department of Communication Systems
Lund University[†]

Abstract.
The existence of a parameterized family of entropy rate functions is proved. This family is generalized to one dentropy function, explicitly dependent on the continuous complexity parameter. The dentropy function establishes a better bound for the tail of a distribution than Chernoff's bound. The "second law" for complex systems is derived. An increased disorder is counterbalanced by an increased complexity such that the new thermodynamical dentropy is kept constant.

Key words: entropy, k-moment entropy, dentropy, Maxwell demon, complexity, complexity parameter, second law for complex systems.

1. Introduction

The entropy measure plays a fundamental role in many scientific fields, like the information theory, thermodynamics, statistical mechanics, optimization problems and communication networks, to mention only a few of them. In the context of the information theory, entropy is interpreted as a measure of the observer's ignorance about the observed system. Entropy also relates the observer's information to physically significant quantities, as for example the energy of a system.

The interpretation of the entropy measure as the relation between the observer's information[1] and the energy of the system can be applied to describe the behavior of any living structure. Extracting useful energy from the environment is one of the main goals of a living structure in its struggle of surviving. To achieve this goal, a living structure has to utilize the information on the environment and make decisions which lead to a transfer of useful energy from the environment to this structure. Even other social, political or economical structures reveal a similar type of behavior.

[†]P.O. Box 118, S-221 00 Lund, Sweden Phone: int+46-46-2224910, Fax: int+46-46-145823
E-mail: wlodek@tts.lth.se
[1]Information and ignorance are used as corresponding interpretations of the entropy measure. These two interpretations differ only by choice of the sign.

K. M. Hanson and R. N. Silver (eds.), Maximum Entropy and Bayesian Methods, 341–346.

Generally, any abstract structure which behaves as described above, can be considered as an instance of a generalized Maxwell Demon [1]. The Maxwell Demon is an abstract creature which observes a system and tries to extract useful energy[2] from this system by making some type of intelligent decisions.

Could we make the Demon more intelligent, such that it will make better decisions? What does this intelligence mean?

1.1. THERMODYNAMIC MODEL OF COMPLEXITY

In the last decade a new complexity theory has been suggested [2],[3]. This new theory tries to answer such an intriguing question as "what is a mechanism of self organization" observed in complex structures [4],[5],[7],[8].

The most interesting achievement of the complexity theory is the discovery of emergence. The intelligent (e.g. adaptive) behavior of a system is caused by interaction between the members of this system. The collective behavior of all members possesses complex properties, although each member doesn't have such a complex property.The discovery of emergence as a nonlinear interaction between simple individuals is important, since it removes the need for a "mind" which in some mysterious way creates intelligence or life in the system.

Although this new direction in the theory of complex systems seems to be promising, many important questions are still unanswered.

It has been observed that complex systems behave in a similar manner as systems in phase transition state - the "edge of chaos" [6],[2],[3]. There is a believe that this behavior could be described by thermodynamical laws, enhanced to complex systems. The question of fundamental importance is if it is possible to derive a foundation of a thermodynamical theory which could explain the basic properties of complex systems.

One particular question in such a theory is very interesting and should find an explanation in order to make this theory successful: "why do the living systems increase their complexity with the time arrow".

Some call this property an extended second law of thermodynamics for complex systems [2]. Living complex systems seem to develop in a direction which is opposite to the one predicted by the second law of "traditional" thermodynamics. It states that the thermodynamical entropy of an isolated system can only increase, which corresponds to increased disorder in this system. This prediction seems to be in contradiction to our experience that living systems result in increasing order with the time arrow due to their ability to increase the self organization and complexity.

In the remainder of this paper I will make an attempt to introduce a foundation of a thermodynamic model of complex systems. Based on this model I will derive the "second law" in context of this model, i.e. explain why living systems increase

---

[2]The reason why this creature is called Demon is because it extracts more useful energy from the observed system than what is allowed by the second law of thermodynamics, i.e., Demon violates law of physics.

the complexity. This new model is based on a new entropy measure. This new entropy measure is a generalization of the "old" entropy measure.

Firstly, I will prove the existence of a parameterized family of entropy rate functions. Secondly, I will generalize this family of entropy rate functions to one function explicitly dependent on the continuous complexity parameter. I will call this generalized function *dentropy function* (<u>D</u>emon's <u>entropy</u>). The complexity parameter in the dentropy function appears in a similar way as the temperature does in the "old" entropy function.

The most important mathematical property of this new entropy measure with the complexity parameter, is that it establishes a better bound for the tail of a distribution than the "old" entropy rate function. This property has an interesting interpretation. Complex systems which are ruled by this dentropy measure can predict the energy of the system with greater accuracy than systems which are ruled by the old entropy measure and the accuracy of prediction increases with the complexity parameter.

This property of complex systems corresponds to the ability of the Maxwell demon to extract additional energy from a system beyond what is allowed by the "old" second law. This ability is due to the increased knowledge[3] about the observed system. For the infinite value of the complexity parameter the Demon will posses an ultimate knowledge about the observed system.

2. k-moment entropy rate function

The Chernoff's bound [9] denoted as the *zero moment entropy rate function*: $I^0(C)$ can be derived from the Markov inequality $P(X \geq C) \leq \frac{E(X)}{C}$:

$$- \ln P(X \geq C) = - \ln P(e^{sX} \geq e^{sC}) \geq \sup_s (sC - \ln E(e^{sX})) = I^0(C) \quad (1)$$

Another similar bound denoted as the *k-moment entropy rate function* can be derived in a similar way.

$$- \ln P(X \geq C) = - \ln P(X^k e^{sX} \geq C^k e^{sC}) \geq$$
$$\sup_s (\ln E(C^k e^{sC}) - \ln E(X^k e^{sX})) = I^k(C) \quad (2)$$

Theorem 1
$$- \ln P(X \geq C) \geq I^m(C) \geq I^k(C) \quad (3)$$

for every positive integer m and k such that $m > k$.
proof[4] (by induction): Assume that this theorem is valid for $k = i$.

$$- \ln P(X \geq C) = - \ln P(X^i e^{sX} \geq C^i e^{sC}) \geq$$
$$\sup_s (\ln E(C^i e^{sC}) - \ln E(X^i e^{sX})) = I^i(C) \quad (4)$$

---

[3] Knowledge in the meaning as a theory about the system.
[4] Only inductive part of the proof is presented. For full proof see [10].

here the optimal value of s_i which gives the supremum $I^i(C)$ is defined according to:

$$C = \frac{d}{ds} \ln(E(X^i e^{sX})) \mid_{s=s_i} \tag{5}$$

From this assumption the conclusion that the theorem is valid for $k = i + 1$ will be derived.

$$-\ln P(X \geq C) = -\ln P(X^{i+1} e^{sX} \geq C^{i+1} e^{sC}) \geq$$
$$\sup_{s}(\ln E(C^{i+1} e^{sC}) - \ln E(X^{i+1} e^{sX})) = I^{i+1}(C) \tag{6}$$

$$\ln(E(X^{i+1} e^{sX})) = \ln(\frac{d}{ds} E(X^i e^{sX})) = \ln\left(E(X^i e^{sX}) \frac{d}{ds} \ln E(X^i e^{sX})\right) =$$
$$\ln(E(X^i e^{sX})) + \ln(\frac{d}{ds} \ln(E(X^i e^{sX}))) \tag{7}$$

then define:

$$\tilde{I}^{i+1}(C, s) = \ln\left(\frac{C}{\frac{d}{ds} \ln E(X^i e^{sX})}\right) + \ln E(C^i e^{sC}) - \ln(E(X^i e^{sX})) \tag{8}$$

but s_i for $I^i(C)$ is defined according to 5 and thus:

$$\ln\left(\frac{C}{\frac{d}{ds} \ln E(X^i e^{sX}) \mid_{s=s_i}}\right) = 0 \tag{9}$$

which implies the relations:

$$\tilde{I}^{i+1}(C, s_i) = I^i(C) \tag{10}$$

$$I^{i+1}(C) = \sup_{s} \tilde{I}^{i+1}(C, s) \geq I^i(C) \square \tag{11}$$

Theorem 2

$$\frac{d}{dC} I^k(C) = s_k(C) + \frac{k}{C} \tag{12}$$

where $s_k(C)$ is the shift parameter defined by 5. Proof is given in [10].

2.1. GENERALIZATION TO CONTINUOUS COMPLEXITY PARAMETER

Define the *dentropy rate function* $I(\lambda, C)$ as a bound:

$$-\ln P(X \geq C) = -\ln P(X^\lambda e^{sX} \geq C^\lambda e^{sC}) \geq$$
$$\sup_{s}(\ln(C^\lambda e^{sC}) - \ln(E(X^\lambda e^{sX}))) = I(\lambda, C) \tag{13}$$

where $\lambda \geq 0$ is a continuous *complexity parameter*.

Conjecture: 1

$$\forall \lambda \left(\frac{\partial}{\partial \lambda} I(\lambda, C) \geq 0 \right) \tag{14}$$

An example of dentropy rate function is shown in figure 1. s is used as a variable due to the relation $I(\lambda, C(s)) = I(\lambda, s)$ and $C(s)$ is calculated from equation 5 generalized to $i = \lambda$.

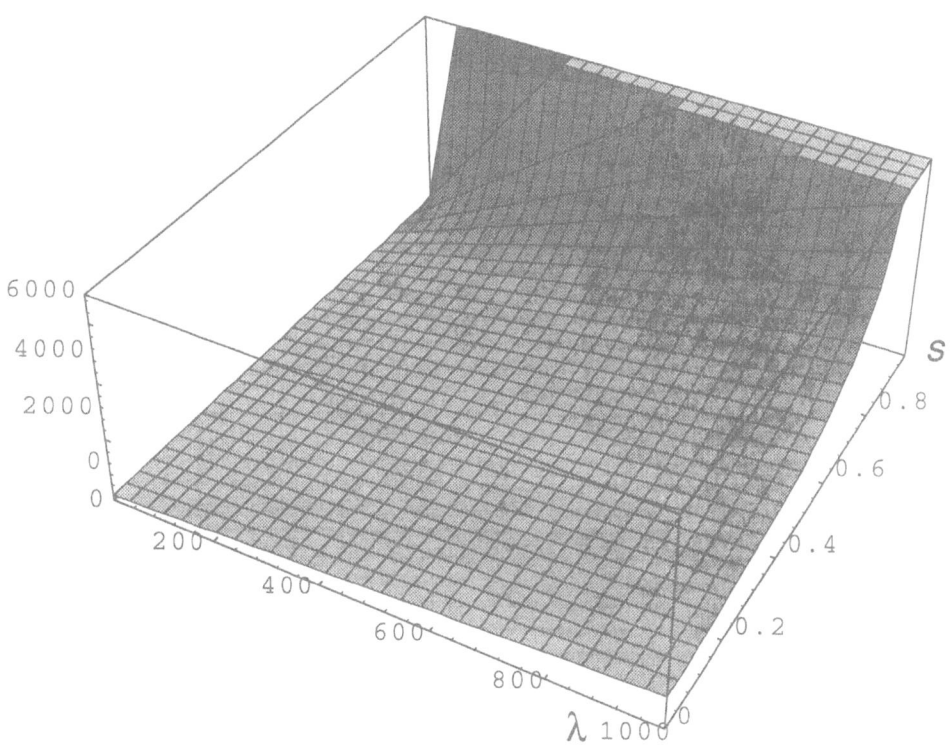

Figure 1.   The dentropy rate function for the exponential distribution $f(x) = e^{-x}$

3.   "Second law" for complexsystems

Assume that the complex system is constrained by the *stationary* condition.

$$dI(\lambda, s) = \frac{\partial}{\partial \lambda} I(\lambda, s) d\lambda + \frac{\partial}{\partial s} I(\lambda, s) ds = 0 \tag{15}$$

then

$$\frac{d\lambda}{ds} = -\frac{\frac{\partial}{\partial s}I(\lambda, s)}{\frac{\partial}{\partial \lambda}I(\lambda, s)} \tag{16}$$

but

$$\frac{\partial}{\partial s}I(\lambda, s) = \frac{\partial}{\partial C}I(\lambda, s)\frac{dC}{ds} = \left(s_\lambda + \frac{\lambda}{C}\right)\frac{d^2}{ds^2}\ln(E(e^{sX})) \tag{17}$$

$$\frac{d\lambda}{ds} = -\frac{(s_\lambda + \frac{\lambda}{C})V_s(X)}{\frac{\partial}{\partial \lambda}I(\lambda, s)} \tag{18}$$

The right side of 18 is negative since all terms are positive. $V_s(X)$ is a variance and $I(\lambda, s)$ increases with λ. Thus, the change of λ must have the opposite sign to the change of s. Particularly when s decreases, which corresponds to increasing temperature and increasing disorder in the system, the complexity parameter must increase.

The complex system responds with increased complexity to the increasing disorder.

References

1. Wojciech H. Zurek "Complexity, Entropy and the Physics of Information" Volume VIII, Santa Fe Institute, Addison-Wesley 1990.
2. Roger Lewin "Complexity, Life at Edge of Chaos" Macmillan 1992.
3. M. Mitchell Waldorp "Complexity the Emerging Science at the Edge of Order and Chaos" Penguin 1992.
4. Christopher G. Langton, Editor "Artificial Life", Volume VI, Santa Fe Institute, Addison-Wesley 1989.
5. Christopher G. Langton, *et al.*, Editors "Artificial Life II", Proceedings Volume X, Santa Fe Institute, Addison-Wesley 1992.
6. Christopher G. Langton "Studying Artificial Life with Cellar Automata", Physica 22D (1986) 120-149, North-Holland, Amsterdam.
7. Stuart A. Kauffman "Antichaos and Adaptation" Scientific American August 1991.
8. Per Bak, Kan Chen "Self-Organized Criticality" Scientific American January 1991.
9. Richard S. Ellis "Entropy, Large Deviations, and Statistical Mechanics" Springer-Verlag New York 1985.
10. Wlodek Holender "A New Entropy Measure with the Explicit Notion of Complexity" (full version) submitted to Physica D, North-Holland, Amsterdam.

MAXIMUM ENTROPY STATES AND COHERENT STRUCTURES IN MAGNETOHYDRODYNAMICS

RICHARD JORDAN

Carnegie Mellon University
Center for Nonlinear Analysis and Department of Mathematics
Pittsburgh, PA 15213

AND

BRUCE TURKINGTON

University of Massachusetts
Department of Mathematics and Statistics
Amherst, MA 01003

Abstract. We review a recently developed model of coherent structures in two-dimensional magnetohydrodynamic turbulence. This model is based upon a constrained maximum entropy principle: Most probable states are determined as maximizers of entropy subject to constraints imposed by the conservation of energy, cross-helicity, and flux under the evolution of an ideal two-dimensional magnetofluid. Predictions of the model are compared with results of high-resolution numerical simulations of magnetofluid turbulence.

Key words: magnetohydrodynamics, coherent structures, turbulence, maximum entropy

1. Introduction and Overview

In this note, we present a model of coherent structures in two-dimensional (2D) magnetohydrodynamic (MHD) turbulence. By a coherent structure, we mean a large-scale organized state that persists amidst the small-scale turbulent fluctuations of the magnetic field and the velocity field. The emergence of such macroscopic states is a dominant feature of large Reynolds number MHD flows, as has been demonstrated by direct numerical simulations. We model these structures as maximizers of an appropriate entropy functional subject to constraints dictated by the conserved quantities of the ideal (nondissipative) MHD equations. Excellent qualitative and quantitative agreement is found with recent numerical simulations of 2D MHD turbulence.

347

K. M. Hanson and R. N. Silver (eds.), Maximum Entropy and Bayesian Methods, 347–353.
© *1996 Kluwer Academic Publishers.*

2. Ideal Magnetohydrodynamics

The equations of ideal, incompressible MHD in appropriately normalized variables are:

$$B_t = \nabla \times (V \times B), \tag{1}$$
$$V_t + V \cdot \nabla V = (\nabla \times B) \times B - \nabla p, \tag{2}$$
$$\nabla \cdot B = 0 \quad , \quad \nabla \cdot V = 0, \tag{3}$$

where $B(x, t)$ is the magnetic field, $V(x, t)$ is the velocity field, and $p(x, t)$ is the fluid pressure. Note that p is determined instantaneously in response to the incompressibility constraint on V. These equations are assumed to hold in a regular bounded spatial domain D in R^2, and $x = (x_1, x_2)$ denotes a generic point in D. The magnetic field and the velocity field take values in R^2. Boundary conditions are given by

$$B \cdot n = 0, \ V \cdot n = 0 \quad \text{on } C, \tag{4}$$

where C is the boundary of D and n is the outward normal to C. The model developed below also applies with minor modifications to the case of a fundamental period domain corresponding to periodicity of B and V in x_1 and x_2.

A 2D ideal magnetofluid conserves energy, flux, and cross-helicity. These quantities are given by, respectively,

$$E = \frac{1}{2} \int_D (B^2 + V^2) \, dx, \tag{5}$$

$$F_f = \int_D f(a) \, dx, \tag{6}$$

$$H_f = \int_D B \cdot V f'(a) \, dx. \tag{7}$$

Here a is the vector potential (or flux function), and is defined by the relation

$$B = (a_{x_2}, -a_{x_1}). \tag{8}$$

The vector potential satisfies the boundary condition

$$a = 0 \quad \text{on } C. \tag{9}$$

The function f in (6) and (7) must satisfy certain regularity (eg. smoothness) conditions, but is otherwise arbitrary. Thus, there are infinite families of conserved flux integrals and cross-helicity integrals. These conserved functionals, which give the dynamics of the 2D magnetofluid its special characteristics, will play a fundamental role in the model sketched below.

3. Macroscopic Description of the MHD System

The high-resolution numerical simulations of Biskamp et al. [1-3] clearly display the turbulent behavior of a slightly dissipative 2D magnetofluid. As the field-flow state $Y = (B, V)$ evolves, it develops rapid fluctuations on very fine spatial scales. After a short period of time, large scale coherent structures emerge in the form of macroscopic magnetic and kinetic islands. These structures persist for a relatively long time period amidst the turbulent fluctuations before the dissipation causes them to decay. In the ideal limit of vanishing dissipation, we expect that the mixing would continue indefinitely, exciting arbitrarily small spatial scales, and that a turbulent relaxed state, consisting of a large-scale coherent structure and infinitesimal-scale local fluctuations, would be approached. Our main goal is to characterize this final turbulent relaxed state.

The field-flow state Y constitutes a *microscopic* description of the MHD system. Due to its highly intricate small-scale behavior, the microstate Y does not furnish a palpable description of the long-time behavior of the magnetofluid. For this reason, we introduce a coarse-grained, or *macroscopic* description of the system. A macrostate $(\rho(x, y))_{x \in D}$ is a family of local probability densities on the values $y \in R^4$ of the microstate Y at each point x in the domain D. That is, for each x in D, $\rho(x, y)$ represents a joint probability density on the values $y = (b, v)$ of the fluctuating field-flow pair $(B(x), V(x))$. By appealing to the methods of nonlinear analysis, it is possible to show that the macroscopic description ρ may be interpreted as a possible long-time $(t \to \infty)$ weak limit of the microscopic field-flow state $Y(x, t)$ (See [4]). We say that $Y(x, t)$ converges weakly to ρ as $t \to \infty$ if for all bounded continuous functions $G(x, y)$ on $D \times R^4$ there holds

$$\int_D \int_{R^4} G(x, y)\rho(x, y)\, dy dx = \lim_{t \to \infty} \int_D G(x, Y(x, t))\, dx.$$

Technically, we may need to pass to a subsequence of times $t_n \to \infty$ in the definition.

4. Constraints on Macrostates

The conservation of energy, flux and cross-helicity under the ideal dynamics translates into corresponding constraints on admissible macrostates. These constraints are formulated in a manner consistent with the above-mentioned weak convergence of $Y(x, t)$ to ρ. They take the forms (see [4,5] for mathematical details):

$$E(\rho) \equiv \frac{1}{2} \int_D \int_{R^4} (b^2 + v^2)\rho(x, y)\, dy dx = E^0, \qquad (10)$$

$$F_f(\rho) \equiv \int_D f(\overline{a}(x))\, dx = F_f^0, \qquad (11)$$

$$H_f(\rho) \equiv \int_D \int_{R^4} b \cdot v f'(\overline{a}(x))\rho(x, y)\, dy dx = H_f^0, \qquad (12)$$

where E^0, F_f^0, and H_f^0 are the values of energy, flux, and cross-helicity fixed by the initial state of the MHD system; the local mean magnetic field $\overline{B}(x)$ is defined by the relation

$$\overline{B}(x) = \int_{R^4} b\rho(x, y)\, dy, \qquad (13)$$

and $\overline{a}(x)$ is the vector potential corresponding to $\overline{B}(x)$. For future reference, we also define the local mean velocity field

$$\overline{V}(x) = \int_{R^4} v\rho(x, y)\, dy \tag{14}$$

We note that both the mean field-flow and the fluctuations contribute to the energy and cross-helicity integrals, whereas only the mean field contributes to the flux integrals. The latter is a consequence of the smoothing property of the operator $B \to a$ [4]. We might say that energy and cross-helicity are cascaded to infinitesimal scales, while flux is cascaded to large scales.

5. Most Probable States: The Maximum Entropy Principle

There have been previous applications of maximum entropy principles to determine most probable states in MHD turbulence. An interesting model along those lines was proposed by Montgomery et al. [6]. Our own research into this subject began as an attempt to build upon the ideas presented in [6]. The classical statistical mechanical theory of MHD turbulence, as set forth by Fyfe and Montgomery [7] utilizes a truncated Fourier series representation of the field-flow state, together with a canonical ensemble on the Fourier amplitudes.

Our approach is inspired in part by the recent maximum entropy model of Robert et al. [8,9] for coherent structures in 2D hydrodynamics. A novelty of our model is that it incorporates the complete list of conserved integrals of the ideal MHD dynamics, unlike the above mentioned theories. Another new feature of our model is that it provides a scheme for determining analytical expressions for both the large-scale mean field-flow and the infinitesimal-scale fluctuations inherent in the long-evolved state.

The entropy functional that we use is essentially the classical Gibbs-Boltzmann-Shannon entropy:

$$S(\rho) = -\int_D \int_{R^4} \rho(x, y) \log \rho(x, y)\, dy dx. \tag{15}$$

As such, S is a measure of (the logarithm) of the number of microstates corresponding to the macrostate ρ. Implicit in its definition as an integral over D is the assumption that fluctuations at two separated points in D are statistically independent. A detailed discussion of the rationale behind this assumption is provided in [10].

In accordance with the principles expounded by Jaynes [11], we now determine the most probable macrostate ρ as a maximizer of the entropy (15) subject to the constraints (10)-(12) on energy, flux, and cross-helicity. That is, we solve the constrained entropy maximization problem

(MEP) $S(\rho) \to \max$, subject to $E(\rho) = E^0$, $F_f(\rho) = F_f^0$, $H_f(\rho) = H_f^0$,

where f varies over all (sufficiently smooth) functions on the invariant range of the flux function a.

In [4], a slightly different entropy functional (a Kullback relative entropy functional with a Gaussian reference measure) was employed. However, identical results are obtained with either the entropy used in [4] or the entropy (15) used in the present note. Our maximum entropy formulation may be partially justified by appealing to the theory of large deviations, as was done in [4], or by the methods of [10], in which a discrete system that satisfies a Liouville property was used to approximate the continuous MHD system.

6. Calculation of Equilibrium States

For sake of economy, we consider here the simplified problem (**SMEP**):

$$S(\rho) \to \max,$$

subject to the constraints

$$E(\rho) = E^0,$$

$$F_i(\rho) = \int_D f_i(\overline{a}) \, dx = F_i^0, \; i = 1, \ldots, M,$$

$$H(\rho) = \int_D \int_{R^4} b \cdot v\rho(x, y) \, dy dx = H^0.$$

Here, $f_i, i = 1, \ldots, M$, may be chosen from some convenient family of basis functions. Such a discretization of the flux constraints approximates quite accurately the infinite family of constraints [12].

In taking into account only the quadratic cross-helicity constraint, we are simplifying considerably the full statistical equilibrium problem (**MEP**). However, this simplified problem does capture the essence of the correlation effects between the field and the flow that result from the conservation of cross-helicity. For an analysis of the consequences of the complete family of cross-helicity integrals, the reader is referred to [10].

The solution ρ of (**SMEP**) follows from the Lagrange multiplier rule:

$$S'(\rho) = \beta E'(\rho) + \sum \alpha_i F_i'(\rho) + \gamma H'(\rho), \tag{16}$$

where β, α_i, and γ are Lagrange multipliers corresponding to the constraints on energy, flux, and cross-helicity, respectively. The derivatives in (16) are functional derivatives. From (16) it follows that

$$\rho = Z^{-1} \exp(-\beta E'(\rho) - \sum \alpha_i F_i'(\rho) - \gamma H'(\rho)),$$

where $Z(x)$ is the partition function which enforces the normalization constraint

$$\int_{R^4} \rho(x, y) \, dy = 1, \text{ for all } x \text{ in } D.$$

After algebraic manipulations, we arrive at the expression

$$\rho = \frac{\beta^2(1 - \mu^2)}{4\pi^2} \exp\left(-\frac{\beta}{2}(1 - \mu^2)(b - \overline{B}(x))^2 - \frac{\beta}{2}(v - \mu b)^2\right), \tag{17}$$

where $\mu = -\gamma/\beta$. We note that $-1 < \mu < 1$ (see [4]).

7. Analysis of Equilibrium States

A glance at equation (17) reveals that the most probable macrostate ρ is for each x in D a Gaussian distribution on the field-flow pair $(B(x), V(x))$. On closer inspection we find that Var $B_i(x)$=Var $V_i(x) = 1/(\beta(1-\mu^2))$, corr $(B_i(x), V_i(x)) = \mu$, for $i = 1, 2$, and for each x in D. The other components are uncorrelated. The mean field-flow can be shown to satisfy the equations (see [4])

$$\overline{V}(x) \ = \ \mu \overline{B}(x), \tag{18}$$

$$\overline{J}(x) \ = \ \sum \lambda_i f_i'(\overline{a}(x)), \tag{19}$$

where

$$\overline{J}(x) = \nabla \times \overline{B}(x) = -\nabla^2 \overline{a}(x),$$

is the current density corresponding to $\overline{B}(x)$, and $\lambda_i = -\alpha_i/(\beta(1 - \mu^2))$.

In particular, it follows from (18)-(19) that the mean field-flow is a stationary solution of the ideal MHD equations (1)-(3). The theory predicts, therefore, that the ideal magnetofluid will evolve to a turbulent relaxed state consisting of a stationary mean field-flow (the coherent structure) and Gaussian fluctuations. We also see from (19) that the mean field $\overline{B}$ is a critical point of the (deterministic) magnetic energy, $\frac{1}{2} \int_D B^2 \, dx$, subject to the flux constraints, $\int_D f_i(a) \, dx = F_i^0$.

8. Comparison with Numerical Simulations

In general, the predictions of our maximum entropy model are in good agreement with the numerical simulations of Biskamp et al. [1-3]. They observe local Gaussian distributions on the magnetic field and velocity field, and a cascade of flux to large scales, which is indicative of the formation of macroscopic magnetic structures. They also report a cascade of energy to small-scales.

A particularly remarkable prediction of our model is that the ratio of kinetic to magnetic energy in statistical equilibrium is less than 1, regardless of the initial ratio. This follows from straightforward calculations and the fact that the correlation μ satisfies $-1 < \mu < 1$. Indeed, we have for the magnetic energy E_m and the kinetic energy E_k the following expressions

$$E_m = \frac{1}{2} \int_D \int_{R^4} b^2 \rho(x, y) \, dy dx = \frac{1}{2} \int_D \overline{B}^2 \, dx + \text{volume}(D)/(\beta(1 - \mu^2)),$$

$$E_k = \frac{1}{2} \int_D \int_{R^4} v^2 \rho(x, y) \, dy dx = \frac{\mu^2}{2} \int_D \overline{B}^2 \, dx + \text{volume}(D)/(\beta(1 - \mu^2)).$$

This prediction is also in accord with the numerical studies of Biskamp et al. [1-3], in which they observed the rapid relaxation of E_k/E_m to an almost constant value less than 1, even for initial ratios as large as 25.

For more detailed discussions of the predictions of our model and for further comparisons with the numerical simulations of Biskamp et al. [1-3], the reader is referred to [4,5,10].

9. Related Results

The maximum entropy model for 2D MHD turbulence proposed above has been derived by Jordan and Turkington [10] as a continuum limit of a discrete model that utilizes a spatial discretization of the field-flow state $Y(x, t)$. This discrete model is based on a discrete Fourier transform together with an *implicit* canonical ensemble on the discretized variables. We also wish to bring to the attention of the reader the very interesting work of Isichenko and Gruzinov [13], who have obtained results similar to those reported here. Their approach utilizes a canonical ensemble for a truncated spectral representation of the MHD system. A clever rescaling of the inverse temperatures enables them to formally pass to a continuum limit, thereby obtaining statistics that respect the complete set of ideal MHD invariants. Their model also predicts that the ideal magnetofluid will evolve to a state consisting of a stationary coherent structure and Gaussian fluctuations.

Acknowledgements

The research of R.J. is supported in part by the ARO and the NSF through grants to the Center for Nonlinear Analysis. The research of B.T. is supported in part by the NSF under grant DMS-9307644.

References

1. D. Biskamp and H. Welter, "Dynamics of decaying two-dimensional magnetohydrodynamic turbulence," *Phys. Fluids B*, 1, p. 1964, 1989.
2. D. Biskamp and H. Welter, "Magnetic field amplification and saturation in two-dimensional magnetohydrodynamic turbulence," *Phys. Fluids B*, 2, p. 1787, 1990.
3. D. Biskamp, H. Welter, and M. Walter, "Statistical properties of two-dimensional magnetohydrodynamic turbulence," *Phys. Fluids B*, 2, p. 3024, 1990.
4. R. Jordan, "A statistical equilibrium model of coherent structures in magnetohydrodynamics," *Nonlinearity*, 8, p. 585, 1995.
5. B. Turkington and R. Jordan, "Turbulent relaxation of a magnetofluid: A statistical equilibrium model," in *Advances in geometric analysis and continuum mechanics*, (Boston), International Press, 1995.
6. D. Montgomery, L. Turner, and G. Vahala, "Most probable states in magnetohydrodynamics," *J. Plasma Phys.*, 21, p. 239, 1979.
7. D. Fyfe and D. Montgomery, "High-beta turbulence in two-dimensional magnetohydrodynamics," *J. Plasma Phys.*, 16, p. 181, 1976.
8. R. Robert, "A maximum entropy principle for two-dimensional perfect fluid dynamics," *J. Stat. Phys.*, 65, p. 531, 1991.
9. R. Robert and J. Sommeria, "Statistical equilibrium states for two-dimensional flows," *J.Fluid Mech.*, 229, p. 291, 1991.
10. R. Jordan and B. Turkington, "Ideal magnetofluid turbulence." submitted to *J. Stat. Phys.*
11. E. T. Jaynes, "Information theory and statistical mechanics," *Phys. Rev.*, 106, p. 620, 1957.
12. B. Turkington, A. Lifschitz, A. Eydeland, and J. Spruck, "Multiconstrained variational problems in magnetohydrodynamics: Equilibrium and slow evolution," *J. Comput. Phys.*, 106, p. 269, 1993.
13. M. B. Isichenko and A. V. Gruzinov, "Isotopological relaxation, coherent structures, and Gaussian turbulence in two-dimensional magnetohydrodynamics," *Phys. Plasmas*, 1, p. 1801, 1994.

Bob Wagner making a point to Richard Silver as Ellyn Santi-Wagner listens.

A LOGNORMAL STATE OF KNOWLEDGE

P.R. DUKES AND E.G. LARSON
Department of Physics and Astronomy
Brigham Young University
Provo, UT 84602 USA[†]

Abstract. The Lognormal distribution is derived as the representation of a particular state of knowledge using a combination of maximum entropy and group invariance arguments.

Key words: Lognormal distribution

1. Introduction

The Bayesian philosophy towards probability theory could be summed up in the phrase, "a probability distribution is a mathematical representation of a state of knowledge." One of the most powerful tools available for obtaining a probabilistic representation of incomplete knowledge is the Maximum Entropy Method, often referred to as MaxEnt. When error probabilities are assigned by MaxEnt, the user enters information about certain properties of the errors that he knows are true, and the entropy or *missing information* is maximized. This situation can be seen as the most desirable when it is observed that other possible properties of the noise, about which the user has no knowledge, will not be relied on in subsequent inferences. In other words, with MaxEnt no unwarranted assumptions or artifacts are introduced into the representation and this, it has been shown, is an essential feature in probability theory as extended logic [1]. The measure of missing information that is used is due to Shannon [2]. Jaynes [3] has since shown that for continuous distributions the Shannon definition involves a measure function. That is, the correct definition for the missing information of a continuous distribution is

$$H = -\int p(x) \ln \frac{p(x)}{m(x)} dx. \tag{1}$$

It is easy to demonstrate that with the measure function $m(x)$ properly taken into account the entropy of a continuous distribution *is* invariant to a change of

---

[†]Email: dukesp@dirac.byu.edu and larson@acoust.byu.edu

K. M. Hanson and R. N. Silver (eds.), Maximum Entropy and Bayesian Methods, 355–359.

the parameter x and that this measure function also plays the roll of a pre-prior distribution which describes complete ignorance of the x parameter [4].

Tribus [5] has shown how various classic or well known distributions can be *derived* as representations of a particular state of knowledge by imposing suitable constraints while maximizing the Shannon entropy. These constraints are identifiable with certain elements of the user's prior state of knowledge and they typically consist of average values or moments. Tribus also notes that because the method of maximum entropy is not very old the number of formally worked out cases is also not very large, so that not *every* distribution that is defined in a statistics text has been fully and properly identified with any particular state of knowledge by MaxEnt or any other Bayesian method. Those examples which have been worked out all seem to involve a uniform measure function in the Shannon entropy. Here, we will describe an example in which the proper measure function/pre-prior is not uniform.

2. The Problem

Inferences based on the Gaussian or normal error law have been found very satisfactory in a great many cases, even when a Gaussian is not a plausible frequency distribution of the errors. Jaynes [1] has explained why this should be so. The Gaussian error law corresponds to a rather minimal state of knowledge about the errors which a user would be in possession of almost universally. But every Bayesian knows that each problem should be considered individually, and sometimes there will be prior information which indicates that something other than a Gaussian would be a better representation of his prior knowledge.

In order to introduce the problem that concerns us in this paper and to illustrate how our concerns may be realized in practice, we consider an example problem in which an experimenter wants to infer the value of a fixed parameter τ given a collection of noisy measurements $D = \{x_1...x_n\}$. The Bayesian approach to this problem is to obtain a posterior probability distribution for the possible values of τ by updating the user's prior distribution with the new information obtained in the data D. This is done by Bayes' theorem which for this problem can be written

$$P(\tau|D, I) = \frac{P(D|\tau, I)P(\tau|I)}{\int P(D|\tau, I)P(\tau|I)d\tau}. \tag{2}$$

$P(\tau|I)$ is the prior probability distribution for τ which is obtained by encoding all the user's prior information about τ. The term $P(D|\tau, I)$ is called the likelihood function and is proportional to the probability of the observed data with an assumed value of τ. I denotes any and all relevant background information which would be called upon. Because there is noise in the measured values while τ is presumably fixed, the likelihood function will depend solely on the probability for the errors induced by the noise. The probability distribution for the errors is also obtained by encoding the user's prior information, and *this* is the aspect of our example problem we wish to focus on.

The prior information in our example will include knowledge that the parameter τ is positive valued and also that the results of each measurement x_i can only have positive values. If it is assumed that the noise is independent of the value of the parameter τ, the most general way to model its effects in the data is by the equation

$$x_i = \tau \epsilon_i. \tag{3}$$

Thus, the error in any one measurement appears as an unknown scale parameter. This situation is applicable to the errors in monthly rainfall measurements [6] and the measurements of many other positive valued parameters.

As indicated above, the measure function in the entropy of a continuous distribution also plays the roll of a noninformative pre-prior. To use MaxEnt and assign a probability distribution for the errors, we first require a pre-prior for the scale parameter ϵ. Jaynes [4] was the first to recognize that complete ignorance of a scale parameter could only mean that a change of scale

$$\epsilon' = a\epsilon \tag{4}$$

would leave that state of knowledge, and therefore its representation, invariant. By imposing the corresponding group invariance, Jaynes showed that the noninformative pre-prior of a scale parameter is the Jeffreys prior

$$m(\epsilon) = \frac{const.}{\epsilon}. \tag{5}$$

With the measure function/pre-prior determined, we can now continue to encode the rest of our prior knowledge about the noise using MaxEnt.

3. The Solution

First, we seek a monotonic transformation through which the noise parameter ϵ, with an interval of possible values in the range $0 < \epsilon < \infty$, could be reversibly transformed into the new parameter e, with an interval of possible values $-\infty < e < \infty$. This change of parameter is to also transform the measure function $m(\epsilon)d\epsilon = \frac{d\epsilon}{\epsilon}$ into

$$m(e) = 1\,de. \tag{6}$$

Our motivation for considering such a transformation is that the user's essential prior information may be more easily recognized and expressed in the new parameter e with its greater symmetry and uniform measure function. The transformation that does what we require is unique to within a meaningless additive constant,

$$e = \ln(\epsilon). \tag{7}$$

With this change of parameter the two measure functions are equivalent, $m(e)de = m(\epsilon)d\epsilon$, and from the pre-prior point of view can be said to represent equivalent states of ignorance. In this change of parameter Eqn. (3) becomes

$$\ln(x_i) = \ln(\tau) + \ln(\epsilon_i) \tag{8a}$$

or

$$y_i = \chi + e_i. \tag{8b}$$

We find the errors in this new parameterization are additive and ignorance of e implies location invariance consistent with the uniform measure function/pre-prior in Eqn. (6) [4].

We now consider what properties we could reasonably ascribe to the distribution of the error parameter e. Let $P(e|I)$ be the probability density for the errors given background information I. Following Bretthorst [7], we know that this probability density will be normalized,

$$\int_{-\infty}^{\infty} P(e|I)de = 1. \tag{9}$$

We also know that the average power carried by the noise would have some finite if unknown value,

$$\int_{-\infty}^{\infty} (e - \bar{e})^2 P(e|I)de = \sigma^2, \tag{10}$$

where the power is understood as the average of the square of the amplitude. In what follows we will assume $\bar{e} = 0$. The distribution that maximizes the entropy with Eqns. (9) and (10) as constraints is found to be a Gaussian

$$P(e|\sigma^2, I) = \frac{1}{\sqrt{2\pi\sigma^2}} \exp\left(-\frac{1}{2\sigma^2}e^2\right), \tag{11}$$

where we indicate the dependency on the power parameter σ^2 explicitly. (If the parameter σ^2 is itself uncertain, it would be treated as a nuisance parameter and marginalized out with respect to the user's prior on σ^2.) Transforming back to the original parameter ϵ we find

$$P(\epsilon|\sigma^2, I) = \frac{1}{J(e)} P(e|\sigma^2, I) = \frac{1}{\epsilon} P(e = \ln(\epsilon)|\sigma^2, I) \tag{12}$$

where $J(e)$ is the Jacobian of the transformation

$$J(e) = \left|\frac{\partial e}{\partial \epsilon}\right|^{-1}. \tag{13}$$

Giving

$$P(\epsilon|\sigma^2, I) = \frac{1}{\sqrt{2\pi\sigma^2\epsilon^2}} \exp\left(-\frac{1}{2\sigma^2}\ln^2(\epsilon)\right). \tag{13}$$

Thus the Lognormal distribution is derived as the correct representation of the user's state of knowledge about the noise parameter ϵ.

References

1. E. Jaynes, *Probability Theory: The Logic of Science*, not yet published, manuscript available at http://omega.albany.edu:8008/JaynesBook.html, 1995.
2. C. Shannon and W. Weaver, *The Mathematical Theory of Communications*, University of Illinois Press, Urbana, Ill., 1949.
3. E. Jaynes, "Brandeis lectures," in *E. T. Jaynes: Papers on Probability, Statistics and Statistical Physics*, R. Rosenkrantz, ed., pp. 40-76, Kluwer Academic, Dordrecht, 1989.
4. E. Jaynes, "Prior probabilities," *IEEE Trans. Sys. Sc. Cy.*, **SCC-4**, pp. 227-241, 1968.
5. M. Tribus, *Rational Descriptions, Decisions and Designs*, Pergamon Press, New York, 1969.
6. W. Evenson, "Climate analysis in ohi'a dieback on the island of hawaii," *Pacific Sci.*, **37**, pp. 375-384, 1983.
7. G. Bretthorst, *Bayesian Spectrum Analysis and Parameter Estimation*, Springer-Verlag, New York, 1988.

Dinner music was provided at Richard's by guitarist Anna Maria Padilla.

PIXON-BASED MULTIRESOLUTION IMAGE RECONSTRUCTION FOR YOHKOH'S HARD X-RAY TELESCOPE

THOMAS R. METCALF AND HUGH S. HUDSON

Institute for Astronomy, University of Hawaii, 2680 Woodlawn Dr., Honolulu, HI 96822 USA

TAKEO KOSUGI

National Astronomical Observatory of Japan, 2-1-1 Osawa, Mitaka, Tokyo 181 JAPAN

AND

R. C. PUETTER AND R. K. PIÑA

Center for Astrophysics and Space Sciences, University of California, San Diego, 9500 Gilman Dr., La Jolla, CA 92093-0111 USA

Abstract. We present results from the application of pixon-based multiresolution image reconstruction to data from *Yohkoh's* Hard X-ray Telescope. The goal of the pixon algorithm is to minimize the number of degrees of freedom used to describe an image within the accuracy allowed by the noise. This leads to a reconstruction which is optimally constrained. We apply the pixon code to two solar flares in the HXT database and compare the results of the pixon reconstruction to the results of a direct, linear, smoothed inversion of the HXT Fourier synthesis data and to a maximum entropy reconstruction. The maximum entropy reconstruction is vastly better than the direct inversion, but the pixon reconstruction gives superior noise suppression and photometry. Further, the pixon reconstruction does not suffer from over-resolution of the images.

1. Introduction

For hard X-ray imaging using the bi-grid modulation collimators of the hard X-ray telescope (HXT; [1]) on board the *Yohkoh* spacecraft, the reconstruction problem is particularly difficult. The HXT data set is sparse, consisting of only 64 collimator outputs from which a relatively large image (typically 64 by 64) is normally reconstructed. Pixon-based reconstruction is ideally suited to this problem since the goal of the method is to minimize the number of degrees of freedom and hence to make the best possible use of the limited data.

K. M. Hanson and R. N. Silver (eds.), Maximum Entropy and Bayesian Methods, 361–365.

Below, we review the theory of pixon-based image reconstruction and describe a pixon-based algorithm used to reconstruct hard X-ray images for HXT. Finally, we apply the algorithm to two solar flare data sets and compare the results to an ME reconstruction and to a direct, linear inversion of the Fourier data.

2. Pixon-Based Image Reconstruction

Most non-linear image reconstruction methods can be understood in terms of a Bayesian estimation scheme in which the reconstructed image is, in some sense, the most probable (e.g. [2]). This implies maximizing the joint probability distribution of the reconstructed image (I) and the model (M) given the data (D): $p(I, M|D)$. The model defines the relationship between the data and the image (e.g. the physics of the image encoding process, the pixel size, etc). Using Bayes' theorem, this probability distribution can be factored to yield

$$p(I, M|D) = \frac{p(D|I, M)p(I|M)p(M)}{p(D)} \propto p(D|I, M)p(I|M) \qquad (1)$$

where $p(X|Y)$ is the probability of X given Y. Here we use $p(I, M|D)$ rather than $p(I|D, M)$ since the model is allowed to vary in the pixon method.

The key idea behind pixon-based image reconstruction is the realization that not all parts of an image require the same spatial resolution. Indeed, for hard X-ray imaging of solar flares, discussed below, most of the image is blank and has no information content. Why then should blank pixels be included as additional degrees of freedom? To avoid this, Piña and Puetter ([3]) introduced the concept of a pixon.

A pixon is a generalized pixel. It is a variable cell in the image representing a single degree-of-freedom (DOF) in the reconstruction. Ideally, the set of pixons used to describe the image would be the minimum set required to describe the information content of the image. Having reduced the reconstruction to the fewest possible degrees of freedom, it makes the best possible use of the available data.

3. A Pixon Algorithm for Fourier Synthesis Hard X-Ray Imaging

The Hard X-Ray Telescope onboard the *Yohkoh* spacecraft uses a set of 64 bi-grid modulation collimators to image the Sun in energies between 13.9 keV and 92.8 keV in 4 energy bands. Each subcollimator is a pair of nearly identical one dimensional grids mounted in parallel planes separated by 1.4 m ([1]). The 64 grid pairs approximately measure the Fourier components of the image. The HXT data set consists of the count rate of hard X-ray radiation observed through each subcollimator.

This is a difficult reconstruction problem since the data are sparse ([4]). The normal reconstruction is onto a 64x64 array of standard 2.46 arc sec pixels. The standard *Yohkoh*/HXT imaging software has been the maximum entropy method of Gull and Daniell ([5]) and Willingdale ([6]), in the implementation described by Sakao ([7]). However, with 4096 degrees of freedom but only 64 data points,

Figure 1. Comparison of three image reconstruction algorithms applied to three HXT energy bands in the solar flare of 1992 January 13. The first column shows the pixon based image reconstruction while the second and third columns show an ME reconstruction and a direct, smoothed inversion of the data, respectively. All images are negative images and use the same greyscale. The images are oriented with solar north up and solar east to the left. Each image is 64 by 64 with 2.46 arc second pixels.

spurious sources are expected. Since the pixon-based image reconstruction minimizes the number of degrees-of-freedom, it will give superior results. For HXT data sets, we typically find at most 100 degrees of freedom sufficient for pixon image reconstruction. Hence, spurious sources are greatly reduced or eliminated.

A related advantage of the pixon-based reconstruction is better photometry. Due to the nature of the HXT data, the number of counts observed in the field-of-view is approximately conserved regardless of the algorithm used in the reconstruction (e.g. [8]). Hence, if there are spurious sources, these counts are removed from the real sources yielding poor photometry. Thus, a pixon-based reconstruc-

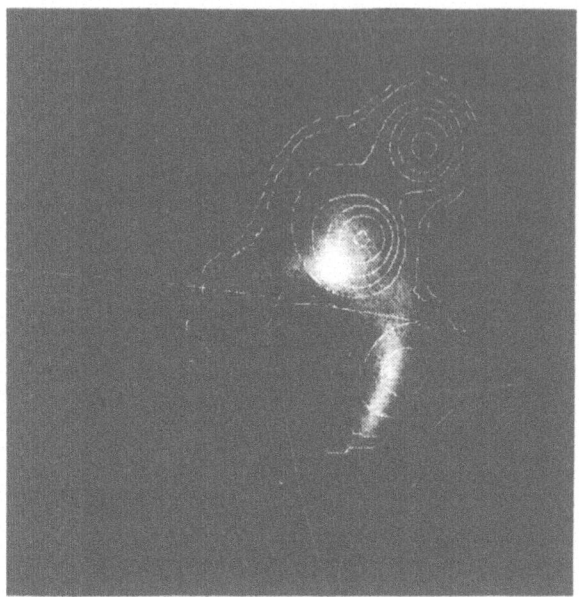

Figure 2. HXT LO channel contours overlaid on an SXT image for the 1992 February 21 limb flare. The HXT contours are at 0.06, 0.08, 0.15, 0.25, 0.4, 0.65, and 0.95 of the maximum value. The SXT image, obtained at 03:20:14 UT, uses the beryllium filter. The HXT data was summed from 03:15:00 through 03:25:04. The image is 64x64 with 2.46 arc second pixels. Solar east is up and solar north is to the right. The curved line passing through the image shows the location of the solar limb.

tion which eliminates spurious sources will have better photometry than other methods.

4. Results

Masuda et al. ([9]) studied hard and soft X-ray images of a flare that occurred near the west solar limb on 1992 January 13. Since this important event is well known, and since it has an interesting combination of diffuse and point emission in the 14-23, 23-33, and 33-55 keV HXT channels (LO, M1, and M2 channels, respectively), we chose it for a detailed comparison of the pixon-based, ME, and direct image reconstruction algorithms.

Figure 1 compares image reconstructions for the 1992 January 13 solar flare.

Each column shows a different reconstruction algorithm while each row shows a different HXT energy channel; all images have a common grey-scale representation. Comparing the images across the three rows, the advantages of the pixon algorithm are clear: the pixon reconstructions show less background structure and, hence, are brighter, implying better photometry.

Figure 2 shows an FPB image reconstruction for the solar limb flare of 1992 February 21 ([10]). This was a gradual event with weak but long-enduring hard X-ray emission. The underlying image in Figure 2 is from the Soft X-ray Telescope (SXT) on *Yohkoh* ([11]) using the Be filter and the contours are from the FPB reconstruction of the LO channel on HXT. We expect a good (but not perfect) correspondence between the two images; the pixon-based reconstruction adequately reproduces the expected image. For this 1992 February 21 data set, the ME reconstruction has not yet been successful in producing a meaningful image.

The research at the University of Hawaii was supported through Lockheed under NASA contract NAS8-37334 with with the Marshall Space Flight Center. R. Puetter and R. Piña acknowledge support from the NSF, NASA, and the California Association for Research in Astronomy.

References

1. T. Kosugi, K. Makishima, T. Murakami, T. Sakao, T. Dotani, M. Inda, K. Kai, S. Masuda, H. Nakajima, Y. Ogawara, M. Sawa, and K. Shibasaki, "The hard x-ray telescope (hxt) for the solar-a mission'," *Solar Phys.*, **136**, p. 17, 1991.
2. R. C. Puetter, "Pixon-based multiresolution image reconstruction and the quantification of picture information content," *The International Journal of Image Systems and Technology*, in press, 1995.
3. R. K. Piña and R. C. Puetter, "Bayesian image reconstruction: The pixon and optimal image modeling," *P.A.S.P.*, **105**, p. 630, 1993.
4. T. R. Metcalf, H. S. Hudson, T. Kosugi, R. C. Puetter, and R. K. Piña, "Pixon-based multiresolution image reconstruction for yohkoh's hard x-ray telescope," *Astrophys. J. (in press)*, 1996.
5. S. F. Gull and G. J. Daniell, "Image reconstruction from incomplete and noisy data," *Nature*, **272**, pp. 686-690, 1978.
6. R. Willingale, "Use of the maximum entropy method in x-ray astronomy," *Mon. Not. R. Astr. Soc.*, **194**, pp. 359-364, 1981.
7. T. Sakao, *Characteristics of Solar Flare Hard X-Ray Sources as Resolved with the hard X-Ray Telescope aboard the Yohkoh Satellite*, Ph.D. Thesis, University of Tokyo, 1994.
8. T. R. Metcalf, R. C. Canfield, E. H. Avrett, and F. T. Metcalf, "Flare heating and ionization of the low solar chromosphere. i. inversion methods for mgi 4571 and 5173," *Astrophys. J.*, **350**, p. 463, 1990.
9. S. Masuda, T. Kosugi, H. Hara, S. Tsuneta, and Y. Ogawara, "A loop-top hard x-ray source in a compact solar flare as evidence for magnetic reconnection," *Nature*, **371**, p. 495, 1994.
10. S. Tsuneta, "Observation of a solar flare at the limb with the *johkoh* soft x-ray telescope," *P.A.S.J.*, **44**, p. L63, 1992.
11. S. Tsuneta, et al., "The soft x-ray telescope for the solar-a mission," *Solar Phys.*, **136**, p. 37, 1991.

Julian Besag, William Fitzgerald, Sibusiso Sibisi, and Miao-Dan Wu at Richard Silver's home.

BAYESIAN METHODS FOR INTERPRETING
PLUTONIUM URINALYSIS DATA

G. MILLER AND W. C. INKRET

Los Alamos National Laboratory
Los Alamos, NM 87545[†]

Abstract. We discuss an internal dosimetry problem, where measurements of plutonium in urine are used to calculate radiation doses. We have developed an algorithm using the MAXENT method. The method gives reasonable results, however the role of the entropy prior distribution is to effectively fit the urine data using intakes occurring close in time to each measured urine result, which is unrealistic. A better approximation for the actual prior is the log-normal distribution; however, with the log-normal distribution another calculational approach must be used. Instead of calculating the most probable values, we turn to calculating expectation values directly from the posterior probability, which is feasible for a small number of intakes.

Key words: prior, lognormal, plutonium, Bayesian, bioassay

1. Introduction

In the field of health physics, exposure to alpha-emitting radionuclides like plutonium is usually monitored by periodic urinalysis. We have developed a new algorithm to calculate the radiation dose to an individual from internal depositions of plutonium (by inhalation or via contaminated wounds) based on their urinalysis data. The code is described in more detail in Ref. [1]. The mathematical method is to maximize the Bayesian posterior probability using an entropy function as the prior probability distribution. The MAXENT method as implemented in the MEMSYS software package is used. Some advantages of the new code are that it ensures positivity of the calculated doses, it smooths out fluctuating data, and it provides an estimate of the propagated error in the calculated doses. This method is generally applicable to the internal dosimetry problem, and we plan to implement it also for tritium and uranium, which are also monitored by urinalysis.

The MAXENT method, although an advance over previously used data unfolding methods, is still not definitive, since the entropy form of the prior probability

---

[†]Email: guthrie@lanl.gov

K. M. Hanson and R. N. Silver (eds.), Maximum Entropy and Bayesian Methods, 367–373.

distribution is not realistic. The entropy function has the property that its standard deviation divided by its mean is always less than $\sqrt{2}$, which is the limiting value for small α, as will be discussed. A more realistic prior would have a larger value of this quantity, as is the case, for example, with the log-normal distribution. We believe the log-normal distribution to be a better approximation to the actual prior distribution as determined by statistical studies of historical data. The relative narrowness of the entropy prior leads to an underestimate of radiation doses in the internal dosimetry problem. We will discuss a Bayesian calculation for a log-normal prior distribution.

The current approach to internal dosimetry is to interpret bioassay measurements in terms of radionuclide intake quantities. Let x_i for $i = 1, M$ denote the intake that occurred during time interval i, where M is the total number of time intervals. For example, in plutonium dosimetry x_i is the activity of plutonium taken into the body on the ith day by inhalation or via a contaminated wound.

The bioassay data are denoted by y_j for $j = 1, N$, with uncertainty estimates (standard deviations) σ_j. For example, y_j is the jth measurement of plutonium activity excreted per day. The biokinetic response is assumed linear and known, so that

$$f_j = \sum_{i=1}^{M} x_i u_{ij}$$

is the predicted bioassay result at time j given intakes x_i, where u_{ij} is the biokinetic response at time j for unit intake at time i.

The problem is to determine the "best fit" values of $\{x_i\}$ given $\{y_j\}$ (read $\{x_i\}$ as "the set of x_i for all i"). The MEMSYS method is to define "best fit" as the x_i values that give a maximum of the Bayesian posterior probability of $\{x_i\}$, given the data and an entropy prior probability distribution.

2. MAXENT

The entropy form of the prior distribution is given by

$$P(x_i) = C \exp\left[\alpha S(x_i)\right], \tag{1}$$

where

$$S(x_i) = x_i - m_i - x_i \log \frac{x_i}{m_i},$$

α and m_i are parameters, and C is a normalization constant such that

$$\int P(x_i) \frac{dx_i}{x_i^{1/2}} = 1$$

(note the metric factor $1/x_i^{1/2}$).

In Ref. [1] a simplified approximation for the entropy function is obtained for the case of small α. For small α,

$$P(x_i) = \sqrt{\frac{\lambda}{\pi}} \exp(-\lambda x_i), \tag{2}$$

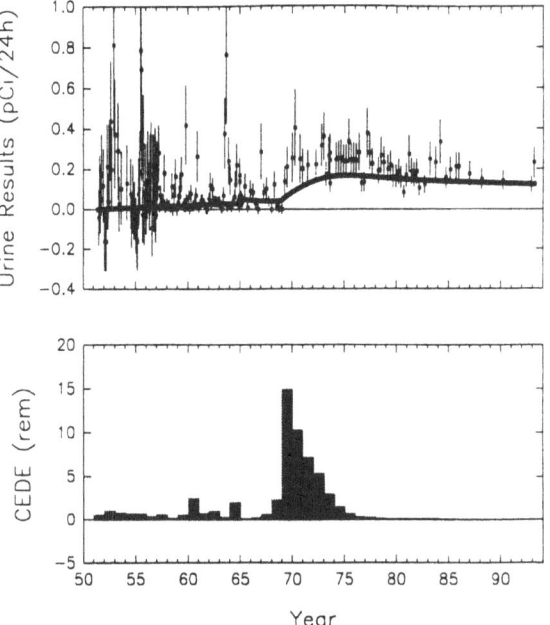

Figure 1.   Example of a MAXENT code dose calculation.

where $\lambda = 2\alpha \log(\zeta)$. The quantity ζ is determined by solving the equation

$$\zeta = \sqrt{\frac{1}{2\alpha m_i \log \zeta}},$$

iteratively for ζ starting with $\zeta = e$ as the initial guess. This requires that α not be too large. Using this approximation, the mean and standard deviation about the mean of x_i are given by

$$\langle x_i \rangle = \frac{1}{2\lambda}$$
$$\sqrt{\langle (\Delta x_i)^2 \rangle} = \sqrt{2} \langle x_i \rangle \qquad (3)$$

The entropy distribution has the property that it's standard deviation divided by mean is always less than $\sqrt{2}$, which is the limiting value for small α. A more realistic prior probability distribution would have a larger value of standard deviation divided by mean, as will be discussed.

A test of the MAXENT code using real ^{239}Pu urine excretion data is shown in Fig. 1. The MAXENT code reconstructed the intake scenario using a noninformative prior probability distribution (uniform small values of m_i).

Basically, the intake scenario calculated in Fig. 1 is reasonable. However, on closer inspection, one sees that an upward excursion of a urine excretion data point is fit by assuming intakes occurring very close in time to the time of the usually assume an intake occurring further away in time, and giving a higher dose estimate.

3. Log-Normal Prior Distribution

The log-normal distribution has the form

$$P(x)dx = \frac{1}{\sqrt{2\pi}\sigma_0} \exp\left[-\frac{\left(\log\frac{x}{m}\right)^2}{\sigma_0^2}\right]\frac{dx}{x}. \tag{4}$$

The log-normal distribution is known empirically to describe many cases of interest[2]. We suspect, and have some empirical evidence, that it gives a reasonable description of the distribution of radionuclide intakes for a population of exposed workers. The log-normal distribution has mean and standard deviation given by [3]

$$\langle x \rangle = m e^{\sigma_0^2/2}$$
$$\sqrt{\langle (\Delta x)^2 \rangle} = \langle x \rangle \sqrt{e^{\sigma_0^2} - 1} \tag{5}$$

Thus, in contrast to the entropy distribution, the log-normal distribution can be arbitrarily broad for large σ_0.

The log-normal distribution does not have the inverse property of the entropy distribution that allows solution of the maximum probability equations by data space mapping. That is, if

$$f(x) = \frac{\partial}{\partial x}\left[\log P(x)\right],$$

not all values of y allow solutions of $y = f(x)$. This means that the maximum posterior probability problem with M variables (M might be 10000) and N data points (N might be 100) will involve M equations in M unknowns rather than N equations in N unknowns–a very important practical difference.

Perhaps using the log-normal prior distribution fundamentally changes the mathematical character of the problem. To begin to understand whether this might be true, we consider a very simple case, with one intake x and one measurement y. It turns out that the problem easily generalizes to multiple measurements with one intake.

The posterior probability is then given by

$$P(x|y)dx \sim \exp\left[-\frac{(xu-y)^2}{2\sigma^2} - \frac{1}{2\sigma_0^2}\left(\log\frac{x}{m}\right)^2\right]\frac{dx}{x},$$

where u is the fraction of intake excreted at the sampling time after an intake, x is the intake amount, y is the measured activity of plutonium excreted per day, σ is the measurement error standard deviation, and σ_0 and m are log-normal parameters that are considered known. Simplifying the notation,

$$P(x|y)dx \sim \exp\left[-\alpha(z-z_0)^2 - \beta(\log z)^2\right]\frac{dz}{z}, \tag{6}$$

where

$$z = \frac{x}{m},$$

$$z_0 = \frac{y}{um},$$

$$\alpha = \frac{(um)^2}{2\sigma^2},$$

$$\beta = \frac{1}{2\sigma_0^2}. \tag{7}$$

If we ask the question "What is the most probable value of x?", the first problem is to decide "With respect to what variable?". If the probability density is plotted versus x, the maximum will occur at a different x than if plotted versus $\log x$. The second problem for Eq. 6 is that there are sometimes multiple local maxima.

The maxima of the exponential in Eq. 6 occur when

$$\frac{\partial}{\partial z} \left[\alpha(z - z_0)^2 + \beta(\log z)^2 \right] = 0,$$

or, when

$$z - z_0 = -\frac{\beta}{\alpha} \frac{\log z}{z} = -\gamma \frac{\log z}{z}, \tag{8}$$

where we have defined γ as

$$\gamma \equiv \frac{\beta}{\alpha}. \tag{9}$$

Figure 2 shows that Eq. 8 always has solutions. However, for large γ there are multiple solutions as shown in Fig. 3. The critical γ turns out to be

$$\gamma_{crit} = 2e^3 = 40.2, \tag{10}$$

and, for $\gamma > \gamma_{crit}$, there are multiple solutions.

For these two reasons, 1) ambiguity about metric factors, and 2) possible multiple local maxima, we propose that finding the maximum of the posterior probability may not be the optimum approach.

Another approach is to directly use the posterior probability distribution to evaluate expectation values. This is conceptually simpler.

For the problem at hand,

$$\langle x^n \rangle = \frac{\int_0^\infty x^n P(x|y)dx}{\int_0^\infty P(x|y)dx}, \tag{11}$$

with $P(x|y)dx$ given by Eq. 6. We therefore can easily calculate $\langle x \rangle$, $\langle (\Delta x)^2 \rangle$, $\langle \chi^2 \rangle$ and other quantities of interest.

If there are multiple measurements, the generalizations of Eqs. 7 are

$$z_0 = \frac{\sum_{j=1}^N w_j \frac{y_j}{mu_j}}{\sum_{j=1}^N w_j}, \tag{12}$$

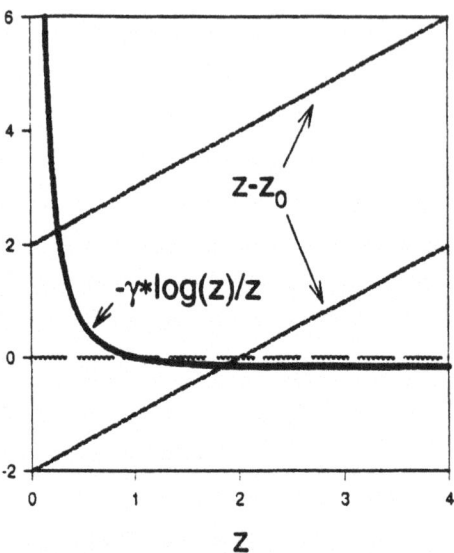

Figure 2.   Graphical solution of the equation determining the most probable z values.

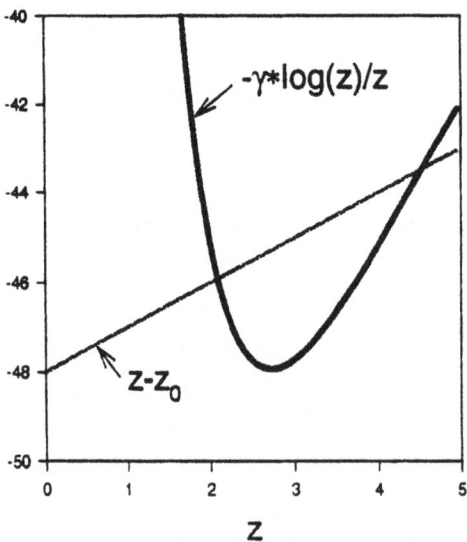

Figure 3.   A case where multiple local maxima occur.

and

$$\alpha = \frac{1}{2} \sum_{j=1}^{N} w_j,$$ (13)

where the weighting factor w_j is given by

$$w_j = \left(\frac{mu_j}{\sigma_j} \right)^2.$$

4. Conclusions

The form of the prior distribution is important. We see this clearly in the internal dosimetry calculation where the MAXENT doses are smaller than those calculated "by hand" for simple cases. Unfortunately, the entropy form of the prior distribution is not able to approximate what we believe is the actual broad form of the prior for cases of interest. Thus, we are led to investigate application of the log-normal prior distribution.

With the log-normal prior, 1) data space mapping fails, and 2) there may be multiple local maxima of the posterior probability. Thus, we cannot simply change MAXENT to use a log-normal prior, and are led to consider alternate approaches.

An elegant approach is to directly evaluate average values using the posterior probability distribution. Straightforward application of this method is practical for determining small numbers of intakes and quite trivial for single intakes. The single-intake method can be used in a data unfolding procedure where multiple intakes are successively determined by one or more subsequent data points. The challenge for the future is to find methods to evaluate very large dimension integrals in order to be able to simultaneously determine large numbers of multiple intakes.

References

1. G. Miller and W. C. Inkret, "Bayesian maximum posterior probability method for interpreting urinalysis data," *Rad. Prot. Dosimetry*, 1995. submitted.
2. H. J. Gale, "The lognormal distribution and some of its applications in the field of radiation protection," Tech. Rep. AERE-R4736, Atomic Energy Research Establishment, 1965.
3. V. Rothschild and N. Logothetis, *Probability Distributions*. New York: John Wiley, 1986.

MaxEnt old timers Glenn Heidbreder, Paul Fougere, and John Berg at Richard's portal.

THE INFORMATION CONTENT OF SONAR ECHOES

RICHARD PITRE
Code 7144, Acoustics Division[†]
Naval Research Laboratory
Washington D. C., 20375

Abstract. An information statistic is defined which quantifies sonar localization performance. The evolution of the *a posteriori* distribution and the localization information statistic are evaluated as functions of the number of pings for an idealized sonar model.

1. Introduction

Sonar target echoes do not always contain enough information to localize a target after a single ping or even after several pings. Information rates achievable with a sonar in the ocean are limited by clutter, reverberation, and background noise. Signal inversion in this context is a dynamic process wherein data are combined with other or older data in order to refine an *a posteriori* distribution[1,2]. At any stage of the evolution the information statistic indicates how many possibilities or cells have been eliminated. Rather than a basis for estimation, the *a posteriori* distributions are a basis for improving the odds of correct decisions. Signal and parameter estimation[3] can be justified in the high information limit(many pings).

The addition of data from different sources, places and times is referred to as "data fusion". Averaging, and the multiplication of probabilities for independent observations are common representations of data fusion[1]. In the sonar example presented, ouputs of several sensors over a band of frequencies from multiple sonar pings will be combined to overcome the noisy ocean background.

2. Information, Information-Bases, and Priors

Information can be interpreted as the reduction in the number of possibilities[4]. An expression for Shannon entropy which generalizes to continuous spaces is given by[1,2]

$$\log \aleph \equiv - \int dP \, \log \left[\frac{dP}{d\mathcal{N}} \right] = - \sum_i p_i \, \log \, p_i , \qquad (1)$$

[†]Email: pitre@n5160d.nrl.navy.mil

K. M. Hanson and R. N. Silver (eds.), Maximum Entropy and Bayesian Methods, 375–380.

where the probability measure P is defined by the probabilities p_i and where $\left[\frac{dP}{dN}\right]$ is the Radon-Nikodym[5] derivative of the measure P with respect to the counting measure $\mathcal{N}$. An interpretation of $\aleph$ as the number of possibilities follows from the limit[6,7]

$$\log \aleph = \lim_{n \to \infty} \frac{1}{n} \log \mathcal{N}^{(n)}[T_P^{(n)}] , \tag{2}$$

where $T_P^{(n)}$ is the typical set of the measure P on the n-sample product space and $\mathcal{N}^{(n)}$ is the product measure defined by $\mathcal{N}$. On a continuous space $\mathcal{N}$ can be a scaled physical measure. If $V[\sigma]$ is the physical measure of a set σ and the selected cell-scale is v, then a counting or *information-base* measure can be defined as

$$\mathcal{N}[\sigma] \equiv \frac{V[\sigma]}{v} . \tag{3}$$

Densities are Radon-Nikodym derivatives relative to a coordinate volume

$$P(x) \equiv \left[\frac{dP}{dx}\right] , \quad \text{and} \quad V(x) \equiv \left[\frac{dV}{dx}\right] . \tag{4}$$

In the case of a metric volume $V(x)$ is the square root of the determinant of the coordinate representation of a metric tensor $\mathbf{g}$. The Radon-Nikodym derivative of P with respect to $\mathcal{N}$ is equal to the coordinate independent ratio of densities

$$\left[\frac{dP}{d\mathcal{N}}\right] = v \frac{P(x)}{V(x)} .$$

Localization performance will be quantified in terms of a volumetric information-base. A similar analysis of classification performance requires an information-base on the particular classification parameters of the target model. Absent an obvious choice, unique measures can sometimes be obtained by imposing more symmetry on the absolute objects of a physical model[8,9]. The space may be isomorphic to a locally compact group in which case it has a unique invariant measure[5].

3. Information Channels and Channel Gain

An information channel is a conditional distribution, $P(\alpha_{out}|\alpha_{in})$, for outputs given the inputs. For a known output, α_{out}, and an arbitrary *a priori* distribution, the *a posteriori* entropy is

$$\log \aleph_{\alpha_{in}|\alpha_{out}} = -\int d\alpha_{in} \, P(\alpha_{in}|\alpha_{out}) \, \log \frac{P(\alpha_{in}|\alpha_{out})}{\mathcal{N}(\alpha_{in})} .$$

The *channel gain*, $\mathcal{G}$, is defined to be the average reduction in entropy obtained from receiver outputs *when the a priori distribution is uniform relative to* $\mathcal{N}$,

$$\mathcal{G} \equiv \left\langle \log \frac{\aleph_{\alpha_{in}}}{\aleph_{\alpha_{in}|\alpha_{out}}} \right\rangle_{\mathcal{N}} . \tag{5}$$

The gain equals the mutual information between receiver inputs and outputs. The output consists of outputs from different sensors or from the same sensor at different times and places,

$$\alpha_{out} = \left(\alpha_{out}^{(1)}, \alpha_{out}^{(2)}, \ldots, \alpha_{out}^{(n)}\right) .$$

When the model implies statistically independent outputs then

$$P(\alpha_{out}|\alpha_{in}) = P(\alpha_{out}^{(1)}|\alpha_{in}) \; P(\alpha_{out}^{(2)}|\alpha_{in}) \; \ldots \; P(\alpha_{out}^{(n)}|\alpha_{in}),$$

and the information gain is the sum of the individual gains minus the redundancy

$$\mathcal{G} = \sum_i \mathcal{G}^{(i)} - \mathcal{R} , \tag{6}$$

where the redundancy is defined as the mutual information

$$\mathcal{R} \equiv \int d\alpha_{out} \; P(\alpha_{out}) \; \log \frac{P(\alpha_{out})}{P\left(\alpha_{out}^{(1)}\right) \; P\left(\alpha_{out}^{(2)}\right) \; \ldots \; P\left(\alpha_{out}^{(n)}\right)} . \tag{7}$$

INFORMATION GAIN AND SYSTEM PERFORMANCE

For optimally encoded channel inputs, average cost is a monotonic function of the information rate[10]. This rate distortion theoretic result is not directly applicable[11] to sonar because target parameters are not optimally encoded. In general, information gain and average operational costs are unrelated[12]. Information gain is more relevant to optimal sonar system design than to optimal operation. Individual sonars are operated in many different contexts each of which have different cost functions so that system design for minimal operational cost is ambiguous. On the other hand all of the cost functions are likely to rely on the ability of the sonar to localize so that the localization information gain can be usefully optimized.

4. A Sonar Model

In the context of sonar[13], the ocean is a waveguide with a reflective rough moving air-sea interface and a layered ocean bottom with rough layer interfaces. Within the water and sediment layers there are random inhomogeneities which, along with the rough surface and interface boundaries, cause clutter. Noise due to shipping, ocean surface dynamics, and biologics can also be relevant factors. Specular reflections from interfaces and the surface[14] result in multiple propagation paths. The target response[15], when convolved with the waveguide response can result in a complex echo. Realistic performance prediction models are becoming possible but do not yet exist. In order to illustrate some of the basic points discussed herein it is only necessary to capture some basic features of an actual system. The multimodal *a posteriori* distributions are to be interpreted as the inversion of limited amounts of data from an exact model of an idealization of a sonar.

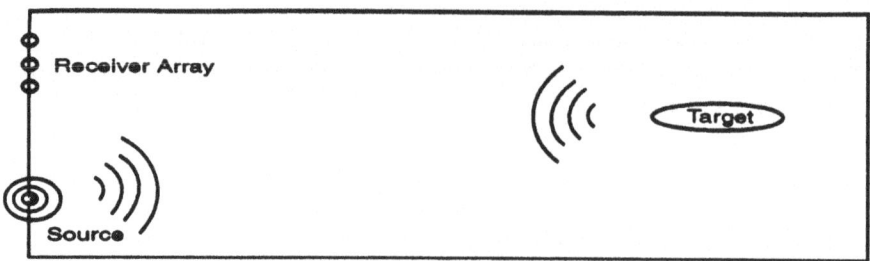

Figure 1. Simulation Geometry

The waveguide, depicted in figure 1 is a single 150 meter thick layer of infinite horizontal extent. Three point receivers and a point source are located on a vertical axis at depths of 20, 50, 80 and 90 meters respectively. The noise is Gaussian, spatially uncorrelated and temporally correlated by effective bandpass filtering to the bandwidth of the pings. The pings were composed of 10 frequency lines over a 1.25 Hz bandwidth centered at 100 Hz. The ping length was approximately 10 seconds. The noise level is roughly the same intensity at the receiver as the target echo. A point target echo is generated by weighting the target incident field with a target strength and propagating the weighted result back to the receivers. The Helmholtz equation for this waveguide separates in range, r, and depth, z,

$$G(r_1, z_1 | r_2, z_2, \omega) \approx Q \sum_n \sin \gamma_n z_1 \ \sin \gamma_n z_2 \ \frac{e^{ik_n|r_1 - r_2|}}{\sqrt{k_n|r_1 - r_2|}} \, ,$$

where the eigen-wavenumbers γ_n and k_n are determined by the boundary reflection conditions. The boundaries are perfect reflectors except that the bottom absorbs all energy impinging on it at grazing angles greater than 10 degrees.

The echo from the q-th ping on the i-th hydrophone at frequency ω_j is denoted by $\alpha_{ij}^{(q)}$. The modeled target echo for a target at (r, z) in the absence of any noise is denoted by $\beta_{ij}(r, z)$ and the noise is denoted by n_{ij}. If the noise autocovariance matrix is denoted by $K_{ij,kl}$ and a dot, $\cdot$, is used to denote the sum, over frequency and receiver indices, of a product then the forward model for n pings is

$$P(\alpha | r, z) = \prod_{q=1}^{n} \frac{e^{-\frac{1}{2}(\beta(r,z) - \alpha^{(q)})^{\dagger} \cdot K^{-1} \cdot (\beta(r,z) - \alpha^{(q)})}}{(2\pi)^{M/2} \sqrt{K}} \, , \qquad (8)$$

where M is the product of the number of receivers times the number of frequency samples. The samples are assumed to be separated in space and time by more than a sufficient multiple of a noise coherence interval and the environment is static so that K is modeled as diagonal. Because K is diagonal, each factor in the product is itself a product with one factor for each frequency sample and each hydrophone. The resulting distribution represents data fusion in the form of a product of contributions from many independent data samples at different times, frequencies, and locations. Figure 2 illustrates the *a posteriori* distribution for a single ping. The dark areas are the regions of low probability and have been effectively eliminated

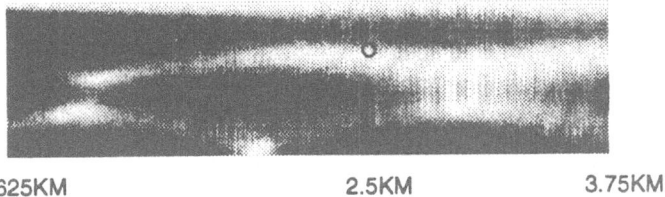

.625KM   2.5KM   3.75KM

Figure 2. The *a posteriori* distribution after inverting a single ping.

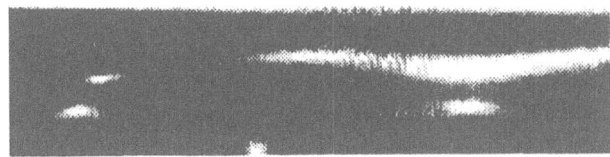

Figure 3. The *a posteriori* distribution after inverting eight pings.

by the information in the ping. The small circle indicates the location of the target
at a range 2.5 kilometers and a depth of 50 meters. This *a posteriori* distribution is
the best that can be done with the available information, when this model applies.
While it is not a choice basis for making decisions, a finite percentage of the possibilities have been eliminated and the chances of a good decision are better than
the chances of a good decision based on a uniform distribution. Note that the received signal has a nonlinear dependence on target location so that the *a posteriori*
distribution is not even remotely Gaussian. Figure 3 illustrates the considerable
improved *a posteriori* distribution after eight(8) pings. After thirty two(32) pings
the noise has been largely overcome so that all but a few likely possibilities remain
for the source location as illustrated in figure 4. The reduction in the number of
likely target locations is reflected by the information gain as a monotonic function
of the number of pings illustrated in figure 5.

5. Conclusion

In many applications Bayesian inversion is a dynamic process of elimination.
As more data is accumulated the *a posteriori* distribution evolves from a uni-

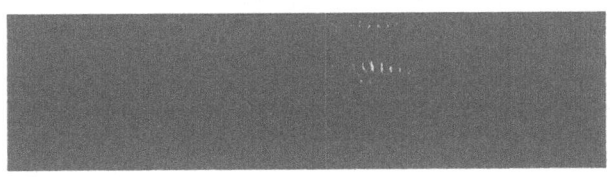

Figure 4. The *a posteriori* distribution after inverting 32 pings.

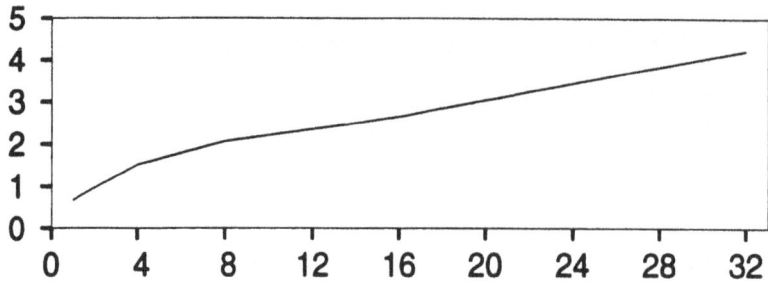

Figure 5. Bits of localization information versus number of pings.

form(relative to the information-base) *a priori* distribution, towards some limiting distribution(usually a Dirac delta function). The number of possibilities decreases from the number of possibilities in the entire hypothesized set to a small number of possibilities near the peak. Between the two extremes the distribution can be multimodal due to model nonlinearity. Information statistics provide a quantitative interpretation of "number of possibilities" at any stage of the evolution.

ACKNOWLEDGMENT. The author thanks Nolan Davis for fruitful discussions.

References

1. R. Pitre and N. R. Davis, "Application of probability modeling and bayesian inversion to passive sonar signal processing. i. concepts and formulation," *J. Acoust. Soc. Am.*, **97**, pp. 978–992, 1995.
2. N. R. Davis and R. Pitre, "Application of probability modeling and bayesian inversion to passive sonar signal processing. ii. application to ocean acoustic source localization," *J. Acoust. Soc. Am.*, **97**, pp. 993–1005, 1995.
3. H. L. V. Trees, *Detection, Estimation, and Modulation Theory*, Wiley, New York, 1968.
4. C. E. Shannon, "A mathematical theory of communication," *Bell Syst. Tech. J.*, **27**, pp. 379–423, 1948.
5. P. R. Halmos, *Measure Theory*, Van Nostrand, Princeton, 1966.
6. T. M. Cover and J. A. Thomas, *Elements of Information Theory*, Wiley, New York, 1991.
7. I. Csiszár and J. Körner, **Information Theory** *Coding Theorems for Discrete Memoryless Systems*, Academic, New York, 1981.
8. E. T. Jaynes, "Prior probabilities," *IEEE Trans. Systems Sci. Cybernet*, **SSC-4**, pp. 227–241, 1968.
9. J. L. Anderson, *Principles of Relativity Physics*, Academic, New York, 1968.
10. T. Berger, Rate Distortion Theory *A Mathematical Basis for Data Compression*, Printice-Hall, New Jersey, 1971.
11. S. D. Briles, "Information-theoretic performance bounding of bayesian identifiers," in *Automatic Object Recognition III*, pp. 255–266, SPIE Proceedings, 1993.
12. D. F. Mela, "Information theory and search theory as special cases of decision theory," *JORSA*, **9**, pp. 907–909, 1961.
13. R. J. Urick, *Principles of Underwater Sound*, McGraw-Hill, New York, third ed., 1983.
14. C. S. Clay and H. Medwin, **Acoustical Oceanography:** *Principles and Applications*, Wiley, New York, 1977.
15. G. T. Schuster and L. C. Smith, "A comparison among four direct boundary integral methods," *J. Acoust. Soc. Am.*, **77**, pp. 850–864, 1985.

OBJECTIVE PRIOR FOR COSMOLOGICAL PARAMETERS

GUILLAUME EVRARD
GRAAL†, URA 1368 UMII/CNRS
Université Montpellier II
34095 Montpellier Cedex 05, France

Abstract. Jaynes' transformation group method is applied to derive the uninformative prior forthe parameters of the standard cosmological models. Scale invariance and invariance under a change of cosmic epoch, imply that in the usual (H, Ω) parametrisation, this improper prior takes the form $\mu(H, \Omega) \propto (H\Omega|\Omega-1|)^{-1}$. The interpretation of this solution, and particularly of its singularity at the critical density $\Omega = 1$, negates the existence of the so-called "flatness problem" of standard cosmology.

Key words: cosmology, uninformative prior, group invariance

1. Introduction: the cosmological inverse problem

Today's most generally accepted description of the universe is supplied by the simplest relativistic cosmological models: the Friedman models. These models have two "free parameters" whose values have to be determined by empirical means. Cosmologists therefore have to solve a generalized inverse problem [1]: from observational data, they have to retrieve information on theoretical parameters. This process can be done via the classical cosmological tests [2], which relate observed quantities (apparent magnitudes, angular diameters, redshifts, number-counts, etc...) to the model parameters.Since the available data are incomplete or noisy, this estimation problem usually requires probabilistic reasoning. Indeed, in absence of certainty, expressing a state of knowledge on such parameters (theoretical or observational) can be done by the mean of a probability density defined over the corresponding parameter space. By accepting this principle, one explicitly acknowledges the Bayesian interpretation of probability.

The main question addressed in this paper is: if any state of knowledge on the cosmological parameters is to be represented by a probability density,which one describes total ignorance, or rather, minimal knowledge? This problem can alternatively be posed in equivalent terms ofmeasure on the parameter space [3].

---

†e-mail: evrard@graal.univ-montp2.fr

K. M. Hanson and R. N. Silver (eds.), Maximum Entropy and Bayesian Methods, 381–386.
© 1996 *Kluwer Academic Publishers.*

The answer is not trivial, essentially because probability densitiesdepend on the chosen parametrisation of the parameter space. Therefore, it must be emphasized that, contrary to common belief, minimal knowledge is *not* necessarily represented by a constant probability density. Consequently, a general rule for assigning uninformative priors/measures must be sought. Jaynes' transformation group method [4] provides such a principle based on invariance arguments. In the following, this technique is applied to the parameters of the Friedman world models to find the prior for the cosmological parameters.

2. Parametrization of the cosmological models

The classical quantities used to characterize a Friedman model are H_0, the so-called *Hubble constant*, which relates the rate of expansion of the universe, and Ω_0, the *density parameter*, which describes the ratio of the present density of the universe to its critical density (the suffix "0" indicates that the value of the parameter is taken at the present epoch). The relative evolution of these parameters is described by the Friedman equations.

There is no preferred parametrisation for the two-dimensional parameter space of Friedman cosmologies, any set (x, y) in a one-to-one correspondence with (H, Ω) equivalently characterizes the model. In the following, I use the (a, χ) parametrisation defined by

$$a = \frac{c}{H \sqrt{|1 - \Omega|}} \,, \tag{1}$$

and

$$\chi = \frac{c}{2H} \frac{\Omega}{|1 - \Omega|^{3/2}} \,. \tag{2}$$

In this parametrisation, the Friedman equations simply read

$$a \left(k + \frac{\dot{a}^2}{c^2} \right) = 2\chi, \tag{3}$$

where the dot denotes derivation with respect to cosmic time and k is the curvature of the spatial sections of the model, normalised to take the values 0 if $\Omega = 1$ ("spatially flat" model), 1 if $\Omega > 1$ ("spatially closed" model) and -1 if $\Omega < 1$ ("spatially open model").

In Eq.(3), a is the *scale factor*, relating how the relative distance of comoving objects evolve. The quantity χ can be seen as an absolute scale parameter, it remains constant during the evolution of the system described by the dynamical equation (3). Its value is given by the "initial value equation"

$$\chi = \frac{4\pi G \rho a^3}{c^2}, \tag{4}$$

where ρ is the matter density (decreasing with time), in a dust-filled, zero-pressure universe.

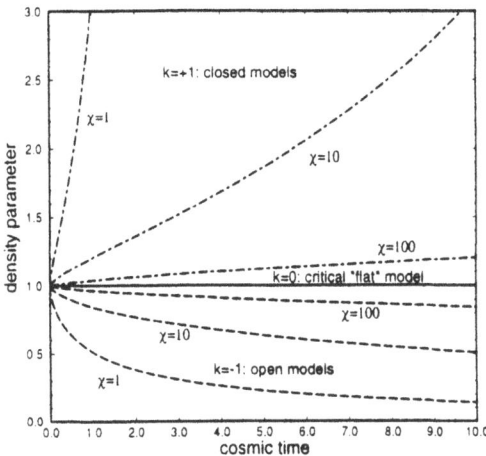

Figure 1.   The evolving universe: density parameter Ω vs cosmic time t (in arbitrary units). Each line is a solution of the Friedman equation (3) depending on the values of k and χ.

The description of the system could have been equivalently performed in terms of cosmic time t, temperature T, Hubble time T_H, or any other quantity monotonically related to a, but the equations turn out to be simpler in the (χ, a) parametrisation.

3. Uninformative prior based on group invariance

There are some changes of parametrisation which just transform the problem into an equivalent one: the system under study is described by different parameters, which however keep the same physical sense. Jaynes's transformation group method is then based on the principle that the uninformative prior must be invariant under such transformations.

The two parameters a and χ defined in Eqs. (1) and (2) are scale parameters, having the dimension of a length. As there is no preferred length scale of reference in the universe, the Friedman equations (3) are invariant under changes of scale of the form $a \to \alpha a$ and $\chi \to \alpha\chi$.

Moreover, as a consequence/description of the expansion of the universe, a is an evolving scale parameter. Still, there is no preferred epoch of reference[1]. Therefore, the Friedman equations are invariant under changes of cosmic epoch, which read $a \to \beta a$ (χ remains constant during the evolution). Stated differently, this means that just waiting does no bring you any information about the cosmological parameters, only observations do so!

In application of Jaynes' principle, the uninformative prior (or the corresponding measure) $\mu_{a\chi}\, da\, d\chi$ must be invariant under the general transformations $\chi' = \alpha\chi$ and $a' = \beta a$. This leads to the well-known Jeffreys' prior [5] for scale parame-

---

[1] This statement can be seen as a consequence of a generalized Copernican principle: we do not occupy a special (spatial or temporal) place in the universe.

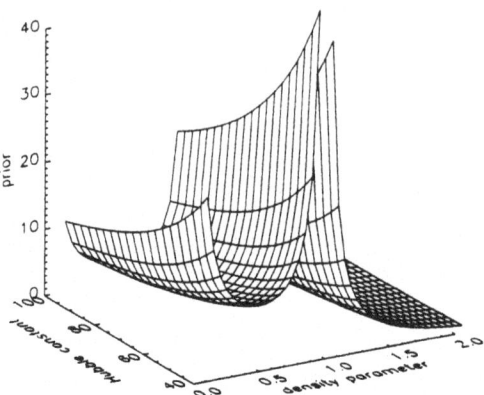

Figure 2. The uninformative prior (6) obtained from Jaynes' transformation group method. The singularity at $\Omega = 1$ negates the existence of the "flatness problem" of standard cosmology.

ters, namely

$$\mu_{a\chi}(a, \chi) \propto \frac{1}{a\,\chi}. \tag{5}$$

Notice that in terms of cosmic time t (the age of the universe), the transformation $a' = \beta a$ reads $t'(\beta, t) = a^{-1}[\beta a(t)]$. This invariance law is different from the classical time translation $t' = t + \Delta t$. However, a series expansion of $t'(\beta, t)$ about the neutral element $\beta = 1$ shows that the classical time translation invariance law is a valid first order approximation of the correct $t'(\beta, t)$ relation, as long as the quantity Δt is small compared to the age t of the universe.

From Eq. (5), it is straightforward to obtain the corresponding form of this measure in the usual (H, Ω) parametrisation. Equations (1) and (2), and the Jacobian of this transformation yield

$$\mu_{H\Omega}(H, \Omega) \propto \frac{1}{H\,\Omega\,|\Omega - 1|}. \tag{6}$$

This improper prior is represented on Fig.(2).

It must be recalled that the density parameter Ω is dimensionless. In the (H, Ω) parametrisation, only H is scale dependent, therefore the $1/H$ form of $\mu_{H\Omega}$ is simply the Bayesian rule of thumb for scale parameters.

The singularities of $\mu_{H\Omega}$ at $\Omega = 1$ and $\Omega = 0$ are of a different nature. As a result of the Friedman equations, the critical value $\Omega = 1$ is the starting point of every model (whatever the value of χ). For non spatially-flat models, Ω evolves away from the critical value 1, more or less rapidly, depending on the value of χ (see Fig. 1). If the density exceeds the critical value (i.e. $\Omega > 1$), the expansion is eventually followed by a recontraction phase. On the contrary, if $\Omega < 1$, the models evolve towards a state of free expansion with $\Omega \to 0$. The critical model, with $\Omega = 1$ exactly, remains in that state forever. Therefore, $\Omega = 1$ and $\Omega = 0$ are stable states, all other values of Ω are transient. Consequently, in absence of any other information, it seems reasonable to infer that the system should be in one of these two fixed states (hence the singularities of the prior).

4. The flatness problem connection

The prior (6) sheds a new light, and hopefully, from an objective viewpoint, on the so-called *flatness problem* of standard cosmology.

This problem can be summarized as follows: the value of Ω today is not known with a precision better than, say, $0.1 < \Omega_0 < 1.5$. Getting back to an arbitrary early epoch, this gives only a very "small" range of acceptable values for Ω_{early}, which appears "extraordinary close" to 1. This situation is then declared very "improbable". In other words, had this "fine-tuning" not occurred, the universe would have recontracted or diffused ($\Omega = 0$) long before the present epoch.

It must first be stressed that this problem is obviously of a probabilistic nature. It is also clearly related to the definition of a measure on the parameter space, and especially on its Ω-sections. However, such a measure is not explicitly specified. Yet, it can be suspected[2] that an underlying and untold measure/prior is assumed to yield the above statements of fine-tuning, closeness and improbability

Indeed, the above reasoning is valid if one accepts that the measure for the cosmological parameters at an early epoch is $\mu_{H\Omega} \, dH \, d\Omega \propto dH \, d\Omega$, or equivalently, that the prior describing minimal knowledge is constant in the (H, Ω) parametrisation. If this was the case, the interval for Ω_{early} corresponding to the actual values of Ω_0, would appear then extremely small, and therefore improbable.

However, as it was shown in the previous section, such a measure is not invariant, especially under a change of cosmic epoch. The constant prior assumption therefore particularizes a given epoch (the t_{early} time). On the contrary, the time-invariance requirement yields the form (6) for the uninformative prior. In this sense, assuming a constant prior consequently represents a considerable prejudice compared to the least-informative and, therefore, least-prejudiced measure (6). In this case, because of the singularity at $\Omega = 1$, the acceptable ranges for both Ω_{early} and Ω_0 have an infinite measure, and the fine-tuning problem disappears.

Moreover, the critical $\Omega = 1$ case is always infinitely distant from any non-flat model. It is therefore extremely dangerous to speak of "Ω close to 1".

The measure (6) enhances the fact that the three geometries are three different competing models, and that the conclusion Ω stricly equals 1 cannot be inferred from measurements. Indeed, because of observational errors, it is impossible to obtain experimentally the absolute accuracy necessary to yield the precise equality $\Omega = 1$. Such a choice must be *decided*, possibly after a physical principle (the theory of inflation [7] can perhaps play this role), and this kind of postulate can only be invalidated by experimental evidence, or alternatively, accepted as compatible with observation, but it can never be rigourously *proved*.

---

[2] "... there are still dangerous hypotheses; they are first, and above all, those that are tacit and unconscious.", from Ref [6].

5. Summary and conclusion

The quest for the cosmological parameters from observational quantities is an example of a generalized inverse problem. In the application of a Bayesian inference process to solve such a problem, the first step is the assignement of a prior. However, in the case of cosmology, the prior knowledge is very close to mere ignorance. Yet, ignorance does not imply "no prior", it just means that we have to argue from a state of minimal information.

The corresponding (probability) density function is found by applying Jaynes' transformation group method, and considering the invariance properties derived from the very definition of the chosen parameters. These invariance transformations are changes of scale, and changes of cosmic epoch.

In the usual (H, Ω) parametrisation, the resulting measure/prior (Eq. 6) is singular at the critical value $\Omega = 1$ of the density parameter, separating those models that recollapse in the future from those that expand for ever. Analysis of this singularity negates the existence of the "flatness problem" of standard cosmology. Indeed, once this apparent enigma has been recognised as essentially probability-based, and closely related to the measure assigned to the parameter space of cosmological models, the former result brings a natural answer to the conundrum of today's observed value of $\Omega_0 \simeq 1$.

Since present observational constraints on the cosmological parameters, and especially on Ω are very weak [8], they provide a very flat likelihood in the vicinity of $\Omega_0 = 1$. Hence, the posterior for the cosmological parameters is dominated by the prior. Consequently, the actual knowledge on the geometry of the universe, described by Ω, can be summarized by: assuming the validity of the Friedman models, the observations are compatible with every geometrical issue. In particular, the spatially-flat $k = 0$ case is not ruled out. However, on no account is this situation special or extraordinary, on the contrary, the prior (6) shows that it corresponds to what should be expected from a lack of observational information.

References

1. A. Tarantola, *Inverse problem theory. Methods for data fitting and model parameters estimation.*, Elsevier Science Publishers B.V., Amsterdam, 1987.
2. A. Sandage, "Observational tests of world models," *Ann. Rev. Astron. Astrophys.*, **26**, pp. 561–630, 1988.
3. G. Evrard and P. Coles, "Getting the measure of the flatness problem," *Class. Quantum Grav.*, 1995 (in press).
4. E. T. Jaynes, "Prior probabilities," *IEEE Transactions on Systems Science and Cybernetics*, *SSC-4*, **3**, pp. 227–240, 1968.
5. H. Jeffreys, *Theory of Probability.*, Clarendon Press, Oxford, 1939.
6. H. Poincaré, *La science et l'hypothèse*, Flammarion, Paris, 1902.
7. J. V. Narlikar and T. Padmanabahn, "Inflation for astronomers," *Ann. Rev. Astron. Astrophys.*, **29**, pp. 325–362, 1991.
8. B. Guiderdoni, "High-redshift tests of Ω_0," in *Observational Tests of Cosmological Inflation*, T. Shanks, A. J. Banday, R. S. Ellis, C. Frenk, and A. Wolfendale, eds., pp. 217–241, Kluwer Academic Publishers, Dordrecht, The Netherlands, 1991.

MEAL ESTIMATION: ACCEPTABLE-LIKELIHOOD EXTENSIONS OF MAXENT

PETER S. FAYNZILBERG
Carnegie Mellon University
Graduate School of Industrial Administration
Schenley Park
Pittsburgh, Pennsylvania 15213-3890[†]

Abstract. An inference scheme that extends and generalizes both MaxEnt and the Maximum Likelihood Principle is proposed. It incorporates both prior knowledge and sample data but, unlike the Bayesian inference, does not involve distributions on the parameter space. In contrast to the Maximum Likelihood Principle, it produces meaningful estimates even when the size of the sample is small.

Key words: Statistical inference, satisficing, likelihood, MaxEnt, Maximum Likelihood.

1. Introduction

It is often pointed out (see for example, Csiszár 1995, this volume) that the Maximum Entropy Principle (MaxEnt) may be used in both Bayesian and classical contexts. In the former, MaxEnt is usually invoked to infer a reasonable prior. And in the latter case, one infers a maximally unbiased probability distribution that is consistent with all available knowledge — the distribution that "speaks the truth and nothing but the truth" (Katz [3]). Thus, when faced with a coin-tossing mechanism with unknown properties, MaxEnt suggest to think that the coin is fair. But what should the observer think after he or she tosses the coin a few times?

In this paper, we describe Maximum-Entropy Acceptable-Likelihood (MEAL) estimation scheme that has been recently proposed in Faynzilberg [2]. MEAL extends and generalizes MaxEnt: in addition to prior deterministic knowledge, it incorporates sample data. The fact that it utilizes both types of information makes MEAL similar to Bayesian inference. In contrast to the latter, however, MEAL remains an essentially classical construct and requires no elicitation of the prior from

---

[†] Internet: petersf@andrew.cmu.edu

K. M. Hanson and R. N. Silver (eds.), Maximum Entropy and Bayesian Methods, 387–392.

the decision-maker. As such, it may be viewed as an alternative to the Bayesian inference scheme.

Another motivation for introducing MEAL may be given if, instead of viewing MaxEnt as a starting point, we concentrate first on the sample data. The classical Maximum Likelihood Principle works well with large samples: under certain regularity conditions, the maximum likelihood estimate (MLE) is consistent, that is, it converges to the true value of the parameter (Lehmann [4], Bickel and Doksum [1]). The main deficiency of the MLE is, however, that it performs poorly in small samples. Moreover, the Maximum Likelihood Principle does not prescribe to the decision-maker how to utilize available prior knowledge.

In fact, the MLE dictates the observer to disregard prior beliefs in the cases when such beliefs may be reasonably formed. In the absence of prior knowledge, for example, an observer who faces an unfamiliar coin-tossing mechanism may believe that the coin is fair — in accordance with both the Principle of Insufficient Reason of Bernoulli and MaxEnt of E.T. Jaynes. Even after the first coin toss, however, the MLE forces the observer to adopt an opposite, rather extreme view that the coin is one-sided — both faces contain heads or both contain tails. [1]

The proposed MEAL inference scheme alleviates the above difficulties of the MLE and extends the MLE by incorporating prior knowledge into the inference process. In doing so, MEAL becomes applicable to inference from small samples, *i.e.*, where the MLE fails. The extension is made possible by the notion of *acceptable likelihood*, to which we now turn.

2. Acceptable Likelihood

Once maximization of the sample likelihood is carried out, it is impossible to incorporate any other information, including prior knowledge, since the result of this optimization is, in most cases, unique. In order to utilize prior knowledge, we suggest to relax the requirement of likelihood maximization to that of likelihood satisficing. [2] That is, we suggest to exclude from consideration all parameter values that are not sufficiently close to the MLE. The Maximum Likelihood Principle is stronger: it excludes all the parameter values *other* than the MLE. The degree of satisficing is idiosyncratic to the decision-maker and described by a (real) parameter α.

Turning to specifics, let Ω be an observation space, Θ a parameter space, [3] and $p : \Omega \times \Theta \to \mathbf{R}_+$ a likelihood function. The prior knowledge available to the decision-maker may be formulated as explicit reduction of Θ or given implicitly in

---

[1] If the probability of heads θ is an unknown parameter of the coin, the likelihood value is either θ or $1 - \theta$ depending on whether the first toss produced heads or tails, respectively. Maximizing these on the interval $[0, 1]$ leads to the above stated conclusion.

[2] The reader familiar with the literature on rationality may find that the relationship between the maximum and acceptable likelihoods is similar in flavor to that between rationality (in which preferences are maximized) and bounded rationality (in which only a certain satisfactory level of preferences is attained). In contrast to bounded rationality, however, we do not make any assertions with regard to the processing and capacity constraints of the individual.

[3] To ease exposition, we consider the parametric case.

the form of constraints on the choice of θ. Let Θ^f be the set of feasible parameter values, *i.e.*, those consistent with prior knowledge.

For a real number α and data $\omega \in \Omega$, we shall say that a parameter $\theta \in \Theta$ makes likelihood of ω acceptable at the level α, or α-acceptable, if

$$p(\omega, \theta) \geq \alpha p^{ML}(\omega), \tag{1}$$

where $p^{ML}(\omega) = \sup_{\tau \in \Theta} p(\omega, \tau)$ is the likelihood of ω under the MLE.

Given the observed sample ω, the requirement of likelihood acceptability reduces the set of parameter values under consideration from Θ to a subset $\Theta_\alpha(\omega) \subseteq \Theta$ whose elements satisfy (1). In contrast to the MLE, this subset is not a singleton, thus enabling the decision-maker to combine the notion of acceptable likelihood with any other inference scheme.

More specifically, suppose that the decision-maker solves the probability assignment problem — the problem of making an inference on the basis of the prior knowledge only — by means of a decision rule δ : for each feasibility set Θ^f, this rule yields an inferred value $\delta(\Theta^f) \in \Theta$. We shall say that a decision rule δ_α, defined by $\delta_\alpha(\Theta^f, \omega) = \delta\left(\Theta^f \cap \Theta_\alpha(\omega)\right)$, is an *acceptable-likelihood extension* of δ. Observe that δ_α involves both the prior knowledge and the sample data. In the next section, we specialize to the case where δ is represented by MaxEnt and obtain MEAL — an acceptable-likelihood extension of MaxEnt.

3. Acceptable-Likelihood Extensions of MaxEnt

The MEAL inference rule may be formulated thus:

> For a given acceptance level α and data ω, select that probability measure which maximizes entropy subject to all available knowledge *and* makes the likelihood of ω acceptable at the level α.

Whereas MaxEnt prescribes maximization over Θ^f, MEAL estimate is obtained by entropy maximization over $\Theta^f \cap \Theta_\alpha(\omega)$. Thus, both knowledge elements — prior knowledge via Θ^f and the sample data via $\Theta_\alpha(\omega)$ — influence the result. In this sense, MEAL extends both the MLE (by incorporating prior knowledge) and MaxEnt (by incorporating the sample information).

Note that with $\alpha = 0$ any parameter yields an acceptable likelihood of the data. In other words, acceptability of likelihood is not binding in maximization of entropy, and the resulting MEAL estimate coincides with the MaxEnt value. This situation describes a maximally conservative decision-maker who ignores any and all the sample evidence in favor of the prior knowledge. At the other extreme, with $\alpha = 1$, only the MLE is acceptable. [4] This case corresponds to the decision-maker who may be said to "lack convictions:" even when faced with a small sample, he abandons prior knowledge entirely and relies exclusively on the sample data. Thus, MEAL not only extends but generalizes both MaxEnt and the MLE as well. In fact both MaxEnt and the Maximum Likelihood Principle may now be viewed as

---

[4] MEAL may still be preferred here over the MLE in this case: it resolves a non-uniqueness of the MLE should it be present.

points on a continuum parameterized by the weighting "conservatism" parameter α.

As the sample size grows larger, the likelihood function becomes more and more peaked, and the requirement (1) becomes more restrictive. In the limit, MEAL estimate coincided with the MLE, thus retaining good large-sample properties of the latter. At the other limit, when no data are available, (1) is vacuously satisfied, and MEAL reproduces MaxEnt. Thus, sequential estimation with MEAL describes a learning process in which the decision-maker changes his estimate *continuously* from a value determined primarily by the prior knowledge to the a value (usually true) derived from the sample data. In particular, it provides a meaningful estimate when the sample is small.

In addition to the above mentioned normative features, MEAL seems to be realistic in describing the real-world decision-makers. Typically, the condition (1) of likelihood acceptability is not binding in small samples. As a consequence, MEAL realistically describes the tendency of the decision-maker to ignore evidence contained in small samples even when it contradicts his prior knowledge (see the coin-toss example of the next section). Developing this line of reasoning further, Faynzilberg [2] outlines a statistical theory of social change and discusses other relationships with behavioral research.

4. MEAL Inference in Bernoulli trials

As an illustration of MEAL inference scheme, consider a coin with unknown parameter θ tossed n times. The MLE in this case coincides with the relative frequency, s, of occurrence of heads in the sample. The MEAL estimate solves the following optimization problem: maximize entropy

$$-\theta \log \theta - (1 - \theta) \log \theta$$

subject to acceptability of the likelihood:

$$\theta^s (1 - \theta)^{1-s} \geq \alpha^{1/n} s^s (1 - s)^{1-s}. \tag{2}$$

Inequality (2) is obtained from (1) by raising both sides to the power $1/n$ — a monotone transformation. In a more general case when some prior knowledge is available, additional constraints on the in addition to this inequality, constraints on the choice representing available prior knowledge are present as well.

Simulation results for the true parameter $\theta^* = 1/4$ and $n \leq 250$ are presented in Figure 1. The lowest line in the figure represents the MLE, and the other two lines show the behavior of MEAL estimates with $\alpha = 0.01$ (upper) and $\alpha = 0.1$ (middle), respectively.

As we can see from the picture, both of the MEAL estimates begin at the MaxEnt value $\theta_0 = 1/2$ and progress toward the true value θ^* (consistency). As we mentioned earlier, when n is relatively small (up to 20 in the present case), likelihood is acceptable, and the decision-maker adhere to the MaxEnt value. As the sample grows larger, he begins to give weight to the sample data and "drifts" toward the MLE. The departure from the MaxEnt value occurs sooner with larger

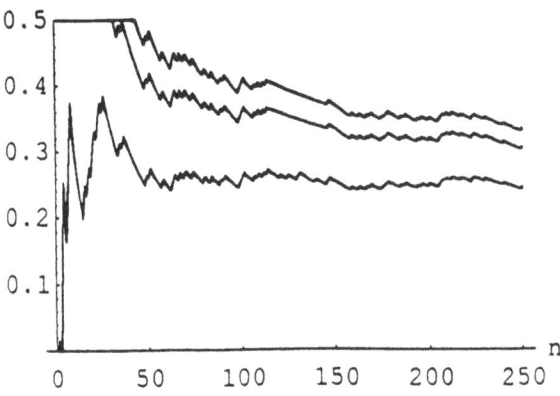

Figure 1. Bernoulli Trials

values of α. Thus, the decision-maker modifies his beliefs rather smoothly — from the initial $1/2$ to the true value $1/4$. Note also that the behavior of the MEAL estimates is much more satisfactory at small values of n than that of the MLE.

5. Informational Content of Likelihood Acceptability

Interestingly, the requirement (1) of likelihood acceptability may itself be expressed in terms of entropy — at least in the case of multinomial distributions. For a multinomial distribution with unknown probabilities $p = (p_1, p_2, \ldots, p_n)$, the MLE is given by the observed relative frequencies $p^e = (N_1/N, \ldots, N_n/N)$ in a sample of size N. Inequality (1) takes the form

$$\prod_{i=1}^{n} p_i^{N_i} \geq \alpha \prod_{i=1}^{n} \left(\frac{N_i}{N}\right)^{N_i},$$

which by taking the logarithm may be transformed into the following inequality:

$$\sum_{i=1}^{n} (N_i/N) \log \frac{(N_i/N)}{p_i} \leq \frac{1}{N} \log \frac{1}{\alpha}. \tag{3}$$

The reader will recognize that the left hand side of this inequality is equal to the Kullback-Leibler number, *i.e.*, relative entropy from the empirical distribution p^e to the estimate p. Thus, our requirement of likelihood acceptability amounts to the restriction that the estimate be *informationally* close to the empirical one. In other words, while the decision-maker need not accept the empirically observed frequency as the MLE would dictate, he should not extensively — in terms of the information contained — deviate from it either.

Besides the interpretation of the parameter α as the degree of conservatism, another one may be given. Namely, if the decision-maker has a limited discriminating ability among the values of the parameter, then the right hand side of (3) turns out to represent the resulting granularity of the parameter space Θ. Then the acceptability of likelihood condition merely states that if the decision-maker accepts a given estimate, he should be consistent and accept those values that he cannot distinguish from the posited value of θ.

6. MEAL as Retrospection

As we mentioned earlier, MEAL is a classical construct. Interestingly, it may be used in a Bayesian setting as well, namely, in *Bayesian inference with retrospection* (Faynzilberg [2]) that allows for *endogenous* determination of the priors.

The reader will recall that in the traditional Bayesian framework, the priors are determined exogenously and, as a consequence, remain unchanged by observations. It may be argued, however, that after observing a sample, the decision-maker has a different view of the world than *ex ante*; he or she may, then, formulate *ex post* a prior that differs from that which may be used in the absence of the sample. The process by which the priors are modified on the basis of the sample data are referred to as retrospection, or backward inference; the usual Bayesian updating is a forward inference in this scheme.

Retrospection, of course, may be implemented in numerous ways. In particular, MEAL may be utilized for this purpose. Given the data and prior knowledge, the decision-maker solves the MEAL optimization problem and views the solution as a prior distribution, which is subsequently updated to posterior in the usual fashion. The resulting scheme exhibits both classical (the sample likelihood is used in MEAL) and Bayesian (updating) features and occupies an intermediate position between the two inference schemes. Furthermore, Bayesian inference with MEAL retrospection also generalizes them: with $\alpha = 0$ it coincides with the traditional Bayesian framework, and with $\alpha = 1$ it reverts to the Maximum Likelihood Principle.

Thus, a decision-maker with retrospection need not decide *whether* to be a classical or a Bayesian statistician. Rather, by selecting the level of likelihood acceptability α, he chooses *to what extent* he wants to be a classical and a Bayesian decision-maker simultaneously.

References

1. P.J. Bickel and K.A. Doksum. *Mathematical Statistics. Basic Ideas and Selected Topics.* Holden-Day, San Francisco, 1977.
2. P.S. Faynzilberg. *Acceptable Likelihood and Bayesian Inference with Retrospection.* Working Paper 1995–02, Graduate School of Industrial Administration, Carnegie Mellon University, Pittsburgh, Pennsylvania, March 1995.
3. Amnon Katz. *Principles of Statistical Mechanics.* W.H. Freeman and Company & Sons, San Francisco, 1967.
4. E.L. Lehmann. *Theory of Point Estimation. Wiley Series in Probability and Mathematical Statistics,* John Wiley & Sons, New York, 1983.

ON CURVE FITTING WITH TWO-DIMENSIONAL UNCERTAINTIES

F.H. FRÖHNER

Forschungszentrum Karlsruhe
Institut für Neutronenphysik und Reaktortechnik
Postfach 3640, D-76021 Karlsruhe
Germany

Abstract. Fitting of theoretical curves to uncertain data, by adjustment of the curve parameters, is an everyday task in science and technology. The case where merely the ordinates of the data points are uncertain is well understood theoretically, whereas only heuristic recipes seem to be available if also the abscissae are uncertain. Usually the given information consists of best values, uncertainties and correlations of the point coordinates and, perhaps, of similar prior information about the curve parameters (the parameters of a given model). Likelihood function and prior are then determined unambiguously by the principle of maximum entropy but the unknown true point coordinates must be integrated out as nuisance parameters. For linear models this can be done analytically. For nonlinear models one needs suitable approximations to derive analytical expressions. It is shown that saddle point integration yields a generalized least-squares algorithm involving iterative solution of the normal equations, with a manifestly covariant form of the resulting likelihood function. The result differs from current ad-hoc recipes. Straight-line fitting to independently measured points with two-dimensional uncertainties, without prior information about the parameters, serves as illustration.

1. Introduction: Curve Fitting with Uncertain Ordinates

In most scientific applications curve fitting is restricted to point data with uncertain ordinates, abscissa uncertainties being neglected. Let us write the curve equation in Cartesian coordinates as $y = f(x, \theta)$, where θ is the vector of curve parameters. For given prior density $p(\theta|\,)$ and uncorrelated data $\xi_j = x_j$, $\eta_j = \langle y(x_j) \rangle$, without abscissa uncertainties but with mean-square errors $\sigma_j^2 = \langle (y_j - \eta_j)^2 \rangle$ of the ordinates, the maximum entropy posterior for the curve parameters is

$$p(\theta|\{\xi_j, \eta_j, \sigma_j\})\, d(\theta) \propto \Big\{ \prod_j \exp\Big[-\frac{1}{2}\Big(\frac{\eta_j - f(\xi_j, \theta)}{\sigma_j}\Big)^2 \Big] \Big\} p(\theta|\,)\, d(\theta)\,, \quad (1)$$

K. M. Hanson and R. N. Silver (eds.), Maximum Entropy and Bayesian Methods, 393–405.
© *1996 Kluwer Academic Publishers.*

where $d(\theta) = \prod_\mu d\theta_\mu$ denotes the volume element in the space of parameter vectors (to be distinguished from the infinitesimal vector $d\theta$). Here and in the sequel we denote true observables by Latin letters (x_j, y_j, ...), the experimentally observed quantities by Greek letters (ξ_j, η_j, ...), with Latin subscripts distinguishing data points ($j, k = 1, 2 ..., N$), Greek subscripts labeling curve parameters ($\mu = 1, 2, ..., M$). Matrix notation will be used to exhibit the covariance of the equations.

Any estimate of the curve parameters depends on the loss function, i. e. on the penalty that must be expected if parameter values other than the unknown true values are used in a given application. If no particular application is envisaged it is common practice to use quadratic loss as a fair approximation – at least in the neighborhood of the true values – to any reasonable loss function. The best estimate (in the sense of minimizing expected losses) consists then of the mean parameter vector $\langle\theta\rangle$ and the covariance matrix $\langle(\theta - \langle\theta\rangle)(\theta - \langle\theta\rangle)^\dagger\rangle$, where $\langle...\rangle$ indicates averaging over the posterior distribution of possible parameters. The dagger, usually denoting Hermitean conjugation, simply means transposition in what follows since all our vectors and matrices will be real.

If the model is linear in the parameters, i. e. if a straight line is to be fitted, the likelihood function is multivariate Gaussian. If the prior is Gaussian too, expressing prior information about the parameters, their uncertainties and, perhaps, their correlations, or if it is uniform (least informative), the posterior is Gaussian again. This leads to the conventional least-squares formalism, with σ_j^{-2} weighting of the data and exact analytical expressions for the mean vector and the covariance matrix (cf. e. g. Dragt et al. 1977, Schmittroth 1979, Perel et al. 1994)

Most models in science and technology are nonlinear, however, which means that posteriors are rarely pure Gaussians with readily calculated means and (co) variances. It is true that means and (co)variances can always be found by numerical integration, with Monte Carlo methods if necessary, but in order to get more transparent analytical results one must employ approximations. Particularly well adapted to typical maximum entropy distributions is the approximation obtained if the exact posterior is replaced by an osculating Gaussian with the same peak location and the same curvatures at the peak, a technique often referred to as method of steepest descent or saddle point integration (cf. e. g. Mathews and Walker 1965). We write the posterior density in the form

$$p(\theta|\{\xi_j, \eta_j, \sigma_j\}) \equiv e^{-Q(\theta)/2} \tag{2}$$

and determine the peak location by minimising the exponent. This means we must solve the "normal equations"

$$[\nabla Q]_{\theta=\hat\theta} = 0 , \tag{3}$$

for instance by Newton-Raphson iteration starting from a prior estimate if available, or from some first guess. Expanding the exponent about the solution $\hat\theta$ and truncating after the quadratic term we get a quadratic form,

$$Q(\theta) \simeq Q(\hat\theta) + \frac{1}{2}(\theta - \hat\theta)^\dagger [\nabla\nabla^\dagger Q]_{\theta=\hat\theta}(\theta - \hat\theta) , \tag{4}$$

where $\nabla_\mu \equiv \partial/\partial\theta_\mu$. In this approximation the posterior is a Gaussian again, with mean vector and covariance matrix

$$\langle \theta \rangle = \hat{\theta} \,, \qquad (5) \qquad\qquad \langle \delta\theta\,\delta\theta^\dagger \rangle = \left[\tfrac{1}{2}\nabla\nabla^\dagger Q \right]^{-1}_{\theta=\hat{\theta}} \qquad (6)$$

In effect the saddle point approximation replaces integrations by differentiations, and the mean vector by the most probable vector. The resulting formalism can be considered as a generalisation of the conventional least-squares method to non-linear models and to Bayesian inclusion of prior information (see Fröhner 1986, 1993).

2. Uncorrelated Data Points with Two-Dimensional Uncertainties

Only ad hoc recipes seem to be available for data with two-dimensional uncertainties. Usually they are derived from some kind of least-squares postulate, and in analogy to the case of one-dimensional errors, but without clear foundation in probability theory. Let us therefore look at the problem from a Bayesian point of view. We write (unknown) true and (reported) observed data points as two-dimensional vectors,

$$\mathbf{r}_j = \begin{pmatrix} x_j \\ y_j \end{pmatrix} \,, \qquad (7) \qquad\qquad \rho_j = \begin{pmatrix} \xi_j \\ \eta_j \end{pmatrix} \,, \qquad (8)$$

and (reported) uncertainties and correlations as 2×2 covariance matrices

$$\mathbf{C}_{jk} = \langle (\mathbf{r}_j - \rho_j)(\mathbf{r}_k - \rho_k)^\dagger \rangle \,, \qquad (9)$$

and interpret the observed coordinates and the covariance matrix as the measurer's best estimates under quadratic loss. The given model or curve equation will also be written in covariant form,

$$F(\mathbf{r}, \theta) = 0 \,, \qquad (10)$$

with $\mathbf{r}$ indicating an arbitrary point on the curve and θ being again the vector of model or curve parameters. For straight line fitting one would have $F(\mathbf{r}, \theta) = y - mx - b$, $\theta = (m\ \ b)^\dagger$, for example.

In preparation for the general case we shall begin with the simpler problem of independently measured data points, with correlations only between abscissa and ordinate of each point, $\mathbf{C}_{jk} = \mathbf{C}_{jj}\delta_{jk}$. In this situation one has, instead of the usual error bar corresponding to the mean square error σ_j^2 of the ordinate, an error ellipse corresponding to the matrix

$$\mathbf{C}_{jj} \equiv \begin{pmatrix} \tau_j^2 & \tau_j\sigma_j\rho_j \\ \tau_j\sigma_j\rho_j & \sigma_j^2 \end{pmatrix} \,, \qquad (11)$$

where τ_j, σ_j are the standard deviations of abscissa and ordinate and ρ_j is the correlation coefficient. (If the abscissa error is independent of the ordinate error there is no correlation at all, $\rho_j = 0$, and the error ellipse has its axes oriented parallel to the coordinate axes.) Given the true vector $\mathbf{r}_j$ and the covariance matrix $\mathbf{C}_{jj}$ the sampling distribution of the observed vector ρ_j is

$$p(\rho_j|\mathbf{r}_j, \mathbf{C}_{jj})\,d(\rho_j) \propto \exp\left[-\frac{1}{2}(\rho_j - \mathbf{r}_j)^\dagger \mathbf{C}_{jj}^{-1}(\rho_j - \mathbf{r}_j)\right]d(\rho_j) \equiv e^{-Q_j/2}\,d(\rho_j)\,.$$
$$(12)$$

Actually $\mathbf{r}_j$ is constrained by the curve equation $F(\mathbf{r}, \theta) = 0$. We represent the constraint by (the limit of) a distribution $p(\mathbf{r}|\theta)d(\mathbf{r})$ that vanishes everywhere except on (a narrow strip along) the curve. This allows us to write the joint distribution for all unknown true points and curve parameters as

$$p(\theta, \{\mathbf{r}_j\}|\{\rho_j, \mathbf{C}_{jj}\})d(\theta)\prod_j d(\mathbf{r}_j) \propto \left\{\prod_j p(\rho_j|\mathbf{r}_j, \mathbf{C}_{jj})\,p(\mathbf{r}_j|\theta)\,d(\mathbf{r}_j)\right\}p(\theta|\,)\,d(\theta)\,,$$
$$(13)$$

by virtue of Bayes' theorem and the basic multiplication rule for joint probabilities. We need the distribution of the curve parameters alone, so we must integrate out the possible postitions of all true points as uninteresting "nuisance parameters". With Eq. 12 we get

$$p(\theta|\{\rho_j, \mathbf{C}_{jj}\})\,d(\theta) \propto \left\{\prod_j \int e^{-Q_j/2}\,p(\mathbf{r}_j|\theta)\,d(\mathbf{r}_j)\right\}p(\theta|\,)\,d(\theta)\,,\qquad(14)$$

which shows that we must calculate line integrals of the type $\int e^{-Q/2}p(\mathbf{r}|\theta)d(\mathbf{r})$, one for each data point. (We shall drop the point subscript j in what follows.) The integrand is a Gaussian if the curve is a straight line, and approximately Gaussian if the curvature is not too strong in the relevant region near the observed point (see Fig. 1). We shall therefore approximate the curve by its tangent at the point where the integrand has its maximum.

The location of the maximum is determined by the condition $Q(\mathbf{r}) = \min$ under the constraint $F(\mathbf{r}, \theta) = 0$ or, with the Langrange multiplier λ, by the unconstrained condition

$$\frac{1}{2}(\mathbf{r} - \rho)^\dagger \mathbf{C}^{-1}(\mathbf{r} - \rho) + \lambda F(\mathbf{r}, \theta) = \min.\qquad(15)$$

Differentiating with respect to $\mathbf{r}$ and λ we get the "normal equations" for the position of the maximum at $\mathbf{r} = \hat{\mathbf{r}}$,

$$\mathbf{C}^{-1}(\hat{\mathbf{r}} - \rho) + \lambda\hat{\mathbf{s}} = 0\,,\qquad(16)\qquad\qquad F(\hat{\mathbf{r}}, \theta) = 0\,.\qquad(17)$$

Here we introduced the gradient $\mathbf{s} \equiv \partial F/\partial \mathbf{r}$ and the abbreviation $\hat{\mathbf{s}} \equiv \mathbf{s}(\hat{\mathbf{r}}, \theta)$. Note that the gradient vectors are normal to the curve specified by $F = 0$ which implies that $\hat{\mathbf{s}}^\dagger(\mathbf{r} - \hat{\mathbf{r}}) = 0$ is the equation of the tangent at the maximum $\hat{\mathbf{r}}$. The normal equations can be solved for λ and $\hat{\mathbf{r}}$ by iteration. Let $\mathbf{r}_n$ be the n-th approximation, and (in obvious notation) $\hat{F} = F_n + \mathbf{s}_n^\dagger(\hat{\mathbf{r}} - \mathbf{r}_n) + \ldots = 0$ the Taylor expansion of (17) about it. Neglecting higher-order terms one finds from the normal equations

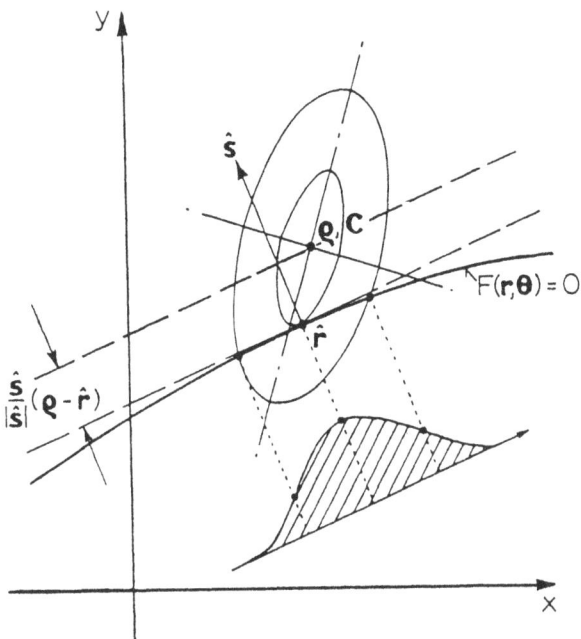

Fig. 1. Illustration of integration over possible true points showing given curve, measured point with error ellipse, smaller contour ellipse through most probable curve point, normal and tangent at this point. The nearly Gaussian integrand (hatched) is also shown.

$$\lambda \hat{s}^\dagger C \hat{s} = -\hat{s}^\dagger (\hat{r} - \rho) \simeq -s_n^\dagger (\hat{r} - \rho) \simeq \lambda s_n^\dagger C s_n , \tag{18}$$

$$F_n + s_n^\dagger (\rho - r_n) \simeq -s_n^\dagger (\hat{r} - \rho) . \tag{19}$$

The resulting approximation for λ can be inserted in (16), whereupon multiplication by C yields the iteration recipe

$$\hat{r} \simeq \rho - \frac{F_n + s_n^\dagger (\rho - r_n)}{s_n^\dagger C s_n} C s_n \equiv r_{n+1} . \tag{20}$$

Starting with $r_0 = \rho$ one can calculate F_0, s_0 and r_1, then reinsert the improved value r_1, and so on. Eventually the curve equation will be satisfied, $F_n \to \hat{F} = 0$ if $n \to \infty$, hence

$$r_n \to \hat{r} = \rho - \frac{\hat{s}^\dagger (\rho - \hat{r})}{\hat{s}^\dagger C \hat{s}} C \hat{s} . \tag{21}$$

We can now write the exponent in the line integral $\int e^{-Q/2} p(r|\theta) d(r)$ in terms of $\hat{r}$ and $\hat{s}$,

$$Q = \hat{Q} + (\mathbf{r} - \hat{\mathbf{r}})^\dagger \mathbf{C}^{-1}(\mathbf{r} - \hat{\mathbf{r}}) \,, \tag{22}$$

where

$$\hat{Q} = (\hat{\mathbf{r}} - \rho)^\dagger \mathbf{C}^{-1}(\hat{\mathbf{r}} - \rho) = \frac{[\hat{\mathbf{s}}^\dagger(\rho - \hat{\mathbf{r}})]^2}{\hat{\mathbf{s}}^\dagger \mathbf{C} \hat{\mathbf{s}}} \tag{23}$$

as follows from (21).

Next we replace the given curve by its tangent at the maximum of the integrand. It is convenient to approximate the density $p(\mathbf{r}|\theta)$ that constrains the integration to the curve by an extremely elongated bivariate Gaussian centred at $\hat{\mathbf{r}}$ and oriented along the tangent to the curve at this point,

$$p(\mathbf{r}|\theta) = \frac{1}{\sqrt{\det(2\pi\mathbf{D})}} e^{-(\mathbf{r} - \hat{\mathbf{r}})^\dagger \mathbf{D}^{-1}(\mathbf{r} - \hat{\mathbf{r}})/2} \,. \tag{24}$$

In this approximation the integrand is a Gaussian again and thus readily integrated,

$$\int e^{-Q/2} p(\mathbf{r}|\theta) d(\mathbf{r}) = \frac{e^{-\hat{Q}/2}}{\sqrt{\det(2\pi\mathbf{D})}} \int \exp\left[-\frac{1}{2}(\mathbf{r} - \hat{\mathbf{r}})^\dagger (\mathbf{C}^{-1} + \mathbf{D}^{-1})(\mathbf{r} - \hat{\mathbf{r}})\right] d(\mathbf{r})$$

$$= e^{-\hat{Q}/2} \sqrt{\frac{\det \mathbf{C}}{\det(\mathbf{C} + \mathbf{D})}}. \tag{25}$$

In order to calculate $\det(\mathbf{C} + \mathbf{D})$ we transform to the principal-axis system of $\mathbf{D}$ by an orthogonal transform $\hat{\mathbf{s}}' = \mathbf{O}\hat{\mathbf{s}}$, $\mathbf{C}' = \mathbf{O}\mathbf{C}\mathbf{O}^\dagger$ such that

$$\hat{\mathbf{s}}' = \begin{pmatrix} 0 \\ |\hat{\mathbf{s}}| \end{pmatrix}, \qquad \mathbf{D}' = \begin{pmatrix} \epsilon^{-2} & 0 \\ & \epsilon^2 \end{pmatrix}, \qquad \mathbf{C}' = \begin{pmatrix} C'_{11} & C'_{12} \\ C'_{12} & C'_{22} \end{pmatrix}, \tag{26}$$

which means that in the transformed system we want the gradient vector to be parallel to the y' axis, and the elongated Gaussian to be parallel to the x' axis, tending towards the tangent for $\epsilon \to 0$ in such a way that the correct normalisation is preserved. It is easy to check that the required orthogonal transformation matrix is

$$\mathbf{O} = \frac{1}{|\hat{\mathbf{s}}|} \begin{pmatrix} \hat{s}_1 & -\hat{s}_2 \\ \hat{s}_2 & \hat{s}_1 \end{pmatrix}. \tag{27}$$

For sufficiently small ϵ one can use the approximation

$$\det(\mathbf{C} + \mathbf{D}) = \det(\mathbf{C}' + \mathbf{D}') \simeq \epsilon^{-2} C'_{22} \,. \tag{28}$$

Transforming back to the original coordinate system one finds $C'_{22} = \hat{\mathbf{s}}^\dagger \mathbf{C} \hat{\mathbf{s}} / (\hat{\mathbf{s}}^\dagger \hat{\mathbf{s}})$. The line integral in "tangent approximation" is therefore

$$\int e^{-\hat{Q}/2} p(\mathbf{r}|\theta) d(\mathbf{r}) \simeq \epsilon \sqrt{\det \mathbf{C}} \sqrt{\frac{\hat{\mathbf{s}}^\dagger \hat{\mathbf{s}}}{\hat{\mathbf{s}}^\dagger \mathbf{C} \hat{\mathbf{s}}}} e^{-\hat{Q}/2} \tag{29}$$

The iterative determination of $\hat{\mathbf{r}}$ and $\hat{\mathbf{s}}$ and the integration must be performed for all data points.

Reintroducing the point subscript j we can now write the posterior distribution for the curve parameters in saddle point approximation. We need not keep θ-independent constants such as ϵ and $\sqrt{\det \mathbf{C}}$ because they cancel upon normalisation. So we note with interest that the transition $\epsilon \to \infty$ is harmless. The resulting posterior is

$$p(\theta|\{\rho_j, \mathbf{C}_{jj}\})d(\theta) \propto \prod_j \left[\sqrt{\frac{\hat{\mathbf{s}}^\dagger \hat{\mathbf{s}}}{2\pi \hat{\mathbf{s}}^\dagger \mathbf{C}\hat{\mathbf{s}}}} \exp\left(-\frac{[\hat{\mathbf{s}}^\dagger(\rho - \hat{\mathbf{r}})]^2}{2\hat{\mathbf{s}}^\dagger \mathbf{C}\hat{\mathbf{s}}}\right) \right]_j p(\theta|\,)d(\theta)$$

$$\simeq \prod_j \left[\frac{|\hat{\mathbf{s}}|}{\sqrt{2\pi\omega^2}} \exp\left(-\frac{F(\rho,\theta)^2}{2\omega^2}\right) \right]_j p(\theta|\,)d(\theta) \,. \qquad (30)$$

The last expression is obtained with $F(\rho,\theta) \simeq \hat{\mathbf{s}}^\dagger(\rho - \hat{\mathbf{r}})$, $\omega^2 \equiv \hat{\mathbf{s}}^\dagger \mathbf{C}\hat{\mathbf{s}}$ and, as before, $|\hat{\mathbf{s}}| \equiv \sqrt{(\hat{\mathbf{s}}^\dagger \hat{\mathbf{s}})}$. The main θ-dependence of the likelihood function, that of the deviations $F(\rho_j, \theta)$ ($=\eta_j - f(\xi_j, \theta)$ for instance), is displayed explicitly. Noting (see Fig. 1) that

$$\frac{\hat{\mathbf{s}}^\dagger}{|\hat{\mathbf{s}}|}(\rho - \hat{\mathbf{r}}) \qquad \text{is the shortest distance of the observed point from the tangent,}$$

and

$$\frac{\hat{\mathbf{s}}^\dagger}{|\hat{\mathbf{s}}|}\mathbf{C}\frac{\hat{\mathbf{s}}}{|\hat{\mathbf{s}}|} \qquad \text{the mean square error of this distance, we recognise that the}$$

likelihood function with the nuisance parameters integrated out is approximately a product of univariate Gaussian distributions. The variate of each Gaussian is the shortest distance between an observed point and the tangent at the most probable curve point. This result might have been guessed: The most probable curve is as close to the given points as the prior permits, closeness being defined in terms of the size and orientation of the error ellipses.

The mean and the covariance matrix of the model parameter vector θ for given prior density $p(\theta|\,)$ and given curve equation can always be computed numerically. Analytical expressions, on the other hand, require approximations again, such as saddle point integration (finding mean vector and covariance matrix of the osculating Gaussian in parameter space).

3. Correlated Data Points with Two-Dimensional Uncertainties

Correlations are usually induced by unknown common ("systematic") errors that affect entire data sets, such as errors of standards, detector efficiencies, instrumental resolution, sample specifications and geometry (cf. Fröhner 1986, 1994). Generalisation of our equations to correlated data points is straightforward if we gather all abscissae and all ordinates in the vectors $\mathbf{x} = (x_1, \ldots x_N)^\dagger$, $\mathbf{y} = (y_1, \ldots y_N)^\dagger$ of (possible) true coordinates and $\xi = (\xi_1, \ldots \xi_N)^\dagger$, $\eta = (\eta_1, \ldots \eta_N)^\dagger$ of observed coordinates and define the $2N$-dimensional vectors

$$\mathbf{r} = \begin{pmatrix} \mathbf{x} \\ \mathbf{y} \end{pmatrix} , \qquad (31) \qquad \rho = \begin{pmatrix} \xi \\ \eta \end{pmatrix} \qquad (32)$$

with the $2N \times 2N$ covariance matrix

$$\mathbf{C} = \begin{pmatrix} \langle(\mathbf{x}-\xi)(\mathbf{x}-\xi)^\dagger\rangle & \langle(\mathbf{x}-\xi)(\mathbf{y}-\eta)^\dagger\rangle \\ \langle(\mathbf{y}-\eta)(\mathbf{x}-\xi)^\dagger\rangle & \langle(\mathbf{y}-\eta)(\mathbf{y}-\eta)^\dagger\rangle \end{pmatrix} . \qquad (33)$$

Similarly, we combine the N individual conditions $F(\mathbf{r}_j, \theta) = 0$ into one vector equation

$$\begin{pmatrix} F(\mathbf{r}_1, \theta) \\ \vdots \\ F(\mathbf{r}_N, \theta) \end{pmatrix} \equiv \mathbf{F}(\mathbf{r}, \theta) = 0 \qquad (34)$$

which constrains all possible true points to the curve specified by the parameter vector θ. Instead of the gradient vectors $\mathbf{s}_j = (F_{xj} \ F_{yj})^\dagger$ there is now a $2N \times N$ matrix of gradients

$$\mathbf{S} \equiv \begin{pmatrix} \mathbf{F_x} \\ \mathbf{F_y} \end{pmatrix} , \qquad (35)$$

where $\mathbf{F_x}$, $\mathbf{F_y}$ are diagonal $N \times N$ matrices with elements $\partial F(\mathbf{r}_j, \theta)/\partial x_j$, $\partial F(\mathbf{r}_j, \theta)/\partial y_j$. In exact analogy with the case of uncorrelated points the likelihood function is to be taken as the maximum entropy density for given $\mathbf{r}$ and $\mathbf{C}$,

$$p(\rho|\mathbf{r}, \mathbf{C}) \propto \exp\left[-\frac{1}{2}(\rho-\mathbf{r})^\dagger \mathbf{C}^{-1}(\rho-\mathbf{r})\right] \equiv e^{-Q/2} , \qquad (36)$$

and the posterior distribution for the curve parameters is

$$p(\theta|\rho, \mathbf{C}) \, d(\theta) \propto \left\{\int e^{-Q/2} \, p(\mathbf{r}|\theta) \, d(\mathbf{r})\right\} p(\theta| \,) \, d(\theta) . \qquad (37)$$

To get rid of the nuisance parameter vector $\mathbf{r}$ we employ saddle point integration as before, but now in $2N$ instead of 2 dimensions. The maximum of the integrand is found as the solution of the variational problem

$$\frac{1}{2}Q + \mathbf{F}^\dagger \lambda = \min , \qquad (38)$$

hence as the solution of the normal equations

$$\mathbf{C}^{-1}(\hat{\mathbf{r}}-\rho)+\hat{\mathbf{S}}\lambda = 0 , \qquad (39) \qquad \hat{\mathbf{F}} = \mathbf{F}_n + \mathbf{S}_n^\dagger(\hat{\mathbf{r}}-\mathbf{r}_n)+\ldots = 0 , \qquad (40)$$

with the vector of Lagrange multipliers $\lambda = (\lambda_1, \ldots \lambda_N)^\dagger$. The iteration recipe is now

$$\hat{\mathbf{r}} \simeq \rho - \mathbf{C}\mathbf{S}_n(\mathbf{S}_n^\dagger \mathbf{C}\mathbf{S}_n)^{-1}[\mathbf{F}_n + \mathbf{S}_n^\dagger(\rho-\mathbf{r}_n)] \equiv \mathbf{r}_{n+1} , \qquad (41)$$

which yields eventually

$$\mathbf{r}_n \to \hat{\mathbf{r}} = \rho - \mathbf{C}\hat{\mathbf{S}}(\hat{\mathbf{S}}^\dagger \mathbf{C}\hat{\mathbf{S}})^{-1}\hat{\mathbf{S}}^\dagger(\rho-\hat{\mathbf{r}}) \qquad (42)$$

and

$$\hat{Q} = (\hat{\mathbf{r}} - \rho)^{\dagger}\hat{\mathbf{S}}(\hat{\mathbf{S}}^{\dagger}\mathbf{C}\hat{\mathbf{S}})^{-1}\hat{\mathbf{S}}^{\dagger}(\hat{\mathbf{r}} - \rho) . \tag{43}$$

The integral over the nuisance parameters is

$$\int e^{-Q/2} p(\mathbf{r}|\theta) d(\mathbf{r}) \simeq e^{-\hat{Q}/2} \int \exp\left[-\frac{1}{2}(\mathbf{r} - \hat{\mathbf{r}})^{\dagger}\mathbf{C}^{-1}(\mathbf{r} - \hat{\mathbf{r}}) \right] p(\mathbf{r}|\theta) d(\mathbf{r}) , \tag{44}$$

where the density function $p(\mathbf{r}|\theta)$ constrains all possible true points to the curve specified by θ. As before we approximate it by an elongated Gaussian, now in $2N$ instead of 2 dimensions, oriented along the tangent hyperplane $\hat{\mathbf{S}}\dagger(\mathbf{r} - \hat{\mathbf{r}})$. In exact analogy to the case of uncorrelated data points one finds eventually in "tangent approximation"

$$p(\theta|\rho, \mathbf{C}) \, d(\theta) \propto \sqrt{\frac{\det \hat{\mathbf{S}}^{\dagger}\hat{\mathbf{S}}}{\det \hat{\mathbf{S}}^{\dagger}\mathbf{C}\hat{\mathbf{S}}}} \, \exp\left[-\frac{1}{2}(\rho - \hat{\mathbf{r}})^{\dagger}\hat{\mathbf{S}}(\hat{\mathbf{S}}^{\dagger}\mathbf{C}\hat{\mathbf{S}})^{-1}\hat{\mathbf{S}}^{\dagger}(\rho - \hat{\mathbf{r}}) \right] p(\theta|) \, d(\theta) \simeq$$

$$\sqrt{\frac{\det \hat{\mathbf{S}}^{\dagger}\hat{\mathbf{S}}}{\det \Omega^2}} \, \exp\left[-\frac{1}{2}\mathbf{F}(\rho, \theta)^{\dagger}\Omega^{-2}\mathbf{F}(\rho, \theta) \right] p(\theta|) \, d(\theta) , \tag{45}$$

where we have displayed the essential θ-dependence by means of $\mathbf{F}(\rho, \theta)^{\dagger} \simeq \hat{\mathbf{S}}^{\dagger}(\rho - \hat{\mathbf{r}})$ and $\Omega^2 \equiv \hat{\mathbf{S}}^{\dagger}\mathbf{C}\hat{\mathbf{S}}$.

This looks like a rather general posterior for parameter estimation from given data points and given associated covariance matrix but two further generalisations are clearly possible. First, the covariant form of the last equations shows that they apply also to curves and data given in more than two dimensions. Second, the parameter adjustment can involve more than one theoretical curve. So far we assumed that the same curve equation, $F(\hat{\mathbf{r}}_j, \theta) = 0$, is valid for all the data, whereas in many situations one has data on different observables all pertaining to the same model parameters. In nuclear physics the data could be neutron scattering, capture and fission cross sections measured for the same isotope, related theoretically in different ways to the same model – quantum-mechanical resonance theory, for example, or the optical ("cloudy crystal ball") model of the nucleus. In such cases one has different model functions and different curve equations $F_j(\mathbf{r}_j, \theta) = 0$ for different observations, but all depending on the same parameters – resonance energies and widths in one case, radii and depths of potential wells in the other. With the modified set of model equations, $\mathbf{F} = (F_1(\mathbf{r}_1, \theta), \dots F_N(\mathbf{r}_N, \theta))^{\dagger} = \mathbf{0}$, the derivation of Eq. 45 goes through as before. Different data types pertaining to the same model can thus be utilised in the same "generalised least-squares" fit. The main differences between our Bayesian approach and current heuristic least-squares recipes for two-dimensional uncertainties (e. g. Deming 1943, Guest 1961, York 1966, Zijp 1984, Lyons 1986) are as follows.

- We do not start from an ad hoc least-squares principle but from more basic premises, viz. the multiplication rule for joint probabilities, Bayes' theorem, and the maximum entropy principle which prescribes Gaussian error distributions if only observed data and their uncertainties or their covariance matrix are given (whatever the unknown true distributions might happen to be). A special least-squares principle is not needed.

- For nonlinear models an iteration scheme emerges in a natural way if the unknown true observables (point coordinates) are integrated out in saddle point approximation. Most current adjustment recipes do not go beyond the first step of this scheme.

- Prior information is included right from the beginning. Mean vector and covariance matrix of the posterior are readily obtained by saddle point integration especially if uniform or Gaussian priors are applicable.

- Another difference to current recipes are the square roots in Eqs. 30 and 45. For $F_j(\mathbf{r}_j,\theta) = y - f(x_j,\theta)$ and uncorrelated data points Eq. 30 contains factors $\sqrt{(1+f_j'^2)}$ (with $f' \equiv \partial f/\partial x$) modifying the sensitivity to parameter changes in steep portions of the curve, for instance in the flanks of sharp peaks (Fröhner 1993). The square root is missing if abscissa uncertainties are not considered at all as in Eq. 1 but it appears, as a consequence of the line integral over possible true points, as soon as they are introduced – no matter how small they are. We conclude that Eq. 1 is deficient: The curve concept is not fully exploited. If the curve is represented by joint probabilities for ordinates and abscissae, $p(\mathbf{r}_j|\theta)d(\mathbf{r}_j) = p(y_j|x_j,\theta)dy_j p(x_j|\theta)dx_j$, there is not only the factor $p(y_j|x_j,\theta)dy_j = \delta(y_j - f_j))dy_j$ saying a true point must lie on the curve, but also a factor $p(x_j|\theta)dx_j \propto dx_j\sqrt{(1+f_j'^2)}$ saying the probability for the abscissa to lie in dx_j must be proportional to the length of the line element there, $\sqrt{(dx_j^2 + dy_j^2)}$.

4. Example: Straight-Line Fit to Uncorrelated Data Points

Application of the generalised Bayesian least-squares technique requires computers except with very simple models. The simplest model is a straight line. Various parametrisations are possible. We choose

$$F(\mathbf{r},\theta) = \mathbf{s}^\dagger(\mathbf{r} - r_0) = r\sin(\varphi - \varphi_0) - r_0 \,, \qquad (46)$$

where r_0 is the shortest distance from the origin to the straight line and φ_0 the angle relative to the abscissa axis (see Fig. 2). The gradient vector, $\mathbf{r}$-independent for straight lines,

$$\nabla F = \begin{pmatrix} -\sin\varphi_0 \\ +\cos\varphi_0 \end{pmatrix} = \mathbf{s} \,, \qquad (47)$$

is then a unit vector, $\mathbf{s}^\dagger\mathbf{s} = 1$. Furthermore, our choice of parameters makes the form of the "least informative" prior rather obvious. Without any information about the parameters our prior probability assignment is not changed if we go from one possible line to another one by parallel translation, $r_0 \to r_0 + c$, or by rotation, $\varphi_0 \to \varphi_0 + \chi$. If no translation and no rotation is favored over any other, so that the possible lines can be considered as covering the plain evenly (see Fig. 2), the prior must be uniform in both parameters,

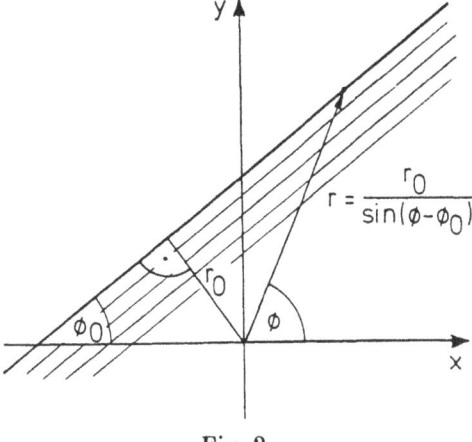

Fig. 2.

$$p(\theta|\ )d(\theta) \propto dr_0\, d\varphi_0\ , \qquad (-\infty < r_0 < \infty,\ 0 \le \varphi_0 < \pi)\ . \tag{48}$$

Jaynes (1976) found this prior in the form $p(\theta|\ )d(\theta) \propto (1+b^2)^{-3/2} db\, dm$ as the solution of a functional equation, where $m = \tan \varphi_0$ is the slope and $b = r_0 / \cos \varphi_0$ the intercept of the straight line $y = mx + b$. Gull (1989), noticing that this was not the only solution to the functional equation, expressed some doubts about its applicability to straight-line fitting. It would seem, however, that our derivation in terms of r_0 and φ_0 provides a rather compelling confirmation of Jaynes' result.

We shall now assume that the data points are uncorrelated so that Eq. (30) applies. With the definition $\omega_j^2 \equiv \mathbf{s}^\dagger \mathbf{C}_{jj}\mathbf{s}$ we can write, for N points,

$$\sum_{j=1}^{N} \frac{(r_0 - \mathbf{s}^\dagger \rho_j)^2}{\mathbf{s}^\dagger \mathbf{C}_{jj}\mathbf{s}} = \frac{N}{\overline{\omega^2}}\left[(r_0 - \mathbf{s}^\dagger \overline{\rho})^2 + \mathbf{s}^\dagger(\overline{\rho\rho^\dagger} - \overline{\rho}\ \overline{\rho}^\dagger)\mathbf{s}\right], \tag{49}$$

where overbars denote ω^{-2}-weighted sample averages, e. g. $\overline{\rho} \equiv \sum_j \omega_j^{-2}\rho_j / \sum_j \omega_j^{-2}$ and $\overline{\omega^2} = \mathbf{s}^\dagger \overline{\mathbf{C}}\mathbf{s} \equiv N / \sum_j \omega_j^{-2}$. With the abbreviation $\mathbf{V} \equiv \overline{\rho\rho^\dagger} - \overline{\rho}\ \overline{\rho}^\dagger$ for the sample covariance matrix one can write the posterior in the product form

$$p(r_0, \varphi_0|\{\rho_j, \mathbf{C}_{jj}\})\, dr_0\, d\varphi_0 = p(r_0|\varphi_0, \{\rho_j, \mathbf{C}_{jj}\})\, dr_0\, p(\varphi_0|\{\rho_j, \mathbf{C}_{jj}\})\, d\varphi_0 \tag{50}$$

with

$$p(r_0|\varphi_0, \{\rho_j, \mathbf{C}_{jj}\}) = \frac{1}{\sqrt{2\pi\overline{\omega^2}/N}} \exp\left[-\frac{(r_0 - \mathbf{s}^\dagger \overline{\rho})^2}{2\overline{\omega^2}/N}\right], \tag{51}$$

$$p(\varphi_0|\{\rho_j, \mathbf{C}_{jj}\}) = \frac{1}{Z} \exp\left[-\frac{N}{2}\frac{\mathbf{s}^\dagger \mathbf{V}\mathbf{s}}{\mathbf{s}^\dagger \overline{\mathbf{C}}\mathbf{s}}\right]. \tag{52}$$

The sample averages here depend on φ_0 but if the data define the slope fairly well, as is usually the case in straight-line fitting, the distribution (52) for φ_0 is extremely narrow, and the dependence on φ_0 is quite weak. In the simplest possible case (Fig. 3) where instead of error ellipses one has error circles,

$$\mathbf{C}_{jj} = \sigma_j^2 \begin{pmatrix} 1 & 0 \\ 0 & 1 \end{pmatrix}, \tag{53}$$

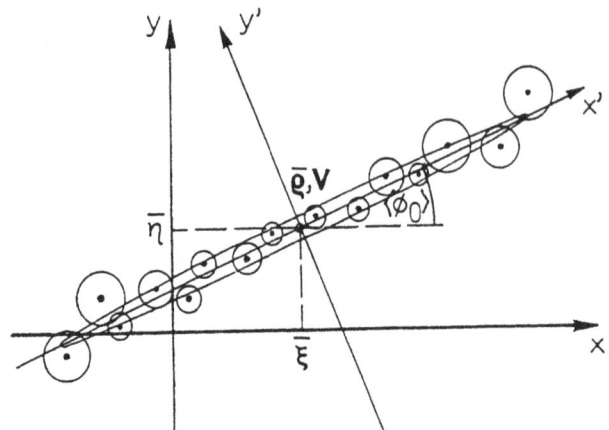

Fig. 3 Typical straight-line fitting situation (schematical, for error circles instead of ellipses): The angle φ_0 is defined very accurately as the ellipse corresponding to **V** is extremely elongated.

the sample averages, with weights σ_j^{-2}, do not depend on φ_0 at all. One can then obviously simplify by

- a shift of the origin to the "centre of mass" $\bar{p}$ of the points,
- a rotation to the principal-axes system of the matrix **V** through the angle

$$\varphi = \frac{1}{2}\arctan\frac{2V_{12}}{V_{11} - V_{22}} = \frac{1}{2}\arctan\frac{\overline{2(\xi - \bar{\xi})(\eta - \bar{\eta})}}{\overline{(\xi - \bar{\xi})^2} - \overline{(\eta - \bar{\eta})^2}} . \tag{54}$$

In the new coordinate system r_0' and φ_0' are uncorrelated, and their distributions are

$$p(r_0|\{\rho_j, \mathbf{C}_{jj}\})dr_0 = \frac{1}{2\pi\overline{\sigma^2}/N}\exp\left[-\frac{r_0^2}{2\overline{\sigma^2}/N}\right]dr_0, \tag{55}$$

$$p(\varphi_0|\{\rho_j, \mathbf{C}_{jj}\})d\varphi_0 = \frac{1}{Z}\exp\left[-\frac{V_{11}' - V_{22}'}{2\overline{\sigma^2}/N}\sin^2\varphi_0\right]d\varphi_0, \tag{56}$$

with normalisation constant

$$Z \simeq \sqrt{2\pi\overline{\sigma^2}/N} \tag{57}$$

and estimates under quadratic loss

$$\langle r_0' \rangle = 0, \tag{58}$$

$$\operatorname{var} r_0' = \frac{\overline{\sigma^2}}{N}, \tag{59}$$

$$\langle \varphi_0' \rangle = 0, \tag{60}$$

$$\text{var}\,\varphi_0' \simeq \frac{\overline{\sigma^2}}{N}\frac{1}{V_{11}' - V_{22}'} = \frac{\overline{\sigma^2}}{N}\frac{1}{\overline{\xi'^2} - \overline{\eta'^2}}. \tag{61}$$

The approximations are valid for the usual case, $\overline{\xi'^2} \gg \overline{\eta'^2}$ (see Fig. 3). Exact expressions can be given in terms of confluent hypergeometric (Kummer) functions but the resulting formulae are not particularly illuminating. So we find that the distribution of r_0' is Gaussian and that of φ_0' approximately Gaussian. The most probable straight line goes through the "centre of gravity" $\overline{p}$ under an angle $\langle\varphi_0\rangle = \varphi$ given by (54).

ACKNOWLEDGEMENTS

Support by Prof. G. Kessler, Dr. H. Küsters and the Nuclear Fusion Project (PKF) at Forschungszentrum Karlsruhe is gratefully acknowledged.

References

1. W.E. Deming, *Statistical Adjustment of Data*, New York (1943).
2. J.B. Dragt, J.W.M. Dekker, H. Gruppelaar and A.J. Janssen, "Methods of Adjustment and Error Evaluation of Neutron Capture Cross Sections", Nucl. Sci. Eng. **62** (1977) 117.
3. F.H. Fröhner, "Principles and Techniques of Data Evaluation", Kernforschungszentrum Karlsruhe, Report KfK 4099 (1986).
4. F.H. Fröhner, "Modern Principles of Data Fitting and Evaluation", in *Proc. Symp. Nucl. Data Eval. Methodol., Brookhaven, 1992*, C.L. Dunford (ed.), World Scientific, Singapore (1993) p. 209.
5. F.H. Fröhner, "Assignment of Uncertainties to Scientific Data", in *Reactor Phys. and Reactor Comput.*, Y. Ronen and E. Elias (eds.), Ben-Gurion U. of the Negev Press, Beer-Sheva (1994) p. 287.
6. P.G. Guest, *Numerical Methods of Curve Fitting*, Cambridge U. Press (1961).
7. S.F. Gull, "Bayesian Data Analysis: Straight-Line Fitting", in *Maximum Entropy and Bayesian Methods*, J. Skilling (ed.), Kluwer, Dordrecht (1989) p. 511.
8. E.T. Jaynes, "Confidence Intervals vs. Bayesian Intervals", in *Foundations of Probability Theory, Statistical Inference, and Statistical Theories of Science*, W.L. Harper and C.A. Hooker (eds.), Reidel, Dordrecht (1976), reprinted in: E.T. Jaynes, *Papers on Probability, Statistics and Statistical Physics*, R. Rosenkrantz (ed.), Reidel, Dordrecht (1983) cf. p. 196.
9. L. Lyons, *Statistics for Nuclear and Particle Physics*, Cambridge U. Press (1986).
10. R.L. Perel, J.J. Wagschal, Y. Yeivin, "Analysis of Experimental Data: a General Review with Particular Reference to Monte Carlo Evaluation of Sensitivities", in *Reactor Phys. and Reactor Comput.*, Y. Ronen and E. Elias (eds.), Ben-Gurion U. of the Negev Press, Beer-Sheva (1994).
11. J. Mathews and R.L. Walker, *Mathematical Methods of Physics*, Benjamin, New York (1965).
12. F. Schmittroth, "A Method for Data Evaluation with Lognormal Distributions", *Nucl. Sci. Eng.* **72** (1979) 19.
13. D. York, "Least-Squares Fitting of a Straight Line", *Can. J. Phys.* **44** (1966) 1079.
14. W.L. Zijp, "Generalized Least Squares Principle for Straight Line Fitting", ECN Petten, Report ECN-154 (1984).

John Skilling, Gary Erickson (facing the other way), Larry Bretthorst, John Stutz, and Sibusiso Sibisi on the balcony at St. John's College.

BAYESIAN INFERENCE IN SEARCH FOR THE *IN VIVO* T_2 DECAY-RATE DISTRIBUTION IN HUMAN BRAIN

I. GIDEONI
University of British Columbia
Dept. of Physics
6224 Agricultural Rd.
Vancouver, B.C.
Canada V6T 1Z1 [†]

Abstract.
A Bayesian procedure for model selection, parameter estimation, and classification, was applied to the problem of *In Vivo* T_2 decay rate distributions in brain tissues. The work determined, for the first time, the probability of existence of short (5-15ms) T_2 components associated with myelin water in brain tissues, and the number of decaying components in the T_2 distributions. The Bayesian aspects of the procedure will be discussed elsewhere [7].

Key words: MRI, Decay rate ,T_2, spin-spin interaction, model selection, classification

1. Introduction

The problem of inferring the distribution of decay rates, given some measurements in time, appears in many forms in MRI related data analysis. Accordingly, it has been discussed in elaborated traditional, frequentist, literature [2,3,6]. Unfortunately, the pathology of this problem leads the traditional methods to manifest their weaknesses: inverse theory methods fail to recognize the problem as an inference problem, leading to solutions that lack any measure of the results' credibility. One of the common methods is the Non-Negative-Least-Square (NNLS) [3], where one fits (using χ^2 criterion) the data to a large number (100-200) of decaying exponents in fixed decay rates. The large number of free parameters leads to *fitting the noise*. Taking care of this by additional constraints, like minimizing the solution norm, is not a consistent description of the prior knowledge, leading to imposed artifacts, such as components' widths.

[†] Email: iftah@physics.ubc.ca.
The data for this work was supplied by Alex MacKay and Ken Whittal, Radiology dept., UBC

K. M. Hanson and R. N. Silver (eds.), Maximum Entropy and Bayesian Methods, 407–412.
© *1996 Kluwer Academic Publishers.*

The traditional non-linear method is to fit (again, using the χ^2 criterion) the data to models with a growing number of decaying exponents [6]. Since there is no consistent frequentist method for comparingmodels with different number of parameters, the method is regarded by many frequentist as 'unreliable'; failing to find the "right" number of decaying exponents leads to wrong estimation of the parameters.

In this paper, we demonstrate the Bayesian treatment of such a problem. We explicitly pose the data, the background information, and the questions asked. We then demonstrate the results of applying the Bayesian procedure for selecting a model and estimating its parameters.

We also apply a Bayesian classification procedure for classifying new data sets to known tissues, leading to syntheticreconstruction of tissue-classified brain images. We hope that in the future, by incorporating additional prior information, this procedure will become clinically useful.

Computational and theoretical aspects of these Bayesian procedures are discussed in [7].

2. Posing the Problem

2.1. THE DATA

Using a 32 echo CPMG MR imaging sequence on a 1.5T GE clinical MR scanner, spin-spin relaxation (T_2) decay curves were acquired from the brains of 11 normal humans[2]. A decay curve is the collection of the amplitudes of the same pixel in the image through 32 consecutive images. Accordingly, each decay curve contains amplitudes of 32 points in time, from 10 to 320ms (Figure 1). The partition to specific tissues was assigned by a neurologist on the images of ten brains. For each of twelve tissues, five of white matter and seven of gray matter, 800-5000 decay curves had been collected. The eleventh brain's data provides the new data for the classification stage.

It is clear from the Bayesian standpoint, that we lose information by using this data instead of the raw signal collected by the scanner. The available data is the result of the imaging reconstruction, including FFT, and thus suffers from FFT artifacts. Moreover, after assigning the specific tissues, the localization of the pixels' data is lost, preventing us from taking possible tissue assignment errors and inter-tissue contamination into account. Nevertheless, in our formulation of the problem, we regard this data as the raw data. The information loss will contribute to the "noise".

2.2. BACKGROUND INFORMATION

2.2.1. *The Set of Models*
We model the mechanism that produced the decay curves by:

$$d_i = \int_0^\infty f(T_2) \cdot e^{-\frac{t_i}{T_2}} dT_2 + [a + [bt_i + [ct_i^2]]] + d \cdot (-1)^i \cdot e^{-\frac{t_i}{76}} + Noise$$

Raw Data

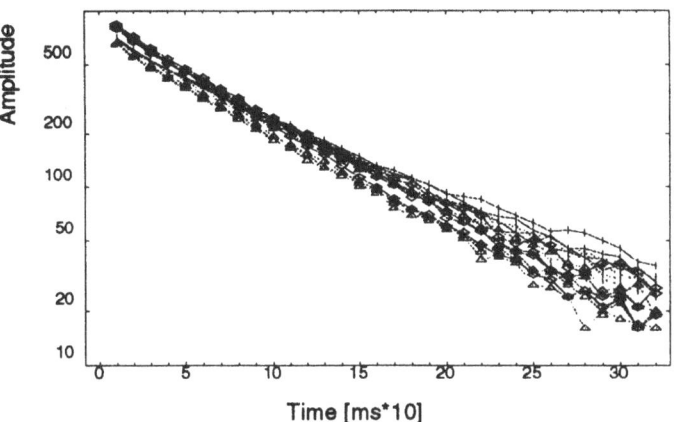

Figure 1.   A Sample of the Raw Data. Different symbols signify data taken from different humans.

Where i=1...32

The T_2 distribution, $f(T_2)$, is expected to be composed of a few discrete components. These components are parameterized by their amplitude, decay rate, and possibly width:

$$f(T_2) = \sum_{m}^{j-k} a_m \delta(T_{2m} - T_2) + \sum_{n}^{k} a_n \frac{1}{\sqrt{2\pi(width)^2}} \cdot e^{-\frac{(T_{2n} - T_2)^2}{2(width)^2}}$$

Polynomial components (a, bt_i, ct_i^2) may exist, and their amplitudes are not known. The third term, an Alternating-Echo-Refocusing (AER), exists, but its amplitude is not known.

2.2.2. *Noise*

The noise, in the Bayesian interpretation, is not a 'random' process, but merely a process in which our ignorance regarding the producing mechanism is such that its effect on the data can not be anticipated exactly. Any known behavior of the producing mechanism can, in principle, be extracted from the noise and incorporated into the model. The noise will include any effects, systematic or otherwise, that are present in the data but not in the model.

In our case, the non-systematic effects are expected to be of finite variance but are otherwise not known. Accordingly, we assign the noise a Gaussian distribution with unknown variance.

2.2.3. *Prior Information*

We assume no preference for any model in the model space, and ignorance regarding the values of the models' parameters. The numerical representation of this ignorance is assigned by considering the amplitudes of any of our base functions as location parameters (leading to flat priors), and the decay rates and widths of components as scale parameters (leading to Jeffreys' priors). Differences in the representation of the ignorance will have, in our case, a negligibleinfluence on the results, as long as we keep the assigned priors uninformative.

The priors for all the models are assigned equal, by indifference.

2.3. THE QUESTIONS

The questions we ask are always regarding the posterior probability (or posterior *pdf*) of the proposition we are interested in.

2.3.1. *Model Selection and Parameter Estimation*

For each tissue, the following questions are asked:

- Which model M_{jkl} has the highest posterior probability $p(M_{jkl}/D_{tissue}I)$?
- Given that model, what are the the most probable values of its parameters? What is the credible region for the value of each of the parameters?

2.3.2. *Classification*

After inferring the T_2 distribution of each of the tissues, using a collection of data sets, and given a new set of data d_{new}, we want to find the tissue from which it came.We are looking for the tissue i, that will maximize the probability $p(C_i/d_{new}D_iI)$, where $C_i \equiv$ "The new data was produced by the same mechanism that produced the old data set, D_i." This mechanism is described by a model (functional form) and model's parameters' values.

3. Results

3.1. MODEL SELECTION AND PARAMETER ESTIMATION

In Figure 2 we give a sample of the results of the model selection and parameter estimation. The figure shows the modelthat had been selected for each tissue, and the estimated parameters of that model for the tissue. The credible regions for the parameters cannot be seen in the figure.

3.2. CLASSIFICATION

The posterior probability of C_i is found to be proportional to the sum, over all possible models and parameters' sets, of the likelihood of the model (and parameters' set)in light of the new data, weighted by the probability of that model (and parameters' set) given the old data set D_{tissue}.

We use this result to classify each pixel in a 32-image-set, generating a synthetic image of classified tissues (Figure 3).

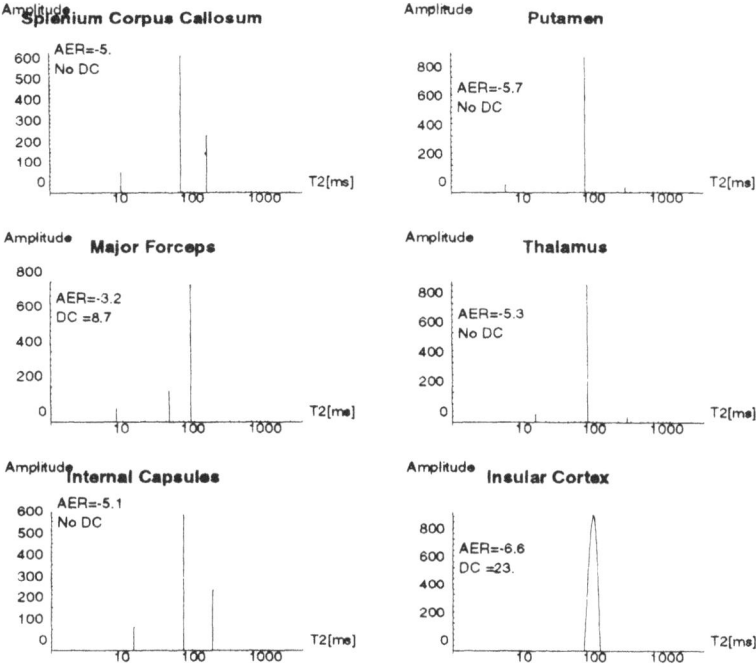

Figure 2. Results of Model Selection and Parameter Estimation for White Matter Tissues (left) and Gray Matter Tissues (right).

Determining the number of decaying components is crucial for the estimation of the decay rates. Base functions describing decaying exponents are always non-orthogonal, leading to wrong estimations in case of wrong assumption regarding the number of exponents [3]. This work infers, for the first time in a consistent manner, the number of components for each tissue. A component of short (5-15ms) T_2 in the brain tissues is assigned to water confined within myelin bilayers[2]. The amplitude of this component distinguishes white from gray matter tissues and may serve as an age indicator for lesions of white matter diseases such as Multiple Sclerosis [5]. This work determines the probability of existence of such low T_2 component and infers its amplitude. Previous frequentist works[2], given the same data and prior information, determined that the main component of the T_2 distribution in most of the tissues is continuously distributed. This work determines that for most of the tissues, the evidence in the data does not justify the additional complexity of wide T_2 components. These frequentist works, from the Bayesian standpoint, used information which is neither in the data, nor part of our background knowledge.

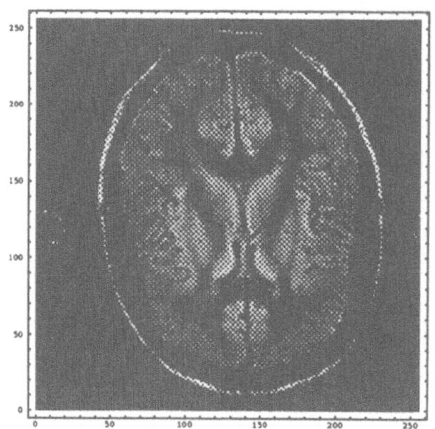 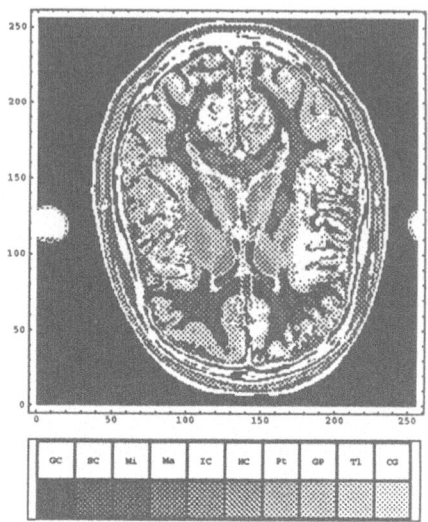

Figure 3. Results of Classification: The MRI Image (left) and the Classified Synthetic Image (right). Each gray level in the synthetic image corresponds to certain tissue. In the data used for this work, the variance among the T_2 distribution from brains of different humans was larger than the difference between the T_2 distribution of different tissues in the same human's brain. The resulting image fails to classify closely behaving gray matter tissues, though the discrimination between white and gray Matter is reasonable.

References

1. G. Larry Bretthorst, *Bayesian Spectrum Analysis and Parameter Estimation.* Lecture Notes in Statistics:**48**. Springer-Verlag, 1988.
2. A.MacKay, K.Whittall, J. Adler, D. Li, D. Paty, D. Graeb, *In Vivo visualization of myelin water in brain bymagnetic resonance* Mag. Res. in Med. **31**:673-677, 1994.
3. Kenneth P.Whittall and Alexander L. MacKay, *Quantitative interpretation of NMR relaxation data.* J. of Mag. Res. **84**:134-152, 1989.
4. R.S.Menson and P.S. Allen, *Application of continuous relaxation time distributions to the fitting of data from model systems and excited tissue.* Mag. Res. in Med. **20**:214-227, 1991.
5. H.B.W.Larsson, J.Frederiksen, L.Kjaer, O.Henriksen, J.Olesen, *In vivo determination of T_1 and T_2 in the brain of patients with severe but stable multiple sclerosis.* Mag. Res. in Med. **7**:43-55, 1988.
6. Wendy A.Stewart, Alexander L.MacKay, Kenneth P. Whittall, G.R.Wayne Moore, Donald W. Paty, *Spin-spin relaxation in experimental allergic encephalomyelitis. Analysis of CPMG data using a nonlinear least squares method and linear inverse theory.* Mag. Res. in Med. **29**:767-775, 1993.
7. I. Gideoni, *Toward general solutions to time-series problems: Notes on obstacles and noise,* In the Bayesian Analysis E-Print Archive[1]

---

[1] Email: bayes-an@xxx.lanl.gov   WEB: http://xxx.lanl.gov/archive/bayes-an

BAYESIAN COMPARISON OF FIT PARAMETERS
APPLICATION TO TIME–RESOLVED X-RAY SPECTROSCOPY

V. KASHYAP

Department of Astronomy & Astrophysics, Univ. of Chicago,
5640 S.Ellis Ave, AAC, Chicago, IL 60637 [†]

Abstract. Analysis of X-ray data of the stars AD Leo and Wolf 630, obtained with the *Röntgen Satellit* (ROSAT) provide important clues to the structure of the coronae on these low-mass, main-sequence (i.e., young) stars. In particular, time-resolved X-ray spectroscopy of these stars allow us to derive estimates for the low- and high-temperature components of the plasma emission measures. Using Bayes' theorem, we show that these temperature components are physically distinct and that the high-T components are correlated with the X-ray light-curves of the stars, while the low-T components are steady. Thus we are able to model the low-T emission as relatively compact, quiescent, static coronal loops, and the high-T emission as unstable flaring components. The work reported here was carried out in collaboration with Mark Giampapa[1], Robert Rosner[2], Tom Fleming[3], Jurgen Schmitt[4], and Jay Bookbinder[5] (Giampapa *et al.*, 1995).

Key words: X-ray Astronomy, stars, solar analogy, χ^2, frequentist, Bayes' Theorem, Odds ratio

1. Introduction

In many instances, analysis of astronomical data requires inter-comparison of the results of similar forward-fit processes applied to different data sets. There is currently no reliable standard method by which such results may be statistically

[†] Email: kashyap@ockham.uchicago.edu
[1] NOAO, National Solar Observatory
[2] Dept. of Astronomy & Astrophysics, University of Chicago
[3] Steward Observatory, University of Arizona
[4] Max Planck Institut fur Astrophysik
[5] Harvard-Smithsonian Center for Astrophysics

K. M. Hanson and R. N. Silver (eds.), Maximum Entropy and Bayesian Methods, 413–418.

compared, because of the generally asymmetrical confidence ranges of the fit parameters. We have developed a Bayesian method to carry out such a comparison,
and apply it to the fit parameters obtained from the time–resolved X-ray spectroscopy of 2 nearby dMe (low-mass, young) stars. In §2 we describe the astronomical problem which motivated the observations, and the statistical problem which
developed upon data analysis. In §3 we describe the manner in which we adapt
the problem to solution by Bayes' Theorem, and in §4 we summarize the results
and list some continuing problems.

2. Coronal Emission from Late-type Stars

We have known for more than a decade now that most stars with masses of order of
the Sun or lower, and of comparable age (in other words, late-type, main-sequence
stars) are X-ray emitters. The emission material is thermal plasma at temperatures
ranging from 10^6 to 10^7 K (Rosner et al., 1985). The paradigm which best explains
this emission is the 'solar analogy,' according to which stellar X-ray emission is
due to the same physical processes that operate on the Sun: A convection-driven
dynamo generates a magnetic field which controls the topology and temperature
of the corona.

X-ray observation of M dwarfs assume an added importance in the context of
the solar analogy: For such low-mass stars, the convective region of heat transport
reaches all the way to the core of the star, and the stellar dynamo is expected to be
fundamentally different in nature than in the case of the Sun. Such a change in the
dynamo should affect the character of the X-ray emission and could potentially
constrain many parameters in models describing stellar activity. Recent survey
observations made with the *Röntgen Satellit* (ROSAT) however show that the efficiency of coronal heating does *not* change in this region for flaring M stars(Fleming
et al., 1993). In order to better understand the structure of the coronae on these
stars, we obtained longer duration observations of 2 of the brightest such stars: AD
Leo (dM4e, $R_* = 0.49R_\odot$, $M_* = 0.44M_\odot$) and Wolf 630 (dM4.5e, $R_* = 0.68R_\odot$,
$M_* = 0.45M_\odot$).

X-ray data as obtained with ROSAT are simply a list of photon events at the
detector (a position sensitive proportional counter; Trümper *et al.* 1991), and contain information on the photon position, arrival time, and photon energy. ROSAT's
detectors are sensitive to photons in the energy range 0.1-2.4 keV, and a total of
∼ 56,000 and ∼ 26,000 photon counts were detected from AD Leo and Wolf 630
in 15.242 and 8.780 ksec respectively in this range.

The resulting count spectra were fit with model spectra of emission from opti-
cally thin thermal plasma (Raymond & Smith 1977, Raymond 1988; no other type
of emission gives reasonable fits to the data). Fitting was carried out by minimiz-
ing the χ^2 statistic, $\sum_{i=1}^{N} \frac{(C_i - \mu_i)^2}{\sigma_i^2}$, where the data are binned into N bins with

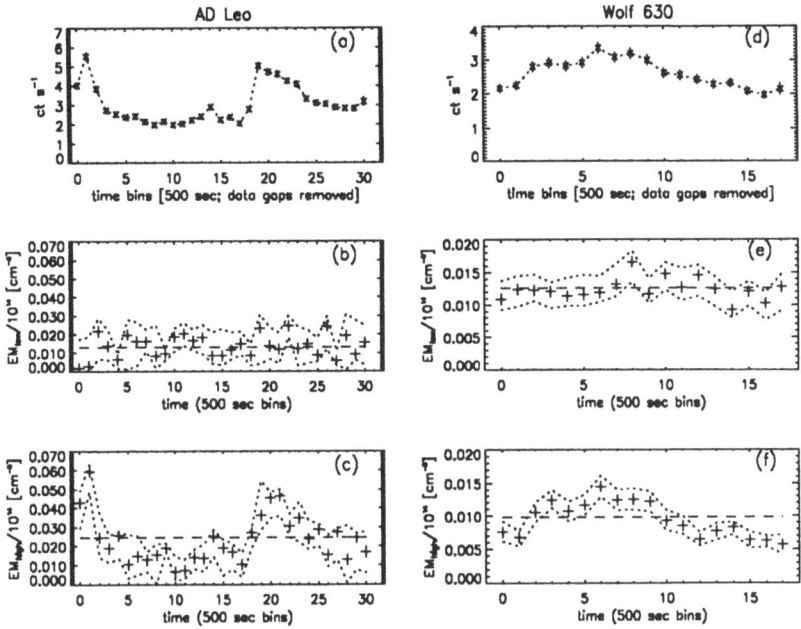

Figure 1. Comparison of EM with count-rate light-curve. (a) Light-curve of AD Leo with the data gaps removed. The length of each vertical bar denotes the 2σ error on the count rate in each bin. (b) EM_L for AD Leo at different times. The X-axis is the same as [a]; '+'s denote the best-fit values of EM_L; dotted lines enclose the 90% confidence interval of the best-fits; and the horizontal dashed line denotes the average value of EM_L over the entire observation. (c) As [b], but for EM_H. (d) As [a], but for Wolf 630. (e) As [b], but for Wolf 630. (f) As [c], but for Wolf 630.

C_i counts, with corresponding errors σ_i in each[6], and the model predicted counts are μ_i. The model best describing the data is one with two spectral components with differing temperatures, and 5 free parameters – an absorption column density to the source (N_H), the two temperatures (T_L and T_H), and the corresponding emission measures[7] (EM_L and EM_H). The best-fit values of these parameters (which result in χ^2_{min}), may then be used to construct static coronal loops which would then define the coronae on these stars (Rosner *et al.* 1978, Serio *et al.* 1981, Giampapa *et al.* 1995). However, the light-curves of these stars show significant variability (Figures 1a, 1d). It is therefore necessary to determine the variation of the plasma parameters with time. To this end, we fit 2-T thermal plasma spectra to count spectra accumulated over 500 sec bins. The results of this time-resolved X-ray spectroscopy are shown in Figure 1, where the count-rate light-curve is compared with the low- and high-T EM components of AD Leo and Wolf 630. In both stars, it appears that EM_H is correlated with the light-curve, while EM_L is rel-

---

[6] There are enough counts in each bin to justify a normal error distribution, an assumption that allows us to use deviations from χ^2_{min} to obtain confidence levels below.

[7] The emission measure (EM) is a quantity describing the "amount" of material that is emitting. Formally, $EM = n_e^2 V$, where n_e is the electron number density and V is the volume of the emitting region.

atively constant. This (unexpected) result suggests the existence of two distinct emitting components on the surfaces of these stars: a low-T component that may be consistently modeled with static coronal loops, and a high-T component that likely arises from a superposition of transient flaring events. Unfortunately, the errors on the fit parameters are not normally distributed, and the above impression cannot be reliably confirmed using standard statistical methods.

In the following section, we present a conceptually simple application of Bayes' Theorem to prove the above impressions.

3. A Bayesian Solution

The error in a given model parameter a may be estimated by minimizing the $\chi^2(a)$ statistic for different (fixed) values of a (a_x, say). Assuming that the data errors are normal (which is a good approximation in our case), the deviation, $\Delta\chi^2(a_x) = \chi^2(a_x) - \chi^2_{min}$, is distributed as the χ^2-distribution with 1 degree of freedom[8] (Lampton et al. 1976, Avni 1976, Press et al. 1992). This allows us to generate cumulative probability distributions describing the statistical variation in each fit parameter, and hence differential 'confidence' distributions for the fit parameters.

Our analysis thus far has been frequentist in approach. Now we shall assign the differential distribution computed above to the true probability distribution of the fit parameters, and proceed along a Bayesian route. It may be objected that this assignment is ad hoc and unwarranted, especially because the frequentist analysis does not allow us to claim that the fit parameters are "real" (there are an infinite number of quite plausible models that we cannot rule out based on χ^2 minimization alone – a maximum entropic reconstruction of the spectrum is perhaps the best direction to concentrate on in the future). However, our aim is to interpret the fit results; recognizing the limitations of instrumentation and analysis, we assume the validity of the fits and the resulting probability distributions. Ideally, posterior probability distributions of EM should be calculated using a Bayesian "spectral-fit" process, but because of the mathematical similarity between the two processes, the results would be similar.

We use the resulting probability distributions of the low- and high-T emission measures ($p_L(EM;t)$ and $p_H(EM;t)$) for the two program stars to first estimate the parameters in two models describing $EM(t)$ and to then compare the two models themselves. For the sake of simplicity[9], we confine our attention to the following models

---

[8] $\Delta\chi^2$ is in general distributed as the χ^2-distribution with the same degrees of freedom as the number of free parameters. However, this multi-dimensional surface may always be projected on to the appropriate subspace by minimizing the χ^2 w.r.t. all the other parameters.

[9] More complicated models may be easily constructed, but the following models are (i) sufficient to understand the underlying physics, and (ii) satisfy Ockham's razor by having just one adjustable parameter.

- **M_1**: The emission measure remains constant with time; $EM(t) = N$, where N is a normalization parameter.
- **M_2**: The emission measure varies with time exactly as the count-rate light-curve, $C(t)$; $EM(t) = N \cdot \ell(t)$, where N is a normalization parameter and $\ell(t) = C(t)/<C>$, where $<C>$ is the average count-rate.

Identifying the various values of the normalization parameter N above with rival, independent, hypotheses (i.e., N represents the hypothesis "the value of the normalization parameter is N"), we can write the posterior probability distribution of N given the model $M_{i=1,2}$ and the data D (representing EM),

$$p(N|D, I) = p(N|I) \cdot \frac{p(D|N, I)}{p(D|I)} . \qquad (1)$$

Here I represents all the background information available, $p(N|I)$ is the prior probability distribution of N (which we assume to be flat), $p(D|N, I)$ is the likelihood of obtaining D for the given N, and $p(D|I)$ is the global likelihood for (M_i, I), and serves as a normalization factor (our notation is similar to that used by Gregory & Loredo 1992). The likelihood function $p(D|N, I)$ is derived for $(N|M_i)$ by determining $EM(t)$ and obtaining the corresponding value of $p(EM; t)$. The posterior probability function $p(N|D, I)$ may then be used to compute the mean and confidence levels of N.

The "odds ratio" (Gregory & Loredo, 1992) of model M_1 over model M_2, which compares the probabilities of two models,

$$O_{12} \equiv \frac{p(M_1|D, I)}{p(M_2|D, I)} = \frac{p(M_1|I)}{p(M_2|I)} \frac{p(D|M_1, I)}{p(D|M_2, I)} , \qquad (2)$$

where $p(M_i|I)$ are the prior probabilities of the models (here assumed equal to each other) and $p(D|M_i, I)$ are simply the global likelihoods for the appropriate models from eqn (1) above[10]. The odds ratio serves to determine which model describes the data better.

4. Results

Applying the method described in §3 to time-resolved X-ray data from AD Leo and Wolf 630, we find that the odds in favor of the "constant" model over the "light-curve" model are $\sim 8600 : 1$ and $\sim 40 : 1$ respectively for the low-T components, and $1 :\sim 10^{19}$ and $1 :\sim 10^{21}$ respectively for the high-T components. These results confirm our initial impression of EM_L being relatively constant and EM_H varying as the light-curve. We are therefore justified in treating the two components as physically distinct. Modeling the emission components with static coronal loops then results in loop solutions (Giampapa et al., 1995) where

---

[10] The proposition (M_i, I) is true if and only if M_i is true. Hence $p(D|M_i, I)$ is equivalent to $p(D|M_i)$, which is the global likelihood for model M_i.

1. for the low-T component, the loop lengths are small ($l \lesssim s_p$, the pressure scale height at the appropriate temperature), and high base pressures ($p_0 \gtrsim p_{0_\odot}$, the corresponding value on the Sun); and

2. for the high-T component, no solutions are possible unless it is assumed that the loops cover a small fraction (< 0.1) of the surface area of the star; then two kinds of solutions occur: large loop lengths ($l \lesssim s_p$) and high base pressures ($p_0 \gtrsim p_{0_\odot}$) at small surface filling fractions (~ 0.1), or small loops ($l \ll s_p$) and very high base pressures ($p_0 \gg p_{0_\odot}$) at very small surface filling fractions ($\lesssim 0.01$).

This type of behavior allows us to identify the low-T component with quasi-static loops such as those found in Quiet Sun and quiescent active regions, and the high-T component with a superposition of transient or flare events arising due to magnetic flux structures presumably highly active due to field reconnection during flux emergence (Giampapa et al., 1995).

It must be noted that more work is needed in order to validate the procedure outlined in §3. Simulations are necessary to verify that the departures from normal errors one encounters in the count spectra are indeed ignorable, and also to characterize the confidence levels in the case of non-normal errors which would obtain for weaker sources. Further, the above discussion assumes that $\Delta\chi^2(a_x)$ is well-behaved, viz., that there are no local minima close to the global minimum. The interpretation of confidence intervals is uncertain in such cases (the cases we encountered were all shallow and could be attributed to the limited number of fit iterations).

Acknowledgments: This work was supported by NASA grants, and LANL & SFI at MaxEnt95.

References

Avni, Y. 1976, *Astrophysical Journal*, **210**, 642.

Fleming, T.A., Giampapa, M.S., Schmitt, J.H.M.M., & Bookbinder, J.A. 1993, *Astrophysical Journal*, **410**, 387.

Giampapa, M.S., Rosner, R., Kashyap, V., Fleming, T.A., Schmitt, J.H.M.M., & Bookbinder, J.A. 1995, submitted to the *Astrophysical Journal*.

Gregory, P.C. & Loredo, T.J. 1992, *Astrophysical Journal*, **398**, 146.

Lampton, M., Margon, B., & Bowyer, S. 1976, *Astrophysical Journal*, **208**, 177.

Press, W.H., Teukolsky, S.A., Vetterling, W.T., & Flannery, B.P. 1992, *Numerical Recipes: The Art of Scientific Computing*, 2nd ed. (Cambridge), 695.

Raymond, J.C. 1988, in *Hot thin plasmas in astrophysics: Proceedings of a NATO Advanced Study Institute*, ed. R.Pallavicini (Dordrecht/Boston:Kluwer), 3.

Raymond, J.C. & Smith, B.W. 1977, *Astrophysical Journal Supplements*, **35**, 419.

Rosner, R., Golub, L., & Vaiana, G.S. 1985, *Annual Reviews of Astronomy & Astrophysics*, **23**, 413.

Rosner, R., Tucker, W., & Vaiana, G.S. 1978, *Astrophysical Journal*, **220**, 643.

Serio, S., Peres, G., Vaiana, G.S., Golub, L., & Rosner, R. 1981, *Astrophysical Journal*, **243**, 288.

Trümper, J., et al. 1991 *Nature*, **349**, 579.

EDGE ENTROPY AND VISUAL COMPLEXITY

P. MOOS AND J.P. LEWIS
Kerner Optical Research
3160 Kerner Blvd., San Rafael, California[‡]

AND

Industrial Light & Magic
P.O.Box 2459, San Rafael, California[§]

Abstract. This philosophical paper discusses visual complexity as distinct from more general notions such as algorithmic complexity and depth. The discussion is supported by a simple experiment that can be interpreted as illustrating a "transition to texture": a set of patterns was ranked both by subjective visual complexity and an objective but perceptually oriented complexity measure ("edge entropy"); the subjective and objective rankings have a significant association at low and medium objective complexity but no association exists at high objective complexity. This relationship is unexpected if one assumes that subjective complexity is some unknown but simple function of algorithmic complexity.

Key words: algorithmic complexity, visual complexity, Shannon entropy, texture

1. Introduction: Algorithmic and Visual Complexity

Computational vision is properly addressed from a Bayesian formulation, and the maximum entropy principle is clearly applicable to the inverse problems arising in vision. Although neurophysiological research results are often interpreted with information theoretical principles, there is as yet little biological or psychological evidence concerning the extent to which the maximum entropy principle may be employed in human vision. We suspect that if it is not employed on a large scale, it is due to 'finite hardware' rather than lack of applicability.

This paper considers a more modest issue—whether subjective visual complexity can be related to possible objective complexity measures such as entropy and algorithmic complexity. After reviewing relevant experimental psychology literature, we describe a simple experiment which suggests that subjective visual complex-

---

[‡]On leave during 1995
[§]Email: zilla@kerner.com

K. M. Hanson and R. N. Silver (eds.), Maximum Entropy and Bayesian Methods, 419–424.
© *1996 Kluwer Academic Publishers.*

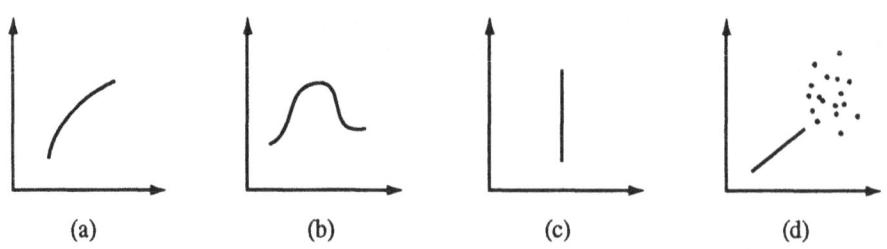

Figure 1.   The unknown relationship between subjective visual complexity (vertical axis) and algorithmic complexity

ity is decoupled from objective complexity measures at high objective complexity. This result is speculatively interpreted as resulting from finite hardware.

The concept of algorithmic complexity (A.C.) [1] provides a useful framework for discussing complexity issues. This concept is probably not in itself a suitable model for visual complexity (V.C.), however. In viewing a complex image people may perceive some of the structure as being texture or noise. For example, a true random image has maximal A.C. yet can be perceptually summarized as being "just noise". The popular Mandelbrot and Julia set images on the other hand are evidently interesting (visually complex?) despite their low A.C. (although large 'logical depth' [1]).

While our thesis that A.C. and V.C. are different things is plausible, it also seems almost meaningless, since V.C. is subjective, while A.C. is not computable!

The subjectivity issue is easier to address: Kendall's coefficient of concordance establishes that most of the variation in subjective judgement of the "complexity" of the patterns in Figure 2 is accounted for by the patterns themselves rather than being due to differing judges, i.e., there is a common notion of V.C. across people on at least some types of image.

Although the algorithmic complexity of the patterns in Figure 2 is unknown, we assume that A.C. ranks the patterns in some simple way—either increasing from left to right, or perhaps constant, or (implausibly) decreasing from left to right. We argue that if V.C. is simply related to A.C. (i.e. is some simple function of it) then we would expect that the subjective complexity rankings versus A.C. of these patterns would have a form such as Figure 1(a), (b), or (c).

In fact, the results of our experiment appear to show a relationship similar to Figure 1 (d). While this outcome does not demonstrate our thesis, the relationship shown in Figure 1 (d) requires explanation if one believes the alternative view that V.C. is simply related to A.C.

2. Study of Visual Complexity

Attenave and Arnoult wrote in 1956 that "there is virtually no psychophysics of shape or pattern" [2]. There have been a number of studies since that time; some of the relevant work will be briefly surveyed in this section.

Figure 2.   Edge-only patterns

An experiment by Attneave directly addresses the question of objective correlates of subjective V.C. [3]. In this experiment a set of simple patterns was ranked
by subjective complexity. The patterns were constructed by randomly choosing a
number of locations on graph paper, taking the convex hull of these points, and
introducing non-convexities using an algorithm operating on the interior points.
Attneave then defined possible objective measures on the number and configuration of vertices. Multiple regression on these measures resulted in

$$J = 5.46 \, log_{10} \, T + .41 \, S - 2.30$$

where J is the judged complexity, T is the "number of turns" (i.e. the number of
extrema of the curvature vs. arc length function), and S is a symmetry measure.
Thus, the log of the number of vertices explains most of the perceived complexity.

Granovskaya, Bereznaya, and Grigorieva [4] propose a specific algorithmic complexity measure (including factors such as mirror versus rotational symmetry) that
accounts for the subjective complexity ranking of a variety of simple synthetic patterns. Granovskaya. et. al. also contains a comprehensive survey of experimental
research on shape including references to some (apparently untranslated) Russian
research on this subject. In describing their experimental results, Granovskaya et.
al. write, "all three theoretical estimates correlated worst of all with a subjective one for the set with the least number of regular figures and a high density
of lines". This statement strongly anticipates the result obtained below, although
the authors did not report statistical tests of this effect.

Other relevant work that cannot be described in the available space includes
"quantitative aesthetics" experiments (which find an inverted-U shaped relation
between objective complexity measures and subjective aesthetic preference for
patterns, e.g. [5]) and the Julesz preattentive texture discrimination experiments
(e.g. [6], which for our purposes show that human visual processing has limits that
can be experimentally demonstrated).

3.   Edge-Only Patterns and Edge Entropy

The research described in the previous section establishes objective correlates of
V.C. on several types of simple synthetic patterns. Stepping back from this, do we
have any expectations on what V.C. might relate to for images in general? If we
knew how images are encoded in the brain, we might propose that subjective visual
complexity relates somehow to the cost of encoding particular images. The "visual
code" is of course unknown, but the role of edges in early visual processing has
been established from both biological and computational viewpoints (e.g. [7,8]).

Psychological research indicates that derivative information is also important in perception, e.g. [9]. Thus, patterns in which the edge information is clear, separate, and measurable might provide suitable stimuli with which to probe subjective visual complexity.

The patterns in Figure 2 are distinguished only by their distribution of edges, that is, edges are (intended to be) the only significant thing in these patterns. The patterns are generated by the following procedure. First, histograms with specified Shannon entropy are synthesized. These histograms are interpreted as histograms of "edge quanta". The quanta in each histogram bin are summed to provide an edge magnitude, and an edge polarity is chosen at random. The resulting edge process is integrated to yield a piecewise constant function, and Brownian drift is removed. Lastly, this function is interpreted as $[0, 2\pi) \rightarrow$ radius because it is anticipated that people might find one-dimensional functions to be a somewhat abstract class of "image".

The "edge entropy" (entropy of the generating histogram) serves as an objective complexity measure on these patterns. Although other nonlinear functions of edge length would differentiate these patterns, the choice of Shannon entropy is partially motivated by its information theoretical interpretation.[1] Also, since we are considering the relation between V.C. and A.C., we note in passing the connection between entropy and A.C. that resulted from resolution of the Maxwell's demon paradox (i.e., that the demon can postpone erasure and hence entropy generation by storing memory of its past actions in a compressed form) [10]. We surmise that for the patterns in Figure 2 the relation between the edge entropy and 'typical' A.C. (averaged across patterns of a particular H) is some "simple" function.

4. Experiment

We have suggested that subjective V.C. may be related to representation cost, and that edge entropy is a plausible objective representation cost for the class of patterns in Figure 2. The relation between this objective measure and subjective V.C. is explored with the following simple experiment.

Subjects were instructed with an operational definition of visual complexity:

> You will see pairs of shapes that resemble evil spaceships. Imagine that you have to pick one of each pair and describe it to a police artist over the telephone. For each pair, quickly pick the shape that would be easier to describe.

Subjects were shown randomly generated patterns from each of 10 edge entropy levels, resulting in 45 pair comparisons. A global subjective ranking was obtained for each person from the pair comparisons.

Twenty-three people completed an initial version of the experiment. The chi-square statistic was used to test the significance of the association between subjective rankings and edge entropy. Statistics are shown in Table 1. The column

---

[1] NH is the expected number of bits needed to specify a pattern generated using N quanta from a distribution having entropy H.

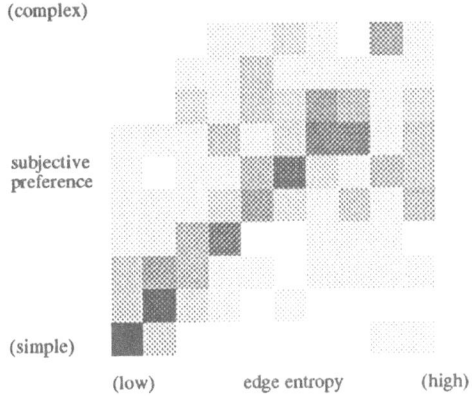

Figure 3. Plot of subjective complexity ranking (vertical axis) versus objective "edge entropy". Darkness is proportional to the number of people who produced the pattern ranking represented by the particular cell.

labeled 'low complexity' reports the association when only the three lowest edge entropy patterns were considered, and likewise for the high complexity column.

The main result is this: while there is a statistically significant association between subjective and objective complexity across all the patterns, and this association holds up when only the low objective complexity patterns are considered (with increased strength in fact), the association fails at high objective complexity.

Our results differ from [4,3] in that

1. stimuli are limited to patterns for which we can assign a plausible encoding cost,
2. the strength of the subjective/objective relation is tested independently at high and low complexity,
3. we demonstrate the failure of this relation at high complexity.

TABLE 1. Subjective/objective chi-square significance and association strength

	all patterns	low complexity	high complexity
χ^2 probability	.0000	.0061	0.14
Cramer V	0.33	0.49	(0.39)

5. Discussion

The argument presented in this paper is circuitous. We summarize it here for clarity:

- We propose that V.C. may relate somehow to the cost of encoding images in the brain.
- Although we do not know how images are coded, edges are known to be an important primitive in early visual processing.
- "Edge-only patterns" are simple patterns distinguished only by their distribution of edges.
- "Edge entropy" is an objective measure of the representation cost of these patterns. Edge entropy probably relates to the A.C. of the patterns in some simple monotonic way.
- We inquire whether edge entropy relates to perceived V.C. In Figure 3 there appears to be an almost linear relationship between V.C. and edge entropy in the lower edge entropy region of the table. The range of edge entropy was such that this relationship clearly fails among the high entropy patterns, however. (The entropy range was intentionally selected to illustrate this effect).
- We conclude with the following speculation: A "transition to texture" occurs as complexity increases beyond our ability to encode it; structure is summarized as texture rather than being directly perceived. Defending this speculation, we expect that some images may exceed our finite visual coding ability, yet we need to be able to process these images in some way — thus, we split the continuum of structural complexity into shape and texture. We hope to explore this speculation in future work.

6. Acknowledgments

Vladik Kreinovich introduced the authors to the important reference [4]. The Common Lisp programming environment (Franz implementation on SGI computers) allowed rapid and painless experimentation.

References

1. C. H. Bennett, "How to define complexity in physics, and why," in *Complexity, Entropy, and the Physics of Information*, W. H. Zurek, ed., pp. 137–148, Addison-Wesley, Redwood City, Ca, 1990.
2. F. Attneave and M. D. Arnoult, "The quantitative study of shape and pattern perception," *Psychological Bulletin*, 53, pp. 453–471, 1956.
3. F. Attneave, "Physical determinants of the judged complexity of shapes," *J. Experimental Psychology*, 53, pp. 221–227, 1957.
4. R. M. Granovskaya, I. Y. Bereznaya, and A. N. Grigorieva, *Perceptions of Form and Forms of Perception*, Lawrence Erlbaum, Hillsdale New Jersey, 1987.
5. D. E. Berlyne, ed., *Studies in the New Experimental Aesthetics*, Wiley, New York, 1974.
6. B. Julesz, "Visual pattern discrimination," *IEEE Trans. Info. Theory*, 8, pp. 84–92, 1962.
7. D. H. Hubel and T. N. Wiesel, "Receptive fields of single neurons in the cat's striate cortex," *J. Physiol.*, 148, pp. 574–591, 1959.
8. D. Marr, *Vision*, W. H. Freeman, San Francisco, 1982.
9. F. Attneave, "Some informational aspects of visual perception," *Psychological Review*, 61, pp. 183–193, 1954.
10. C. H. Bennett, "The thermodynamics of computation—a review," in *Maxwell's Demon: Entropy, Information, Computing*, H. Leff and A. Rex, eds., pp. 213–248, Princeton U. Press, Princeton, 1990.

MAXIMUM ENTROPY TOMOGRAPHY

C. T. MOTTERSHEAD
Los Alamos National Laboratory
AOT-1, MS H808, Los Alamos, NM, USA 87545 [†]

Abstract. Tomography is the process of estimating an unknown two-dimensional density distribution from a set of its one-dimensional projections (views). The ability of Minerbo's [1] maximum entropy tomography (MENT) algorithm to work with a very small number of views makes it especially useful in industrial radiography, accelerator beam diagnostics, and other potential applications. The MENT algorithm is reviewed using a simple continuum derivation of the factored product form for the two dimensional maximum entropy reconstruction function. The projection coordinates for the various views are related by arbitrary linear transformations. The unknown one dimensional Lagrange factor functions that constitute the maximum entropy interpolant are sampled identically to the digital projection data arrays, thus avoiding the geometric issues of pixel projection and underdetermined null-spaces. An iterative procedure is described for solving the large system of coupled non-linear integral equations that result when the product form is substituted into the projection integral constraints. Some of the advantages and costs of using this pure maximum entropy formalism for tomographic reconstruction are illustrated and discussed.

Key words: tomography, diagnostics, integral equations

1. The Geometry of Projection

The projection of a two dimensional function $f(x, y)$ on the x-axis is defined by the integral

$$p(x) = \int dy f(x, y) \tag{1}$$

The limits of integration are finite because we consider only functions of compact support (i.e. those that vanish outside a finite circle). The input data for tomographic reconstruction is a set of such projection integrals onto N different axes (s, t) defined by a set of transformation matrices R_n, for $n = 1, 2, ..., N$:

[†] Email: mottershead@lanl.gov

K. M. Hanson and R. N. Silver (eds.), Maximum Entropy and Bayesian Methods, 425–430.

$$\begin{pmatrix} s \\ t \end{pmatrix} = R_n \begin{pmatrix} x \\ y \end{pmatrix} = \begin{pmatrix} a_n & b_n \\ c_n & d_n \end{pmatrix} \begin{pmatrix} x \\ y \end{pmatrix} = \begin{pmatrix} \cos\theta_n & \sin\theta_n \\ -\sin\theta_n & \cos\theta_n \end{pmatrix} \begin{pmatrix} x \\ y \end{pmatrix} \quad (2)$$

The last form is a simple rotation matrix through angle θ_n, used for spacial reconstruction. The general form could also be beam transport matrices for reconstruction of accelerator phase-space densities from profile monitor data. The projection sample coordinate is s, and the transverse integration coordinate t. The Jacobian of the transformation, giving the area of the (s,t) plane corresponding to a unit area in the (x,y) plane, is $J_n = \det|R_n| = a_n d_n - b_n c_n$. The Jacobian $J_n = 1$ for both rotation and beam transport matrices, but is carried here for the sake of generality. The inverse transformation from the nth view integration coordinates back to the (x,y) reconstruction plane is specified by the pair of functions

$$x = x_n(s,t) = (d_n s - b_n t)/J_n = s\cos\theta_n - t\sin\theta_n \quad (3)$$

$$y = y_n(s,t) = (a_n t - c_n s)/J_n = s\sin\theta_n + t\cos\theta_n \quad (4)$$

In general, the projections may be thought of as a set of one-dimensional views through the unknown two-dimensional function $f(x,y)$. In this notation, the nth projection integral, or "view", is defined by

$$p_n(s) = \int dt\, f[x_n(s,t), y_n(s,t)]. \quad (5)$$

In the reconstruction plane, the sample axis for the nth view maps into the line $t = c_n x + d_n y = 0$, and the integration directions are along lines of constant $s = a_n x + b_n y$.

2. Variational Maximization of Entropy

Inversion of a set of projection data $\{p_n(s), R_n\}$, $n = 1, 2, ...N$, to find $f(x,y)$ is only unique[2] in the limit $N \to \infty$. For some industrial applications[3–5], N may be quite small, perhaps only 3 or 4, so an additional principle is needed to select a unique inverse. The maximum entropy principle argues that the most reasonable distribution function $f(x,y)$ to choose is the one that maximizes the entropy

$$H(f) = -\iint dx\,dy\, f(x,y) \log f(x,y) \quad (6)$$

subject to the projection data constraints of Eq.5.

The rationale[6–9] is that the maximum entropy distribution can be produced in the greatest number of ways, so has the highest *a-priori* probability, and contains the least information, incorporating everything we know, and nothing else. Whatever the rationale, the calculus of variations leads directly to a unique, simple product form for the 2-D function $f(x,y)$ of maximum entropy. Introduce N Lagrange multiplier functions $\lambda_n(s)$ to form the functional

$$\psi(f,\lambda) = H(f) + \sum_{n=1}^{N} \int ds\,\lambda_n(s)\left\{\int dt\, f[x_n(s,t), y_n(s,t)] - p_n(s)\right\}, \quad (7)$$

and demand that it be stationary with respect to variations in both the unknown two-dimensional function $f(x, y)$ and the N unknown one-dimensional functions $\lambda_n(s)$. Remember that $p_n(s)$ is fixed data, independent of the variations in $\lambda_n(s)$ or $f(x, y)$. By construction, the condition $\delta\psi = 0$ under the variation $\lambda_n(s) \rightarrow \lambda_n(s) + \delta\lambda_n(s)$ only reproduces the projection constraint Eqs.5. The variation due to $\delta f(x, y)$ in the entropy term defined by Eq. 6 is straightforward. Collecting the coefficients of $\delta f(x, y)$ from the summation term requires the observation that each of the double integrals on s and t is over the entire plane, and can be mapped back to the entire (x, y) plane by Eq. 2. For the nth mapping, we denote by $s_n(x, y)$ the function giving the value of s corresponding to the reconstruction point (x, y):

$$s = s_n(x, y) = a_n\, x + b_n\, y = x\, \cos\theta_n + y\, \sin\theta_n. \tag{8}$$

After these coordinate transformations, the variation due to $f(x, y) \rightarrow f(x, y) + \delta f(x, y)$ may be written

$$\delta\psi = -\int\!\!\int dx dy \{1 + \log f(x, y) - \sum_{n=1}^{N} J_n \lambda_n[s_n(x, y)]\}\delta f(x, y) \tag{9}$$

The condition that $\delta\psi = 0$ for arbitrary $\delta f(x, y)$ therefore implies

$$\log f(x, y) = \sum_{n=1}^{N} J_n \lambda_n[s_n(x, y)] - 1 \quad \text{or} \quad f(x, y) = \prod_{n=1}^{N} h_n[s_n(x, y)], \tag{10}$$

where the unknown Lagrange multipliers $\lambda_n(s)$ have been replaced by the equally unknown one-dimensional functions $h_n(s) \equiv \exp(J_n \lambda_n(s) - 1/N)$. The arguments of these functions are $s_n = a_n\, x + b_n\, y$, completely specified by the geometry, but their shapes are still to be determined by fitting the projection data.

In computer implementations, the projection data is received as an array of samples for each view. The MENT algorithm has the almost unique virtue that the solution is represented by an identically sampled array of h-factors. The continuous functions $h_n(s)$ are evaluated by simple linear interpolation in s. Decisions about the number of pixels in the final display of $f(x, y)$ are postponed until after the solution is found. There are no undetermined null-spaces.

3. Solving the Constraint Equations

Substitution of the maximum entropy product form Eq.10 into the projection constraint integrals Eqs.5 results in a formidable set of N coupled nonlinear integral equations to be solved for the N unknown one dimensional functions $h_n(s)$:

$$p_n(s) = \int dt \prod_{k=1}^{N} h_k[s_k(x_n, y_n)] \qquad n = 1, 2, ...N. \tag{11}$$

The product form means N views produces an Nth order problem. In general the arguments of the $h_k(s)$ functions inside the integral are

$$s_k[x_n(s, t), y_n(s, t)] = [(a_k d_n - b_k c_n)s + (a_n b_k - a_k b_n)t]/J_n \tag{12}$$

This reduces to $s_k(x_n, y_n) = s\cos(\theta_k - \theta_n) + t\sin(\theta_k - \theta_n)$ in the rotational case. Since this mapping is the identity for $k = n$, we may always factor $h_n(s)$ out of the constraint integral for the nth view, and write

$$p_n(s) = h_n(s)Q_n(s), \quad \text{where} \quad Q_n(s) \equiv \int dt \prod_{k \neq n} h_k[s_k(x_n, y_n)] \qquad (13)$$

is the integral over the other views. This form is the basis of the iterative solution schemes used in the computer implementations of this algorithm. The process begins with an initial guess for the h-factors, usually $h_n(s) = 1$. The $Q_n(s)$ integrals for each view are then computed, and the error $\Delta_n(s) \equiv h_n(s)Q_n(s) - p_n(s)$ is evaluated for all the views. The integration domain is a circle large enough to cover the projections. The $h_n(s)$ are renormalized each iteration to force the mean error, averaged over all views, to zero: $< \Delta > = 0$. A full update step would then be $h_n(s) \rightarrow p_n(s)/Q_n(s)$, but this tends to overcorrect $h_n(s)$. Therefore a damping factor $\gamma < 1$ is used to accept only a fraction of the indicated change in $h_n(s)$, on the grounds that for each iteration, especially the early ones, $Q_n(s)$ is only an estimate:

$$h_n(s) \rightarrow h_n(s) - \gamma\frac{\Delta_n(s)}{Q_n(s)} \qquad (14)$$

At present, trial and error is used to find the critical damping factor sufficient for convergence. More damping seems to be needed for larger N.

Convergence is judged, and the iterations stopped, when either the RMS error residual of the current estimate $<\Delta^2>^{\frac{1}{2}}$, or its rate of decrease, drops below a preset tolerance. This damped iterative procedure differs from Minerbo's original MENT code, which used a fast Gauss-Seidel update scheme where the new h_n's for each view are used immediately to compute the Q_n's for the next view.

The product form Eq.10 is the direct consequence of maximizing entropy in the noiseless limit. But it gives all views a kind of veto power over any given view that could be exploited for noise rejection logic. For example, if $Q_n(s) = 0$ for a particular sample, it means that the current estimate of $h_n(s)$ for the other views exclude the possibility of a non-zero signal in $p_n(s)$. So if $p_n(s) \neq 0$, it must either be a noise spike, or the other views are wrong.

4. Performance of the Algorithm

Tomographic reconstruction produces images $f(x, y)$ that may be represented by a set of M pixels $\{p_i | i = 1, M\}$. If these are treated as a probability distribution, normalized to $\sum p_i = 1$, the image entropy is $H = -\sum p_i \log p_i$. This entropy is a measure on the grey-level histogram of the image, and reaches the maximum possible value of $H = \log M$ for the uniform distribution $p_i = 1/M$. The pixels of a given grey-level could be scattered randomly over the image, or all swept onto one corner, without affecting the entropy. An M-pixel image with only a fraction F of the pixels non-zero can have a maximum entropy of $H = \log FM$ when all the non-zero pixels are the same. Solving for the effective occupation fraction $F \equiv \exp H/M$ provides a convenient, M-independent, measure of the image entropy H. The unconstrained maximization of entropy of the full M-pixel image always leads to a uniform distribution with F=100%. The constrained problem leads to distributions of lesser entropy with shapes determined by the observed projections.

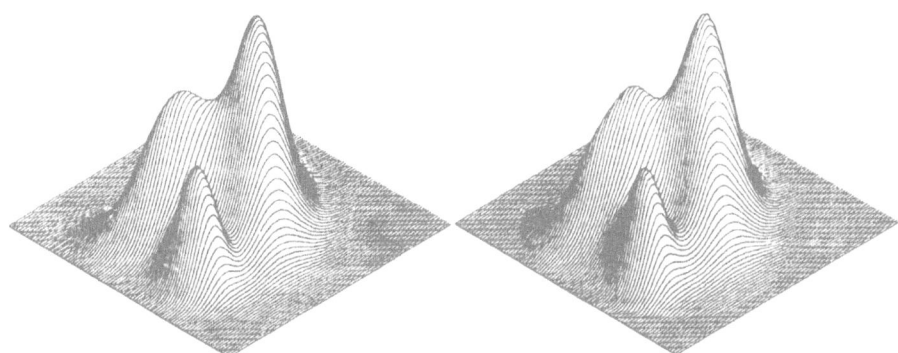

Figure 1.   Original and Reconstructed Distributions for a Smooth Multigaussian Example.

Figure 1 shows isometric plots of a simple smooth multigaussian distribution (left), and its reconstruction from six views (right). The reconstruction fidelity is very good in this easy case, with F=40% for both the original and the reconstruction.

Figure 2 shows the original distribution, and a representative projection, for a more difficult post and ring problem. The reconstructions shown in Fig. 3 recover the main features of the original. Their rippled surface is not simply noise, nor failure to converge. It is an artifact of the product form when the projections have the sharp peaks shown in Fig. 2. The six view reconstruction (left) has F=67.3%, and the more constrained twelve view (right) has F=64.7%, both higher than the F=58.8% of the original, which has a very narrow, single grey-level, histogram. Whether this higher entropy really implies greater *a-priori* probability is left to the speculation of the reader.

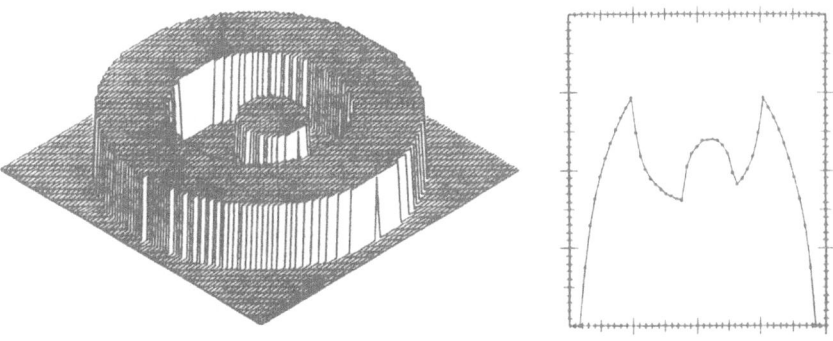

Figure 2.   Original Image, and Representative Projection, for Post and Ring Problem.

Figure 3.   Six-view and Twelve-view Reconstructions of the Post and Ring.

Acknowledgments

This work was supported by a series of programs at Los Alamos National Laboratory. At the beginning of these, G. Minerbo provided copies of his code and many helpful discussions.

References

1. G. N. Minerbo, "MENT: A Maximum Entropy Algorithm for Reconstructing a Source from Projection Data," Comp. Graphics Image Proc. **10**, pp. 48-68, 1979.
2. Johann Radon, "Uber die Bestimmung von Funktionen durch irhe Integralwerte langs gewisser Mannigfaltigkeiten", Ber. Verh. Sachs. Akad. Wiss. Leipzig, Kl. **69**, pp. 262-277, 1917.
3. O. R. Sander, G. N. Minerbo, R. A. Jameson, and D. D. Chamberlin, "Beam Tomography in Two and Four Dimensions," Proc. 1979 Linac Conf., Brookhaven National Laboratory report BNL-51134, pp. 314-318, 1980.
4. C. T. Mottershead, "A Tomographic Streak Camera for Ultrahigh-Speed Movies," IEEE Trans. Nucl. Sci. **29**, pp. 900-903, 1982.
5. C. T. Mottershead, " Maximum Entropy Beam Diagnostic Tomography", IEEE Trans. Nuc. Sci. **NS-32**, pp. 1970-1972, 1985.
6. R. E. Turner and D. S. Betts, Chap. 3, pp. 16-26 in *Introductory Statistical Mechanics*, Sussex University Press, 1974.
7. E. T. Jaynes, "On the Rationale of Maximum-Entropy Methods," Proc. IEEE **70**, pp. 939-952, 1982.
8. E. T. Jaynes, "Prior Probabilities," IEEE Trans. Systems Sci. and Cybernetics, **SSC-4**, pp. 227-241, 1968.
9. E. T. Jaynes, "Where Do We Go From Here?", pp. 21-58, in *Maximum-Entropy and Bayesian Methods in Inverse Problems*, C. Ray Smith and W.T. Grandy,Jr. (eds.), Reidel 1985.

BAYESIAN REGULARIZATION OF SOME SEISMIC OPERATORS

M. D. SACCHI AND T. J. ULRYCH

Dept. of Geophysics and Astronomy
The University of British Columbia
2219 Main Mall, Vancouver, Canada, V6T 1Z4 [†]

Abstract. We present a Bayesian approach to estimate the discrete Fourier and Radon transforms. We show that limited aperture can be compensated by introducing a prior that mimics a sparse distribution of parameters.

Key words: Regularization, inversion, Cauchy-Gauss model, 2-D DFT, Radon transform, resolution

1. Introduction

A common approach to signal analysis and decomposition is based on mapping the data into a new domain where the support of each signal is considerably reduced and, consequently, decomposition can be easily attained. This is applicable to the discrete Fourier transform and to the Radon transform [1], operators which are often used in seismic processing. Since we are always concerned with a finite amount of data the correct decomposition of seismic events is complicated by sidelobe artifacts [2].

We show that, in general, these type of problems can be written down as linear inverse problems where the correct regularization is crucial to the resolution of seismic signals when the aperture of the array of receivers is below the resolution limit. The regularization is obtained by incorporating into the problem a "long tailed" prior using Bayes' rule.

2. Linear inversion of the discrete Fourier transform

2.1. DEFINITION OF THE PROBLEM

For simplicity we start with the 1-D DFT since extensions to higher dimensions are straightforward. Consider a N-sample time or spatial series $x_0, x_1, x_2, \ldots, x_{N-1}$. The DFT of the discrete series is given by

---

[†]Email: sacchi@geop.ubc.ca , ulrych@geop.ubc.ca

K. M. Hanson and R. N. Silver (eds.), Maximum Entropy and Bayesian Methods, 431–436.

$$X_k = \sum_{n=0}^{N-1} x_n e^{-i2\pi nk/N} \quad k = 0, \ldots, N-1, \tag{1}$$

and similarly, the inverse DFT is given by

$$x_n = \frac{1}{N} \sum_{k=0}^{N-1} X_k e^{i2\pi nk/N} \quad n = 0, \ldots, N-1. \tag{2}$$

We wish to estimate M spectral samples where $M > N$. A standard approach to solving this problem is by means of zero padding. Defining a new time series consisting of the original series plus a zero extension for $n = N, \ldots, M-1$, we can estimate M spectral samples using the DFT. This procedure helps to remove ambiguities due to discretization of the Fourier transform but, as is well know, it does not reduce the sidelobes created by the temporal/spatial window or improve the resolution. Let us therefore consider the estimation of M spectral samples without zero padding. In other words we want to estimate the DFT using only the available information. The underlying philosophy is similar to Burg's maximum entropy method (MEM) [3] except for the fact that in the MEM the target is a PSD estimate, a phase-less function.

To avoid biasing our results by the discretization we also impose the condition $M \gg N$. Rewriting equation (2) as

$$x_n = \frac{1}{M} \sum_{k=0}^{M-1} X_k e^{i2\pi nk/M} \quad n = 0, \ldots, N-1, \tag{3}$$

gives rise to a linear system of equations

$$\mathbf{y} = \mathbf{F}\mathbf{x}, \tag{4}$$

where the vectors $\mathbf{y} \in \mathbf{R}^N$ and $\mathbf{x} \in \mathbf{C}^M$ denote the available information and the unknown DFT, respectively. Equation (4) represents an ill-posed problem. Uniqueness is imposed by defining a regularized solution, $\hat{\mathbf{x}}$, which is obtained solving the problem [4] expressed by

$$min\{J(\mathbf{x}) = \Phi(\mathbf{x}) + ||\mathbf{y} - \mathbf{F}\mathbf{x}||_2^2\}. \tag{5}$$

In the next section we explore a Bayesian approach to compute the regularizer.

2.2. ZERO ORDER QUADRATIC REGULARIZATION

We consider data contaminated with noise which is distributed as $N(0, \sigma_n^2)$. We also assume that the samples of the DFT may be modelled with a gaussian prior. After combining the likelihood with the prior probability of the model by means of Bayes' rule, the MAP solution is computed by minimizing the following objective function:

$$J_{gg}(\mathbf{x}) = \lambda ||\mathbf{x}||_2^2 + ||\mathbf{y} - \mathbf{F}\mathbf{x}||_2^2, \tag{6}$$

where the subscript gg stands for the Gauss-Gauss model (Gaussian model and Gaussian errors). The scalar is the ratio of variances, $\lambda = \sigma_n^2/\sigma_x^2$. Equation (6) is the objective function of the problem. The first term represents the model norm while the second term is the misfit function. Taking derivatives and equating to zero yields

$$\hat{\mathbf{x}} = (\mathbf{F}^H \mathbf{F} + \lambda \mathbf{I}_M)^{-1} \mathbf{F}^H \mathbf{y}. \tag{7}$$

We can write equation (7) in another form using the following identity

$$(\mathbf{F}^H \mathbf{F} + \lambda \mathbf{I}_M)^{-1} \mathbf{F}^H = \mathbf{F}^H (\mathbf{F}\mathbf{F}^H + \lambda \mathbf{I}_N)^{-1}, \tag{8}$$

where $\mathbf{I}_M$ and $\mathbf{I}_N$ represent $M \times M$ and $N \times N$ identity matrices, respectively. Recalling equation (11) and that $\mathbf{F}\mathbf{F}^H = \frac{1}{M}\mathbf{I}_N$,

$$\hat{\mathbf{x}} = (\frac{1}{M} + \lambda)^{-1} \mathbf{F}^H \mathbf{y}. \tag{9}$$

The result is simply the DFT of x_n modified by a scale factor. The solution expressed by equation (9) becomes

$$\hat{X}_k = \frac{1}{1 + \lambda M} \sum_{n=0}^{N-1} x_n e^{-i2\pi nk/(M-1)}. \tag{10}$$

Equation (10) represents the DFT of the windowed time series and is equivalent to padding with zeros in the range $n = N, \ldots, M - 1$. It is clear that the zero order regularization yields a scaled version of the conventional DFT and the associated periodogram will show the classical sidelobes artifacts due to truncation.

2.3. REGULARIZATION BY THE CAUCHY-GAUSS MODEL

We propose the use of a regularization derived from a pdf that mimics a sparse distribution of spectral amplitudes. A "long tailed" distribution, like the Cauchy pdf, will induce a sparse model consisting of only a few elements different from zero. If the data consist of a few number of harmonics (1-D case) or a limited number of plane waves (2-D case) a sparse solution will help to attenuate sidelobes artifacts. The Cauchy pdf is given by

$$p(X_k|\sigma_c) \propto \frac{1}{(1 + \frac{X_k X_k^*}{2\sigma_c^2})}, \tag{11}$$

where σ_c is a scale parameter. When we combine the Cauchy prior with the data likelihood the cost function becomes

$$J_{cg}(\mathbf{x}) = S(\mathbf{x}) + \frac{1}{2\sigma_n^2}(\mathbf{y} - \mathbf{F}\mathbf{x})^H(\mathbf{y} - \mathbf{F}\mathbf{x}), \tag{12}$$

where the subscript cg stands for the Cauchy-Gauss model. The function, $S(\mathbf{x})$, which is expressed by

$$S(\mathbf{x}) = \sum_{k=0}^{M-1} \log(1 + \frac{X_k X_k^*}{2\sigma_c^2}) \tag{13}$$

is the regularizer imposed by the Cauchy distribution and is a measure of the sparseness of the vector of spectral powers $P_k = X_k X_k^*$, $k = 0, \ldots, M-1$. The constant σ_c controls the amount of sparseness that can be attained by the inversion.

Taking derivatives of $J_{cg}(\mathbf{x})$ and equating to zero yields the following result

$$\hat{\mathbf{x}} = (\lambda \mathbf{Q}^{-1} + \mathbf{F}^H \mathbf{F})^{-1} \mathbf{F}^H \mathbf{y}, \tag{14}$$

where $\lambda = \sigma_n^2/\sigma_c^2$ and $\mathbf{Q}$ is a $M \times M$ diagonal matrix with elements given by

$$Q_{ii} = 1 + \frac{X_i X_i^*}{2\sigma_c^2}, \quad i = 0, \ldots, M-1. \tag{15}$$

Using an identity similar to equation (8), the solution may be written as

$$\hat{\mathbf{x}} = \mathbf{Q} \mathbf{F}^H (\lambda \mathbf{I}_N + \mathbf{F} \mathbf{Q} \mathbf{F}^H)^{-1} \mathbf{y}. \tag{16}$$

We stress that, although from the theoretical point of view of uniqueness and convergence, the operators given by equations (14) and (16) are equivalent, from the point of view of computational advantages, the following observations apply. 1)- Whereas equation (14) demands the inversion of a $M \times M$ matrix, equation (16) requires the inversion of a $N \times N$ matrix. 2)- The operator $(\lambda \mathbf{I}_N + \mathbf{F} \mathbf{Q} \mathbf{F}^H)$ in equation (19) is Toeplitz Hermitian provided that the time series is uniformly discretized and a fast solver like Levinson's recursion can be used.

The hyper-parameters of the problem are fitted by means of a χ^2 criterion. Equation (16) is iteratively solved. The initial model is the DFT of the truncated signal. In general, about 10 iteration are necessary to minimize J_{cg}.

2.4. HIGH RESOLUTION 2-D SPECTRAL ANALYSIS

We present a hybrid procedure based on standard Fourier analysis in the temporal variable while, for the spatial variable, we use the Cauchy-Gauss model to retrieve the wavenumbers.

The algorithm is applied to estimate the spatio-temporal spectrum of a signal received by a passive array of receivers.

We model two sinusoids with unit amplitude and with normalized wavenumbers of 0.30 and 0.25 units and normalized frequency of 0.20 units, respectively. The third wave ($f = 0.35, k = -0.25$) has an amplitude which is 25% below the amplitude of the first and second waves. The temporal extension of each channel is 150 samples, which represents one order of magnitude above the aperture of the array (15 receivers). Gaussian noise with standard deviation, $\sigma_n = 0.1$ was added to the composite record. The noise represents 40% of the amplitude of the third wave. The spatio-temporal spectrum computed using the periodogram is illustrated in Figure 1a. The $f - k$ plane is dominated by sidelobes due to truncation in space and time. This is more noticeable for the wavenumber, since

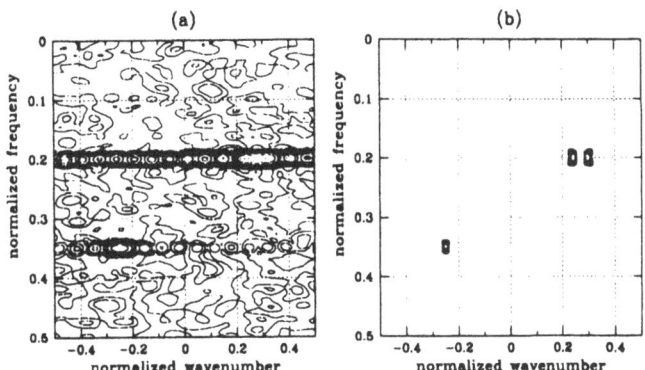

Figure 1. 2-D spectrum of three narrow band signals of normalised frequency-wavenumber pairs: (0.2, 0.3), (0.2, 0.25) and (0.35, −0.25). a) Conventional 2-D estimator obtained with DFT. b) 2-D estimator obtained with the Cauchy-Gauss model. The contour lines correspond to normalised amplitudes ranging from 0 to −40db, with an interval of −5db.

the aperture of the array is one order smaller that the length of the time series. The data were processed with the hybrid procedure based on the Cauchy-Gauss model. The parameters σ_n and σ_c were chosen to reject the noise. The resulting high resolution $f - k$ panel is portrayed in Figure 1b. There is a clear enhancement of the spatial resolution and a suppression of the background noise.

3. Inversion of the Radon operator

Given a wavefield, $d(x, t)$, composed by a linear superposition of plane waves. The Radon transform, $m(p, \tau)$, where τ denotes time and p the ray parameter, may be obtained by solving the following problem

$$d(x, t) = \sum_p m(p, t = \tau + x\, p),\qquad(17)$$

equation (17) may be broken down into several problems in the space-frequency domain $(x - \omega)$,

$$d(x, \omega) = \sum_p m(p, \omega)e^{-i\omega\, p\, x}.\qquad(18)$$

In matrix form (ω is omitted),

$$\mathbf{d} = \mathbf{Lm}.\qquad(19)$$

A sparse prior can also be used in this context. The reason is very simple. Since a limited number of wavefields is expected, a sparse regularisation will retrieve a solution consisting mainly of isolated points in the $\tau - p$ domain. The latter helps to sharpen the seismic image, contrary to the zero order quadratic regularization that introduces amplitude smearing.

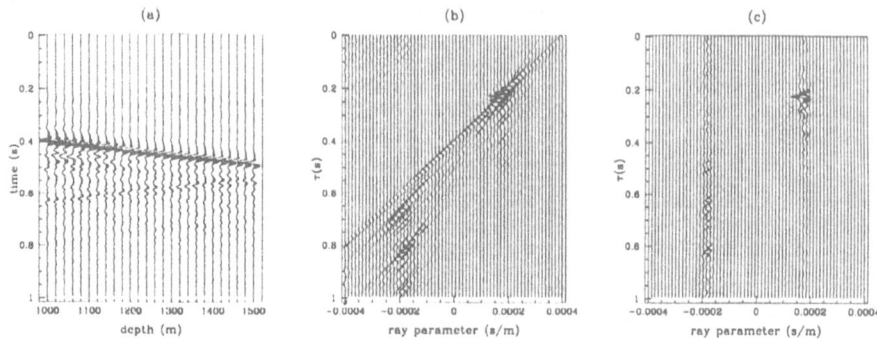

Figure 2. Wavefield (a) acquired in a VSP experiment. The Radon operator is used to decompose the wavefield. The operator is derived using zero order quadratic regularisation (a) and the Cauchy-Gauss regularisation (b).

3.1. APPLICATION TO VERTICAL SEISMIC PROFILING (VSP)

In vertical seismic profiling, the receivers are located in a vertical well log. A seismic source generates a broad-band signal that excites the receivers. The goal in VSP is to estimate the variation of some geophysical parameters, e.g., compressional and shear waves velocities, using the waveform as primary information. The VSP data are composed of two principal linear wavefields: the down-going (wave traveling from the source to the receivers) and the up-going waves (reflected from geological interfaces). The data in Figure 2a correspond to a spatio-temporal window of a VSP experiment. We first compute the Radon transform using zero-order quadratic regularization and then by means of the Cauchy regularization. The results are portrayed in Figures 2b and 2c, respectively. It is clear from a comparison of the $\tau-p$ panels that the Cauchy regularization has managed to attenuate the smearing introduced by the limited aperture.

4. Summary

The algorithm described here provides an interesting approach to limited aperture problems. The sparse prior, incorporated by means of Bayes' theorem, allows us to construct a solution consisting of a few nonzero events, a feature that is consistent with the expected model which consists of a limited number or harmonics or plane-waves.

References

1. S. R. Deans, *The Radon transformation and its applications*, John Wiley and Sons, 1983.
2. S. M. Marple, *Digital spectral analysis, with applications*, Prentice-Hall, NY, 1987.
3. J. P. Burg, *Maximum entropy spectral analysis*, Ph.D thesis, Stanford University, 1975.
4. A. H. Tikhonov and A. V. Goncharsky, *Ill-posed problems in the natural sciences*, MIR, Moscow, 1987.

MULTIMODALITY BAYESIAN ALGORITHM FOR IMAGE RECONST
IN POSITRON EMISSION TOMOGRAPHY

SRIKANTH SASTRY
Physical Sciences Laboratory
Division of Computer Research and Technology †

JOHN W. VANMETER
Laboratory of Neuroscience, National Institute of Aging

AND

RICHARD E. CARSON
Positron Emission Tomography Department, Clinical Center
National Institutes of Health, Bethesda, MD 20952

Abstract. Bayesian reconstruction algorithms offer a method for improving the image quality for Positron Emission Tomography. We present a new method which follows the overall strategy of Bayesian image reconstruction with priors, but without assumptions of global or piecewise smoothness, which are not satisfied in practice. Specifically, we use segmented Magnetic Resonance (MR) images to assign tissue composition to each reconstruction pixel. The reconstruction is performed as a constrained expection–maximization of activities of each tissue type, applying a prior based on the MR segmentations. The algorithm is tested in realistic simulations employing a full physical scanner model.

Key words: Positron Emission Tomography, Magnetic Resonance images, Multimodality Image Reconstruction, Expectation Maximization, image segmentation

1. Introduction

The Expectation Maximization (EM) algorithm for Maximum Likelihood (ML) image reconstruction in Positron Emission Tomography (PET) has been widely studied [1]. While the EM algorithm is flexible in its ability to incorporate the physical and statistical model of projection measurements [2], the algorithm is slow to converge and suffers further from an increase in noise with increasing number of iterations (see, e.g., [3]). Bayesian reconstruction algorithms based on a variety of priors have been studied to extend and regularize the EM-ML method.[1]

---

† Email: sastry@nih.gov
[1] In the more general context of maximum *a posteriori* estimation, see [4].

K. M. Hanson and R. N. Silver (eds.), Maximum Entropy and Bayesian Methods, 437–441.
© *1996 Kluwer Academic Publishers.*

Most of these methods impose smoothing priors or model the image as composed of uniform regions separated by well-defined edges. Neither of these idealizations is satisfactory in practice, since tissue composition and hence activity levels vary on a scale comparable to typical pixel sizes.

An appealing approach to extend the EM method to suppress noise and enhance resolution is multimodal image reconstruction, i.e., PET reconstruction using data from other modalities, such as Magnetic Resonance Imaging (MRI) (see e.g., [5–7]). The use of MR images is particularly desirable because of the high resolution of these anatomical images.

2. The Algorithm

We have developed an algorithm designed to exploit prior information in the form of segmented MR images of the brain. In addition to the PET projections, the algorithm requires MR segmentations registered with the PET image volume, assigning to each pixel its tissue composition (gray matter (GM), white matter (WM), cerebrospinal fluid (CSF) or other (Other)). The algorithm is based on the model that the activity levels at each PET pixel can be represented as a weighted sum of activity levels of the constituent tissue types. Thus,

$$\lambda_j = \sum_{n=\text{GM,WM,CSF,Other}} \lambda_{jn} p_{jn} \tag{1}$$

where λ_j is the total activity at pixel j, λ_{jn} are the activity levels of tissue types n, and p_{jn} are weights derived from probabilistic MR segmentations, with $\sum_n p_{jn} = 1$.

In the unconstrained EM algorithm, the posterior density for pixel activities λ_j, given the projection measurements Y_i is equal to the the likelihood function, since no prior is assumed. The log-likelihood function, given in terms of the Poisson distribution for the projection counts, is maximized iteratively (see e.g., [8]). In a Bayesian reconstruction, the sum of the log-likelihood and the logarithm of the specified prior are maximized iteratively.

The activity levels of gray matter, white matter and the cerebrospinal fluid are known to be very different. Hence, in defining our prior, we assume that the activity levels of each tissue type are narrowly distributed compared to the overall variability. Since the MR segmentations offer information regarding the tissue composition of each image pixel, we perform the reconstruction for the activity levels λ_{jn} for each tissue type n, constraining the individual values to be close to the global mean activity level $\bar{\lambda}_n$. The appropriate prior has the form $exp(-U(\lambda))$ where,

$$U(\lambda) = \sum_{j,n} \frac{\left(\lambda_{jn} - \bar{\lambda}_n\right)^2}{\sigma_{jn}^2}. \tag{2}$$

Here, $\sigma_{jn}^2 = c_{jn}\sigma_n^2$, where c_{jn} is a scaling factor. In general, σ_n^2 should be a function of the mean activity values $\bar{\lambda}_n$. In the present study, we have choosen

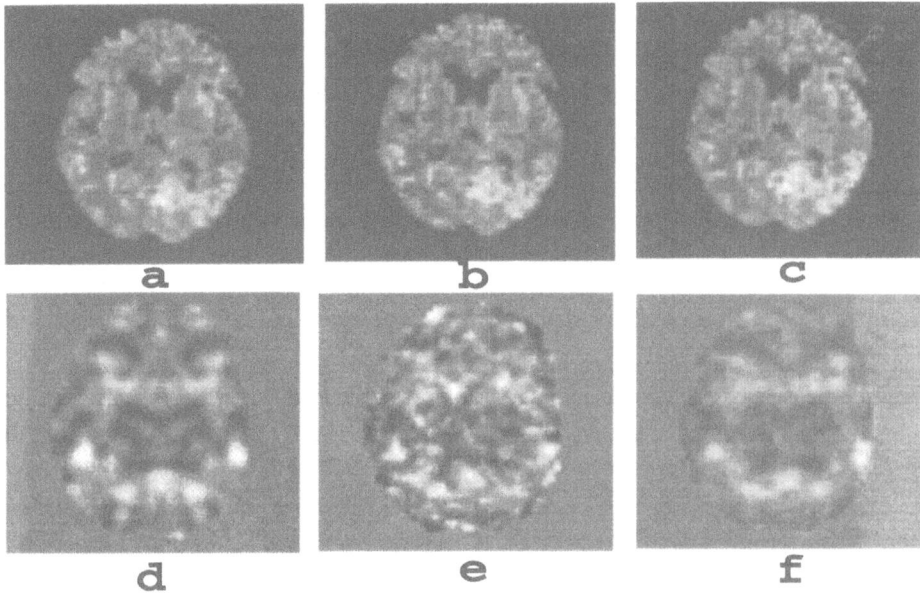

Figure 1.   Comparison of true and reconstructed images. (a) True image, (b) Reconstruction for one projection sample, (c) Average of reconstructed images for 10 independent projection samples, (d) True gray matter profile, (e) Reconstructed grey matter profile for one projection sample, and (f) Average of reconstructed gray matter profiles for 10 independent projection samples.

$\sigma_n^2 = K\overline{\lambda}_n$, so that the degree of variability allowed to each tissue type is set through an arbitrary parameter K. By maximizing the quantity,

$$E' \equiv E\left(lnf|Y,\lambda^{(m)}\right) - U(\lambda) \tag{3}$$

where $E\left(lnf|Y,\lambda^{(m)}\right)$ is the log-likelihood which is maximized in the unconstrained EM-ML algorithm, with Y representing the projections, $\lambda^{(m)}$ the reconstructed activities at the mth iteration, we obtain the iterative equations,

$$\lambda_{jn}^{(m)} = \overline{\lambda}_n(1 - K/2) + \left(\overline{\lambda}_n^2(1 - K/2)^2 + 2K\overline{\lambda}_n\lambda_{jn\ EM}^{(m+1)}\right)^{1/2}, \tag{4}$$

where $\lambda_{jn\ EM}^{(m+1)}$ is the updated activity one obtains using the EM algorithm. This iterative equation clearly displays how K determines the balance between the activities obtained without a prior and the constraint imposed by the prior. The above equation yields the updated values of activity levels for individual tissue types. The overall activity at a pixel is obtained by using Eq. (1). Thus, during the reconstruction, we obtain the overall image as well as the activity profiles of each tissue type.

3. Simulations

We have tested the algorithm in realistic simulations of the brain where the "true" image was constructed from segmented human MR images by assigning activity profiles to each tissue type, (with the ratio of activities GM:WM:CSF:Other = 4:1:.005:0) and applying the full projection model of the Scanditronix PC2048-15B PET scanner incorporating finite resolution, attenuation, scatter and random counts, and generating projections with roughly one million counts per slice. In the simulations, the activity profiles of individual tissue types are not assumed to be constant, but instead the activity varies from pixel to pixel as a smoothed random field with a standard deviation of $\equiv 10\%$.

The reconstructed images as well as individual tissue activity level profiles compare favorably to the known "true" images. Figure 1 shows the true image and the reconstructed images (both total and gray matter) in a simulation with K = 60. Figure 2(a) shows reconstructed gray matter activity levels plotted against true activity levels for a series of K values. These data are obtained from the reconstruction of a single projection from the "true" image, by grouping image pixels according to the true activity level and averaging the activities in those pixels in the reconstructed image. It is seen that the two sets are strongly correlated and within the range of K values shown, the correlation increases with K. This is an expected result since a lower K value implies a stronger constraint, resulting in a flatter activity profile within each tissue type and greater bias.

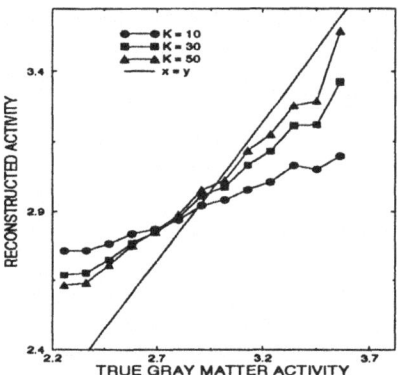 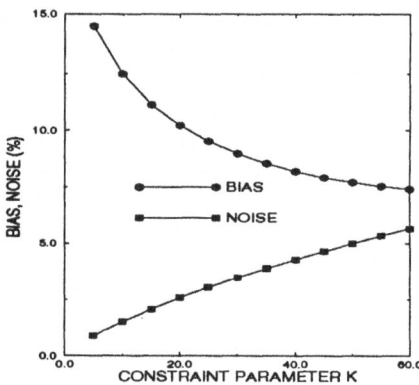

Figure 2. (a) Reconstructed gray matter activity levels *vs.* true activity levels for K=10, 30, 50. The straight line (x = y) is plotted for easy comparison. (b) Bias of the averaged image (of 10 samples) with respect to the true image (open circles) and the noise (open squares).

In Figure 2(b) we show the bias of the reconstructed image from the true image

$$d_t = \left[(1/N_{pixels}) \sum_j (\lambda_j^{rec} - \lambda_j^{true})^2 \right]^{(1/2)} (100\%/\lambda_{ave}^{true}). \qquad (5)$$

These data are generated by simulating a sample of 10 noisy realizations from the same "true" image, performing reconstruction on each of these projections, and obtaining the average of the reconstructed activity at each pixel. It is seen that the bias decreases with increasing K. In Figure 2(b) we also show, for the sample of 10 realizations, the *rms* deviation of the sample, d_s, from the mean reconstructed activities (d_s is defined similarly to d_t). The *rms* deviation of the sample increases with increasing K indicating the increasingly noisy nature of the reconstructed images as the strength of the constraint is reduced. Thus, a proper choice of the K value has to be made to achieve an appropriate tradeoff between bias and noise.

4. Summary

We have presented a Bayesian reconstruction algorithm for PET which utilizes segmented MR anatomical data to define a prior density for the image activities. The defined prior does not make any assumptions about the smoothness of the expected images and instead imposes a constraint of the overall variability of activity levels for individual tissue types. The algorithm performs very satisfactorily in terms of the reconstructed images, but it is found that the convergence of the images is quite slow.

Goals presently being pursued in developing and applying this algorithm are: (i) study and optimization of convergence, (ii) dependence of image quality, bias, and variability on the global activity level constraint, (iii) reliability of the algorithm when abnormal activity levels exist in restricted parts of the image, and (iv) the effect of registration errors between MR and PET image volumes. We have also applied the algorithm to human PET projection data. An issue presently under study is the processing of segmentations from the MR image volumes to account for the differences in axial resolution between the two modalities.

References

1. L. A. Shepp and Y. Vardi, *IEEE Trans. Med. Imag.* **MI-1**, pp. 113-122, 1982.
2. R. E. Carson, Y. Yan, B. Chodowski, T. K. Yap and M. E. Daube-Witherspoon, *IEEE Trans. Med. Imag.* **13**, pp. 526-537, 1994.
3. D. Snyder, M. Miller, L. Thomas, and D. Politte, *IEEE Trans. Med. Imag.* **6**, pp. 98-105, 1987.
4. S. Geman and D. Geman, *IEEE Trans. Pattern Anal. Machine Intel.* **6**, pp. 721-740, 1984.
5. G. Gindi, M. Lee, A. Rangarajan, and I. G. Zubal, *IEEE Trans. Med. Imag.* **12**, pp. 670-680, 1993.
6. X. Ouyang, W. H. Hong, V. E. Johnson, X. Hu and C.-T. Chen, *IEEE Trans. Med. Imag.* **13**, pp. 627-40, 1994.
7. R. Leahy and X. Yan, pp. 105-120, in *Information processing in medical imaging*, A. C. F. Colchester and D. J. Hawkes (eds.), New York, Springer, 1991.
8. Martin A. Tanner, *Tools for Statistical Inference*, Springer-Verlag (1993).

Sibusiso Sibisi's wife, Kay, with her baby girl, Nomvelo, and Jennifer Skilling on the balcony.

EVIDENCE INTEGRALS

W. VON DER LINDEN, R. FISCHER AND V. DOSE
Max-Planck-Institut für Plasmaphysik, EURATOM Association
D-85740 Garching b. München, Germany ¶

Abstract. Evidence integrals are a key ingredient of quantified maximum entropy (QME). They allow to evaluate hyper-parameters which in turn determine the amount of regularization, the noise level, confidence intervals etc. Consequently, the exact evaluation of these multi-dimensional integrals is of central importance to the theory. Since the conventional 'steepest descent' (SD) approximation fails frequently a more refined approach is needed. Using diagrammatic techniques we derive a correction factor to the SD result which is readily incorporated in existing QME codes and which provides significantly improved results.

Key words: Quantified MaxEnt, evidence integral, regularization parameter

1. Introduction

From the theoretical point of view QME [1,2] is a straight forward application of Bayesian probability theory. The numerical implementation, however, is not that simple since multi-dimensional *evidence integrals* are required. We consider an inferential problem summarized by $\mathcal{I}$ and characterized by certain *hyper-parameters* λ, such as the regularization parameter α, the width of the preblur [2], and the noise level. In QME the *prior* probability $p(f|\lambda\mathcal{I}) \propto \exp(\alpha S)$ for the image f depends on the entropy S. We restrict the discussion to a Gaussian *likelihood* function $p(D|f\lambda\mathcal{I}) \propto \exp(-\frac{1}{2}\chi^2)$, where χ^2 measures the discrepancy between data D and fit. Consequently, the joint posterior of the image and the hyper-parameters is of the form $p(\lambda f|D\mathcal{I}) = c(\lambda)\exp\phi(f,\lambda)$. According to the Bayesian sum-rule the evidences for the data $p(D|\mathcal{I})$ and for the hyper-parameters $p(\lambda|D\mathcal{I})$, respectively, are proportional to the multi-dimensional evidence integrals in image space

$$I = \int_0^\infty e^{\phi(f,\lambda)}\mathcal{D}f = \int_0^\infty e^{\phi(f,\lambda)} \prod_i \frac{df_i}{\sqrt{f_i}} \qquad . \tag{1}$$

In general such integrals cannot be determined analytically. Several numerical approaches to 'add up' have been proposed and analyzed by Skilling[3]; so far,

---

¶e-mail: wvl@ibmop5.ipp-garching.mpg.de

K. M. Hanson and R. N. Silver (eds.), Maximum Entropy and Bayesian Methods, 443–448.

however, with limited success. The commonly employed SD comprises three approximations. a) ϕ is expanded to second order about its maximum f^* (Gaussian approximation), b) it is assumed that $\exp(f)$ is sharply peaked at positive f^* so that the lower integration limits can be extended to $-\infty$, and c) the integration measure is replaced by a constant and extracted from the integral

$$I^{\text{SD}} := \frac{e^{\phi(f^*,\lambda)}}{\prod_i \sqrt{f_i^*}} \int_{-\infty}^{\infty} e^{\Delta f^T (A+B)\Delta f} df^N = \frac{e^{\phi(f^*,\lambda)} \pi^{N/2}}{\prod_i \sqrt{f_i^*}} \det^{-\frac{1}{2}}(A+B) \qquad . \quad (2)$$

We introduced the abbreviation $\Delta f = f - f^*$ and split the Hessian $-\frac{1}{2}\nabla\nabla^T \phi|_{f^*}$ into the diagonal matrix A and the residual trace-zero matrix B. The SD approximation has the advantage of providing an analytic expression for the evidence integrals that can be evaluated with little numerical effort. While the Gaussian approximation has proven to be reliable, tampering with the integration limits and measure has the following negative impact. Due to the divergent prefactor $(\prod_i f_i^{*-1/2})$ and the modified integration limits, images with values close to zero are preferred which is accomplished by less regularization, i.e. small α. For the same reason confidence intervals are in general strongly overestimated.

2. Diagrammatic Approach

To overcome these shortcomings we propose a diagrammatic approach providing a correction factor $C = I/I^{\text{SD}}$ to the SD result which can be implemented into existing QME codes with little effort. The impatient reader who is merely interested in a reliable correction formula for his QME code can directly jump to Eq.13,15. We start out by writing the Taylor expansion of the exponent ϕ as

$$\phi = \phi|_{f^*} - \Delta f^T B \Delta f + \sum_i [R(f_i, f_i^*) - A_{ii}\Delta f_i^2] \qquad . \quad (3)$$

Only the second order terms, to which the Gaussian approximation is restricted, introduce correlations between image points f_i and f_j. Higher order contributions originate from the entropy and are 'diagonal' in f, expressed by the special form $\sum_i R(f_i, f_i^*)$. The evidence integral is recast into

$$I(B) = e^{\phi(f^*,\lambda)} \int_0^{\infty} e^{-\Delta f^T B \Delta f} \prod_i e^{R(f_i,f_i^*)-A_{ii}\Delta f_i^2} \mathcal{D}f \qquad . \quad (4)$$

The analytic evaluation of this integral is hampered by the integration limits and the image-space measure. The correction factor reads

$$C := \frac{I(B)}{I^{\text{SD}}} = \frac{I(B)}{I(0)} \det^{\frac{1}{2}}(\mathbb{1} + BA^{-1}) \prod_i \kappa_i \quad (5)$$

$$\kappa_i := \sqrt{\frac{A_{ii}f_i^*}{\pi}} \int_0^{\infty} \mathcal{D}f_i \, e^{R(f_i,f_i^*)-A_{ii}(f_i-f_i^*)^2} \quad (6)$$

The SD approximation becomes progressively reliable as $f_i^* \to \infty$ and $C \to 1$ in this limit. The one-dimensional integrals in Eq.6 can be determined easily by numerical means or even analytically as we will see below. The difficult part is

$$\frac{I(B)}{I(0)} = \int_0^\infty e^{-\Delta f^T B \Delta f} \underbrace{\prod_i \frac{e^{R(f_i, f_i^*) - A_{ii} \Delta f_i^2}}{\int_0^\infty \mathcal{D}f'_i \, e^{R(f'_i, f_i^*) - A_{ii} \Delta f_i^2}} \mathcal{D}f}_{\rho(f)} =: \langle e^{-\Delta f^T B \Delta f} \rangle_\rho \quad (7)$$

The distribution function $\rho(f)$ is positive definite and normalized to one. Furthermore, due to the product form of $\rho(f)$, $\langle \Delta f_i \Delta f_j \rangle_\rho = \langle \Delta f_i \rangle_\rho \langle \Delta f_j \rangle_\rho, \forall i \neq j$. The particular form of Eq.7 prompted us to employ diagrammatic techniques [1]. The Taylor expansion of the exponential yields an N-th order contribution

$$\left\langle \left(\sum_{\substack{ij \\ =1}}^{L} \Delta f_i B_{ij} \Delta f_j \right)^N \right\rangle_\rho = \left\langle \sum_{\substack{i_1, j_1, \ldots, i_N, j_N \\ =1}}^{L} \Delta f_{i_1} B_{i_1 j_1} \Delta f_{j_1} \cdots \Delta f_{i_N} B_{i_N j_N} \Delta f_{j_N} \right\rangle_\rho \quad (8)$$

For a graphical representation we introduce N vertical pairs of sites labeled i and j and enumerate from 1 to N, as illustrated in fig.1 for $N = 7$. Sites (circles)

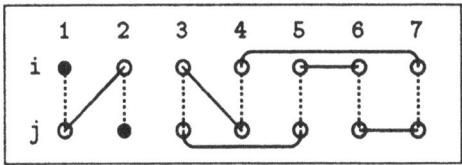

Figure 1.   *Graphical representation of the 7-th order term as explained in the text.*

with coordinates (ν) stand for factors Δf_ν, where ν is a compound index for i_l or j_l, respectively. All vertical pairs are connected by dashed lines which for the sites (i_l) and (j_l) carry a factor B_{i_l, j_l}. Next we split Δf_ν into expectation value $\Delta_\nu := \langle \Delta f_\nu \rangle_\rho$ and fluctuation $\delta_\nu := \Delta f_\nu - \Delta_\nu$ which will simplify the diagrammatic rules since $\langle \delta_\nu \rangle = 0$. Consequently, the N-th order product in Eq.8 decomposes further into a sum over 4^N terms consisting of all possible distributions of Δ-s and δ-s on the 2N sites. Graphically we represent expectation values (Δ) by full circles and fluctuations (δ) by open circles. All possible arrangements of open and full circles, respectively, have to be generated and summed over. A typical arrangement is depicted in fig.1. The Δ-s in the product can be taken out of the expectation value in Eq.8 leaving behind expectation values of products of δ factors only. So far, the entire derivation is exact. Next we propose a pair-fluctuation approximation (PA) which corresponds to a partial summation of diagrams to infinite order. [2] In PA higher order correlation functions are decomposed into all

---

[1] An exact but much less efficient approach is suggested by Monte Carlo importance sampling.

[2] A systematic improvement to higher order correlations is straight forward, although it becomes increasingly cumbersome.

possible products of pair-fluctuations

$$\langle \delta_{\nu_1} \delta_{\nu_2} \cdots \delta_{\nu_{2k}} \rangle = \sum_{\mathcal{P}} \langle \delta_{\mathcal{P}_1^{(1)}} \delta_{\mathcal{P}_2^{(1)}} \rangle \cdots \langle \delta_{\mathcal{P}_1^{(k)}} \delta_{\mathcal{P}_2^{(k)}} \rangle \quad . \tag{9}$$

The sum runs over all $(2k-1)!!$ different complete pairings of the indices $\nu_1, \ldots, \nu_{2k}$. Expectation values for an odd number of δ-s vanish in PA since $\langle \delta \rangle = 0$. The PA is exact for a Gaussian ρ with unrestricted support like in Eq.2. Due to the product-form of Eq.7 it is also exact in the general case for δ-s with unequal indices. The PA approach is closely related to Wick's theorem [4] used in quantum-field theory. Graphically, pair-fluctuations $\langle \delta_\nu \delta_{\nu'} \rangle$ are represented by bonds (solid lines) connecting the open circles at ν and ν'. The complete pairing of δ-s demands that precisely two bonds are attached to each open circle, while there are no bonds connected to full circles. The diagrams are generally composed of *connected diagrams*, in which all sites are successively connected by solid or dashed lines, respectively. E.g. the diagram in fig.1 consists of 2 connected diagrams, one of order 2 and one of order 5. The order of a connected diagram is the number of B-factors, or rather dashed lines. Since $B_{\nu\nu} = 0$ and $< \delta_\nu \delta_{\nu'} >= 0$ for $\nu \neq \nu'$ there are no bonds between vertical nearest neighbor sites. Generally the N-th order term can be decomposed into a set of connected diagrams, specified by $\{n_1, \ldots, n_N\}$, with n_l being the number of connected diagrams of order l. We introduce the constraint $\sum_l l\, n_l = N$ to enforce the correct expansion order N. All connected diagrams of order N contribute one and the same factor $\mathcal{F}_N := \langle (\Delta^T B\Delta)^N \rangle_\rho^c$ to the product Eq.8 so that we merely have to count the number of topologically distinct connected diagrams. The superscript c indicates that only connected diagrams are taken into account. The number of ways to group N different objects into subsets defined by the arrangement $\{n_1, \ldots, n_N\}$ is $N!/\prod_{l=1}^{N}(n_l!\, l!^{n_l})$. E.g. for the integers $(1, 2, 3, 4)$ there are 3 possibilities to form the arrangement specified by $\{n_1 = 0, n_2 = 2, n_3 = 0, n_4 = 0\}$, namely $(12)(34)$, $(13)(24)$, and $(14)(23)$. Eventually, Eq.7 is recast into the remarkable expression

$$\langle e^{-\Delta f^T B\Delta f} \rangle_\rho = \sum_{N=0}^{\infty} \sum_{\substack{n_1, \ldots, n_\infty \\ =0}}^{\infty} \delta(\sum_l l n_l - N) \frac{\prod_l (-1)^{l n_l}}{\prod_l (n_l!\, l!^{n_l})} \prod_{l=1}^{\infty} (f_l)^{n_l}$$

$$= \exp(\langle e^{-\Delta f^T B\Delta f} \rangle_c - 1) \quad , \tag{10}$$

according to which it is sufficient to evaluate connected diagrams only. It is harmless to extend the number of terms in the second sum to n_∞ due to the delta function constraint. Obviously, connected diagrams form either one closed loop containing only open circles or one open loop with full circles at the ends. An example of a closed loop of order 5 and an open loop of order 2 has already been encountered in fig.1 . To generate all closed loops of order N we start at site i_1 and place a bond to the upper or lower site of one of the remaining $(N-1)$ vertical pairs. With the alternative site of this pair we proceed to form a bond to one of the now remaining $(N-2)$ vertical pairs and so on until the loop is closed. In total there are therefore $(N-1)!2^{N-1}$ possible closed loops contributing

$$\mathcal{F}_N^c = (N-1)!\mathrm{tr}[(B\tilde{A}^{-1})^l]/2 \tag{11}$$

to Eq.8. The definition of $\tilde{A}$ is given in Eq.14, where we exploit the fact that the integral-correction vanishes if the regularization parameter is such that $\tilde{A} + B$ is not positive definite. The definition of $\tilde{A}$ accelerates the numerical procedure by enforcing positivity from the outset. Open loops are constructed from closed loops by cutting one of the N bonds which yields a total number of $N!2^{N-1}$. Their contribution to Eq.8 amounts

$$\mathcal{F}_l^o = l! \Delta^T (B\tilde{A}^{-1})^l \tilde{A} \Delta \tag{12}$$

Combining Eq.11,12,10, and 5 we obtain the desired correction factor

$$\mathcal{C} \;=\; \sqrt{\frac{\det(\mathbb{1} + BA^{-1})}{\det(\mathbb{1} + B\tilde{A}^{-1})}} e^{\Delta^T((\mathbb{1}+B\tilde{A}^{-1})^{-1}-\mathbb{1})\tilde{A}\Delta} \prod_i \kappa_i \tag{13}$$

$$\tilde{A}_{\nu\nu'} \;:=\; \delta_{\nu,\nu'} \cdot \max[(2\langle \Delta f_\nu^2 \rangle)^{-1}, A_{\nu\nu}] \tag{14}$$

The expectation values of Δf_i and Δf_i^2, entering $\tilde{A}_{\nu\nu'}$ and Eq.13, are easily determined numerically. An even simpler and yet fairly reliable approach is the Gaussian approximation, i.e. $R(\Delta, f^*) = 0$ in Eq.3 and Eq.6. In this case κ_i can be expressed in terms of the modified Bessel functions $I_{\pm 1/4}$ [5] while Δ_i and Δf_i^2 follow from κ_i upon differentiation with respect to f_i^* and A_{ii}, respectively

$$\kappa_i \;=\; f_i^* \sqrt{\frac{z\pi}{2}} e^{-z} \left(I_{1/4}(z) + I_{-1/4}(z) \right) \Big|_{z=A_{ii} f_i^{*2}/2}$$

$$A_{ii}\langle \Delta f_i^2 \rangle \;=\; z\left(1 - \frac{I'_{1/4}(z) + I'_{-1/4}(z)}{I_{1/4}(z) + I_{-1/4}(z)}\right)\Big|_{z=A_{ii} f_i^{*2}/2} \tag{15}$$

$$A_{ii}\langle \Delta f_i \rangle \;=\; (f_i^*)^{-1}\left(1/4 - A_{ii}\langle \Delta f_i^2 \rangle \right)$$

In summary the diagrammatic approach treats the diagonal part of ϕ exactly, while the off-diagonal part is approximated by partial summation of diagrams to infinite order. This approximation includes all terms up to second order completely.

3. Putting the integral correction to the test

In order to assess the importance of the proper treatment of the evidence integrals and the reliability of the integral correction formula proposed in the present paper we studied a variety of problems. One example will be given below others can be found elsewhere in these proceedings [7]. We compared the results obtained by the diagrammatic approach (PA) with those of the steepest-descent approximation (SD). In both cases α is determined upon maximizing the evidence $P(\alpha|D\mathcal{I})$. A conceptually different approach was proposed recently by Strauss, Wolpert and Wolf[6] (SWW) in which α is marginalized over, making evidence integrals obsolete as long as no further hyper-parameters or confidence intervals are required. The PA results agree remarkably well with those obtained by the SWW approach in all cases where the SD approximation failed badly. As a specific example we present here an ill-posed inversion problem encountered in Quantum Monte Carlo simulations. Without digressing into the underlying physics the inversion problem reads

$$g_m = \frac{1}{\pi} \int_0^\infty \frac{\omega}{1 - e^{-\beta\omega}} (e^{-\tau_m\omega} + e^{-(\beta-\tau_m)\omega}) f(\omega) \, d\omega \qquad . \qquad (16)$$

The parameters β and τ_m, $m = 1, \ldots, 40$ are known. Quantum Monte Carlo simulations provide values for g_m with a statistical error of about 1%. QME is employed to infer $f(\omega)$, which is displayed in fig.2 . Obviously SD strongly overfits the data.

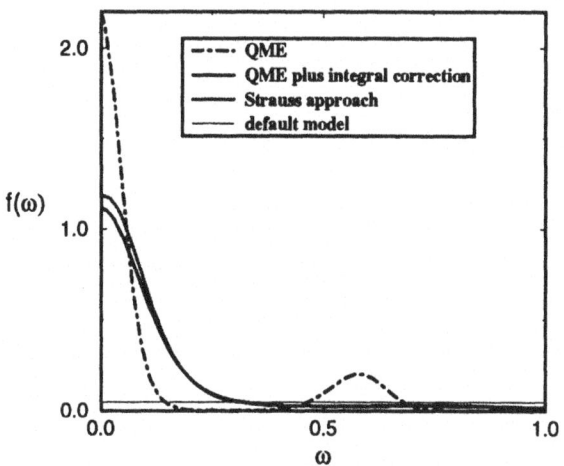

Figure 2. Comparison of QME results obtained by different approximations.

In fact, the associated regularization parameter is an order of magnitude too small. SD would not just lose a 'beauty contest' but it yields the **wrong** physics. We note in closing that confidence intervals are overestimated correspondingly.

References

1. S. F. Gull, "Developments in maximum entropy data analysis," in *Maximum Entropy and Bayesian Methods*, J. Skilling, ed., pp. 53–71, Kluwer Academic, Dordrecht, 1989.
2. J. Skilling, "Quantified maximum entropy," in *Maximum Entropy and Bayesian Methods*, P. F. Fougère, ed., pp. 341–350, Kluwer Academic, Dordrecht, 1990.
3. J. Skilling, "How to add up," in *Maximum Entropy and Bayesian Methods 1993*, G. Heidbreder, ed., Kluwer Academic, Dordrecht, 1995.
4. R. D. Mattuck, *A Guide to Feynman Diagrams in the Many-Body Problem*, Dover Publications, Inc., New York, 1992.
5. I. Gradshteyn and I. Ryzhik, "formulae 3.462, 9.240, 9.215," in *Tables of Integrals, Series, and Products*, Academic Press, San Diego, 1980.
6. C. E. M. Strauss, D. H. Wolpert and D. R. Wolf, "Alpha, evidence, and the entropic prior," in *Maximum Entropy and Bayesian Methods*, G. Heidbreder, ed., Kluwer Academic, Dordrecht, 1995.
7. R. Fischer, W. von der Linden and V. Dose, "On the importance of α marginalization," *In these proceedings*, 1995.

Rose Vigil at the buffet table at Fuller Lodge.

At the dinner at Fuller Lodge; Ali Mohammad-Djafari, Callie Silver, Leda Silver, Jim Press, Rose Vigil, Imre Csiszár, Richard Silver, and Bob Snapp.

Fundamental Theories of Physics

Series Editor: Alwyn van der Merwe, *University of Denver, USA*

1. M. Sachs: *General Relativity and Matter.* A Spinor Field Theory from Fermis to Light-Years. With a Foreword by C. Kilmister. 1982   ISBN 90-277-1381-2
2. G.H. Duffey: *A Development of Quantum Mechanics.* Based on Symmetry Considerations. 1985   ISBN 90-277-1587-4
3. S. Diner, D. Fargue, G. Lochak and F. Selleri (eds.): *The Wave-Particle Dualism.* A Tribute to Louis de Broglie on his 90th Birthday. 1984   ISBN 90-277-1664-1
4. E. Prugovečki: *Stochastic Quantum Mechanics and Quantum Spacetime.* A Consistent Unification of Relativity and Quantum Theory based on Stochastic Spaces. 1984; 2nd printing 1986   ISBN 90-277-1617-X
5. D. Hestenes and G. Sobczyk: *Clifford Algebra to Geometric Calculus.* A Unified Language for Mathematics and Physics. 1984
          ISBN 90-277-1673-0; Pb (1987) 90-277-2561-6
6. P. Exner: *Open Quantum Systems and Feynman Integrals.* 1985   ISBN 90-277-1678-1
7. L. Mayants: *The Enigma of Probability and Physics.* 1984   ISBN 90-277-1674-9
8. E. Tocaci: *Relativistic Mechanics, Time and Inertia.* Translated from Romanian. Edited and with a Foreword by C.W. Kilmister. 1985   ISBN 90-277-1769-9
9. B. Bertotti, F. de Felice and A. Pascolini (eds.): *General Relativity and Gravitation.* Proceedings of the 10th International Conference (Padova, Italy, 1983). 1984
          ISBN 90-277-1819-9
10. G. Tarozzi and A. van der Merwe (eds.): *Open Questions in Quantum Physics.* 1985
          ISBN 90-277-1853-9
11. J.V. Narlikar and T. Padmanabhan: *Gravity, Gauge Theories and Quantum Cosmology.* 1986   ISBN 90-277-1948-9
12. G.S. Asanov: *Finsler Geometry, Relativity and Gauge Theories.* 1985
          ISBN 90-277-1960-8
13. K. Namsrai: *Nonlocal Quantum Field Theory and Stochastic Quantum Mechanics.* 1986   ISBN 90-277-2001-0
14. C. Ray Smith and W.T. Grandy, Jr. (eds.): *Maximum-Entropy and Bayesian Methods in Inverse Problems.* Proceedings of the 1st and 2nd International Workshop (Laramie, Wyoming, USA). 1985   ISBN 90-277-2074-6
15. D. Hestenes: *New Foundations for Classical Mechanics.* 1986
          ISBN 90-277-2090-8; Pb (1987) 90-277-2526-8
16. S.J. Prokhovnik: *Light in Einstein's Universe.* The Role of Energy in Cosmology and Relativity. 1985   ISBN 90-277-2093-2
17. Y.S. Kim and M.E. Noz: *Theory and Applications of the Poincaré Group.* 1986
          ISBN 90-277-2141-6
18. M. Sachs: *Quantum Mechanics from General Relativity.* An Approximation for a Theory of Inertia. 1986   ISBN 90-277-2247-1
19. W.T. Grandy, Jr.: *Foundations of Statistical Mechanics.*
    Vol. I: *Equilibrium Theory.* 1987   ISBN 90-277-2489-X
20. H.-H von Borzeszkowski and H.-J. Treder: *The Meaning of Quantum Gravity.* 1988
          ISBN 90-277-2518-7
21. C. Ray Smith and G.J. Erickson (eds.): *Maximum-Entropy and Bayesian Spectral Analysis and Estimation Problems.* Proceedings of the 3rd International Workshop (Laramie, Wyoming, USA, 1983). 1987   ISBN 90-277-2579-9

Fundamental Theories of Physics

22. A.O. Barut and A. van der Merwe (eds.): *Selected Scientific Papers of Alfred Landé.*
[*1888-1975*]. 1988   ISBN 90-277-2594-2
23. W.T. Grandy, Jr.: *Foundations of Statistical Mechanics.*
Vol. II: *Nonequilibrium Phenomena.* 1988   ISBN 90-277-2649-3
24. E.I. Bitsakis and C.A. Nicolaides (eds.): *The Concept of Probability.* Proceedings of the
Delphi Conference (Delphi, Greece, 1987). 1989   ISBN 90-277-2679-5
25. A. van der Merwe, F. Selleri and G. Tarozzi (eds.): *Microphysical Reality and Quantum
Formalism, Vol. 1.* Proceedings of the International Conference (Urbino, Italy, 1985).
1988   ISBN 90-277-2683-3
26. A. van der Merwe, F. Selleri and G. Tarozzi (eds.): *Microphysical Reality and Quantum
Formalism, Vol. 2.* Proceedings of the International Conference (Urbino, Italy, 1985).
1988   ISBN 90-277-2684-1
27. I.D. Novikov and V.P. Frolov: *Physics of Black Holes.* 1989   ISBN 90-277-2685-X
28. G. Tarozzi and A. van der Merwe (eds.): *The Nature of Quantum Paradoxes.* Italian
Studies in the Foundations and Philosophy of Modern Physics. 1988
                                                                             ISBN 90-277-2703-1
29. B.R. Iyer, N. Mukunda and C.V. Vishveshwara (eds.): *Gravitation, Gauge Theories
and the Early Universe.* 1989   ISBN 90-277-2710-4
30. H. Mark and L. Wood (eds.): *Energy in Physics, War and Peace.* A Festschrift
celebrating Edward Teller's 80th Birthday. 1988   ISBN 90-277-2775-9
31. G.J. Erickson and C.R. Smith (eds.): *Maximum-Entropy and Bayesian Methods in
Science and Engineering.*
Vol. I: *Foundations.* 1988   ISBN 90-277-2793-7
32. G.J. Erickson and C.R. Smith (eds.): *Maximum-Entropy and Bayesian Methods in
Science and Engineering.*
Vol. II: *Applications.* 1988   ISBN 90-277-2794-5
33. M.E. Noz and Y.S. Kim (eds.): *Special Relativity and Quantum Theory.* A Collection of
Papers on the Poincaré Group. 1988   ISBN 90-277-2799-6
34. I.Yu. Kobzarev and Yu.I. Manin: *Elementary Particles. Mathematics, Physics and
Philosophy.* 1989   ISBN 0-7923-0098-X
35. F. Selleri: *Quantum Paradoxes and Physical Reality.* 1990   ISBN 0-7923-0253-2
36. J. Skilling (ed.): *Maximum-Entropy and Bayesian Methods.* Proceedings of the 8th
International Workshop (Cambridge, UK, 1988). 1989   ISBN 0-7923-0224-9
37. M. Kafatos (ed.): *Bell's Theorem, Quantum Theory and Conceptions of the Universe.*
1989   ISBN 0-7923-0496-9
38. Yu.A. Izyumov and V.N. Syromyatnikov: *Phase Transitions and Crystal Symmetry.*
1990   ISBN 0-7923-0542-6
39. P.F. Fougère (ed.): *Maximum-Entropy and Bayesian Methods.* Proceedings of the 9th
International Workshop (Dartmouth, Massachusetts, USA, 1989). 1990
                                                                             ISBN 0-7923-0928-6
40. L. de Broglie: *Heisenberg's Uncertainties and the Probabilistic Interpretation of Wave
Mechanics.* With Critical Notes of the Author. 1990   ISBN 0-7923-0929-4
41. W.T. Grandy, Jr.: *Relativistic Quantum Mechanics of Leptons and Fields.* 1991
                                                                             ISBN 0-7923-1049-7
42. Yu.L. Klimontovich: *Turbulent Motion and the Structure of Chaos.* A New Approach
to the Statistical Theory of Open Systems. 1991   ISBN 0-7923-1114-0

Fundamental Theories of Physics

Fundamental Theories of Physics

KLUWER ACADEMIC PUBLISHERS – DORDRECHT / BOSTON / LONDON

The manufacturer's authorised representative in the EU is Springer
Nature Customer Service Centre GmbH, Europaplatz 3, 69115 Heidelberg,
Germany. If you have any concerns regarding our products, please
contact ProductSafety@springernature.com

Printed and bound by CPI Group (UK) Ltd, Croydon, CR0 4YY
29/04/2026
02099472-0010